Edible Insects as Innovative Foods

Nutritional, Functional and Acceptability Assessments

Editors

Chuleui Jung
Victor Benno Meyer-Rochow

MDPI • Basel • Beijing • Wuhan • Barcelona • Belgrade • Manchester • Tokyo • Cluj • Tianjin

Editors
Chuleui Jung
Andong National University
Korea

Victor Benno Meyer-Rochow
Oulu University
Finland

Editorial Office
MDPI
St. Alban-Anlage 66
4052 Basel, Switzerland

This is a reprint of articles from the Special Issue published online in the open access journal *Foods* (ISSN 2304-8158) (available at: https://www.mdpi.com/journal/foods/special_issues/edible_insects_innovative_food).

For citation purposes, cite each article independently as indicated on the article page online and as indicated below:

LastName, A.A.; LastName, B.B.; LastName, C.C. Article Title. *Journal Name* **Year**, *Article Number*, Page Range.

ISBN 978-3-03943-076-5 (Hbk)
ISBN 978-3-03943-077-2 (PDF)

Cover image courtesy of Victor Benno Meyer-Rochow and Chuleui Jung.

Contents

About the Editors

Chuleui Jung studied Biology at Seoul National University in Korea and finished his PhD from Oregon State University with "Dispersal behaviors of predatory mites", a group of acari which are important natural enemies of spider mites in agriculture. Since being accepted in Andong National University, he has taught entomology, ecology, acarology, honey bee science and pollination, and ecosystem modeling. His research activities are orient toward the ecological processes and interactions of beneficial insects and arthropods with mutualistic ecosystem services such as biocontrol and pollination. At the same time, he believes that developing insects as generic food or functional sources could alleviate global food-related problems. Prof. Jung has published more than 200 scientific papers and several book chapters. He has also served many scientific journals as Editor or Editorial Board Member. Publications are available from https://www.researchgate.net/profile/Chuleui_Jung.

Victor Benno Meyer-Rochow is a New Zealander with a PhD in Neurobiology and a DSc in Ethnobiological Studies (both from the Australian National University in Canberra). Dr. Meyer-Rochow's neurobiological research took him several times to Antarctica and the Arctic, but his ethnobiologial work focused on Central and Western Australia (1973), Papua New Guinea (1972, 1975, 2002, 2005), North-East India (1990, 2012, 2014, 2015, 2016) and the DPRK (North Korea: 2012). He was Managing Director of the Research Institute of Luminous Organisms on the Japanese Pacific island of Hachijojima from 2014 to 2019, and is currently working as Visiting Professor at Andong National University's Department of Plant Medicals in the Republic of Korea. He is attached to the Department of Genetics and Physiology, Oulu University, Oulu, Finland. Information on his research and publications is available from www.meyer-rochow.com and https://www.researchgate.net/profile/Victor_Meyer-Rochow.

Preface to "Edible Insects as Innovative Foods"

Will insects be the food of the future?

The answer is that we aren't sure yet, but we can be certain that insects in the past were indeed a food item appreciated by humankind worldwide (Bequaert 1921, Bergier 1941; Bodenheimer 1951). Somehow and for reasons not completely understood, but perhaps related to the spread of Christianity that forbade the consumption of insects with the exception of four species of locust, the awareness that insects can be vectors of some fatal diseases, the emotional separation from Nature including entomophobia, and an increasingly greater variety of foodstuffs reaching the consumer, the use of insects as human food became less and less common.

However, there is no doubt that since time immemorial humans have almost certainly consumed some insects either by ingesting them more or less accidentally with fruit and other items or seeking to eat them deliberately. Even our closest animal relatives, the monkeys, have been observed to actively collect insects and other arthropods in order to eat them (e.g. Marshall 1902; Carpenter 1921; Nickle and Heymann 1996; Sanz et al. 2009).

Over the last 30 years, there has been a renewed interest in insects as human food. Conferences have begun to focus on edible insects since the XVI International Pacific Science Congress in Seoul in August 1987 and the International Conference on Minilivestock in Beijing in September 1995 brought this topic to a wider audience. Numerous scientific publications in the last 20 years or so have praised the advantages of insect-based diets over conventional diets containing meats like poultry and especially ruminants, and highlighted the environmentally advantageous aspects of farming of insects over traditionally farmed animals. Various edible insect species have had their acceptability as a novelty food (or feed in animal husbandry and fish culture) examined and their potential risk of carrying diseases or undesired microbes scrutinized. Although insects should not be seen as a food item for humans merely to survive periods of starvation, there is no way to deny that the global food security situation for the human population is becoming increasingly precarious, and that global food production has to increase by at least 50% to meet demand by 2050. Despite earlier reports of people who live traditionally in different parts of the world engaging in entomophagy (the consumption of insects), until Meyer-Rochow (1975) none of these reports had thought to link global food security to the universal and extensive use of insects as a possible and potent way to ease global food shortages.

We now possess a considerable amount of information on the kinds of insect that serve as food to various people in the world; we know that most insects are nutritious, consist of valuable protein, easily digestible fatty acids, and contain important minerals and vitamins, and recommendations exist regarding how to breed the most lucrative species optimally. However, there are still gaps to be filled with regard to the processing of cultured insects, the preparation and conservation of insect-based foods, economics and marketing, and the potential of insects as suppliers of health-promoting drugs and medicines.

It is with these thoughts in mind that we accepted the task of serving as Guest Editors for this Issue of the journal Foods on edible insects and their role as food and as raw material for a variety of products.

1. Bequaert, J. Insects as food: How they have augmented the food supply of mankind in early and recent years. *Nat. Hist. J.* **1921**, *21*, 191–200.

2. Bergier, E. *Peuples Entomophages et Insectes Comestibles: Étude sur les Moeurs de L'homme et de L'insecte*; Imprimérie Rullière Frères: Avignon, France, 1941.

3. Bodenheimer, F.S. *Insects as Human Food*; W. Junk Publishers: The Hague, The Netherlands, 1951.

4. Carpenter, G.D.H. Experiment on the relative edibility of insects with special reference to their coloration. *Trans. Roy. Ent. Soc. London* **1921**, *45*, 1–105.

5. Marshall, G.A.K. Five years' observations and experiments (1896–1901) on the bionomics of South African insects, chiefly mimicry and warning colours. *Trans. Entomol. Soc. Lond.* **1902**, *50*, 287–584.

6. Nickle, D.A.; Heymann, E.W. Predation on orthoptera and other orders of insects by tamarind monkeys, *Saguinus mystax mystax* and *S. fuscicollis nigrifrons* (Primates: Callitrichidae) in north-eastern Peru. *J. Zool. (London)* **1996**, *239*, 799–819.

7. Sanz, C.; Schöning, C.; Morgan, D. Chimpanzees prey on army ants with specialized tool set. *Am. J. Primatol.* **2009**, *71*, 1e8.

8. Meyer-Rochow, V.B. Can insects help to ease the problem of world food shortage? *Search* **1975**, *6*, 261–262.

Chuleui Jung, Victor Benno Meyer-Rochow
Editors

Editorial

Insects Used as Food and Feed: Isn't That What We All Need?

Victor Benno Meyer-Rochow [1,2,*] and Chuleui Jung [1,3,*]

[1]   Agricultural Science and Technology Research Institute, Andong National University, Andong 36729, Korea
[2]   Department of Genetics and Ecology, Oulu University, SF-90140 Oulu, Finland
[3]   Department of Plant medicals, Andong National University, Andong 36729, Korea
*   Correspondence: meyrow@gmail.com (V.B.M.-R.); cjung@andong.ac.kr (C.J.)

Received: 14 July 2020; Accepted: 21 July 2020; Published: 27 July 2020

Abstract: This Special Issue of *Foods* explores different aspects of how insects can be used as a novel resource for food and feed. Some contributions deal with questions of acceptability and legality, others tackle problems related to innovative techniques in processing and marketing food, and yet another group of papers highlights the use of insects and their bio-active products in the context of promoting human health. The collective aim of the contributions by the researchers from at least 20 countries is to examine whether the use of insects—be it for food, feed, or therapeutic purposes—has a future. We conclude that positive aspects undoubtedly exist regarding the nutritional and pharmacological value of various insect species but that environmental and bio-functional issues could even outweigh the nutritional value of food insects.

Keywords: entomophagy; insect edibility; food shortage; acceptance; bio-active compounds; nutrients

---

1. Introduction

The terms "entomophagy"—a combination of the ancient Greek words "entomon" (notched anima) and "phagein" (eating of something specific)—and "insectivorous" and "insectivore", which use a suffix of Latin origin (derived from "vorare", to devour), have been examined and compared with each other by Evans et al. [1] with regard to their suitability and applicability in connection with humans. That study not only demonstrates the uncertainty in how to describe the consumption of insects by humans, whether habitual or accidental, but also highlights that the use of insects by humans as food and feed can be approached from different angles in different disciplines by experts from different locations asking different questions and using a variety of tools. This Special Issue of *Foods* is a case in point because authors located in at least 20 different countries have contributed their findings to it.

The papers chosen to form part of this Special Issue range from those addressing legal matters and questions of acceptability to those that are primarily concerned with the processing, chemical composition, and bioactivity of the insects. The range of papers in this issue underscores the fact that a greater understanding of the potential of insects as food and feed depends on interactions between fields and experts of both the humanities and the sciences. Whether, however, insects will ever be the "food of the future" is difficult to say. They were undoubtedly "food of the past" for our ancestors, and they still are a "food of the present" for many inhabitants of the world today [2–5].

In the past, insects were indeed a food item appreciated by humankind worldwide [6–8]. Somehow—and for reasons not completely understood but probably related to the spread of Christianity that forbade the consumption of insects with the exception of four species of locust, or the awareness of the fact that insects can be vectors of some fatal human diseases, the emotional separation from nature giving rise to entomophobia, and an increasingly greater variety of food stuffs reaching the consumer—the use of insects as human food became continually less popular. Our closest animal relatives, the monkeys, have been observed to actively collect insects and other arthropods in order to eat them—e.g., [9–12]—or, in the case of millipedes, to use them therapeutically [13].

Over the last 30 years, there has been a renewed interest in insects as human food, and the literature dealing with entomophagy and entomotherapy has become so vast that it is seemingly impossible to review it in its entirety. International conferences have begun to focus more and more on edible insects since the XVI International Pacific Science Congress in Seoul in August 1987 and the International Conference on Minilivestock in Beijing in September 1995, and the subsequent book by Maurizio Paoletti [14] brought this topic to a wider audience. Scientific publications, too many to mention, have appeared in the last 20 years or so, praising the advantages of an insect-based diet over a diet consisting of conventional meats like poultry and especially ruminants and highlighting the environmentally advantageous farming practices for mini-livestock such as insects over those for traditionally farmed animals [15–17]. Various edible insect species have had their farming potential assessed, their acceptability as a novelty food (or feed in animal husbandry and fish culture) examined, and their potential risk of carrying diseases or undesired microbes scrutinized. Nevertheless, there remain many open questions, which is why Special Issues such as this are still important.

Although insects should not be seen as a food item for humans merely to survive times of dietary hardship and periods of starvation [18], there is no way of denying that the global food security situation for the human population is becoming increasingly precarious and that food production worldwide has to increase by at least 50% to meet the demand by 2050 [19]. Despite the earlier reports of traditionally living people in different parts of the world engaging in entomophagy (mentioned above), none of these reports until Meyer-Rochow [20] had thought to link global food security with insects being more universally and extensively used as a possible and potent way to ease global food shortages. It was largely due to DeFoliart [21] and Van Huis et al. [22] that the idea was taken up and more widely promoted.

We now possess a considerable amount of information on the kinds of insect that serve as food to various people in the world; we know that most insects are nutritious, consist of valuable protein and easily digestible fatty acids, and contain important minerals and vitamins [23–25], and recommendations exist on how to breed the most lucrative species optimally [26,27]. However, gaps to be filled still exist with regard to the processing of cultured insects, the preservation of insect-based foods and their nutrients, the preparation of insect-containing dishes, the economics and marketing, and the potential of insects as suppliers of health-promoting drugs and medicines.

2. Results

In this Special Issue of *Foods*, Dorothy Nyangena and colleagues [28] stress that quality and safety concerns call for simple, actionable hazard control mechanisms. They tested the extent to which different processing methods affected the proximate composition and microbiological characteristics of the edible cricket *Acheta domesticus* and the grasshopper *Ruspolia differens*. The researchers concluded that traditional processing can alter the nutritional value of the insects but improves microbial safety.

Microbial safety was also an issue that the paper by Nils Grabowski et al. [29] dealt with. The authors pointed out that edible insects, like other foodstuffs, in order to guarantee consumer safety and product quality, should be covered by governmental regulations. Since nations vary with regard to their approaches concerning the inclusion of insects in foodstuffs and commercial farming, the situation is complex. This paper reviews the situation from a hygiene point of view and provides some worthwhile recommendations.

The Korean scientists Su-Jin Pyo and colleagues [30] focused on six species of traditionally used insects in Korea and demonstrated that, in particular, *Allomyrina dichotoma* and *Teleogryllus emma* showed considerable antioxidant activities. *Tenebrio molitor* and honeybee extracts, more so than the other species, exhibited strong haemolytic activities, suggesting that these insects not only possessed potential as a food source but could be used therapeutically as well.

The paper by Sampat Ghosh and colleagues [31] shows that drones and the developmental stages of identical species (in this case, the honeybee *Apis mellifera*), but representing different strains from widely different localities (here: "Buckfast" drones from Denmark and "Italian" drones from Korea) can differ with regard to their composition. Buckfast drones, for example, exhibited the highest antioxidant activity.

Results on the basis of data obtained from 310 Italian consumers were analyzed by Rocco Roma and colleagues [32] in order to ascertain the acceptability of edible insects as food or food supplements. Consumers fell into five groups and the acceptance of insects as feed material amongst them was widespread, but acceptance as a food item for humans was not. Reasons such as neophobia, disgust, assumed health risk, unfamiliarity, and socio-cultural backgrounds are thought to be responsible, although age-dependent differences were noticed.

Giulia Leni and co-workers [33] present a novel method tested on the lesser mealworm *Alphitobius diaperinus* that can be used in conjunction with other insect species to extract high quality protein for food and feed purposes. Enzymatic hydrolysis with *Bacillus licheniformis* for five hours resulted in a protein hydrolysate with a 15% degree of hydrolysis, oil-holding capacity, and excellent solubility.

Da-Young Lee and colleagues [34], who investigated the extent to which freeze-dried mature larval silkworm powder (SMSP) affected ethanol-induced fatty liver disease in rats, found that mature larval silkworm powder reduced triglyceride levels by as much as 35% in the livers of the rats that had received daily doses of 25% ethanol (3 g/kg body weight) for 4 weeks. In ethanol-treated rats given SMSP, the serum levels of triglycerides and various other compounds were lower, suggesting that SMSP can protect against alcoholic liver disease.

In view of the fact that so-called mirror neurons can influence a person's decisions and/or opinions, Meyer-Rochow and Kejonen [35] analyzed a list of commonly used idioms in the Finnish language that make reference to insects in the context of food and eating. The vast majority of these idioms referred to insects in a negative way, which could lead to or reinforce feelings of disgust and prevent people from trying insect-containing food items.

That the nutrient composition of mealworms can be influenced by fresh plant supplements in their diet during the growing phase of the insects was demonstrated by Changqi Liu and co-workers [36]. When mealworm larvae were reared on wheat bran enriched with carrot, orange, or red cabbage for four weeks, the larvae grew better and bigger, being 40–46% heavier than the controls that did not receive the food supplements.

Yang-Ju Son and colleagues [37] used mealworms to produce a defatted powder and mealworm oil. While the powder was rich in protein, minerals, and bioactive compounds and had a pleasant taste, the mealworm oil contained bioactive nutrients, especially gamma tocopherol, and had physicochemical properties that were similar to those of vegetable oil. Mealworm products contain antioxidants and glucosamine derivatives such as chitin and chitosan and possess anti-inflammatory properties.

Inhabitants of the island of Madagascar, as Joost Van Itterbeeck and colleagues [38] could show, have a long history of using especially insects belonging to the Orthoptera as a traditional food. Thirty-seven edible species were recorded. The insects were mostly eaten by humans as snacks and part of the main meal and were primarily collected by children. With the exception of *Brachytrupes membranaceus colosseus*, the insects were not marketed, and using insects as feed for animals was rare.

The microbiological and bioactive properties of powders made from *Acheta domesticus* and *Tenebrio molitor* were examined by Concetta Maria Messina and co-workers [39], who found a higher number of viable bacteria, mainly *Citrobacter* and Enterobacteriaceae, in mealworm flour than in cricket powder, which, on the other hand, contained more Porphyromonadaceae. The protein hydrolysates of both powders were rich in antioxidants.

To what extent an aversion to insects and rejection of insects as human food exists in different countries was studied by Mauricio Castro and Edgar Chambers [40] with a survey launched in 13 countries. Most of the 630 participants in the study were reluctant to accept insects and insect-containing food items, with the exception of respondents from Mexico and Thailand. Obviously, that educating consumers that food insects need not be unhygienic should be one goal in making food insects more widely acceptable, was their conclusion.

The latter is also the recommendation that came from the results obtained by Karin Wendin and colleagues [41], who could show that sensory attributes are a very important factor. Particle size, for example, had a great influence on the appearance and viscosity, colour and odour, as well as texture and taste of the insect-containing food but not the flavour. On the other hand, salt content did affect the taste and flavour. An addition of antioxidants reduced the colour change and rancidity in the food, thereby improving its acceptability.

An interesting test was carried out by Simone Mancini and psychology colleagues [42], who asked participants to evaluate by taste two identical bread samples, one labelled as having been supplemented with insect powder. In reality, however, both types of bread were identical. Even though the participants gave higher scores for flavour, texture, and overall liking to the insect-labelled bread, they nevertheless declared that they were less likely to consume the insect-labelled bread in the future, which suggests deep-seated feelings of insect rejection overriding sensory experience.

The idea that the acceptability of insects as a component of food items depends to a large degree on the way in which the edible insects are processed and presented but also on the packaging has been suggested before, but Wendin, Mårtensson, Djerf, and Langton [43] confirmed this through sensory analyses with probands. The results showed that following mixing and chopping (e.g., of mealworms), the particle size, oil content, salt content, and antioxidants influenced the product more than the storage time. Packed in Tetra Recart cartons, the product also stayed stable for at least 6 months at room temperature.

That there are still big gaps regarding the use and acceptability of insects as human food in various regions of the world has been shown in an interesting paper by Mozhui et al. [44]. In this beautifully illustrated investigation of entomophagic practices by the inhabitants of Nagaland (North-East India), the authors not only provide details on how the 106 edible species (representing nine orders of insects and 32 families) are obtained but also explain the various ways in which local residents turn the insects into delicious dishes.

An important aspect of processing insects as human food is tackled by the Malaysian/Thai team headed by Anuduang [45]. In their contribution they investigated the impact of a range of temperatures and durations on the quality of silkworm powder and found that shorter times of exposure to hot water and a low drying temperature preserve the antioxidant activities of the product best.

Jeong et al. [46] point out that despite their long history of utilization as food and pharmaceutical bioresources, the value of social wasps as a food resource is poorly appreciated and researched. Especially, not only are the larvae of the invasive *Vespa velutina nigrithorax* rich in nutrition and safe to eat but their saliva's amino acid composition has been shown to be worthy of further study in the context of amino acid supplementation and exercise enhancement.

3. Discussion

Coming back to the original question as to whether insects as food and feed have a future, we conclude that there can be no doubt that the many positive aspects in terms of environmental issues, especially when commercially bred species are considered, will outweigh the fact that the nutritional value of food insects is not significantly greater than that of conventional foods [47]. Insects bred to replace or supplement feedstuffs required in fish culture, poultry, and other animal husbandry will encounter fewer acceptability problems than insects and insect products aimed at human consumers [48]. Moreover, although numerous species of insects have been used therapeutically in the past [49], modern chemical analytical and isolating techniques have made it possible to identify potent compounds in them, and, as also some papers in this issue have shown, insects and insect products with medicinal properties are likely to play an increasingly greater role in the future to combat infections and other maladies.

The main advantages of food insects over conventional meat-supplying species are that the former can be reared on much less food and water and require significantly less space than the latter. Furthermore, the rate of reproduction of insects is considerably higher than that of conventional food animals, and a significantly greater proportion of an insect's body weight is utilizable as food than is the case with regard to conventional meat animals. Most people and not just animal lovers would also find the "harvesting" of insects more acceptable than the slaughter of livestock. Finally, the so-called carbon footprint of farmed insects is deemed to be appreciably lower than that of farmed conventional food animals [15–17,22].

Obstacles to overcome are, on the one hand, the declining interest in edible insects by citizens of countries in which insects used to be consumed widely [50] and, on the other hand, the still widespread rejection of insect-containing foods by citizens of countries in which edible insects and insect-supplemented foods were not traditionally accepted but have now become available [51–53]. To understand what governs the selection and acceptance of edible insect species and what role taboos play in this has to be one important goal [54]; education, clever marketing, tapping into the trend of buying healthy food, and highlighting the traditional nutritional and medical benefits are additional goals that could help to make consumers consider insects as food [55,56]. Using catchy slogans such as "Forget about the pork and put a cricket on your fork" or "Mealworms and spaghetti is food that makes you happy" could also be expected to help!

Author Contributions: Conceptualization: V.B.M.-R. and C.J.; Methodology: V.B.M.-R. and C.J.; Validation: V.B.M.-R. and C.J.; Formal analysis: V.B.M.-R. and C.J.; Investigation: V.B.M.-R. and C.J.; Resources: V.B.M.-R. and C.J.; Writing—original draft preparation: V.B.M.-R.; Writing—review and editing: V.B.M.-R. and C.J.; Visualization: V.B.M.-R. and C.J.; Project administration, V.B.M.-R. and C.J. All authors have read and agreed to the published version of the manuscript.

Funding: This research received no external funding.

Conflicts of Interest: The authors declare that there is no conflict of interests.

References

1. Evans, J.; Alemu, M.H.; Flore, R.; Frost, M.B.; Halloran, A.; Jensen, A.B.; Maciel-Vergara, G.; Meyer-Rochow, V.B.; Münke-Svendsen, C.; Olsen, S.B.; et al. 'Entomophagy': An evolving terminology in need of review. *J. Insects Food Feed.* **2015**, *1*, 293–305. [CrossRef]

2. Mitsuhashi, J. *The World's Edible Insects (Sekai Konchuu Shokkou Taizen)*; Yasaka Shobou: Tokyo, Japan, 2008.

3. Jongema, Y. *List of Edible Insects of the World*; Wageningen UR: Wageningen, The Netherlands, 2015; Available online: https://tinyurl.com/mestm6p (accessed on 22 June 2020).

4. Van Huis, A. Chapter 11. Insects as human food. In *Ethnozoology: Animals in Our Lives*; Alves, R., Paulino de Albuquerque, U., Eds.; Elsevier: Amsterdam, The Netherlands, 2018; pp. 195–213.

5. Meyer-Rochow, V.B. Insects (and other non-crustacean arthropods) as human food. In *Encyclopedia of Food Security and Sustainability, Vol. 1*; Ferranti, P., Berry, E.M., Anderson, J.R., Eds.; Elsevier: Amsterdam, The Netherlands, 2019; pp. 416–421. ISBN 9780128126875.

6. Bequaert, J. Insects as food: How they have augmented the food supply of mankind in early and recent years. *Nat. Hist. J.* **1921**, *21*, 191–200.

7. Bergier, E. *Peuples Entomophages et Insectes Comestibles: Étude sur les Moeurs de L'homme et de L'insecte*; Imprimérie Rullière Frères: Avignon, France, 1941.

8. Bodenheimer, F.S. *Insects as Human Food*; W. Junk Publishers: The Hague, The Netherlands, 1951.

9. Marshall, G.A.K. Five years' observations and experiments (1896–1901) on the bionomics of South African insects, chiefly mimicry and warning colours. *Trans. Entomol. Soc. Lond.* **1902**, *50*, 287–584.

10. Carpenter, G.D.H. Experiments on the relative edibility of insects with special reference to their coloration. *Trans. Roy. Entomol. Soc. Lond.* **1921**, *54*, 1–105.

11. Nickle, D.A.; Heymann, E.W. Predation on orthoptera and other orders of insects by tamarind monkeys, *Saguinus mystax mystax* and *Saguinus fuscicollis nigrifrons* (Primates: Callitrichidae) in north-eastern Peru. *J. Zool. (London)* **1996**, *239*, 799–819. [CrossRef]

12. Rothman, J.M.; Raubenheimer, D.; Bryer, M.A.H.; Takahashi, M.; Gilbert, C.C. Nutritional contributions of insects to primate diets: Implications for primate evolution. *J. Hum. Evol.* **2014**, *71*, 59–69. [CrossRef]

13. Weldon, P.; Aldrich, J.R.; Klun, J.A.; Oliver, J.E.; Debboun, M. 2003 Benzoquinones from millipedes deter mosquitoes and elicit self-anointing in capuchin monkeys (*Cebus* spp.). *Sci. Nat.* **2003**, *90*, 301–304. [CrossRef]

14. Paoletti, M.G. *Ecological Implications of Minilivestock: Potential of Insects, Rodents, Frogs, and Snails*; Science Publishers: Enfield, EN, USA, 2005.

15. Oonincx, D.G.A.B.; De Boer, I.J.M. Environmental impact of the production of mealworms as a protein source for humans—a life cycle assessment. *PLoS ONE* **2012**, *7*, e51145. [CrossRef]

16. Abbasi, T.; Abbasi, T.; Abbasi, S.A. Reducing the global environmental impact of livestock production: The minilivestock option. *J. Clean. Prod.* **2015**, *112*, 1754–1766. [CrossRef]

17. Halloran, A.; Caparros Megido, R.; Oloo, J.; Weigel, T.; Nsevolo, P.; Francis, F. Comparative aspect of cricket framing in Thailand, Cambodia, Lao People's Democratic Republic, Democratic Republic of the Congo, and Kenya. *J. Insects Food Feed* **2018**, *4*, 101–114. [CrossRef]

18. Gahukar, R.T. Edible insects collected from forests for family livelihood and wellness of rural communities-a review. *Global Food Secur.* **2020**. [CrossRef]

19. Tilman, D.; Balzer, C.; Hill, J.; Befort, B.L. Global food demand and the sustainable intensification of agriculture. *Proc. Natl. Acad. Sci. USA* **2011**, *108*, 20260–20264. [CrossRef] [PubMed]

20. Meyer-Rochow, V.B. Can insects help to ease the problem of world food shortage? *Search* **1975**, *6*, 261–262.

21. DeFoliart, G.R. Insects as food: Why the western attitude is important. *Ann. Rev. Entomol.* **1999**, *41*, 21–50. [CrossRef]

22. Van Huis, A.; Van Itterbeeck, J.; Klunder, H.; Mertens, E.; Halloran, A.; Muir, G.; Vantomme, P. *Edible Insects: Future Prospects for Food and Feed*; Food and Agriculture Organisation of the United Nations: Rome, Italy, 2013; pp. 1–190.

23. Bukkens, S.G.F. The nutritional value of edible insects. *Ecol. Food Nutr.* **1997**, *36*, 287–319. [CrossRef]

24. Chakravorty, J.; Ghosh, S.; Meyer-Rochow, V.B. Chemical composition of *Aspongopus nepalensis* Westwood 1837 (Hemiptera; Pentatomidae), a common food insect of tribal people in Arunachal Pradesh (India). *Int. J. Vitam. Nutr. Res.* **2011**, *81*, 1–14. [CrossRef]

25. Rumpold, B.A.; Schlüter, O.K. Nutritional composition and safety aspects of edible insects. *Mol. Nut. Food Res.* **2013**, *57*, 802–823. [CrossRef]

26. Mlcek, J.; Rop, O.; Borkovcova, M.; Bednarova, M. A comprehensive look at the possibilities of edible insects as food in Europe-a review. *Pol. J. Food Nutr. Sci.* **2014**, *64*, 147–157. [CrossRef]

27. Sun-Waterhouse, D.; Waterhouse, G.I.N.; You, L.; Zhang, J.; Liu, Y.; Ma, L.; Gao, J.; Dong, Y. Transforming insect biomass into consumer wellness food: A review. *Food Res. Int.* **2016**, *89*, 129–151. [CrossRef]

28. Nyangena, D.N.; Mutungi, C.; Imathiu, S.; Kinyuru, J.; Affognon, H.; Ekesi, S.; Nakimbugwe, D.; Fiaboe, K.K.M. Effects of traditional processing techniques on the nutritional and microbiological quality of four edible insect species used for food and feed in East Africa. *Foods* **2020**, *9*, 574. [CrossRef]

29. Grabowski, N.T.; Tchibozo, S.; Abdulmawjood, A.; Acheuk, F.; Guerfali, M.M.; Sayed, W.A.A.; Plötz, M. Edible insects in Africa in terms of food, wildlife resource, and pest management legislation. *Foods* **2020**, *9*, 502. [CrossRef] [PubMed]

30. Pyo, S.-J.; Kang, D.-G.; Jung, C.; Sohn, H.-Y. Anti-thrombotic, anti-oxidant and haemolysis activities of six edible insect species. *Foods* **2020**, *9*, 401. [CrossRef] [PubMed]

31. Ghosh, S.; Sohn, H.-Y.; Pyo, S.-J.; Jensen, A.B.; Meyer-Rochow, V.B.; Jung, C. Nutritional composition of Apis mellifera drones from Korea and Denmark as a potential sustainable alternative food source: Comparison between developmental stages. *Foods* **2020**, *9*, 389. [CrossRef] [PubMed]

32. Roma, R.; Palmisano, G.O.; De Boni, A. Insects as novel food: A consumer attitude analysis through the dominance-based rough set approach. *Foods* **2020**, *9*, 387. [CrossRef]

33. Leni, G.; Soetemans, L.; Caligiani, A.; Sforza, S.; Bastiaens, L. Degree of hydrolysis affects the techno-functional properties of lesser mealworm protein hydrolysates. *Foods* **2020**, *9*, 381. [CrossRef]

34. Lee, D.-Y.; Hong, K.-S.; Song, M.-Y.; Yun, S.-M.; Ji, S.-D.; Son, J.-G.; Kim, E.-H. Hepatoprotective effects of steamed and freeze-dried mature silkworm larval powder against ethanol-induced fatty liver disease in rats. *Foods* **2020**, *9*, 285. [CrossRef]

35. Meyer-Rochow, V.B.; Kejonen, A. Could western attitudes towards edible insects possibly be influenced by idioms containing unfavourable references to insects, spiders and other invertebrates? *Foods* **2020**, *9*, 172. [CrossRef]

36. Liu, C.; Masri, J.; Perez, V.; Maya, C.; Zhao, J. Growth performance and nutrient composition of mealworms (*Tenebrio molitor*) fed on fresh plant materials-supplemented diets. *Foods* **2020**, *9*, 151. [CrossRef]

37. Son, Y.-J.; Choi, S.Y.; Hwang, I.-K.; Nho, C.W.; Kim, S.H. Could defatted mealworm (*Tenebrio molitor*) and mealworm oil Be used as food ingredients? *Foods* **2020**, *9*, 40. [CrossRef]

38. Van Itterbeeck, J.; Andrianavalona, I.N.R.; Rajemison, F.I.; Rakondrasoa, J.F.; Ralantiarinaivo, V.R.; Hugel, S.; Fisher, B.L. Diversity and use of edible grasshoppers, locusts, crickets, and katydids (Orthoptera) in Madagascar. *Foods* **2019**, *8*, 666. [CrossRef]

39. Messina, C.M.; Gaglio, R.; Morghese, M.; Tolone, M.; Arena, R.; Moschetti, G.; Santulli, A.; Francesca, N.; Settanni, L. Microbiological profile and bioactive properties of insect powders used in food and feed formulations. *Foods* **2019**, *8*, 400. [CrossRef]

40. Castro, M.; Chambers, E.I.V. Consumer avoidance of insect containing foods: Primary emotions, perceptions and sensory characteristics driving consumers considerations. *Foods* **2019**, *8*, 351. [CrossRef] [PubMed]

41. Wendin, K.; Olsson, V.; Langton, M. Mealworms as food ingredient -sensory Investigation of a model system. *Foods* **2019**, *8*, 319. [CrossRef] [PubMed]

42. Mancini, S.; Sogari, G.; Menozzi, D.; Nuvoloni, R.; Torracca, B.; Moruzzo, R.; Paci, G. Factors predicting the intention of eating an insect-based product. *Foods* **2019**, *8*, 270. [CrossRef] [PubMed]

43. Wendin, K.; Mårtensson, L.; Djerf, H.; Langton, M. Product quality during the storage of foods with insects as an ingredient: Impact of particle size, antioxidant, oil content and salt content. *Foods* **2020**, *9*, 885. [CrossRef]

44. Mozhui, L.; Kakati, L.N.; Kiewhuo, P.; Changkija, S. Traditional knowledge of the utilization of edible insects in Nagaland, North-East India. *Foods* **2020**, *9*, 852. [CrossRef]

45. Anuduang, A.; Loo, Y.Y.; Jomduang, S.; Lim, S.J.; Wan Mustapha, W.A. Effect of thermal processing on physico-chemical and antioxidant properties in mulberry silkworm (*Bombyx mori* L.) powder. *Foods* **2020**, *9*, 871. [CrossRef]

46. Jeong, H.; Kim, J.M.; Kim, B.; Nam, J.-O.; Hahn, D.; Choi, M.B. Nutritional value of the larvae of the alien invasive wasp *Vespa velutina nigrithorax* and amino acid composition of the larval saliva. *Foods* **2020**, *9*, 885. [CrossRef]

47. Payne, C.L.R.; Scarborough, P.; Rayner, M.; Nonaka, K. Are edible insects more or less "healthy" than commonly consumed meats? A comparison using two nutrient profiling models developed to combat over- and undernutrition. *Europ. J. Clin. Nutr.* **2016**, *70*, 285–291. [CrossRef]

48. Sogari, G.; Amato, M.; Biasato, I.; Chiesa, S.; Gasco, L. The Potential role of insects as feed: A multi-perspective review. *Animals* **2019**, *9*, 119. [CrossRef]

49. Meyer-Rochow, V.B. Therapeutic arthropods and other, largely terrestrial, folk-medicinally important invertebrates: A comparative survey and review. *J. Ethnobiol. Ethnomed.* **2017**, *13*, 9. [CrossRef] [PubMed]

50. Müller, A. Insects as food in Laos and Thailand: A case of "Westernisation"? *Asian J. Soc. Sci.* **2019**, *47*, 204–223. [CrossRef]

51. Shelomi, M. Why we still don't eat insects: Assessing entomophagy promotion through a diffusion of innovations framework. *Trends Food Sci. Technol.* **2015**, *45*, 311–318. [CrossRef]

52. Megido, R.C.; Gierts, C.; Blecker, C.; Brostaux, Y.; Haubruge, E.; Alabi, T.; Franci, F. Consumer acceptance of insect-based alternative meat products in western countries. *Food Qualit. Pref.* **2016**, *52*, 237–243. [CrossRef]

53. Berger, S.; Bärtsch, C.; Schmidt, C.; Christandl, F.; Wyss, A.M. When utilitarian claims backfire: Advertising content and the uptake of insects as food. *Front. Nutr.* **2018**, *5*. [CrossRef] [PubMed]

54. Meyer-Rochow, V.B. Food taboos: Their origins and purposes. *J. Ethnobiol. Ethnomed.* **2009**, *5*, 18. [CrossRef]

55. Ghosh, S.; Jung, C.; Meyer-Rochow, V.B. What governs selection and acceptance of edible insect species? In *Edible Insects in Sustainable Food Systems*; Halloran, A., Vantomme, P., Roos, N., Eds.; Springer: Cham, Switzerland, 2018; pp. 331–351.

56. Mishyna, M.; Chen, J.; Benjamin, O. 2020 Sensory attributes of edible insects and insect-based foods–Future outlooks for enhancing consumer appeal. *Trends Food Sci. Technol.* **2020**, *95*, 141–148. [CrossRef]

 *foods*

Editorial

Insects Used as Food and Feed: Isn't That What We All Need?

Victor Benno Meyer-Rochow [1,2,*] **and Chuleui Jung** [1,3,*]

[1] Agricultural Science and Technology Research Institute, Andong National University, Andong 36729, Korea
[2] Department of Genetics and Ecology, Oulu University, SF-90140 Oulu, Finland
[3] Department of Plant medicals, Andong National University, Andong 36729, Korea
[*] Correspondence: meyrow@gmail.com (V.B.M.-R.); cjung@andong.ac.kr (C.J.)

Received: 14 July 2020; Accepted: 21 July 2020; Published: 27 July 2020

Abstract: This Special Issue of *Foods* explores different aspects of how insects can be used as a novel resource for food and feed. Some contributions deal with questions of acceptability and legality, others tackle problems related to innovative techniques in processing and marketing food, and yet another group of papers highlights the use of insects and their bio-active products in the context of promoting human health. The collective aim of the contributions by the researchers from at least 20 countries is to examine whether the use of insects—be it for food, feed, or therapeutic purposes—has a future. We conclude that positive aspects undoubtedly exist regarding the nutritional and pharmacological value of various insect species but that environmental and bio-functional issues could even outweigh the nutritional value of food insects.

Keywords: entomophagy; insect edibility; food shortage; acceptance; bio-active compounds; nutrients

1. Introduction

The terms "entomophagy"—a combination of the ancient Greek words "entomon" (notched anima) and "phagein" (eating of something specific)—and "insectivorous" and "insectivore", which use a suffix of Latin origin (derived from "vorare", to devour), have been examined and compared with each other by Evans et al. [1] with regard to their suitability and applicability in connection with humans. That study not only demonstrates the uncertainty in how to describe the consumption of insects by humans, whether habitual or accidental, but also highlights that the use of insects by humans as food and feed can be approached from different angles in different disciplines by experts from different locations asking different questions and using a variety of tools. This Special Issue of *Foods* is a case in point because authors located in at least 20 different countries have contributed their findings to it.

The papers chosen to form part of this Special Issue range from those addressing legal matters and questions of acceptability to those that are primarily concerned with the processing, chemical composition, and bioactivity of the insects. The range of papers in this issue underscores the fact that a greater understanding of the potential of insects as food and feed depends on interactions between fields and experts of both the humanities and the sciences. Whether, however, insects will ever be the "food of the future" is difficult to say. They were undoubtedly "food of the past" for our ancestors, and they still are a "food of the present" for many inhabitants of the world today [2–5].

In the past, insects were indeed a food item appreciated by humankind worldwide [6–8]. Somehow—and for reasons not completely understood but probably related to the spread of Christianity that forbade the consumption of insects with the exception of four species of locust, or the awareness of the fact that insects can be vectors of some fatal human diseases, the emotional separation from nature giving rise to entomophobia, and an increasingly greater variety of food stuffs reaching the consumer—the use of insects as human food became continually less popular. Our closest animal relatives, the monkeys, have been observed to actively collect insects and other arthropods in order to eat them—e.g., [9–12]—or, in the case of millipedes, to use them therapeutically [13].

Over the last 30 years, there has been a renewed interest in insects as human food, and the literature dealing with entomophagy and entomotherapy has become so vast that it is seemingly impossible to review it in its entirety. International conferences have begun to focus more and more on edible insects since the XVI International Pacific Science Congress in Seoul in August 1987 and the International Conference on Minilivestock in Beijing in September 1995, and the subsequent book by Maurizio Paoletti [14] brought this topic to a wider audience. Scientific publications, too many to mention, have appeared in the last 20 years or so, praising the advantages of an insect-based diet over a diet consisting of conventional meats like poultry and especially ruminants and highlighting the environmentally advantageous farming practices for mini-livestock such as insects over those for traditionally farmed animals [15–17]. Various edible insect species have had their farming potential assessed, their acceptability as a novelty food (or feed in animal husbandry and fish culture) examined, and their potential risk of carrying diseases or undesired microbes scrutinized. Nevertheless, there remain many open questions, which is why Special Issues such as this are still important.

Although insects should not be seen as a food item for humans merely to survive times of dietary hardship and periods of starvation [18], there is no way of denying that the global food security situation for the human population is becoming increasingly precarious and that food production worldwide has to increase by at least 50% to meet the demand by 2050 [19]. Despite the earlier reports of traditionally living people in different parts of the world engaging in entomophagy (mentioned above), none of these reports until Meyer-Rochow [20] had thought to link global food security with insects being more universally and extensively used as a possible and potent way to ease global food shortages. It was largely due to DeFoliart [21] and Van Huis et al. [22] that the idea was taken up and more widely promoted.

We now possess a considerable amount of information on the kinds of insect that serve as food to various people in the world; we know that most insects are nutritious, consist of valuable protein and easily digestible fatty acids, and contain important minerals and vitamins [23–25], and recommendations exist on how to breed the most lucrative species optimally [26,27]. However, gaps to be filled still exist with regard to the processing of cultured insects, the preservation of insect-based foods and their nutrients, the preparation of insect-containing dishes, the economics and marketing, and the potential of insects as suppliers of health-promoting drugs and medicines.

2. Results

In this Special Issue of *Foods*, Dorothy Nyangena and colleagues [28] stress that quality and safety concerns call for simple, actionable hazard control mechanisms. They tested the extent to which different processing methods affected the proximate composition and microbiological characteristics of the edible cricket *Acheta domesticus* and the grasshopper *Ruspolia differens*. The researchers concluded that traditional processing can alter the nutritional value of the insects but improves microbial safety.

Microbial safety was also an issue that the paper by Nils Grabowski et al. [29] dealt with. The authors pointed out that edible insects, like other foodstuffs, in order to guarantee consumer safety and product quality, should be covered by governmental regulations. Since nations vary with regard to their approaches concerning the inclusion of insects in foodstuffs and commercial farming, the situation is complex. This paper reviews the situation from a hygiene point of view and provides some worthwhile recommendations.

The Korean scientists Su-Jin Pyo and colleagues [30] focused on six species of traditionally used insects in Korea and demonstrated that, in particular, *Allomyrina dichotoma* and *Teleogryllus emma* showed considerable antioxidant activities. *Tenebrio molitor* and honeybee extracts, more so than the other species, exhibited strong haemolytic activities, suggesting that these insects not only possessed potential as a food source but could be used therapeutically as well.

The paper by Sampat Ghosh and colleagues [31] shows that drones and the developmental stages of identical species (in this case, the honeybee *Apis mellifera*), but representing different strains from widely different localities (here: "Buckfast" drones from Denmark and "Italian" drones from Korea) can differ with regard to their composition. Buckfast drones, for example, exhibited the highest antioxidant activity.

Results on the basis of data obtained from 310 Italian consumers were analyzed by Rocco Roma and colleagues [32] in order to ascertain the acceptability of edible insects as food or food supplements. Consumers fell into five groups and the acceptance of insects as feed material amongst them was widespread, but acceptance as a food item for humans was not. Reasons such as neophobia, disgust, assumed health risk, unfamiliarity, and socio-cultural backgrounds are thought to be responsible, although age-dependent differences were noticed.

Giulia Leni and co-workers [33] present a novel method tested on the lesser mealworm *Alphitobius diaperinus* that can be used in conjunction with other insect species to extract high quality protein for food and feed purposes. Enzymatic hydrolysis with *Bacillus licheniformis* for five hours resulted in a protein hydrolysate with a 15% degree of hydrolysis, oil-holding capacity, and excellent solubility.

Da-Young Lee and colleagues [34], who investigated the extent to which freeze-dried mature larval silkworm powder (SMSP) affected ethanol-induced fatty liver disease in rats, found that mature larval silkworm powder reduced triglyceride levels by as much as 35% in the livers of the rats that had received daily doses of 25% ethanol (3 g/kg body weight) for 4 weeks. In ethanol-treated rats given SMSP, the serum levels of triglycerides and various other compounds were lower, suggesting that SMSP can protect against alcoholic liver disease.

In view of the fact that so-called mirror neurons can influence a person's decisions and/or opinions, Meyer-Rochow and Kejonen [35] analyzed a list of commonly used idioms in the Finnish language that make reference to insects in the context of food and eating. The vast majority of these idioms referred to insects in a negative way, which could lead to or reinforce feelings of disgust and prevent people from trying insect-containing food items.

That the nutrient composition of mealworms can be influenced by fresh plant supplements in their diet during the growing phase of the insects was demonstrated by Changqi Liu and co-workers [36]. When mealworm larvae were reared on wheat bran enriched with carrot, orange, or red cabbage for four weeks, the larvae grew better and bigger, being 40–46% heavier than the controls that did not receive the food supplements.

Yang-Ju Son and colleagues [37] used mealworms to produce a defatted powder and mealworm oil. While the powder was rich in protein, minerals, and bioactive compounds and had a pleasant taste, the mealworm oil contained bioactive nutrients, especially gamma tocopherol, and had physicochemical properties that were similar to those of vegetable oil. Mealworm products contain antioxidants and glucosamine derivatives such as chitin and chitosan and possess anti-inflammatory properties.

Inhabitants of the island of Madagascar, as Joost Van Itterbeeck and colleagues [38] could show, have a long history of using especially insects belonging to the Orthoptera as a traditional food. Thirty-seven edible species were recorded. The insects were mostly eaten by humans as snacks and part of the main meal and were primarily collected by children. With the exception of *Brachytrupes membranaceus colosseus*, the insects were not marketed, and using insects as feed for animals was rare.

The microbiological and bioactive properties of powders made from *Acheta domesticus* and *Tenebrio molitor* were examined by Concetta Maria Messina and co-workers [39], who found a higher number of viable bacteria, mainly *Citrobacter* and Enterobacteriaceae, in mealworm flour than in cricket powder, which, on the other hand, contained more Porphyromonadaceae. The protein hydrolysates of both powders were rich in antioxidants.

To what extent an aversion to insects and rejection of insects as human food exists in different countries was studied by Mauricio Castro and Edgar Chambers [40] with a survey launched in 13 countries. Most of the 630 participants in the study were reluctant to accept insects and insect-containing food items, with the exception of respondents from Mexico and Thailand. Obviously, that educating consumers that food insects need not be unhygienic should be one goal in making food insects more widely acceptable, was their conclusion.

The latter is also the recommendation that came from the results obtained by Karin Wendin and colleagues [41], who could show that sensory attributes are a very important factor. Particle size, for example, had a great influence on the appearance and viscosity, colour and odour, as well as texture and taste of the insect-containing food but not the flavour. On the other hand, salt content did affect the taste and flavour. An addition of antioxidants reduced the colour change and rancidity in the food, thereby improving its acceptability.

An interesting test was carried out by Simone Mancini and psychology colleagues [42], who asked participants to evaluate by taste two identical bread samples, one labelled as having been supplemented with insect powder. In reality, however, both types of bread were identical. Even though the participants gave higher scores for flavour, texture, and overall liking to the insect-labelled bread, they nevertheless declared that they were less likely to consume the insect-labelled bread in the future, which suggests deep-seated feelings of insect rejection overriding sensory experience.

The idea that the acceptability of insects as a component of food items depends to a large degree on the way in which the edible insects are processed and presented but also on the packaging has been suggested before, but Wendin, Mårtensson, Djerf, and Langton [43] confirmed this through sensory analyses with probands. The results showed that following mixing and chopping (e.g., of mealworms), the particle size, oil content, salt content, and antioxidants influenced the product more than the storage time. Packed in Tetra Recart cartons, the product also stayed stable for at least 6 months at room temperature.

That there are still big gaps regarding the use and acceptability of insects as human food in various regions of the world has been shown in an interesting paper by Mozhui et al. [44]. In this beautifully illustrated investigation of entomophagic practices by the inhabitants of Nagaland (North-East India), the authors not only provide details on how the 106 edible species (representing nine orders of insects and 32 families) are obtained but also explain the various ways in which local residents turn the insects into delicious dishes.

An important aspect of processing insects as human food is tackled by the Malaysian/Thai team headed by Anuduang [45]. In their contribution they investigated the impact of a range of temperatures and durations on the quality of silkworm powder and found that shorter times of exposure to hot water and a low drying temperature preserve the antioxidant activities of the product best.

Jeong et al. [46] point out that despite their long history of utilization as food and pharmaceutical bioresources, the value of social wasps as a food resource is poorly appreciated and researched. Especially, not only are the larvae of the invasive *Vespa velutina nigrithorax* rich in nutrition and safe to eat but their saliva's amino acid composition has been shown to be worthy of further study in the context of amino acid supplementation and exercise enhancement.

3. Discussion

Coming back to the original question as to whether insects as food and feed have a future, we conclude that there can be no doubt that the many positive aspects in terms of environmental issues, especially when commercially bred species are considered, will outweigh the fact that the nutritional value of food insects is not significantly greater than that of conventional foods [47]. Insects bred to replace or supplement feedstuffs required in fish culture, poultry, and other animal husbandry will encounter fewer acceptability problems than insects and insect products aimed at human consumers [48]. Moreover, although numerous species of insects have been used therapeutically in the past [49], modern chemical analytical and isolating techniques have made it possible to identify potent compounds in them, and, as also some papers in this issue have shown, insects and insect products with medicinal properties are likely to play an increasingly greater role in the future to combat infections and other maladies.

The main advantages of food insects over conventional meat-supplying species are that the former can be reared on much less food and water and require significantly less space than the latter. Furthermore, the rate of reproduction of insects is considerably higher than that of conventional food animals, and a significantly greater proportion of an insect's body weight is utilizable as food than is the case with regard to conventional meat animals. Most people and not just animal lovers would also find the "harvesting" of insects more acceptable than the slaughter of livestock. Finally, the so-called carbon footprint of farmed insects is deemed to be appreciably lower than that of farmed conventional food animals [15–17,22].

Obstacles to overcome are, on the one hand, the declining interest in edible insects by citizens of countries in which insects used to be consumed widely [50] and, on the other hand, the still widespread rejection of insect-containing foods by citizens of countries in which edible insects and insect-supplemented foods were not traditionally accepted but have now become available [51–53]. To understand what governs the selection and acceptance of edible insect species and what role taboos play in this has to be one important goal [54]; education, clever marketing, tapping into the trend of buying healthy food, and highlighting the traditional nutritional and medical benefits are additional goals that could help to make consumers consider insects as food [55,56]. Using catchy slogans such as "Forget about the pork and put a cricket on your fork" or "Mealworms and spaghetti is food that makes you happy" could also be expected to help!

Author Contributions: Conceptualization: V.B.M.-R. and C.J.; Methodology: V.B.M.-R. and C.J.; Validation: V.B.M.-R. and C.J.; Formal analysis: V.B.M.-R. and C.J.; Investigation: V.B.M.-R. and C.J.; Resources: V.B.M.-R. and C.J.; Writing—original draft preparation: V.B.M.-R.; Writing—review and editing: V.B.M.-R. and C.J.; Visualization: V.B.M.-R. and C.J.; Project administration, V.B.M.-R. and C.J. All authors have read and agreed to the published version of the manuscript.

Funding: This research received no external funding.

Conflicts of Interest: The authors declare that there is no conflict of interests.

References

1. Evans, J.; Alemu, M.H.; Flore, R.; Frost, M.B.; Halloran, A.; Jensen, A.B.; Maciel-Vergara, G.; Meyer-Rochow, V.B.; Münke-Svendsen, C.; Olsen, S.B.; et al. 'Entomophagy': An evolving terminology in need of review. *J. Insects Food Feed.* **2015**, *1*, 293–305. [CrossRef]

2. Mitsuhashi, J. *The World's Edible Insects (Sekai Konchuu Shokkou Taizen)*; Yasaka Shobou: Tokyo, Japan, 2008.

3. Jongema, Y. *List of Edible Insects of the World*; Wageningen UR: Wageningen, The Netherlands, 2015; Available online: https://tinyurl.com/mestm6p (accessed on 22 June 2020).

4. Van Huis, A. Chapter 11. Insects as human food. In *Ethnozoology: Animals in Our Lives*; Alves, R., Paulino de Albuquerque, U., Eds.; Elsevier: Amsterdam, The Netherlands, 2018; pp. 195–213.

5. Meyer-Rochow, V.B. Insects (and other non-crustacean arthropods) as human food. In *Encyclopedia of Food Security and Sustainability, Vol. 1*; Ferranti, P., Berry, E.M., Anderson, J.R., Eds.; Elsevier: Amsterdam, The Netherlands, 2019; pp. 416–421. ISBN 9780128126875.

6. Bequaert, J. Insects as food: How they have augmented the food supply of mankind in early and recent years. *Nat. Hist. J.* **1921**, *21*, 191–200.

7. Bergier, E. *Peuples Entomophages et Insectes Comestibles: Étude sur les Moeurs de L'homme et de L'insecte*; Imprimérie Rullière Frères: Avignon, France, 1941.

8. Bodenheimer, F.S. *Insects as Human Food*; W. Junk Publishers: The Hague, The Netherlands, 1951.

9. Marshall, G.A.K. Five years' observations and experiments (1896–1901) on the bionomics of South African insects, chiefly mimicry and warning colours. *Trans. Entomol. Soc. Lond.* **1902**, *50*, 287–584.

10. Carpenter, G.D.H. Experiments on the relative edibility of insects with special reference to their coloration. *Trans. Roy. Entomol. Soc. Lond.* **1921**, *54*, 1–105.

11. Nickle, D.A.; Heymann, E.W. Predation on orthoptera and other orders of insects by tamarind monkeys, *Saguinus mystax mystax* and *Saguinus fuscicollis nigrifrons* (Primates: Callitrichidae) in north-eastern Peru. *J. Zool. (London)* **1996**, *239*, 799–819. [CrossRef]

12. Rothman, J.M.; Raubenheimer, D.; Bryer, M.A.H.; Takahashi, M.; Gilbert, C.C. Nutritional contributions of insects to primate diets: Implications for primate evolution. *J. Hum. Evol.* **2014**, *71*, 59–69. [CrossRef]

13. Weldon, P.; Aldrich, J.R.; Klun, J.A.; Oliver, J.E.; Debboun, M. 2003 Benzoquinones from millipedes deter mosquitoes and elicit self-anointing in capuchin monkeys (*Cebus* spp.). *Sci. Nat.* **2003**, *90*, 301–304. [CrossRef]

14. Paoletti, M.G. *Ecological Implications of Minilivestock: Potential of Insects, Rodents, Frogs, and Snails*; Science Publishers: Enfield, EN, USA, 2005.

15. Oonincx, D.G.A.B.; De Boer, I.J.M. Environmental impact of the production of mealworms as a protein source for humans—a life cycle assessment. *PLoS ONE* **2012**, *7*, e51145. [CrossRef]

16. Abbasi, T.; Abbasi, T.; Abbasi, S.A. Reducing the global environmental impact of livestock production: The minilivestock option. *J. Clean. Prod.* **2015**, *112*, 1754–1766. [CrossRef]

17. Halloran, A.; Caparros Megido, R.; Oloo, J.; Weigel, T.; Nsevolo, P.; Francis, F. Comparative aspect of cricket framing in Thailand, Cambodia, Lao People's Democratic Republic, Democratic Republic of the Congo, and Kenya. *J. Insects Food Feed* **2018**, *4*, 101–114. [CrossRef]

18. Gahukar, R.T. Edible insects collected from forests for family livelihood and wellness of rural communities-a review. *Global Food Secur.* **2020**. [CrossRef]

19. Tilman, D.; Balzer, C.; Hill, J.; Befort, B.L. Global food demand and the sustainable intensification of agriculture. *Proc. Natl. Acad. Sci. USA* **2011**, *108*, 20260–20264. [CrossRef] [PubMed]

20. Meyer-Rochow, V.B. Can insects help to ease the problem of world food shortage? *Search* **1975**, *6*, 261–262.

21. DeFoliart, G.R. Insects as food: Why the western attitude is important. *Ann. Rev. Entomol.* **1999**, *41*, 21–50. [CrossRef]

22. Van Huis, A.; Van Itterbeeck, J.; Klunder, H.; Mertens, E.; Halloran, A.; Muir, G.; Vantomme, P. *Edible Insects: Future Prospects for Food and Feed*; Food and Agriculture Organisation of the United Nations: Rome, Italy, 2013; pp. 1–190.

23. Bukkens, S.G.F. The nutritional value of edible insects. *Ecol. Food Nutr.* **1997**, *36*, 287–319. [CrossRef]

24. Chakravorty, J.; Ghosh, S.; Meyer-Rochow, V.B. Chemical composition of *Aspongopus nepalensis* Westwood 1837 (Hemiptera; Pentatomidae), a common food insect of tribal people in Arunachal Pradesh (India). *Int. J. Vitam. Nutr. Res.* **2011**, *81*, 1–14. [CrossRef]

25. Rumpold, B.A.; Schlüter, O.K. Nutritional composition and safety aspects of edible insects. *Mol. Nut. Food Res.* **2013**, *57*, 802–823. [CrossRef]

26. Mlcek, J.; Rop, O.; Borkovcova, M.; Bednarova, M. A comprehensive look at the possibilities of edible insects as food in Europe-a review. *Pol. J. Food Nutr. Sci.* **2014**, *64*, 147–157. [CrossRef]

27. Sun-Waterhouse, D.; Waterhouse, G.I.N.; You, L.; Zhang, J.; Liu, Y.; Ma, L.; Gao, J.; Dong, Y. Transforming insect biomass into consumer wellness food: A review. *Food Res. Int.* **2016**, *89*, 129–151. [CrossRef]

28. Nyangena, D.N.; Mutungi, C.; Imathiu, S.; Kinyuru, J.; Affognon, H.; Ekesi, S.; Nakimbugwe, D.; Fiaboe, K.K.M. Effects of traditional processing techniques on the nutritional and microbiological quality of four edible insect species used for food and feed in East Africa. *Foods* **2020**, *9*, 574. [CrossRef]

29. Grabowski, N.T.; Tchibozo, S.; Abdulmawjood, A.; Acheuk, F.; Guerfali, M.M.; Sayed, W.A.A.; Plötz, M. Edible insects in Africa in terms of food, wildlife resource, and pest management legislation. *Foods* **2020**, *9*, 502. [CrossRef] [PubMed]

30. Pyo, S.-J.; Kang, D.-G.; Jung, C.; Sohn, H.-Y. Anti-thrombotic, anti-oxidant and haemolysis activities of six edible insect species. *Foods* **2020**, *9*, 401. [CrossRef] [PubMed]

31. Ghosh, S.; Sohn, H.-Y.; Pyo, S.-J.; Jensen, A.B.; Meyer-Rochow, V.B.; Jung, C. Nutritional composition of Apis mellifera drones from Korea and Denmark as a potential sustainable alternative food source: Comparison between developmental stages. *Foods* **2020**, *9*, 389. [CrossRef] [PubMed]

32. Roma, R.; Palmisano, G.O.; De Boni, A. Insects as novel food: A consumer attitude analysis through the dominance-based rough set approach. *Foods* **2020**, *9*, 387. [CrossRef]

33. Leni, G.; Soetemans, L.; Caligiani, A.; Sforza, S.; Bastiaens, L. Degree of hydrolysis affects the techno-functional properties of lesser mealworm protein hydrolysates. *Foods* **2020**, *9*, 381. [CrossRef]

34. Lee, D.-Y.; Hong, K.-S.; Song, M.-Y.; Yun, S.-M.; Ji, S.-D.; Son, J.-G.; Kim, E.-H. Hepatoprotective effects of steamed and freeze-dried mature silkworm larval powder against ethanol-induced fatty liver disease in rats. *Foods* **2020**, *9*, 285. [CrossRef]

35. Meyer-Rochow, V.B.; Kejonen, A. Could western attitudes towards edible insects possibly be influenced by idioms containing unfavourable references to insects, spiders and other invertebrates? *Foods* **2020**, *9*, 172. [CrossRef]

36. Liu, C.; Masri, J.; Perez, V.; Maya, C.; Zhao, J. Growth performance and nutrient composition of mealworms (*Tenebrio molitor*) fed on fresh plant materials-supplemented diets. *Foods* **2020**, *9*, 151. [CrossRef]

37. Son, Y.-J.; Choi, S.Y.; Hwang, I.-K.; Nho, C.W.; Kim, S.H. Could defatted mealworm (*Tenebrio molitor*) and mealworm oil Be used as food ingredients? *Foods* **2020**, *9*, 40. [CrossRef]

38. Van Itterbeeck, J.; Andrianavalona, I.N.R.; Rajemison, F.I.; Rakondrasoa, J.F.; Ralantiarinaivo, V.R.; Hugel, S.; Fisher, B.L. Diversity and use of edible grasshoppers, locusts, crickets, and katydids (Orthoptera) in Madagascar. *Foods* **2019**, *8*, 666. [CrossRef]

39. Messina, C.M.; Gaglio, R.; Morghese, M.; Tolone, M.; Arena, R.; Moschetti, G.; Santulli, A.; Francesca, N.; Settanni, L. Microbiological profile and bioactive properties of insect powders used in food and feed formulations. *Foods* **2019**, *8*, 400. [CrossRef]

40. Castro, M.; Chambers, E.I.V. Consumer avoidance of insect containing foods: Primary emotions, perceptions and sensory characteristics driving consumers considerations. *Foods* **2019**, *8*, 351. [CrossRef] [PubMed]

41. Wendin, K.; Olsson, V.; Langton, M. Mealworms as food ingredient -sensory Investigation of a model system. *Foods* **2019**, *8*, 319. [CrossRef] [PubMed]

42. Mancini, S.; Sogari, G.; Menozzi, D.; Nuvoloni, R.; Torracca, B.; Moruzzo, R.; Paci, G. Factors predicting the intention of eating an insect-based product. *Foods* **2019**, *8*, 270. [CrossRef] [PubMed]

43. Wendin, K.; Mårtensson, L.; Djerf, H.; Langton, M. Product quality during the storage of foods with insects as an ingredient: Impact of particle size, antioxidant, oil content and salt content. *Foods* **2020**, *9*, 885. [CrossRef]

44. Mozhui, L.; Kakati, L.N.; Kiewhuo, P.; Changkija, S. Traditional knowledge of the utilization of edible insects in Nagaland, North-East India. *Foods* **2020**, *9*, 852. [CrossRef]

45. Anuduang, A.; Loo, Y.Y.; Jomduang, S.; Lim, S.J.; Wan Mustapha, W.A. Effect of thermal processing on physico-chemical and antioxidant properties in mulberry silkworm (*Bombyx mori* L.) powder. *Foods* **2020**, *9*, 871. [CrossRef]

46. Jeong, H.; Kim, J.M.; Kim, B.; Nam, J.-O.; Hahn, D.; Choi, M.B. Nutritional value of the larvae of the alien invasive wasp *Vespa velutina nigrithorax* and amino acid composition of the larval saliva. *Foods* **2020**, *9*, 885. [CrossRef]

47. Payne, C.L.R.; Scarborough, P.; Rayner, M.; Nonaka, K. Are edible insects more or less "healthy" than commonly consumed meats? A comparison using two nutrient profiling models developed to combat over- and undernutrition. *Europ. J. Clin. Nutr.* **2016**, *70*, 285–291. [CrossRef]

48. Sogari, G.; Amato, M.; Biasato, I.; Chiesa, S.; Gasco, L. The Potential role of insects as feed: A multi-perspective review. *Animals* **2019**, *9*, 119. [CrossRef]

49. Meyer-Rochow, V.B. Therapeutic arthropods and other, largely terrestrial, folk-medicinally important invertebrates: A comparative survey and review. *J. Ethnobiol. Ethnomed.* **2017**, *13*, 9. [CrossRef] [PubMed]

50. Müller, A. Insects as food in Laos and Thailand: A case of "Westernisation"? *Asian J. Soc. Sci.* **2019**, *47*, 204–223. [CrossRef]

51. Shelomi, M. Why we still don't eat insects: Assessing entomophagy promotion through a diffusion of innovations framework. *Trends Food Sci. Technol.* **2015**, *45*, 311–318. [CrossRef]

52. Megido, R.C.; Gierts, C.; Blecker, C.; Brostaux, Y.; Haubruge, E.; Alabi, T.; Franci, F. Consumer acceptance of insect-based alternative meat products in western countries. *Food Qualit. Pref.* **2016**, *52*, 237–243. [CrossRef]

53. Berger, S.; Bärtsch, C.; Schmidt, C.; Christandl, F.; Wyss, A.M. When utilitarian claims backfire: Advertising content and the uptake of insects as food. *Front. Nutr.* **2018**, *5*. [CrossRef] [PubMed]

54. Meyer-Rochow, V.B. Food taboos: Their origins and purposes. *J. Ethnobiol. Ethnomed.* **2009**, *5*, 18. [CrossRef]

55. Ghosh, S.; Jung, C.; Meyer-Rochow, V.B. What governs selection and acceptance of edible insect species? In *Edible Insects in Sustainable Food Systems*; Halloran, A., Vantomme, P., Roos, N., Eds.; Springer: Cham, Switzerland, 2018; pp. 331–351.

56. Mishyna, M.; Chen, J.; Benjamin, O. 2020 Sensory attributes of edible insects and insect-based foods–Future outlooks for enhancing consumer appeal. *Trends Food Sci. Technol.* **2020**, *95*, 141–148. [CrossRef]

Article

Effects of Traditional Processing Techniques on the Nutritional and Microbiological Quality of Four Edible Insect Species Used for Food and Feed in East Africa

Dorothy N. Nyangena [1], Christopher Mutungi [2,*], Samuel Imathiu [1], John Kinyuru [1], Hippolyte Affognon [3], Sunday Ekesi [4], Dorothy Nakimbugwe [5] and Komi K. M. Fiaboe [6]

[1]   Department of Food Science and Technology, Jomo Kenyatta University of Agriculture and Technology, Nairobi 00100, Kenya; dorothynyangena@gmail.com (D.N.N.); samuel.imathiu@jkuat.ac.ke (S.I.); jkinyuru@agr.jkuat.ac.ke (J.K.)

[2]   International Institute of Tropical Agriculture (IITA), Mikocheni Light Industrial Area, Dar es Salaam 34441, Tanzania

[3]   West and Central African Council for Agricultural Research for Development (CORAF), Dakar 3120, Senegal; h.affognon@coraf.org

[4]   International Centre of Insect Physiology and Ecology, Nairobi 00100, Kenya; sekesi@icipe.org

[5]   Department of Food Technology and Nutrition, School of Technology, Nutrition and Bio-Engineering, Makerere University, Kampala 7062, Uganda; dnakimbugwe@gmail.com

[6]   International Institute of Tropical Agriculture (IITA), Yaoundé 2008, Cameroon; k.fiaboe@cgiar.org

*   Correspondence: c.mutungi@cgiar.org

Received: 8 March 2020; Accepted: 3 April 2020; Published: 4 May 2020

Abstract: Edible insects are increasingly being considered as food and feed ingredients because of their rich nutrient content. Already, edible insect farming has taken-off in Africa, but quality and safety concerns call for simple, actionable hazard control mechanisms. We examined the effects of traditional processing techniques—boiling, toasting, solar-drying, oven-drying, boiling + oven-drying, boiling + solar-drying, toasting + oven-drying, toasting + solar-drying—on the proximate composition and microbiological quality of adult *Acheta domesticus* and *Ruspolia differens*, the prepupae of *Hermetia illucens* and 5th instar larvae of *Spodoptera littoralis*. Boiling, toasting, and drying decreased the dry matter crude fat by 0.8–51% in the order: toasting > boiling > oven-drying > solar-drying, whereas the protein contents increased by 1.2–22% following the same order. Boiling and toasting decreased aerobic mesophilic bacterial populations, lowered *Staphylococcus aureus*, and eliminated the yeasts and moulds, Lac+ enteric bacteria, and *Salmonella*. Oven-drying alone marginally lowered bacterial populations as well as yeast and moulds, whereas solar-drying alone had no effect on these parameters. Oven-drying of the boiled or toasted products increased the aerobic mesophilic bacteria counts but the products remained negative on Lac+ enteric bacteria and *Salmonella*. Traditional processing improves microbial safety but alters the nutritional value. Species- and treatment-specific patterns exist.

Keywords: entomophagy; processing; traditional knowledge; food/feed safety; nutrition

---

1. Introduction

Insects are part of the diets of humans and domesticated animals in many parts of the world. They have been consumed by communities for many years, and were suggested as a resource that could be used to ease global food shortages [1]. It has been claimed that insects provide vital nutrients including protein, calories, minerals and useful bioactive compounds to more than 2 billion people worldwide [2]. In Africa, insects contribute to the livelihoods and food security of households by generating significant income, and creating employment for the local communities [3]. Edible insect species that are considered unsuitable for consumption by humans have also been used as ingredients to substitute conventional protein sources, e.g., fishmeal in poultry, fish and pig feeds, thus contributing indirectly to human diets [4–6]. Already in East Africa, edible insect farming initiatives have taken off, and some countries have developed regulatory mechanisms to mainstream their production and use, for example, in animal feeds [7]. Hence, there are opportunities for farmers to produce edible insects, which can be delivered to food and feed factories as a raw material for manufacturing value-added products.

To be able to utilize edible insects on a commercial scale, the production of large quantities of biomass is necessary, either through mass rearing or sustainable harvesting. However, the need for suitable postharvest techniques that can make it possible to accumulate practical quantities that are of high quality is also important. Notwithstanding the fact that many insect species are collected from wild habitats or likely to be reared in environments that potentially originate microbiological hazards [8–11], the rich nutritional profile of insects offers a suitable substrate for the growth of unwanted microorganisms, such as spoilage and pathogenic ones, when the conditions are suitable [12,13]. In Africa, freshly harvested or semi-processed insects find their way to rural open-air markets, with some favourite species consumed by humans reaching urban markets and restaurants [3,14]. The processes leading to delivery of the edible insects to the end-consumer are highly variable, as the techniques and practices in collection or harvesting, aggregation, handling, preliminary processing, packaging, storage, and transportation vary widely.

The harvesting of microbiologically hazard-free insects is difficult to achieve. The contamination of insects with unwanted microorganisms is a consequence of a combination of the substrates, the insect species, and the farming or collection environments and processing steps applied [10]. Wynants [15] examined the microbial dynamics during an industrial production cycle of lesser mealworms (*Alphitobius diaperinus*), and found a direct association between the microbial diversity of the substrate and the harvested insects. The microbial populations were, however, generally lower in the larvae compared to the substrate, feed remnants, faeces and exuviae. In a separate study, the microbial dynamics during the industrial rearing, processing, and storage of the edible house cricket, *Gryllodes sigillatus*, were reported [16]. The microbial diversity of the feed substrate and harvested crickets were similar. However, unlike *A. diaperinus*, the overall microbial population was higher in the crickets. In both studies, food pathogens including *Salmonella* spp., *Listeria monocytogenes*, *Bacillus cereus* or coagulase-positive staphylococci were not detected, but fungal isolates corresponding to the genera *Aspergillus* and *Fusarium* were recovered. Elsewhere [11], the possibility of transmission of *Salmonella* sp. to mealworms (*Tenebrio molitor*) reared on contaminated wheat bran substrate was investigated. Survival of *Salmonella* sp. in the larvae was dependent on the contamination level with little or no retention at the lower levels, possibly because of competitive exclusion by the endogenous larval microbiota and/or because of antibacterial activity of the larvae. These findings indicated that some bacterial species have a competitive advantage and become dominant depending on the insect species. Regardless of these complex dynamics, processing presents a critical line of defence against potential hazards. It also interrupts spoilage processes, thereby improving product quality and minimizing product losses. For this reason, recent reviews have concluded that there is a need to upgrade and standardize processing methods so that the safety and nutritional value of insect-based food/feeds can be assured [10,17–19].

Some popular processing methods of edible insects in Africa (see Supplementary Material S1) include steaming, boiling, roasting, toasting, frying, smoking, and drying, or a combination of these [17]. A significant reduction in microbial hazards can be achieved by approaches such as thermal treatment, but some procedures may fail to adequately achieve the desired results [12]. The evaluation and validation of these methods that build on traditional knowledge would be a strong entry point for developing and implementing a successful food/feed safety mechanism, especially in rural settings where insect rearing and collection can have the greatest impacts in securing livelihoods. Thus, the aim of the present work was to examine the effect of some popular processing methods on the nutritional and microbiological quality of different insect species, Specifically, we asked: (i) how do processing techniques affect the nutritional value and microbiological quality? (ii) for the same processing technique, does the nutritional value, and microbiological quality of the processed products vary with the insect species? (iii) is there interaction between processing technique and insect species? Such knowledge would contribute crucial guidance for technological improvements in hazard control plans targeting rural collectors and the small-scale actors involved in insect rearing and preliminary processing activities.

2. Materials and Methods

2.1. Sources of Edible Insect Samples

Adult house crickets (*Acheta domesticus*) reared on a mixture of brewer's waste and kales, black soldier fly (*Hermetia illucens*) pre-pupae reared on a mixture of brewer's and kitchen waste, and African cotton leafworm (*Spodoptera littoralis*) 5th instar larvae reared on black nightshade leaves, were collected from the insect rearing and containment unit of the International Centre for Insect Physiology and Ecology (ICIPE) in Nairobi, Kenya. Adult grasshoppers (*Ruspolia differens*) were collected from the wild in Kampala Central, Nakawa and Makidye divisions, Uganda, and transported overnight in cool boxes to Jomo Kenyatta University of Agriculture and Technology (JKUAT, Nairobi, Kenya), Food Science laboratory for analysis. All the raw insects were washed with chilled tap-water (4 °C) before any processing. All samples were sieved and separated from undesired debris at the point of collection. The samples for microbiological analysis were analysed within 24 h of collection or processing, during which period they were stored in a refrigerator maintained at 4–8 °C. Samples for chemical analysis were preserved in a deep freezer (−21 °C) in sterile zip-lock bags.

2.2. Experimental Design

A factorial design of two factors—insect species and processing method—was applied. There were four levels of insect species—*A. domesticus*, *R. differens*, *H. illucens*, *S. littoralis*—and nine levels of processing method—toasting, boiling, solar-drying, oven-drying, toasting + solar-drying, toasting + oven-drying, boiling + solar-drying, boiling + oven-drying—and finally, the raw insects washed in chilled water. The experiment was replicated three times.

2.3. Post-harvest Processing

2.3.1. Toasting

A clean, dry, stainless pan was placed over an open flame and heated to about 150 °C. The raw insects (500 g) were then placed on the hot pan and without addition of cooking oil fried for 5 min, with regular turning using a wooden cooking stick to avoid sticking or burning [20]. The toasted products were then transferred on to an aluminium foil and left for 20 min to equilibrate to room temperature (22–25 °C). The product was then subdivided into two lots which were packed separately in polyethylene zip-lock bags and stored in a refrigerator or deep freezer awaiting microbiological and chemical analysis, respectively.

2.3.2. Boiling

The raw insects (500 g) were placed on a wire-mesh kitchen sieve and submerged in a boiling water bath (96 °C) for 5 min [12]. The sieve was then lifted from the water, and the contents were allowed to drain for 1 min. The insects were transferred on to an aluminium foil and left for 20 min cool to room temperature (22–25 °C). The product was then subdivided into two lots which were then packed in polyethylene zip-lock bags and stored in a refrigerator or freezer awaiting microbiological and chemical analysis, respectively.

2.3.3. Solar-Drying and Oven Drying

For the solar-dried products, the raw insects (500 g) were placed in a solar dryer, and left to dry to constant weight in 2–3 days. Details of the dyer design were as described elsewhere [21]. Briefly, the design consisted of a clear plastic (polyethylene) sheet stretched over a wooden box (0.6 m wide × 1.2 m long × 0.2 m high), which was placed longitudinally on a slanting metal frame constructed to a height of 1 m of the ground on the air inlet end, and 1.2 m on the air exit end. The inside of the box was lined with a black polyethylene sheet, and the air entry and exit ends were drilled with closely spaced holes of 1 cm diameter. The temperature and relative humidity in the dryer before introducing the insects ranged between 50–60 °C and 15–25%, respectively, as determined using an EL-USB-2 data logger (Lascar electronics Inc., Erie, PA, USA). For the oven-dried samples, raw insects (500 g) were placed in an air-oven dryer (TD-384KN model Thermotec, Tokyo, Japan), maintained at 60 °C and dried to constant weight in 2–3 days. The samples were regarded dry when the change in weight was less than 1% over three samplings performed at one-hour intervals. The products were removed from the drying chambers and left to cool for 20 min, following which they were subdivided into two lots, which were then packed in polyethylene zip-lock bags and stored in a refrigerator or freezer awaiting microbiological and chemical analysis, respectively.

2.3.4. Combined Processes (Boiling/Toasting and Drying)

Separate lots (500 g) of raw insects were toasted or boiled as described in Sections 2.3.1 and 2.3.2, respectively. These were then dried either in the solar-dryer or in the oven-dryer, as described in Section 2.3.3. Each of the final products was then subdivided into two lots, which were then packed in polyethylene zip-lock bags and stored in a refrigerator or freezer awaiting microbiological and chemical analysis, respectively.

2.4. Determination of Proximate Composition

The AOAC standard methods [22] were used. Moisture content was determined by hot-air drying (Method 925.10), crude protein by semi-micro Kjeldahl method for total nitrogen with 6.25 as the nitrogen-to-protein conversion factor (Method 920.87), crude fat by soxhlet extraction (Method 920.85), crude fibre by the Henneberg–Stohmann method (Method 920.86), and crude ash by incineration in a muffle furnace (Method 923.03). Total available carbohydrate was determined by the difference.

2.5. Assessment of Microbiological Quality

2.5.1. Enumeration of Total Viable Count, Lactose Positive Enteric (Lac+) bacteria, and *Staphylococcus Aureus*

The raw and processed products were each separately pulverized in a Philips HR-2850 mini blender that was pre-rinsed with 70% ethanol. Ten (10) grams of the pulverized material was transferred into a conical flask containing 90 mL of sterile peptone water (HiMedia M028) as a diluent, and the mixture was homogenized in a sterile polyethylene bag using a Stomacher® 400 Circulator (Seward, West Sussex, UK) for 2 min. Aliquots (1 mL) of the homogenate were serially diluted up to 10^{-7} in 9 mL of the diluent. Aliquots (0.1 mL) of individual dilutions were cultured in triplicate Petri dishes. Total viable counts (TVC) were determined by pour-plate technique on Plate Count Agar (HiMedia M091) after incubation at 35 °C for 48 h. The TVC provided an estimate of the total viable aerobic mesophilic microbial population present in the samples. Lac+ enteric bacteria were determined on MacConkey Agar (HiMedia M081) following 24 h incubation of the cultured plates at 37 °C. Typical colonies were distinguished by their pink-red colour. *Staphylococcus aureus* counts were determined on Baird Parker Agar (HiMedia M043) containing 5% Egg Yolk Tellurite Emulsion (HiMedia FD046) following incubation of the cultured plates at 35 °C for 48 h. Typical colonies were distinguished by their circular (2–3 mm diameter), smooth, convex, moist appearance, and confirmed using coagulase test. The latter was performed by emulsifying a portion of isolated colonies in physiological saline on a microscope slide to form a suspension, then mixing with rabbit plasma and examining the clumping behaviour of the cells. For all the determinations, plates having 20–300 colonies were counted, and the number was expressed as log-colony-forming units per gram (Log CFU/g) of sample.

2.5.2. Enumeration of Yeasts and Moulds

The determination of total yeasts and mould counts was performed on Potato Dextrose Agar (PDA) (HiMedia Ref. M096) acidified with sterile 10% tartaric Acid (251380, Sigma-Aldrich, Schnelldorf, Germany) to pH 3.5. Aliquots of sample homogenate prepared as in Section 2.5.1 above were serially diluted. Aliquots (0.1 mL) of individual dilutions were cultured in triplicate PDA plates using the spread plate technique, and the plates incubated at 25 °C for 72 h. Plates having 20–300 colonies were counted, and the number expressed as log-colony-forming units per gram (Log CFU/g) of sample.

2.5.3. Detection of Salmonella

Samples (25 g) of the raw and processed products were enriched with 225 mL nutrient broth containing 5 g peptone, 5 g sodium chloride, 1.5 g beef extract, and 1.5 g yeast extract per 1000 mL of water, pH 7.4 (HiMedia M002) and incubated at 35 °C for 24 h. Selective enrichment was then performed by transferring 25 mL of the enriched homogenate to 225 mL of tetrathionate broth (HiMedia M032) and the culture incubated at 37 °C for 24 h. A loopful of the Tetrathionate broth culture was streaked on Salmonella-Shigella Agar (HiMedia, M108) and plates incubated at 37 °C for 24 h. Typical *Salmonella* colonies (colourless colonies with black centres) were further identified using the Triple Sugar Iron (HiMedia, M021I) test. For this test, a sterilized inoculation needle was used to touch the top of a well isolated colony and then inoculated into the Triple Sugar Iron agar by first stabbing through the centre of the media to the bottom of the tube, followed by streaking on the surface of the agar slant. The slants were incubated at 35 °C for 24 h. *Salmonella* forms a red slope (alkaline) and yellow (acid) butt indicating fermentation of glucose, with or without gas (cracks and bubbles in the medium), and with or without H_2S production (blackening of the medium).

2.6. Statistical Analyses

Data were subjected to analysis of variance (ANOVA) using general linear model (GLM) procedures on IBM® SPSS® Statistics 20 (IBM Corporation, Armonk, NY, USA). Insect species and processing method were the two independent factors, while proximate composition parameters (moisture, crude protein, crude fat, crude fibre, ash, total carbohydrate contents) and microbiological parameters (total viable count, and counts of Lac+ bacteria, *St. aureus*, yeasts and moulds) were the dependent variables. For the experimental part examining the effects of processing on proximate composition, five factor levels of processing technique were considered (toasting, boiling, oven-drying, solar-drying, and the raw insects after washing in chilled water), whereas all nine levels were considered for the experimental part designed to examine the effects on microbiological quality. When interactions between insect species and processing method were significant, a one-way ANOVA was performed to test the significance of differences amongst insect species and processing method combinations. Means were separated using Bonferroni's *t*-test at $p < 0.05$.

3. Results and Discussion

3.1. Proximate Composition of Raw Insects

Dry matter (DM) contents of the fresh *A. domesticus*, *R. differens*, *H. illucens* prepupae, and *S. littoralis* larvae ranged between 26–35% (Table 1). The *A. domesticus* and *R. differens* had higher DM contents than *H. illucens* and *S. littoralis* larvae. The dry matter crude protein and crude fat also differed significantly (see Tables 1 and 2). Protein contents followed the order *A. domesticus* > *R. differens* > *H. illucens* > *S. littoralis*. Except for *S. littoralis*, the protein contents were within the known ranges in the literature: 55–70%, 34.2–45.8%, 38–48% for *A. domesticus* [23], *R. differens* [24], and *H. illucens* [4], respectively.

The protein content of *S. littoralis* was lower by a factor of 0.75 compared to the 51% reported by others [25]. The huge difference may be attributed to the substrates on which the caterpillars were reared, i.e., castor bean (*Ricinus communis*) leaves, which are richer in protein compared to black nightshade leaves (*Solanum nigrum*) in our case. Of the four insects, *H. illucens* and *R. differens* contained more fat by a factor of 1.5 compared to *A. domesticus* and *S. littoralis* (see Table 1). Furthermore, the fat contents of *H. illucens* and *A. domesticus* were within the range reported by other researchers: 9.8–22.8% for *A. domesticus* [23] and 15–35% for *H. illucens* [4]. The fat contents of *R. differens* and *S. littoralis* were approximately half the levels reported elsewhere i.e., 42.2–54.3% for *R. differens* [24] and 33% for *S. littoralis* [25].

Table 1. Moisture, crude protein, and crude fat contents of raw and processed insects.

	Insect Species			
	H. illucens	**A. domesticus**	**R. differens**	**S. littoralis**
	Moisture (g/100g)			
Raw	67.5 Cb	65.5 Ca	65.4 Ca	74.0 Cc
Boiled	74.2 Db (+ 9.9)	71.3 Da (+8.9)	76.5 Dc (+17.0)	87.2 Dd (+17.8)
Toasted	32.7 Ba (−52.5)	44.2 Bb (−32.5)	46.1 Bb (−39.5)	32.6 Ba (−55.9)
Solar-dried	10.1 Ad (−85.0)	9.9 Ac (−84.9)	9.8Ab (−85.0)	9.1 Aa (−87.7)
Oven-dried	8.9 Ac (−86.8)	8.8 Ac (−86.6)	8.3Ab (−87.3)	7.2 Aa (−90.3)
	Crude protein (g/100g DM)			
Raw	36.3 Aa	52.3 Ad	43.8 Ac	38.5 Ab
Boiled	39.6 Ca (+9.1)	57.5 Cd (+9.9)	48.8 Bc (+11.4)	41.9 Cb (+8.9)
Toasted	41.3 Da (+13.7)	59.5 Dd (+13.7)	53.5 Cc (+22.1)	43.8 Db (+13.7)
Solar-dried	37.0 Ba (+1.9)	53.1 ABd (+1.5)	42.3 Ac (+3.4)	39.7 Ab (+3.1)
Oven-dried	37.9 Ca (+4.4)	53.8 Bd (+2.9)	44.4 Ac (+1.4)	41.1 Bb (+7.1)
	Crude fat (g/100g DM)			
Raw	29.6 Db	18.3 Da	28.4 Db	17.4 Ca
Boiled	20.3 Bb (−31.4)	16.5 Ba (−9.8)	24.9 Bc (−12.3)	16.3 Ba (−6.3)
Toasted	14.3 Ab (−51.7)	10.5 Aa (−42.6)	20.1 Ac (−29.2)	9.7 Aa (−44.3)
Solar-dried	28.2 Dc (−4.7)	17.9 CDa (−2.2)	27.5 CDb (−3.3)	17.3 Ca (−0.8)
Oven-dried	26.9 Cb (−9.2)	17.1 BCa (−6.6)	26.7 Cb (−6.1)	16.9 BCa (−2.9)

Means in the same column followed by the same capital letters, and means in the same row followed by the same small letters are not significantly different ($p < 0.05$; $n = 3$). Values in parentheses are the percent change relative to the raw product.

The fibre, ash and available carbohydrate contents of the raw insects are presented in Table 2. Crude fibre levels were generally <10%, and lowest in *R. differens*. The crude fibre contents of *R. differens*, *H. illucens*, and *S. littoralis* corresponded well with those reported by other authors i.e., *R differens*: 1.8–2.7% [24]; *H. illucens*: 7% [4]; *S. littoralis*: 10.7% [25], whereas the levels determined in *A. domesticus* were about half the levels (14.9–22.1%) reported elsewhere [23]. Crude fibre also includes chitin and complex carbohydrates such as cellulose and lignin that might be present in the insects' gut [26], and therefore variability may arise depending on preparation processes such as the degutting, cleaning and removal of parts, or the inclusion of a starvation regime prior to harvesting [15]. Crude ash contents were about two times higher in *S. littoralis* compared to the other insects. The ash content of *R. differens* compared well with the findings reported elsewhere [20], whereas the ash content of *A. domesticus* was within the 3.6–9.1% range reported by other authors [23].

Table 2. Crude fibre, ash and carbohydrate contents of raw and processed insects.

	Insect Species			
	H. illucens	*A. domesticus*	*R. differens*	*S. littoralis*
	Crude Fibre (g/100g DM)			
Raw	8.6 Bc	8.1 BCc	4.2 Ba	6.4 Bb
Boiled	7.9 Ac (−8.1)	7.4 Ac (−8.6)	3.9 Aa (−7.1)	6.5 BCb (+1.6)
Toasted	10.6 Cd (+23.2)	8.2 Cc (+1.2)	4.5 Ca (+7.1)	6.2 Ab (−3.1)
Solar-dried	8.7 Bc (+1.2)	8.1 BCc (0.0)	4.2 Ba (0.0)	6.6 Cb (+3.1)
Oven-dried	8.5 Bc (−1.2)	8.0 Bc (−1.2)	4.1 Ba (−2.4)	6.6 Cb (+3.1)
	Crude Ash (g/100g DM)			
Raw	3.9 Bb	3.6 Bab	3.1 Ba	7.6 Bc
Boiled	2.9 Aa (−25.6)	3.0 Aa (−16.7)	2.7 Aa (−12.9)	6.7 Ab (−11.8)
Toasted	4.3 Ca (+10.3)	4.2 Ca (+16.7)	4.3 Da (+38.7)	9.5 Cb (+25.0)
Solar-dried	3.8 Ba (−0.25)	3.6 Ba (0.0)	3.8 Ca (+22.5)	7.6 Bb (0.0)
Oven-dried	3.9 Ba (0.0)	3.5 Ba (−2.8)	3.7 Ca (+19.3)	7.7 Bb (+1.3)
	Available Carbohydrate (g/100g DM)			
Raw	21.6 Ab	17.7 Da	20.4 Cb	30.1 Cc
Boiled	29.3 Cc (+35.6)	15.6 Aa (−11.9)	19.7 Bb (−3.4)	28.6 Bc (−4.9)
Toasted	29.5 Cb (+36.6)	17.6 CDa (−0.6)	17.7 Aa (−13.2)	30.7 Cb (+2.0)
Solar-dried	22.3 ABb (+3.2)	17.3 Ba (−2.3)	22.2 Eb (+8.8)	28.9 Bc (−4.0)
Oven-dried	22.8 Bb (+5.6)	17.5 BCa (−1.1)	21.2 Db (+3.9)	27.6 Ac (−8.3)

Means on the same column followed by the same capital letters, and means on the same row followed by the same small letters are not significantly different ($p < 0.05$; $n = 3$). Values in parentheses are the percent change relative to the raw product.

Previous studies show that edible insects are highly nutritious, but the nutritional composition varies widely. Protein, fat, fibre, and ash contents range between 4.9–77%, 0.7–77%, 0.9–29%, and 0.3–21% on DM basis, respectively [23]. The wide variability has been associated with insect species and feed substrates [27–30], collection sites and swarming seasons [24], age [31], sex [32], and possibly the use of varying methods or false interpretation. The insects examined in this study belonged to the orders Lepidoptera (*S. littoralis*), Orthoptera (*A. domesticus* and *R. differens*), and Diptera (*H. illucens*), and included adults (*A. domesticus* and *R. differens*), pre-pupae (*H. illucen*) and larvae (*S. Littoralis*). According to Rumpold and Schlüter [23], Orthopterans have, on average, a higher protein content (61.3%) compared to Dipterans (49.5%) and Lepidopterans (45.4%), whereas the Lepidopterans have a higher lipid content (27.7%) compared to Dipterans (22.8%) and Orthopterans (13.4%). Fibre contents are higher in Dipterans (13.6%) than Orthopterans (9.6%) and Lepidopterans (6.6%), while carbohydrates are higher in Lepidopterans (18.8%) compared to Orthopterans (13%) and Dipterans (6%). Ash content is higher in Dipterans (10.3%) than Lepidopterans (4.5%) and Orthopterans (3.9%). Examining our data within these categories, the findings on protein, fibre, and carbohydrate correspond well, while those on fat and ash do not, suggesting that the other factors (age, sex, feed substrates, collection sites and swarming seasons, etc.) also had strong effects. Age was reported to influence the nutritional composition of reared house crickets, with protein, total lipids and mineral content being highest at 9–12 weeks [31]. Regarding sex, females crickets were found to have more lipids and less proteins and chitin than the males [32]. *Hermetia illucens* larvae were shown to have the capacity to accumulate both lipid- and water-soluble nutrients from feed substrates [27]. Elsewhere, the *H. illucens* prepupae reared on digestate were low in lipids and high in ash compared to those reared on vegetable waste [29]. It was also reported that substrates comprising brewery waste or a mixture of fruit and vegetables resulted in higher protein content compared to fruit or winery by-product substrates [33]. Geographical area and swarming season were found to influence the nutritional composition of wild-harvested *R. differens* primarily due to differences in feed substrate, possibly arising from the type and abundance of available vegetation or soil type, which particularly influences the mineral composition of the vegetation [24].

3.2. Effects of Processing on Proximate Composition

The detailed statistical analysis results for effects of processing on proximate composition are presented in the Supplementary Material S2. The moisture contents of the processed products (Table 1) varied with the processing method (F = 628.7; df = 4; $p < 0.001$) and species (F = 28.3; df = 3; $p < 0.001$), and the interaction effect of processing method and insect species was significant (F = 92.1; df = 19; $p < 0.001$). Boiling resulted in moisture gain (10–18%) due to hydration or the increased binding of water by hydrophilic tissue components, whereas toasting resulted in a lower moisture content (30–60%) due to evaporative loss. A similar gain in moisture was reported in boiled *Gryllodes sigillatus* [16]. Drying reduced the moisture content of the raw insects by 85–90%, but oven-drying achieved lower moisture levels by about two percentage points compared to solar-drying. The *S. littoralis* gained more moisture during boiling, but also lost more moisture during toasting and drying, which indicates a deviation in hydration properties compared to the other insects. Earlier work [21] demonstrated dissimilarities in the hydration properties of *A. domesticus* and *H. illucens* powders, and attributed the differences to the compositional factors that influence the number and water binding strength of hydrophilic sites.

The crude protein contents increased by 1.2–22% relative to the raw product (see Table 1). The interaction of processing method and insect species was significant (F = 7.1; df = 19; $p < 0.001$), and the main effects were significant as well (processing: F = 99.4; df = 4; $p < 0.001$; insect species: F = 598.9; df = 3; $p < 0.001$). Boiling and toasting increased the protein contents remarkably (boiling: 8.9–11%; toasting 13–22%) on all the insects compared to solar-drying (1.5–3.5%) and oven-drying (1.2–7.1%). The increase in protein content upon drying was significant in *H. illucens* (both solar- and oven-drying) and in *A. domesticus* and *S. littoralis* (oven-drying) but not in *R. differens*. Other authors [34] reported a decrease in protein content when the beetle *Eulepida mashona* and the cricket *Henicus whellani* were boiled for 30–60 min (*E. Mashona*: 1.2–14.7%; *H. whellani* 9.5–10.1%), but then observed no change when the insects were toasted. The decrease was possibly due to the dissolution of protein or disintegration and loss of connective tissue as colloidal constituents in the boiling water. A decrease in crude protein content was also reported in studies with *Imbrasia belina* [34] and *Hemijana variegata* [35]. Other authors, e.g., [36], found no significant change in protein content when the edible caterpillar, *Imbrasia epimethea*, was thermally processed (boiled or boiled and sun-dried). A decrease in nitrogen content might also occur during thermal treatments due to loss of amides and amines [37] or the formation of complexes with primary and secondary lipid oxidation products [38]. Our results suggest that these effects were overshadowed by other dynamics such as the loss of dry matter components, mainly fat. In fact, the measured increase in protein content was strongly correlated with fat loss (R^2 = −0.692; $p < 0.001$). Solar-drying had minimal effect on protein content.

Processing decreased fat content by 2.0–51% (Table 1). Processing method and insect species were significant (processing method: F = 1209.1; df = 4; $p < 0.001$; insect species: F = 1734.9; df = 3; $p < 0.001$). Likewise, the interaction effect was significant (F = 55.5; df = 19; $p < 0.001$). Examining the main effects independently, toasting, boiling, oven-drying and solar-drying decreased the fat contents by 42.0%, 15.0%, 6.1% and 2.7%, respectively. Similarly, the loss of fat content followed the order *H. illucens* (24.2%) > *A domesticus* (15.3%) > *S. littoralis* (13.6%) > *R. differens* (12.8%). Fat loss correlated positively with the fat content of the raw insects (R^2 = 0.57; $p < 0.001$). The average loss was also higher for the non-orthopterans, i.e., *H. illucens* pre-pupae and *S. littoralis* larvae (18.9%), compared to the orthopterans (14.1%), although the difference was not statistically significant (p = 0.326). The loss of fat during boiling was due to the melting of fat globules into the boiling water. Similarly, toasting melted the fat, which became exuded as a result of tissue contraction, while some of it was possibly also lost through thermal decomposition [39]. The decrease in fat content during oven- and solar-drying suggests that some fat may have transuded alongside water vapor or was oxidized into other compounds [40,41], especially since insect lipids are made up of polyunsaturated fatty acids. Furthermore, the lipids of different insect species comprise the fatty acid profiles of varied physico-chemical properties [42], which might explain some of the variability in fat loss magnitudes during thermal treatment.

Crude fibre (see Table 2) increased or decreased depending on processing method and insect species (F = 17.1; df = 19; $p < 0.001$). Except for *S. littoralis*, boiling diminished the contents (7.6–8.7%), whereas toasting increased them (2–22%). Similar results were reported with cooked and roasted mopane worms [26]. Heating shifts the ratio of soluble to insoluble fibre [43]. The decrease in fibre content during boiling was probably because some of the complex carbohydrates dissolved [44] or were washed off the tissues. Toasting, on the other hand concentrated these polysaccharides, especially with the concomitant loss of fat during the process. However, the thermal treatment may also have resulted in an increase in total fibre by causing the formation of protein–fibre complexes [43]. Unlike boiling and toasting, drying did not affect crude fibre contents.

The crude ash content of processed products (Table 2) was also a function of the interaction of processing method and insect species. The main effects were also significant. The ash contents generally decreased upon boiling due to leaching [34,45] but increased in the toasted products. We attribute this increase to concentration effects, unlike in the findings of others [26] where contamination from ash during hot-ash roasting was implicated. There was no significant change in ash contents when samples were subjected to drying, except for *R. differens*, which was peculiar. The contents of available carbohydrates (Table 2) were dependent on the interaction of insect species and processing (F= 26.3; df = 19; $p < 0.001$). The main effects were also significant (insect species: F = 460; df = 3; $p < 0.001$; processing method: F= 7.9; df = 4; $p > 0.001$). Nonetheless there were no clear trends, as the CHO determination was dependent on the increase or decrease in the other parameters. Generally, for all the proximate parameters, the interaction of insect species with processing method was significant, meaning that each combination of species and processing method produced a different outcome.

3.3. Effect of Processing on Microbial Quality

The detailed results of statistical analyses exploring effects of processing technique on microbial quality are presented in the Supplementary Material S3.

3.3.1. Effects of Boiling and Toasting

Total viable count and YMC of the raw and processed products are presented in Table 3. The TVC determined on the raw insects (7.0–9.1 Log CFU/g) were comparable to those reported elsewhere [12,15,16]. The YMC (6.4–8.2 Log CFU/g) were higher by 2–3 log cycles compared to the YMC reported for industrially reared lesser mealworms (*Alphitobius diaperinus*) [15], but within the limits reported for industrially reared house crickets *Gryllodes sigillatus* [16]. The allowable TVC limit in edible foods is 7 Log CFU/g [46] and, therefore, all the raw insects except *S. littoralis* exceeded this limit. For animal feeds, however, the TVC would be within the allowable limit of 10^9 CFU/g [7]. Similarly, the raw insects had a YMC higher than the recommended limit of 6 Log CFU/g [46].

Boiling and toasting lowered the TVC by 4–6 log cycles and completely eliminated the yeasts and moulds. Similar findings were reported for boiled (5–10 min) *G. sigillatus* [16], boiled (ca. 93 °C; 30 min) *I. belina* [47] and toasted (10 min) *R. differens* [48]. The interaction effect of insect species and processing method was significant for the destruction of TVC (F = 1207; df = 35; $p < 0.001$) and YMC (F = 11739.6; df = 35; $p < 0.001$). Furthermore, both main effects were significant for both TVC (processing method: F = 505.5; df = 8; $p < 0.001$; insect species: F = 10.26; df = 3; $p < 0.001$) and YMC (processing method: F = 526.1; df = 8; $p < 0.001$; insect species: F = 13.1; df = 3; $p < 0.001$). Boiling was more effective than toasting at lowering the TVC, due to better heat transfer through the tissues [12]. However, our results also show differences in the reduction in TVC and YMC across the insect species (see Table 3), which might be due to a combination of factors such as the initial contamination levels and size.

With the exception of *S. littoralis*, all the unprocessed insects were positive for Lac+ enteric bacteria (Table 4). This category comprises bacteria such as *Escherichia coli*, *Klebsiella*, and *Enterobacter* spp., and therefore may indicate contamination with faecal coliforms from unhygienic environments, soil or poor handling [49]. In dried insect products for compounding animal feeds, total coliform count should not exceed 500 CFU/g, while pathogenic *E. coli* should not exceeded the

10 CFU/g [7]. The absence of Lac+ enteric bacteria in *S. littoralis* was because of the hygienic conditions of rearing. Elsewhere, contamination with coliforms including *E. coli* was also reported in wild-harvested unprocessed emperor moth caterpillars [13,47] and grasshoppers [50]. The boiling and toasting procedures applied in the present study potentially eliminated *E. coli* and other enteric pathogens. Other authors, [47,48], made similar observations on boiled *I. belina* larvae and toasted *R. differens*, respectively.

Table 3. Total viable counts and yeast and mould counts on raw and processed products.

	Insect Species			
	H. illucens	*A. domesticus*	*R. differens*	*S. littoralis*
	Total viable count (Log CFU/g)			
Raw	7.7 Eb	8.3 Fc	9.1 Ed	7.0 Ea
Boiled	2.6 Bb (−66.2)	2.3 Ba (−72.3)	2.8 Bc (−69.2)	2.6 Bb (−62.9)
Toasted	2.8 BCa (−63.6)	3.1 Cd (−62.7)	2.9 Bb (−68.1)	3.0 BCc (−57.1)
Solar-dried	7.8 Eb (+1.3)	8.8 Fc (+6.0)	9.2 Ed (+1.1)	7.4 Ea (+5.7)
Oven-dried	6.2 Da (−19.4)	6.8 Eb (−18.1)	6.8 Db (−25.3)	6.4 Dc (−8.6)
Boiled + Solar-dried	3.1 Ca (−59.7)	3.9 Db (−53.0)	3.2 Ba (−64.8)	3.1 Ca (−55.7)
Toasted + Solar-dried	3.0 BCab (−61.0)	3.1 Cb (−62.6)	3.9 Cc (−57.1)	2.9 BC (−58.6)
Boiled + Oven-dried	1.6 Ab (−79.2)	1.0 Aa (−87.9)	1.8 Ac (80.2)	1.7 Abc (75.7)
Toasted + Oven-dried	1.8 Ab (−76.6)	1.3 Aa (−78.3)	2.0Ac (−78.0)	2.1 Ac (−70.0)
	Yeast and mould count (Log CFU/g)			
Raw	7.6 Cb	7.8 Cb	8.2 Cc	6.4 Ba
Boiled	0.0	0.0	0.0	0.0
Toasted	0.0	0.0	0.0	0.0
Solar-dried	6.5 Ba (−14.5)	7.0 Bb (−10.2)	8.0 Bc (−2.4)	6.3 Ba (−1.5)
Oven-dried	4.2 Ab (−44.7)	3.2 Aa (−59.0)	5.5 Ac (−32.9)	3.0 Aa (−53.1)
Boiled + Solar-dried	0.0	0.0	0.0	0.0
Toasted + Solar-dried	0.0	0.0	0.0	0.0
Boiled + Oven-dried	0.0	0.0	0.0	0.0
Toasted + Oven-dried	0.0	0.0	0.0	0.0

Means on the same column followed by the same capital letters, and means on the same row followed by the same small letters are not significantly different ($p < 0.05$; $n = 3$). Values in parentheses are the percent change relative to the raw product.

Table 4. Counts of Lac+ enteric bacteria (Log colony-forming units (CFU)/g) on the raw insects and the processed products.

	Insect Species			
	H. illucens	*A. domesticus*	*R. differens*	*S. littoralis*
Raw	6.4 Db	6.1 Cb	5.9 Cb	0.0 Aa
Boiled	0.0	0.0	0.0	0.0
Toasted	0.0	0.0	0.0	0.0
Solar-dried	2.1 Cb (−67.2)	2.8 Bc (−54.1)	2.3 Bb (−60.1)	0.0 Aa
Oven dried	1.1 Bb (−82.8)	1.5 Bc (−75.4)	1.0 Bb (−83.1)	0.0 Aa
Boiled + Solar-dried	0.0	0.0	0.0	0.0
Toasted + Solar-dried	0.0	0.0	0.0	0.0
Boiled ı Oven-dried	0.0	0.0	0.0	0.0
Toasted + Oven-dried	0.0	0.0	0.0	0.0

Means in the same column followed by the same capital letters, and means on the same row followed by the same small letters are not significantly different ($p < 0.05$; $n = 3$). Values in parentheses are the percent change relative to the raw product.

Staphylococcus aureus was present on all the raw products (Table 5). The counts (7.7–9.1 Log CFU/g) were higher than the recommended limit of 4 Log CFU/g [46]. Habitat and manual collection might have been the cause of a significantly higher *St. aureus* contamination in the *R. differens* compared to the other insect species. Processing lowered the contamination levels by 0.9–4.5 Log CFU/g depending on the insect species and processing method ($F = 3110.7$; df = 35; $p < 0.001$). Boiling and toasting remarkably reduced *St. aureus* by 4.8–6.4 log cycles but unlike the Lac+ enteric bacteria, fungi, and *Salmonella*, complete elimination was not achieved. The results on *Salmonella* are shown in Table 5. Microbiological guidelines for food and feed require no detection of *Salmonella* in 25 g of sample [7,46,51]. All the raw insects were positive for *Salmonella*, but the products subjected to boiling or toasting tested negative; thus, boiling or toasting rendered the products safe.

Table 5. Counts of *St. aureus*, and presence (+) or absence (−) of *Salmonella* on the raw and processed products.

	Insect Species			
	H. illucens	*A. domesticus*	*R. differens*	*S. littoralis*
	St. aureus (Log CFU/g)			
Raw	7.9 [Fa]	8.3 [Fb]	9.1 [Ec]	7.7 [Ea]
Boiled	2.5 [Db] (−68.4)	2.1 [CDa] (−74.7)	3.3 [Bd] (−63.7)	2.9 [Cc] (−62.3)
Toasted	1.5 [Ba] (−80.1)	1.9 [Cb] (−77.1)	2.9 [Bc] (−68.1)	1.8 [Ab] (−76.6)
Solar-dried	3.0 [Ea] (−62.0)	2.9 [Ea] (−65.1)	4.5 [Dc] (−50.5)	3.8 [Db] (−50.6)
Oven-dried	1.7 [BCa] (−78.5)	1.6 [BCa] (−80.7)	3.9 [Cc] (−57.1)	3.2 [Cb] (−58.4)
Boiled + Solar-dried	2.6 [Da] (−67.1)	3.0 [Eb] (−63.8)	4.0 [Cc] (−56.0)	3.1 [Cb] (−59.7)
Toasted + Solar-dried	2.0 [Ca] (−74.9)	2.4 [Db] (−71.1)	3.1 [Bc] (−65.9)	2.1 [Ba] (−72.7)
Boiled + Oven-dried	1.2 [ABa] (−8.8)	1.3 [ABa] (−84.3)	2.1 [Ac] (−76.9)	1.9 [ABb] (−75.3)
Toasted + Oven-dried	0.9 [Aa] (−88.6)	1.0 [Aa] (−87.9)	2.0 [Ac] (−78.0)	1.6 [Ab] (−79.2)
	Salmonella			
Raw	+	+	+	+
Boiled	−	−	−	−
Toasted	−	−	−	−
Solar-dried	+	+	+	+
Oven-dried	+	+	+	+
Boiled + Solar-dried	−	−	−	−
Toasted + Solar-dried	−	−	−	−
Boiled + Oven-dried	−	−	−	−
Toasted + Oven-dried	−	−	−	−

For *St. aureus*, means in the same column followed by the same capital letters, and means in the same row followed by the same small letters are not significantly different ($p < 0.05$; $n = 3$). Values in parentheses are the percent change relative to the raw product.

3.3.2. Effects of Solar- and Oven-Drying

Solar-drying did not affect TVC, while oven-drying lowered the counts by 0.5–2.3 log cycles depending on species (see Table 3). Likewise, drying only lowered the YMC in *H. illucens* and *A. domesticus* by one log cycle (solar) and 2.7–4.6 log cycles (oven). While yeasts are not known to cause food poisoning, strains of moulds that are capable of producing mycotoxins with injurious health consequences on humans and animal have been isolated in dried edible insect products [13,52]. Depending on insect species, solar-drying lowered Lac+ bacteria and *St. aureus* contamination by 3.3–4.3 and 3.9–5.4 log cycles, respectively (see Tables 4 and 5). Nonetheless, this did not bring the contamination to the acceptable limits [46]. Contrastingly, oven-drying lowered the Lac+ bacteria and *St. aureus* contaminations by greater margins (Lac+ bacteria: 4.6–5.3 log cycles; *St. aureus*: 4.5–6.5 log cycles) bringing down the contamination levels to the acceptable limits. Both the oven- and solar-dried products, however, were positive for *Salmonella* (Table 5) and were therefore unsafe. We did not find works that investigated purely the effects of drying the raw insects on microbial contamination levels. However, some authors [53] investigated the effects of oven-heating on *E. coli*, and *St. aureus* contamination of beef chops and showed that these micro-organisms were destroyed at an internal cook temperature of 52–57 °C; the authors demonstrated D-values of 66.7, 4.4 and 0.47 min for *E. coli* and 61.4, 5.3 and 0.48 min for *St. aureus* at 52, 54 and 57 °C, respectively, suggesting that the destruction of these organisms could be expected when drying is done at temperatures close to 60 °C.

Furthermore, the authors reported even lower D-values for *Salmonella* (15.3, 6.2, and 1.7 min at 52, 54 and 57 °C, respectively). The presence of *Salmonella* on the samples subjected to drying was probably due to the rather resistant nature of the pathogen to dry-heat [53] or recontamination.

3.3.3. Effects of Combined Processing (Boiling or Toasting Followed by Solar- or Oven-Drying)

The products subjected to boiling or toasting followed by solar- or oven-drying were free of fungi, Lac+ bacteria and *Salmonella*, (See Table 3, Table 4, Table 5), and the *St. aureus* counts on the boiled or toasted and dried samples were lower. Thus, drying reduced the counts further, although marginally (see Table 5). The absence of fungi, *Salmonella* and coliforms including *E. coli* in boiled and dried *I. belina* was also reported, but recontamination, particularly by moulds and yeasts, was observed when the boiled products were subjected to open sun-drying [47]. Other authors reported the presence of *E. coli*, coliforms, and *Salmonella* on thermally processed and dried insects, for instance, on roasted (5 min) and sun-dried *B. alcinoe* [13], and boiled (30 min), sun-dried, and deep-fried (15–20 min) grasshoppers [50]. Recontamination from unhygienic processing and handling was associated with these observations. A recent investigation on the microbiological quality of processed edible insect products sold in Germany and the Netherlands coming from Europe or Asia also showed that the dried products did not comply with bacterial count recommendations for total bacterial counts, *Enterobacteriaceae*, staphylococci, bacilli, and yeast and mould, although the products were negative for *Salmonellae*, *L. monocytogenes*, *E. coli* and *Staphylococcus aureus* [54]. In a separate exploratory study, crickets (*Gryllus bimaculatus*) and superworms (*Zophobas atratus*) subjected to different combinations of cooking and drying regimes varied in the microbial counts, strongly displaying species- and treatment-specific patterns [55]. In the present study, the boiled or toasted samples subjected to oven-drying had lower TVC by about one order of magnitude compared to those subjected to toasting or boiling alone, whereas those subjected to solar-drying had a higher TVC by up to 1.6 orders of magnitude (Table 3), suggesting some recontamination during solar-drying. Nonetheless, the solar- or oven-drying techniques maintained the good microbiological quality of the boiled/toasted products. An increase in microbial counts was observed elsewhere for smoked crickets after subjection to drying [16].

Raw edible insects generally contain high numbers of mesophilic aerobes, bacterial endospores or spore-forming bacteria, *Enterobacteriaceae*, lactic acid bacteria, psychotropic aerobes, and fungi, and potentially harmful species may be present [56]. Pathogens comprising *Staphylococcus aureus*, *Bacillus cereus* and coliforms (including *E. coli*, and species of *Pseudomonas*, *Klebsiella*, *Aerobacter*, *Proteus*), *Salmonella* spp., *Listeria monocytogenes*, as well as moulds of the genera *Aspergillus*, *Penicillium*, and *Fusarium*, have been isolated [15,52,54]. The presence of *St. aureus* is significant because of the ability to produce enterotoxins, although the vegetative cells are destroyed by cooking, but easily reintroduced during handling [17]. *Pseudomonas* and *Proteus* spp. are proteolytic; sometimes lipolytic, and are therefore implicated in spoilage, undesirable flavour, and loss of nutritional value [52]. Some moulds of the genera *Aspergillus*, *Penicillium*, and *Fusarium* are mycotoxigenic [16,47]. Because bacterial populations are generally higher in the gut as compared to the skin, the destruction is sometimes not complete during processing if heat transfer to the inner tissues is inadequate [12]. Unhygienic handling, storage and retailing can also result in unsafe products [13]. Thus, two groups of microbiota should be of interest when assessing the microbial safety of insects—those that are intrinsically associated with insects and those that are introduced from the environment (rearing, handling, etc.). This study examined the decontamination dynamics of both reared and wild-harvested insects, and has shed some light on the potential variations in contamination when similar processes are applied for different insect species. However, a limitation of the study is that it did not extend the examinations to the broader microbial diversity within species, which should be of interest.

4. Conclusions

The objective of the present work was to evaluate the significance of some local processing methods on the nutritional and microbial quality of edible insects collected from the wild or reared for use in human or animal diets in East Africa. On dry matter basis, the nutritional value of all the processed products remained high but there were significant losses in fat and a subsequent gain in protein contents, especially when the insects were toasted or boiled. These effects have implications for the priorities of farmers who want to use insects as animal feed. Fish farmers may be more interested in protein-rich insects, whereas pig farmers may find high insect fat content desirable. Poultry farmers may desire insects with a high protein and high mineral content. The gain in fibre content when the insects are toasted is especially undesirable for fish, and could mean lower nutrient digestibility when ingested by humans or animals. The interaction of processing technique and insect species was significant; hence, the ultimate nutritional content is a function of insect species and processing method. Further studies should examine this interaction in the context of the qualitative aspects, such as changes in amino acid, fatty acid and micronutrient profiles, of the products. Insects can also contain functional substances that confer antimicrobial [11] and antioxidant activity [57], as well as anti-nutrients that might be affected by processing [58]. These were not investigated and should also form a basis for further investigations in the future. The interaction of processing technique and insect species was significant for microbiological quality. Boiling or toasting followed by drying in the oven were the most effective methods. The thermal processes destroyed active cells or lowered microbial load, and drying significantly lowered the water activity. To this end, further studies should examine shelf-life stability with respect to packaging and storage conditions.

Supplementary Materials: The following are available as http://www.mdpi.com/2304-8158/9/5/574/s1, supplementary material S1: Traditional processing of edible insects in Africa; supplementary material S2: ANOVA tables and graphical representation of the interaction effects of insect species and processing technique on proximate composition parameters; supplementary material S3: ANOVA tables and graphical representation of interaction effects on microbiological quality parameters.

Author Contributions: Conceptualization and funding acquisition—K.K.M.F., D.N., S.E., H.A., and C.M.; design and methodology, C.M., S.I., and J.K.; execution of data collection and statistical analysis—D.N.N.; Supervision J.K., S.I., and C.M.; original manuscript draft preparation—D.N.N.; review and editing of manuscript—S.I. and C.M. All authors have approved the final version for be published.

Funding: This work was supported by the 'INSFEED—Insect feed for poultry and fish production in Kenya and Uganda' project (Cultivate Africa Grant No: 107839-001) co-founded by International Development Research Centre (IDRC) and Australia Centre for International Agricultural Research (ACIAR).

Conflicts of Interest: The authors declare no conflict of interest.

References

1. Meyer-Rochow, V. Can insects help to ease the problem of world food shortage? *Search* **1975**, *6*, 261–262.

2. Van Huis, A.; Van Itterbeeck, J.; Klunder, H.; Mertens, E.; Halloran, A.; Muir, G.; Vantomme, P. *Edible Insects: Future Prospects for Food and Feed Security*; FAO: Rome, Italy, 2013.

3. Baiyegunhi, L.J.S.; Oppong, B.B.; Senyolo, G.M. Mopane worm (*Imbrasia belina*) and rural household food security in Limpopo province, South Africa. *Food Secur.* **2016**, *8*, 153–165. [CrossRef]

4. Makkar, H.P.S.; Tran, G.; Heuzé, V.; Ankers, P. State-of-the-art on use of insects as animal feed. *Anim. Feed Sci. Technol.* **2014**, *197*, 1–33. [CrossRef]

5. Roncolini, A.; Milanović, V.; Cardinali, F.; Osimani, A.; Garofalo, C.; Sabbatini, R.; Clementi, F.; Pasquini, M.; Mozzon, M.; Foligni, R.; et al. Protein fortification with mealworm (*Tenebrio molitor* L.) powder: Effect on textural, microbiological, nutritional and sensory features of bread. *PLoS ONE* **2019**, *14*, e0211747. [CrossRef]

6. Zarantoniello, M.; Zimbelli, A.; Randazzo, B.; Compagni, M.D.; Truzzi, C.; Antonucci, M.; Riolo, P.; Loreto, N.; Osimani, A.; Milanović, V.; et al. Black Soldier Fly (*Hermetia illucens*) reared on roasted coffee by-product and Schizochytrium sp. as a sustainable terrestrial ingredient for aquafeeds production. *Aquaculture* **2020**, *518*, 734659. [CrossRef]

7. KEBS. *Dried Insect Products for Compounding Animal Feeds—Specification KS 2117: ICS 65.120*; Kenya Bureau of Standards: Nairobi, Kenya, 2017; p. 9.

8. Belleggia, L.; Milanović, V.; Cardinali, F.; Garofalo, C.; Pasquini, M.; Tavoletti, S.; Riolo, P.; Ruschioni, S.; Isidoro, N.; Clementi, F.; et al. Listeria dynamics in a laboratory-scale food chain of mealworm larvae (*Tenebrio molitor*) intended for human consumption. *Food Control* **2020**, *114*, 107246. [CrossRef]

9. Mancini, S.; Paci, G.; Ciardelli, V.; Turchi, B.; Pedonese, F.; Fratini, F. Listeria monocytogenes contamination of Tenebrio molitor larvae rearing substrate: Preliminary evaluations. *Food Microbiol.* **2019**, *83*, 104–108. [CrossRef]

10. Van der Fels-Klerx, H.J.; Camenzuli, L.; Belluco, S.; Meijer, N.; Ricci, A. Food Safety Issues Related to Uses of Insects for Feeds and Foods. *Compr. Rev. Food Sci. Food Saf.* **2018**, *17*, 1172–1183. [CrossRef]

11. Wynants, E.; Frooninckx, L.; Van Miert, S.; Geeraerd, A.; Claes, J.; Van Campenhout, L. Risks related to the presence of Salmonella sp. during rearing of mealworms (*Tenebrio molitor*) for food or feed: Survival in the substrate and transmission to the larvae. *Food Control* **2019**, *100*, 227–234. [CrossRef]

12. Klunder, H.C.; Wolkers-Rooijackers, J.; Korpela, J.M.; Nout, M.J.R. Microbiological aspects of processing and storage of edible insects. *Food Control* **2012**, *26*, 628–631. [CrossRef]

13. Braide, W.; Oranusi, S.; IUdegbunam, I.; Oguoma, O.; Akobondu, C.; Nwaoguikpe, R. Microbiological quality of an edible caterpillar of an emperor moth, *Bunaea alcinoe*. *J. Ecol. Nat.* **2011**, *3*, 176–180.

14. Mmari, M.W.; Kinyuru, J.N.; Laswai, H.S.; Okoth, J.K. Traditions, beliefs and indigenous technologies in connection with the edible longhorn grasshopper *Ruspolia differens* (Serville 1838) in Tanzania. *J. Ethnobiol. Ethnomed.* **2017**, *13*, 60. [CrossRef]

15. Wynants, E.; Crauwels, S.; Verreth, C.; Gianotten, N.; Lievens, B.; Claes, J.; Van Campenhout, L. Microbial dynamics during production of lesser mealworms (*Alphitobius diaperinus*) for human consumption at industrial scale. *Food Microbiol.* **2018**, *70*, 181–191. [CrossRef]

16. Vandeweyer, D.; Wynants, E.; Crauwels, S.; Verreth, C.; Viaene, N.; Claes, J.; Lievens, B.; Van Campenhout, L. Microbial dynamics during industrial rearing, processing, and storage of tropical house crickets (*Gryllodes sigillatus*) for human consumption. *Appl. Environ. Microbiol.* **2018**, *84*, e00255-18. [CrossRef]

17. Mutungi, C.; Irungu, F.G.; Nduko, J.; Mutua, F.; Affognon, H.; Nakimbugwe, D.; Ekesi, S.; Fiaboe, K.K.M. Postharvest processes of edible insects in Africa: A review of processing methods, and the implications for nutrition, safety and new products development. *Crit. Rev. Food Sci. Nutr.* **2019**, *59*, 276–298. [CrossRef]

18. Belluco, S.; Losasso, C.; Maggioletti, M.; Alonzi, C.C.; Paoletti, M.G.; Ricci, A. Edible Insects in a Food Safety and Nutritional Perspective: A Critical Review. *Compr. Rev. Food. Sci. Food Saf.* **2013**, *12*, 296–313. [CrossRef]

19. Rumpold, B.A.; Schlüter, O.K. Potential and challenges of insects as an innovative source for food and feed production. *Innov. Food Sci. Emerg. Technol.* **2013**, *17*, 1–11. [CrossRef]

20. Kinyuru, J.N.; Kenji, G.M.; Njoroge, S.M.; Ayieko, M. Effect of processing methods on the in vitro protein digestibility and vitamin content of edible winged termite (*Macrotermes subhylanus*) and Grasshopper (*Ruspolia differens*). *Food Bioprocess Technol.* **2010**, *3*, 778–782. [CrossRef]

21. Kamau, E.; Mutungi, C.; Kinyuru, J.; Imathiu, S.; Tanga, C.; Affognon, H.; Ekesi, S.; Nakimbugwe, D.; Fiaboe, K.K.M. Moisture adsorption properties and shelf-life estimation of dried and pulverised edible house cricket *Acheta domesticus* (L.) and black soldier fly larvae *Hermetia illucens* (L.). *Food Res. Int.* **2018**, *106*, 420–427. [CrossRef]

22. AOAC. *Official Methods of Analysis*, 17th ed.; The Association of Official Analytical Chemists: Gaithersburg, MD, USA, 2000.

23. Rumpold, B.A.; Schlüter, O.K. Nutritional composition and safety aspects of edible insects. *Mol. Nutr. Food Res.* **2013**, *57*, 802–823. [CrossRef]

24. Ssepuuya, G.; Smets, R.; Nakimbugwe, D.; Van Der Borght, M.; Claes, J. Nutrient composition of the long-horned grasshopper *Ruspolia differens* Serville: Effect of swarming season and sourcing geographical area. *Food Chem.* **2019**, *301*, 125305. [CrossRef]

25. Sayed, W.A.A.; Ibrahim, N.S.; Hatab, M.H.; Zhu, F.; Rumpold, B.A. Comparative Study of the Use of Insect Meal from *Spodoptera littoralis* and *Bactrocera zonata* for Feeding Japanese Quail Chicks. *Animals* **2019**, *9*, 136. [CrossRef]

26. Madibela, O.R.; Seitiso, T.K.; Thema, T.F.; Letso, M. Effect of traditional processing methods on chemical composition and in vitro true dry matter digestibility of the Mophane worm (*Imbrasia belina*). *J. Arid Environ.* **2007**, *68*, 492–500. [CrossRef]

27. Liland, N.S.; Biancarosa, I.; Araujo, P.; Biemans, D.; Bruckner, C.G.; Waagbø, R.; Torstensen, B.E.; Lock, E.-J. Modulation of nutrient composition of black soldier fly (*Hermetia illucens*) larvae by feeding seaweed-enriched media. *PLoS ONE* **2017**, *12*, e0183188. [CrossRef]

28. Oonincx, D.G.A.B.; van Broekhoven, S.; van Huis, A.; van Loon, J.J.A. Feed Conversion, Survival and Development, and Composition of Four Insect Species on Diets Composed of Food By-Products. *PLoS ONE* **2015**, *10*, e0144601. [CrossRef]

29. Spranghers, T.; Ottoboni, M.; Klootwijk, C.; Ovyn, A.; Deboosere, S.; De Meulenaer, B.; Michiels, J.; Eeckhout, M.; De Clercq, P.; De Smet, S. Nutritional composition of black soldier fly (*Hermetia illucens*) prepupae reared on different organic waste substrates. *J. Sci. Food Agric.* **2017**, *97*, 2594–2600. [CrossRef]

30. Tschirner, M.; Simon, A. Influence of different growing substrates and processing on the nutrient composition of black soldier fly larvae destined for animal feed. *J. Insects Food Feed* **2015**, *1*, 249–259. [CrossRef]

31. Kipkoech, C.; Kinyuru, J.; Imathiu, S.; Roos, N. Use of house cricket to address food security in Kenya: Nutrient and chitin composition of farmed crickets as influenced by age. *Afr. J. Agric. Res.* **2017**, *12*, 3189–3197. [CrossRef]

32. Kulma, M.; Kouřimská, L.; Plachý, V.; Božik, M.; Adámková, A.; Vrabec, V. Effect of sex on the nutritional value of house cricket, *Acheta domestica* L. *Food Chem.* **2019**, *272*, 267–272. [CrossRef]

33. Meneguz, M.; Schiavone, A.; Gai, F.; Dama, A.; Lussiana, C.; Renna, M.; Gasco, L. Effect of rearing substrate on growth performance, waste reduction efficiency and chemical composition of black soldier fly (*Hermetia illucens*) larvae. *J. Sci. Food Agric.* **2018**, *98*, 5776–5784. [CrossRef]

34. Manditsera, F.A.; Luning, P.A.; Fogliano, V.; Lakemond, C.M.M. Effect of domestic cooking methods on protein digestibility and mineral bioaccessibility of wild harvested adult edible insects. *Food Res. Int.* **2019**, *121*, 404–411. [CrossRef] [PubMed]

35. Egan, B.A.; Toms, R.; Minter, L.R.; Addo-Bediako, A.; Masoko, P.; Mphosi, M.; Olivier, P.A.S. Nutritional Significance of the Edible Insect, *Hemijana variegata* Rothschild (Lepidoptera: Eupterotidae), of the Blouberg Region, Limpopo, South Africa. *Afr. Entomol.* **2014**, *22*, 15–23. [CrossRef]

36. Lautenschläger, T.; Neinhuis, C.; Kikongo, E.; Henle, T.; Förster, A. Impact of different preparations on the nutritional value of the edible caterpillar *Imbrasia epimethea* from northern Angola. *Eur. Food Res. Technol.* **2017**, *243*, 769–778. [CrossRef]

37. Hosseini Bai, S.; Darby, I.; Nevenimo, T.; Hannet, G.; Hannet, D.; Poienou, M.; Grant, E.; Brooks, P.; Walton, D.; Randall, B.; et al. Effects of roasting on kernel peroxide value, free fatty acid, fatty acid composition and crude protein content. *PLoS ONE* **2017**, *12*, e0184279. [CrossRef]

38. Lira, G.M.; Barros Silva, K.W.; Figueirêdo, B.C.; Bragagnolo, N. Impact of smoking on the lipid fraction and nutritional value of seabob shrimp (Xiphopenaeus kroyeri, Heller, 1862). *LWT J. Food Sci. Technol.* **2014**, *58*, 183–187. [CrossRef]

39. Knothe, G.; Dunn, R.O. A comprehensive evaluation of the melting points of fatty acids and esters determined by differential scanning calorimetry. *J. Am. Oil Chem. Soc.* **2009**, *86*, 843–856. [CrossRef]

40. Wu, T.; Mao, L. Influences of hot air drying and microwave drying on nutritional and odorous properties of grass carp (*Ctenopharyngodon idellus*) fillets. *Food Chem.* **2008**, *110*, 647–653. [CrossRef]

41. Akonor, P.T.; Ofori, H.; Dziedzoave, N.T.; Kortei, N.K. Drying characteristics and physical and nutritional properties of shrimp meat as affected by different traditional drying techniques. *Int. J. Food Sci. Technol.* **2016**, *2016*, 5. [CrossRef]

42. Guil-Guerrero, J.L.; Ramos-Bueno, R.P.; González-Fernández, M.J.; Fabrikov, D.; Sánchez-Muros, M.J.; Barroso, F.G. Insects as Food: Fatty Acid Profiles, Lipid Classes, and sn-2 Fatty Acid Distribution of Lepidoptera Larvae. *Eur. J. Lipid Sci. Technol.* **2018**, *120*, 1700391. [CrossRef]

43. Dhingra, D.; Michael, M.; Rajput, H.; Patil, R.T. Dietary fibre in foods: a review. *J. Food Sci. Technol.* **2012**, *49*, 255–266. [CrossRef]

44. Kutoš, T.; Golob, T.; Kač, M.; Plestenjak, A. Dietary fibre content of dry and processed beans. *Food Chem.* **2003**, *80*, 231–235. [CrossRef]

45. Hosseini, H.; Mahmoudzadeh, M.; Rezaei, M.; Mahmoudzadeh, L.; Khaksar, R.; Khosroshahi, N.K.; Babakhani, A. Effect of different cooking methods on minerals, vitamins and nutritional quality indices of kutum roach (*Rutilus frisii* kutum). *Food Chem.* **2014**, *148*, 86–91. [CrossRef]

46. Stannard, C. Development and use of microbiological criteria for foods. *Food Sci. Technol. Today* **1997**, *11*, 137–176.

47. Gashe, B.A.; Mpuchane, S.F.; Siame, B.A.; Allotey, J.; Teferra, G. The microbiology of Phane, an edible caterpillar of the emperor moth, *Imbrasia belina*. *J. Food Prot.* **1997**, *60*, 1376–1380. [CrossRef]

48. Ng'ang'a, J.; Imathiu, S.; Fombong, F.; Ayieko, M.; Vanden Broeck, J.; Kinyuru, J. Microbial quality of edible grasshoppers *Ruspolia differens* (Orthoptera: Tettigoniidae): From wild harvesting to fork in the Kagera Region, Tanzania. *J. Food Saf.* **2019**, *39*, e12549. [CrossRef]

49. Todd, E.C.D. Microbiological safety standards and public health goals to reduce foodborne disease. *Meat Sci* **2004**, *66*, 33–43. [CrossRef]

50. Ali, A.; Adji, M.B.; Saidou, C.; Aoudou, Y.; Tchiegang, C. Physico-chemical properties and safety of grasshoppers: Important contributors to food security in the far north region of Cameroon. *J. Appl. Sci. Res.* **2010**, *4*, 108–111. [CrossRef]

51. Kukier, E.; Goldsztejn, M.; Grenda, T.; Kwiatek, K.; Bocian, L. Microbiological quality of feed materials used between 2009 and 2012 in Poland. *Bull. Vet. Inst. Pulawy* **2013**, *57*, 535–543. [CrossRef]

52. Mpuchane, S.; Gashe, B.A.; Allotey, J.; Siame, B.; Teferra, G.; Ditlhogo, M. Quality deterioration of phane, the edible caterpillar of an emperor moth *Imbrasia belina*. *Food Control* **2000**, *11*, 453–458. [CrossRef]

53. Shigehisa, T.; Nakagami, T.; Taji, S.; Sakaguchi, G. Destruction of *salmonellae*, *Escherichia coli*, and *Staphylococcus aureus* inoculated into and onto beef during dry-oven roasting. *Jpn. J. Vet. Res.* **1985**, *47*, 251–257. [CrossRef]

54. Grabowski, N.T.; Klein, G. Microbiology of processed edible insect products—Results of a preliminary survey. *Int. J. Food Microbiol.* **2017**, *243*, 103–107. [CrossRef]

55. Grabowski, N.T.; Klein, G. Microbiology of cooked and dried edible Mediterranean field crickets (*Gryllus bimaculatus*) and superworms (*Zophobas atratus*) submitted to four different heating treatments. *Food Sci. Technol. Int.* **2017**, *23*, 17–23. [CrossRef]

56. Garofalo, C.; Milanović, V.; Cardinali, F.; Aquilanti, L.; Clementi, F.; Osimani, A. Current knowledge on the microbiota of edible insects intended for human consumption: A state-of-the-art review. *Food Res. Int.* **2019**, *125*, 108527. [CrossRef]

57. Di Mattia, C.; Battista, N.; Sacchetti, G.; Serafini, M. Antioxidant Activities in vitro of Water and Liposoluble Extracts Obtained by Different Species of Edible Insects and Invertebrates. *Front. Nutr.* **2019**, *6*. [CrossRef]

58. Chakravorty, J.; Ghosh, S.; Megu, K.; Jung, C.; Meyer-Rochow, V.B. Nutritional and anti-nutritional composition of *Oecophylla smaragdina* (Hymenoptera: Formicidae) and *Odontotermes* sp. (Isoptera: Termitidae): Two preferred edible insects of Arunachal Pradesh, India. *J. Asia-Pac. Entomol.* **2016**, *19*, 711–720. [CrossRef]

 *foods*

Article

Edible Insects in Africa in Terms of Food, Wildlife Resource, and Pest Management Legislation

Nils Th. Grabowski [1,*], Séverin Tchibozo [2], Amir Abdulmawjood [3], Fatma Acheuk [4], Meriem M'Saad Guerfali [5], Waheed A.A. Sayed [6] and Madeleine Plötz [1]

[1] Institute for Food Quality and Food Safety, Hannover University of Veterinary Medicine, Foundation, Bischofsholer Damm 15, D-30173 Hannover, Germany; Madeleine.Ploetz@tiho-hannover.de

[2] Research Centre for Biodiversity Management, 04 B.P., Cotonou BJ-0385, Benin; tchisev@yahoo.fr

[3] Institute for Food Quality and Food Safety, Research Center for Emerging Infections and Zoonoses (RIZ), Hannover University of Veterinary Medicine, Foundation, Bünteweg 2, D-30559 Hannover, Germany; amir.abdulmawjood@tiho-hannover.de

[4] Laboratory of Valorization and Conservation of Biological Resources, University of M'Hamed Bougara of Boumerdes, Avenue de l'indépendance, Boumerdès DZ-35000, Algeria; fatma.acheuk@yahoo.fr

[5] Laboratory of Biotechnology and Nuclear Technologies LR16CNSTN01, National Centre for Nuclear Sciences and Technology, Technopole de Sidi Thabet, Sidi Thabet T-2020, Tunisia; msaad_tn@yahoo.fr

[6] Biological Application Department, Nuclear Research Centre, Atomic Energy Authority, Cairo ET-11787, Egypt; waheedel@hotmail.com

* Correspondence: Nils.Grabowski@tiho-hannover.de

Received: 4 March 2020; Accepted: 13 April 2020; Published: 16 April 2020

Abstract: Entomophagy is an ancient and actually African tradition that has been receiving renewed attention since edible insects have been identified as one of the solutions to improve global nutrition. As any other foodstuff, insects should be regulated by the government to ensure product quality and consumer safety. The goal of the present paper was to assess the current legal status of edible insects in Africa. For that, corresponding authorities were contacted along with an extensive online search, relying mostly on the FAOLEX database. Except for Botswana, insects are not mentioned in national regulations, although the definitions for "foodstuff" allow their inclusion, i.e., general food law can also apply to insects. Contacted authorities tolerated entomophagy, even though no legal base existed. However, insects typically appear in laws pertaining the use of natural resources, making a permit necessary (in most cases). Pest management regulation can also refer to edible species, e.g., locusts or weevils. Farming is an option that should be assessed carefully. All this creates a complex, nation-specific situation regarding which insect may be used legally to what purpose. Recommendations for elements in future insect-related regulations from the food hygiene point of view are provided.

Keywords: entomophagy; food law; Africa; food hygiene; food policy

---

1. Introduction

1.1. Entomophagy on the African Continent

Some years ago, the FAO (Food and Agricultural Organization of the United Nations) recognized the potential of edible insects as one possibility to mitigate hunger and the effects of the climate change, and as a response to that, the discussion of establishing insect farms in traditionally entomophagous countries rather than increasing the extraction from the wild started [1].

Edible insects have been part of the human diet from the dawn of mankind on. However, food habits changed over the millennia, and while consuming insects was largely lost in Europe after the classical antiquity, the tradition lingered on in Africa. There are hundreds of insect species consumed in Africa as foodstuffs or as traditional medicine [1–6]. The awareness of the benefits of edible insects has also reached non-traditional sectors of the African population, and web-based information sites like LINCAOCNET (http://gbif.africamuseum.be/lincaocnet_dev/) provide searchable information on local species.

Insects are traded in a relatively small to medium level. The economic benefit varies with the species and is seldom accounted for, but one of the most significant ones seems to be the phane caterpillars of a saturniid emperor moth *Gonimbrasia belina* (ex "Imbrasia belina"), reaching a yearly trade value of more than $85 million in Southern Africa.

Like with any other foodstuff, the consumption of edible insects may lead to consumer risks, typically allergens, foodborne diseases, food spoilage agents, and contaminants [7]. Being so, the tradition has developed a set of dos and do nots to ensure food safety to a certain degree. However, as traditions develop over long periods of time and tend to become inflexible, some parts of it may not cover "modern" risks like environmental pollution, or even packaging [3]. In fact, the traditional handling of African insect-based products has become submitted to scientific research, and results show that even processed products may contain pathogens. By means of illustration (and far beyond completeness of data), Table 1 provides a look into the microbiology of fresh and processed products from three African insect species.

Table 1. Selection of microbiological findings in three African edible insect species (African mole cricket (*Gryllotalpa africana*), cabbage tree emperor moth (*Bunaea alcinoe*), and African palm weevil (*Rhynchophorus phoenicis*)); based on [1]. Blank spaces either mean that the sample was negative for that pathogen or was not tested for it.

Species	Product	*Acinetobacter* spp.	*Bacillus* spp.	*Corynebacterium* spp.	*Enterobacter* spp.	*Enterobacter Faecalis*	*Escherichia Coli*	*Klebsiella* spp.	*Micrococcus* spp.	*Proteus* spp.	*Pseudomonas* spp.	*Serratia* spp.	*Staphylococcus* spp.	*Aspergillus* spp.	*Mucor* spp.	*Rhizopus* spp.	*Saccharomyces* spp.
Gryllotalpa-africana	Adults and nymphs, raw		+						+	+			+				
Bunaea alcinoe	Larva, raw	+	+						+				+				
	Skin		+					+			+	+	+				+
Rhynchopho-rus phoenicis	Gut		+		+							+	+				
	Larva, fresh		+		+							+	+				
	Larva, fried		+										+				
	Larva, roasted		+			+	+				+		+				

These results suggest that a stricter control of insect-based products is needed, moreover, if insect entrepreneurs and/or consumers lose this traditional knowledge. Implementing food legislation is a proven method to reduce food-related consumer risks.

1.2. The European Union as a Starting Point for Legal Considerations

In Europe and in terms of food legislation, a division can be made between EU member and non-member states, and within the EU, between EU law and national law. These food laws, along with a large number of related legal texts, represent the base of public health control of foodstuffs. It covers all the productions steps of the food chain from the primary production to the purchase of the product by the consumer. With the EU, community and national legislation was harmonized largely to pursue a maximum of congruency among laws and a minimum of features regulated twice (i.e., on EU and national level). The system has been established for the ordinary animal-derived foodstuffs and is in constant revision and improvement.

Europe is one of those geographical areas with no recent entomophagy tradition (despite some exceptions), and the discussion is moreover addressing the feasibility and the practical and legal framework to establish insect farms. This created a dilemma, because on one hand, entrepreneurs that wish to start an insect business depend on the certification by corresponding authorities. On the other hand, in many countries, insects are neither expressly allowed nor forbidden. So, there is no legal framework by which these authorities could certify this enterprise, even if there is good will to promote this development. Thus, authorities ask the entrepreneurs for more information on risk handling, etc., which in turn can only generate once the business is running. This creates a climate of legal uncertainty, which is perceived as obstacles by the insect business operators [8,9].

However, this is changing. In 2015, the EFSA (European Food Safety Authority) published the *Risk profile related to production and consumption of insects as food and feed*, coming to the overall conclusion that, when produced according to current law requirements, insects do not pose a major risk to consumers. However, knowledge on residues and contaminants was scarce, and EFFSA recommended more research. The second important EU publication is the amendment of the novel food regulation, i.e., *REG (EC) 2015/2283*. In it, insects are clearly classified as potential foodstuffs, but for each species and product, a separate authorization procedure must be followed, leading to the inclusion of the novel foodstuff in the so-called Union List. At present, there are several requests, which are being processed. Although this regulation does not contain specific requirements for insects as foodstuffs (e.g., primary production, processing, quality parameters, etc.), this regulation provides the base for a clearer, EU-wide regulatory frame for all sectors involved in insect production.

Finally, there is draft *Ref. Ares (2019) 382900—23/01/2019*, which proposes to add another Appendix A, specific for edible insects to the *REG (EC) 853/2004*. This regulation contains all regulatory issues about the production of animal-based foodstuffs along the production chain. The draft contains a definition of "insect" and proposals on the choice of allowed feedstuffs in insect farming. It has, however, not been ratified so far.

Until then, several nations issued national guidelines, which should be regarded as recommendations and interim solutions. However, there is a marked degree of heterogeneity in this process on the continent, ranging from utter rejection to a relatively developed legal framework to produce, sell, and monitor insect-based foodstuffs (IBF). In addition, not all European nations have issued official statements regarding edible insects so that for those countries, a grey zone is presumed. This current European situation is described in [7].

1.3. African Countries and Economic Communities

Africa, as all other continents, is a spectacular mixture of ethnic groups, languages (approximately 2000), religions, lifestyles, and cultures, subsuming into more than 1.3 billion humans inhabiting it. All African states are part of the African Union (AU). Within it, the African Economic Community (AEC) is the common organization seeking, ultimately, economic and monetary union of the member states, which vaguely corresponds to the EU. AU/AEC recognize several, so-called Regional Economic Communities (REC), some of which have subgroups, i.e.,

- Arab Maghreb Union (UMA);
- Common Market for Eastern and Southern Africa (COMESA);
- Community of Sahel-Saharan States (CEN-SAD);
- East African Community (EAC);
- Economic Community of Central African States (ECCAS), with CEMAC (Economic and Monetary Community of Central Africa) as subgroup;
- Economic Community of West African States (ECOWAS), with UEMOA (West African Economic and Monetary Union) and WAMZ (West African Monetary Zone) as subgroups; the latter does not address specific food or agriculture issues and is therefore not considered here;
- Intergovernmental Authority on Development (IGAD);
- Southern African Development Community (SADC), with SACU (South African Customs Union) as a subgroup.

However, most subgroups are not recognized by AEC (CEMAC, SACU, and UEMOA), and there are other trade unions that also lack recognition by AEC, i.e., the Economic Community of Great Lakes Countries (CEPGL), Indian Ocean Commission (IOC), and Mano River Union (MRU).

Finally, there are trade blocks that include both African and non-African nations. One example is GAFTA, the Greater Arab Free Trade Area, which originates from the Arab League and includes most Arabic-speaking countries inside and outside Africa. IOC is an organization that links African island nations with France via this country's oversea regions. Table 2 subsumes the affiliation of the African nations to these trading communities. It reveals a complex network of interactions between the different African nations.

Table 2. Affiliations of African nations to the different trading blocks. See text for abbreviations. Blocks not addressing food issues were excluded.

Country	CEMAC	CEN-SAD	CEPGL	COMESA	EAC	ECCAS	ECOWAS	GAFTA	IGAD	IOC	MRU	SACU	SADC	UEMOA	UMA	WAMZ
Algeria								+							+	
Angola						+							+			
Benin		+					+							+		
Botswana												+	+			
Burkina Faso		+					+							+		
Burundi			+	+	+	+										
Cameroon	+															
Cape Verde		+				+	+									
Central African Republic	+	+				+										
Chad	+	+				+										
Comoros		+		+						+			+			
Congo, Democratic Republic (Kinshasa)			+	+		+							+			
Congo, Republic (Brazzaville)	+					+										
Djibouti		+		+					+							
Egypt		+		+				+							+	
Equatorial Guinea	+					+										
Eritrea		+		+					+							
Eswatini				+								+	+			
Ethiopia				+					+							
Gabon	+					+										
Gambia		+					+									+
Ghana		+					+									+
Guinea		+					+				+					+
Guinea-Bissau		+					+							+		
Ivory Coast		+					+				+			+		
Kenya		+		+	+				+							
Lesotho												+	+			
Liberia		+					+				+					+
Libya		+		+				+							+	
Madagascar				+						+			+			
Malawi		+		+									+			
Mali		+					+							+		
Mauritania		+													+	
Mauritius				+						+			+			
Morocco		+						+							+	
Mozambique													+			
Namibia												+	+			
Niger		+					+							+		
Nigeria		+					+									+
Rwanda			+	+	+	+										
São Tomé and Príncipe		+				+										
Senegal		+					+							+		
Seychelles				+						+			+			
Sierra Leone		+					+				+					+
Somalia		+		+					+							
South Africa												+	+			
South Sudan					+				+							
Sudan		+		+				+	+							
Tanzania					+								+			
Togo		+					+							+		
Tunisia		+		+				+							+	
Uganda				+	+				+							
Zambia				+									+			
Zimbabwe				+									+			

Regardless of the official status at AU/AEC, these trade blocks have the goal to improve the trade among the member states, creating free trading zones. The trade also includes foods, and just like many European countries, African countries face the problem of trading foodstuffs with different quality standards. Another of the challenges these blocs have is the fact that many African countries belong to the more than one trading block, and the activity of these blocks varies. Some of these blocks have issued common trade rules that can be read in the internet, and some do not.

Besides international rules, national food law and other regulations also a play a very important role, particularly for foodstuffs produced and consumed inside the corresponding countries. The extent of this legislation varies strongly among them.

The aim of this contribution is to outline the current legal situation regarding edible insects in all African countries.

2. Material and Methods

For this research, all African countries were included (Table 2). Not entering a political dispute, states not recognized by the United Nations (typically "break-away states") were included if five or more UN nations recognized them as independent. In this way, Galmudug, Khaatumo, Puntland, Republic of Azania, and the Republic of Somaliland were not considered in this survey, while the Sahrawi Arab Democratic Republic was. However, no data for the latter could be encountered, so no further mention of this country would be made.

Initially, an approach similar that made for the publication regarding the legal status in Europe [7,10] was intended, i.e., contacting the competent authorities (Table 3) personally by telephone and/or mail. However, the degree of response varied strongly among countries, with the best replies from the most French-speaking and Arabic-speaking countries. Others did not respond (even after contacting the corresponding embassies in Europe), even after several attempts, and so their homepages were searched for corresponding regulations. As this was not successful in some of them, the FAOLEX page (http://www.fao.org/faolex/country-profiles/en/) was consulted, trusting in having the most complete and updated database for FAO-related legislations. Eventually, all nations were checked by this mean. In some cases, additional information was found in secondary sources, e.g., Droit-Afrique.com, but there is a possible bias because these regulations may not be reflecting the actual situation of the given country. Data was gathered between the fall of 2018 and that of 2019.

Table 3. List of African countries and authorities that provided information on the legal status of edible insects including the amount of edible species per countries according to Jongema [11]; besides these country-specific data, 4 species were recorded for Northern, 91 for central, 2 for Western, 9 for Eastern, and 20 for Southern Africa.

Country	Institution	REIS *
Algeria	وزارة الفلاحة و التنمية الريفية و الصيد البحري (Ministry of Agriculture, Rural Development and Fisheries)	0
Angola	MINAGRIF—Ministério da Agricultura e Florestas (Ministry of Agriculture and Forests)	16
Benin	ABSSA—Agence Béninoise de Sécurité Sanitaire des Aliments (Beninese Agency for the Sanitary Security of Foodstuffs)	24
Botswana	Ministry of Health	22
Burkina Faso	Ministère de l'Agriculture et des Aménagements Hydrauliques (Ministry of Agriculture and Water Engineering)	7
Burundi	Ministère de l'Agriculture et de l'Élevage (Ministry of Agriculture and Livestock Breeding)	1
Cameroon	MINADER—Ministère de l'Agriculture et du Développement Rurale (Ministry of Agriculture and Rural Development)	59

Table 3. *Cont.*

Country	Institution	REIS *
Cape Verde	Ministério da Agricultura e Ambiente (Ministry of Agriculture and Environment)	0
Central African Republic	Ambassade de la République Centrafricaine à Paris (Embassy of the Central AfricanRepublic in Paris)	54
Chad	Ministère de la Production, de l'Irrigation et des Equipements Agricoles	1
Comoros	Ministère de l'Agriculture, de la Pêche, de l'Environnement, de l'Aménagement du Territoire et de l'Urbanisme (Ministry of Agriculture, Fishery, Environment, Land Use Planning, and Urbanization)	0
Congo, Democratic Republic (Kinshasa)	Ministère de l'Agriculture de l'Élevage et de la Pêche de la République Démocratique du Congo (Ministry of Agriculture, Livestock Breeding, and Fishery of the Democratic Republic of the Congo)	107
Congo, Republic (Brazzaville)	Ministère de l'Agriculture de l'Élevage et de la Pêche (Ministry of Agriculture, Livestock Breeding, and Fishery)	59
Djibouti	MAEM—Ministère de l'Agriculture, de l'Elevage et de la Mer, Chargé des Ressources Hydrauliques (Ministry of Agriculture, Livestock, & Fisheries, in charge of water resources)	0
Egypt	وزارة الزراعة وإستصلاح الأراضي (Ministry of Agriculture and Reclamation of Land)	1
Equatorial Guinea	Ministerio de Agricultura y Bosques (Ministry of Agriculture and Forests)	1
Eritrea	Botschaft des Staates Eritrea in der Bundesrepublik Deutschland (Embassy of the State of Eritrea in the Federal Republic of Germany)	0
Eswatini	Ministry of Agriculture	20
Ethiopia	ጤና ሚኒስቴር - ኢትዮጵያ (Ministry of Health)	1
Gabon	Ministère de l'Agriculture, de l'Elevage, chargé de la mise en œuvre du programme Graine (Ministry of Agriculture, Livestock Breeding, in charge of carrying out the Graine program)	11
Gambia	FSQA—Food Safety and Quality Authority	1
Ghana	FDA—Food Safety Division	3
Guinea	Ministère de l'Agriculture (Ministry of Agriculture)	9
Guinea-Bissau	Ministério da Agricultura e Desenvolvimento Rural (Ministry of Agriculture and Rural Development)	2
Ivory Coast	Ministère de l'Agriculture et du Dévelopment Rural (Ministry of Agriculture and Rural Development)	5
Kenya	Ministry of Agriculture, Livestock and Fisheries	10
Lesotho	Ministry of Agriculture and Food Security	20
Liberia	MOA—Ministry of Agriculture	2
Libya	وزارة الزراعة والثروة الحيوانية والبحرية (Ministry of Agriculture, Fisheries, Livestock, and Irrigation)	1
Madagascar	MAEP—Ministère de l'Agriculture, de l'Elevage et de la Pêche (Ministry of Agriculture, Animal Breeding and Fishery)	34
Malawi	Ministry of Agriculture, Irrigation, and Water Management	20
Mali	MEADD—Ministère de l'Environnement, de l'Assainissement et du Développement Durable (Ministry of Environment, Water Treatment, and Sustainable Development)	8

Table 3. *Cont.*

Country	Institution	REIS *
Mauritania	وزارة التنمية الريفية/Ministère de Développement Rural (Ministry of Rural Development)	0
Mauritius	Ministry of Agro Industry and Food Security	2
Morocco	وزارة الفلاحة والصيد البحري والتنمية القروية والمياه والغابات/ Ministère d'Agriculture, de la Pêche Maritime, du Développement Rural, et des Eaux et Forêts (Ministry of Agriculture, Fishery, Rural Development, Water and Forests)	2
Mozambique	SETSAN—Secretariado Técnico de Segurança Alimentar e Nutricional (Technical Secretariate of Food and Nutrition Safety)	3
Namibia	Ministry of Agriculture, Water, and Forestry	27
Niger	Ministère de l'Èlevage (Ministry of Animal Breeding)	19
Nigeria	NAFDAC—National Agency for Food and Drug Administration and Control	19
Rwanda	MINAGRI—Ministère de l'Agriculture (Ministry of Agriculture)	0
São Tomé and Príncipe	MADR—Ministério da Agricultura e Desenvolvimento Rural (Ministry of Agriculture and Rural Development)	4
Senegal	Ministère de l'Agriculture et de l'Équipement Rural (Ministry of Agriculture and Rural Supply)	10
Seychelles	Ministry of Fisheries and Agriculture	0
Sierra Leone	Ministry of Agriculture, Forestry and Food Security	7
Somalia	Wasaaradda Beerahaiyo Waraabka (Ministry of Agriculture and Irrigation)	0
South Africa	FACS—Food Advisory Consumer Service	58
South Sudan	Ministry of Agriculture and Forestry	1 **
Sudan	وزارة الثروة الحيوانية و السمكية و المراعي (Ministry of Livestock, Fisheries, and Rangeland)	7 **
Tanzania	Ministry of Livestock and Fisheries	22
Togo	Ministère de l'Agriculture, de la Production Animale et de l'Halieutique (Ministry of Agriculture, Animal Breeding, and Fishery)	14
Tunisia	وزارة الفلاحة (Ministry of Agriculture)	0
Uganda	MAAIF—Ministry of Agriculture, Animal Industry, and Fisheries	8
Zambia	Ministry of Agriculture	75
Zimbabwe	Ministry of Lands, Agriculture, Water, Climate and Rural Resettlement	47

* REIS = recorded edible insect species, ** due to its independence in 2011, some species recorded earlier for Sudan may also occur in South Sudan.

Resources considered in this paper were publications from the AU and other trade blocks detailed in Table 2 and regulations that have been in force. It largely omits national policy papers due to space reasons and because policies may be implemented late, in a changed version, or even be abandoned, not meeting this contribution's title. Bilateral agreements among African states or between African and non-African states (which are basically trade agreements) were also excluded in this analysis, as they are of minor concern for the African citizens.

The use of insects other than food was also not considered here.

Regarding which insect species are edible or not, the list of Jongema [11] was taken as reference. Within the text, law references are marked in italics (sometimes abbreviated) and presented before the ordinary reference list as an Appendix A. Although written differently in the various documents, the abbreviations for "number" were unified according to the corresponding typing conventions, i.e., "№" (English), "n°" (French and Portuguese), and "n.°" (Spanish). Regarding capitalization, the original version of the acts was retained.

3. Results

In Africa, edible insects may be regulated mainly in three different contexts:

- Food law;
- Wildlife resources management;
- Pest management.

The applicability of these regulations will depend fundamentally on whether a definition for a given item ("foodstuff", "wildlife species", etc.) can be extrapolated to edible insects.

3.1. Food Law

As in other regions of the world, the food law of a given African country is determined by both international and national regulations. From the international point of view, Codex Alimentarius, AU and the different African trade blocks are the most important ones. As the Codex is well-known, it will be omitted here.

- AU. The *African convention on the conservation of nature and natural resources* is oriented towards managing the entire African environment, regardless of being farmland or not. Member states are called to "establish, strengthen, and implement specific national standards" (p. 10) for processing respectively production methods and product quality (Article XIII).
- CEMAC. Food security is one of the foci of CEMAC (http://www.cemac.int/node/140). Two institutions are thought to be handling food regulation issues, both localized in Ndjamena, Chad. Access to the homepage of CEBEVIRHA (Commission Économique du Bétail, de la Viande et des Ressources Halieutiques, Economic Commission for Livestock, Meat, and Fishery Resources), was not granted by the authors' internet due to safety concerns. Pôle Régional de Recherche Appliquée au Développement des Systèmes Agricoles d'Afrique Centrale (Regional Growth Market of Applied Research for the Development of Agricultural Systems of Central Africa, PRASAC) is the other authority, conceived as a network for agricultural research regarding the savannah. No documents could be downloaded.
- CEPGL. No food-related documents could be encountered on the union's webpage (http://www.cepgl.org/).
- COMESA. Chapter 18 of the COMESA treaty refers to the "co-operation in agriculture and rural development" (COMESA, 2009, p. 69), aiming towards "common agricultural policy", "regional food sufficiency", productivity increases, and "replacement of imports on a regional basis" (ibid.). In terms of food safety, member states are requested to harmonize their regulation and standards in order to facilitate trade within COMESA borders and cooperate among each other with regard to research incl. that to increase productivity. An early warning system regarding food hazards is also an issue. No more specific documents could be found.
- CEN-SAD. The web presence (https://web.archive.org/web/20080709020642/http://www.uneca. org/cen-sad/reportofthe 6th ordinarysessionoftheEC.htm) of this trade block mentions that during the 2001 meeting in Ouagadougou, Burkina Faso, opinions regarding food security (cave, not food safety) were exchanged. No more details were available.
- EAC. This East African REC issued the *EAC food security action plan (2011–2015)*. ECOWAS In it, they include a FAO definition of 'food security', i.e., "Food security exists when all people, at all times, have physical, social and economic access to sufficient, safe and nutritious food to meet their dietary needs and food preferences for an active and healthy life" (p. 4). In this way, food security also includes food safety. Apart from improving production, quality, and distribution of common foodstuffs, using an alternative food supply of agricultural, aquatic, and forestry systems is one of

the priority areas of this action plan. This would, in fact, include insects, regardless of whether they are caught from the wild or farmed.

- ECCAS. Like EAC, ECCAS member states agreed to cooperate in terms of agriculture and food supply. Chapter VII of the corresponding treaty (*Treaty establishing the Economic Community of Central African States*) contains these intentions. As other trade blocks, food supply is supposed to be originated from agriculture and livestock, aquaculture, and forestry management, so insects are potentially included.

- ECOWAS. Article 22 of the ECOWAS treaty (*Revised treaty*) foresees the establishment of a technical commission for food and agriculture. The corresponding section of cooperation in the food sector is Chapter IV. As in other documents, the focus lies on food security rather than food safety.

- GAFTA. Of this trade union, only the declaration text (قرار المجلس الاقتصادي بإعلان منطقة التجارة الحرة العربية الكبرى والاجتماعي رقم 1317 د.ع 59 بتاريخ 1997/2/19) was available. It focuses on the trading conditions of Arab products claiming that a certificate of origin is mandatory, but that imported GAFTA goods will be treated as local ones throughout the member states. It may be assumed that this also includes foodstuffs, but this product was not mentioned in this document.

- IOC. Like other transnational organizations, food security is one of the primary goals of IOC. Comparable to IGAD, the provided IOC documentation [12] focuses on specific projects to improve food security. However, no legal documents could be encountered.

- IGAD. Food security and food safety are primary targets of IGAD, being part in several policy statements and programs published (*Regional Strategy, Volume 1 & 2*) rather than concise legal bases.

- MRU. Due to technical reasons, the organization's homepage could not be accessed.

- SACU. Founded in 1910, SACU is the oldest customs union of the world. It recognizes the predominance of the agricultural sector, which plays an important, but variable role in the member states [13]. However, no specific documents could be encountered.

- SADC. In its declaration of agriculture-related actions to take (*Dar-es-Salam declaration on agriculture and food security in the SADC region*), only aquaculture and "short cycle stocks" (sic!; poultry, small ruminants, pigs) are mentioned. In fact, many insects would be short-cycle livestock. In view of the time of publication, which was far before the FAO started to promote edible insects, opening this definition to productive insects may be an option to be considered, moreover as some insect products like mopane worms (*Gonimbrasia belina*) are an important product with transregional trading relevance. However, the protocol on wildlife management (*Protocol on wildlife conservation and law enforcement*) does include all animal and plant species and allows a sustainable use of this wildlife, particularly by local communities. However, no specific rules were laid down. The guidelines for the regulation of food safety in SADC states (*Regional guidelines for the regulation of food safety in SADC member states*) contain a comprehensive description on how the national food law should be in the 15 member states. Although it does not mention insects as such, the definition of food (" … any substance […] intentionally incorporated into the food … " with a list of duly prohibited substances) would support the inclusion of insects. Many aspects of this document remind on the EU food law, particularly *REC (EC) 178/2002*, but as that latter, no specific data on the different food types or the evaluation criteria is provided.

- UEMOA. Unlike other trade organizations, UEMOA published a concise regulation of food security and safety. Considering that according to *Règlement 007/2007/CM/UEMOA*, every animal species can be a foodstuff, insects are potentially included. Chapter II deals with the health monitoring of animals and products made thereof. While Section 1 sets the obligations of the involved actors, Section 2 refers to animal health control and inspection, and Section 3 to animal and animal product movements inside and outside member countries. Chapter III attends food safety with a similar layout (Section 1: obligations, Section 2: control and inspection, and Section 3: product movements). In terms of food safety, Article 84 is particularly interesting since it prohibits the production and the placing on the market of foodstuffs that are hazardous to human and animal health, do not respond to the consumer information requirements, do not fulfill the ethics of the international food trade as established by the Codex Alimentarius, or do not cope with the requirements for novel foodstuffs. The latter refers to foodstuffs unknown to the member states' population and consumed marginally and also includes GMOs. Furthermore, the regulation refers to the so-called Secrétariat Régional de la Normalisation, de la Certification et de la Promotion de la Qualité (NORMCERQ) that is in charge of harmonizing the corresponding regulations, but no hint of this activity regarding food could be encountered.

- UMA. The Arab Maghreb Union has a commission for food safety, which, according to its website, met 2007 in Nouakchott, Mauretania. It has ratified several conventions regarding the exchange and quarantine of agricultural products among member countries. However, these documents are not online.

As can be seen, most of these international organizations do not set a concise legal framework for the member states. This is a difference to other trade blocks, e.g., the EU. Instead, policies are proposed. Policy documents are typically placed on consulted ministries' homepages, dealing with ways to increase food security, food safety, intermember states exchange, and food and production-related research. In some cases, policy statements are presented rather than concrete legal frameworks.

Policies seldom address edible insects directly. Some exceptions will be presented on a national level.

On a this level, no data on the food law in Chad and Lesotho could be encountered. Technical and safety reasons made it impossible to access the corresponding web presences in Botswana, Cameroon, Sierra Leone, Sudan, Tanzania, and Togo. FAOLEX was used whenever possible.

No country has explicitly included edible insects in their current food legislation so far. However, direct contact to authorities and consulting the internet provided the following results:

- According to the consulted authorities, edible insects are tolerated in Algeria, Benin, Burkina Faso, Burundi, Cameroon, Central African Republic, Madagascar, Morocco, Namibia, South Africa, Togo, and Tunisia. All these countries lack a specific regulation for edible insects.
- In Congo-Kinshasa, Guinea, and Niger, entomophagy is also tolerated, but during the interviews, an interest in establishing a corresponding legal framework was expressed.
- Benin. The central laboratory for food safety (http://www.lcssa-benin.org/index.php/resultats-analyses/) offers a series of chemical, microbiological and contaminant-related analyses for all foodstuffs. By extension, this would theoretically also include edible insects.
- Cape Verde. In *Lei n° 30/VIII/2013*, "animal" is defined as either a mammal, bird, or bee (Art. 3). From that, "products of animal origin meant for human consumption" derive. In this way, other insects are excluded, making edible insects illegal on the archipelago. The only exception would be bee brood.
- Comoros. Food safety is addressed in the public health code (*Code de la santé publique et de l'action sociale pour le bien être de la population*), providing a definition for food hygiene, but not for food.
- Gabon. Most food legislation is about fishery products.
- Gambia. The FSQA homepage has a section for regulations, but instead of providing the original texts, bullet points to specific issues are presented of which none applies to edible insects. Within the subsection "standards", a Food Safety and Quality Act from 2011 is mentioned, but could not be found neither on the FSQA nor the FAOLEX pages.
- Ghana. The FDA homepage contains a large set of regulations and codes for many sectors of food policy, and there is a list of different foodstuffs and the parameters they should be analyzed for (https://fdaghana.gov.gh/index.php/certificate-of-analysis/), yet omitting the threshold values. None refers to edible insects.
- Guinea. In fact, most regulation concerns fishery products.
- Madagascar. On 901 pages, MAEP has merged all relevant regulations pertaining agriculture, animal breeding, and fishery into one document. There is no specific Malagasy legislation for

- edible insects, but they are accepted, and the Codex Alimentarius is used when determining the hygiene and edibility of insects. However, Codex specifications apply to foodstuffs in general or to specific, non-insect foodstuffs.
- Malawi. A national policy that promotes insect consumption is said to exist, but could not be retrieved on the internet.
- Mozambique. While there is an extensive regulatory framework for fishery products, a general food act could not be encountered.
- Namibia. A new policy for food safety [14] does mention edible insects specifically, postulating that the Ministry of Agriculture, Water, and Forestry should be responsible to elaborate standards for a series of foodstuffs including insects. This policy was submitted to the cabinet by the Minister of Agriculture, Water, and Forestry, but has not been amended yet.

These results show that edible insects are tolerated in many countries, and although there is no regulation for them at the moment, some governments are interested in developing them.

When working with legal texts, it is important to check the definitions provided there. In the case of the definitions for 'foodstuff', all countries would permit considering insects as such (Table 4), at least in the case of the countries for which such a definition could be encountered. In most cases in which countries apparently lack a foodstuff definition, however, they do have food legislation, but either they work without the preliminary definition or address specific foodstuffs, e.g., meat, fishery products, or eggs. For UEMOA member states, the food definition provided there would have an official character, even in the countries apparently without their own definition, i.e., Burkina Faso and Senegal. Another probable reason for information gaps may the fact that some regulations have simply not been published on the internet. To give an example, a central regulation for food in Togo is the *Arrêté interministériel n⁰ 06/08/MAEP/MEF*. It is mentioned in secondary literature (e.g., [15]) but could not be traced back in the net.

Table 4. Definitions of "foodstuff" in African food regulation frameworks; for Botswana, Burundi, Cameroon, Chad, Comoros, Congo-Brazzaville, Congo-Kinshasa, Djibouti, Equatorial Guinea, Eritrea, Gabon, Gambia, Guinea, Lesotho, Malawi, Mauritania, Libya, São Tomé and Príncipe, South Sudan, and Togo, no definitions could be encountered.

Country	Definition (Regulation)
Algeria	" … any treated, partially processed or raw substances intended for human consumption and including beverages, chewing gum and any substances used in the manufacture, preparation and processing of foods, excluding those used only in the form of drugs or cosmetics." (*Décret exécutif n⁰ 91-53*, Art. 2)
Angola	" … any substance or mixture of substances, in the solid, liquid, pasty or any other suitable form, intended to provide the human body with the normal elements essential for its formation, maintenance, and development." (*Decreto Presidencial n⁰ 179/18*, Art. 4 c)
Benin	" … any treated, partially processed or raw material intended for human consumption and includes beverages, chewing gum and all substances used in the manufacture, preparation and processing of foods, excluding those used only in the form of drugs or cosmetics." (*Loi n⁰ 84-009 du 15 mars 1984*, Art. 2)
Botswana	" … means any animal product, fish, fruit, vegetable, condiment, beverage and any other substance whatever, in any form, state or stage of preparation which is intended or ordinarily used for human consumption, and includes any article produced, manufactured, sold or presented for use as food or drink for human consumption, including chewing gum, and any ingredient of such food, drink or chewing gum" (*Food Control Act*, 2)
Burkina Faso	" … any treated, partially treated or unwrought substance intended for human consumption, and includes beverages, chewing gum and all substances used in the manufacture, preparation and processing of foods, excluding substances used only in form of drugs, cosmetics or tobacco" (*Règlement n⁰ 007/2007/CM/UEMOA*, 1)

Table 4. *Cont.*

Country	Definition (Regulation)
Cape Verde	" … any substance or product, processed or partially processed or unprocessed, intended to be ingested by, or reasonably likely to be, human beings" (*Decreito-Legislativo n° 3/2009*, Art. 3.1)
Central African Republic	" … any raw or partially processed substance intended for human consumption" (*Loi n° 03.04*, Art. 24)
Egypt	" … any product or substance intended for human consumption, whether primary, raw, semi-processed, wholly/partially processed or not processed, including beverages and bottled water or food additives and any substance including water and gum, except for fodder and plants and crops before harvest, live animals and birds prior to their transport to slaughterhouses, sea creatures and farm-raised fish prior to fishing, pharmaceutical products and cosmetics" (*Law 1/2007*, (1)/6)
Eswatini	" … means food of animal origin including meat, milk, fish, honey and their products. (*Veterinary Public Health Act*, Art. 2)
Ethiopia	" … means any raw, semi-processed or processed substance for commercial purpose or to be served for the public in any way intended for human consumption that includes water and other drinks, chewing gum, supplementary food and any substance which has been used in the manufacture, preparation or treatment of food, but does not include tobacco and substances used only as medicines" (*Proclamation № 661/2009*, Art. 2, 1/)
Ghana	" … includes water, a food product, a live animal or a live plant, and (a) a substance or a thing of a kind used, capable of being used or represented as being for use, for human or animal consumption whether it is live, raw, prepared or partly prepared, (b) a substance or a thing of a kind used, capable of being used or represented as being for use, as an ingredient or additive in a substance or a thing referred to in paragraph (a), (c) a substance used in preparing a substance or a thing referred to in paragraph (a), (d) chewing gum or an ingredient or additive in chewing gum or a substance used in preparing chewing gum, and (e) a substance or a thing declared by the Minister to be a food under Section 146 (3)" * (*Public Health Act, 2012*, Art. 149)
Guinea-Bissau	" … any substance, whether or not treated, intended for human consumption, by swallowing beverages and products of the type of chewing gum, as well as the ingredients used in its manufacture, preparation and treatment" (*Decreto n° 62-E/92*, Art. 2)
Ivory Coast	" … any food, product, or drink intended for human consumption." (*Décret n° 92-487 du 26 aôut 1992*, Art. 2)
Kenya	" … includes any article manufactured, sold or represented for use as food or drink for human consumption, chewing gum, and any ingredient of such food, drink or chewing gum" (*Food, Drugs and Chemical Substances Act*, 2.)
Liberia	" … articles including liquids used as nutriment for human consumption or use, alcoholic and nonalcoholic beverages, chewing gum, ice and articles used for components of any such article. (*Public Health Law—Title 33—Liberian Code of Laws Revised*, § 26.1)
Madagascar	" … any substance or product, processed, partially processed or unprocessed, intended to be ingested or reasonably likely to be ingested by humans. This term also includes beverages, chewing gums and any substance, including water, intentionally incorporated into food during their manufacture, preparation or processing." (*Loi n° 2017-048*, Art. 2)
Mali	" … any totally processed, partially treated or raw material intended for human consumption and including beverages, chewing gum and all substances used in the manufacture, preparation and processing of foods, excluding cosmetics or tobacco or substances used solely as medicaments" (*Décret n° 06-259/P-RM du 3 juin 2006*, 2)
Mauritius	" … (a) means any article or substance meant for human consumption and includes (i) drinks and bottled water; (ii) chewing gum and other products of similar nature and use; and (iii) articles and substances used or intended for use as ingredients in the composition or preparation of food; (b) does not include (i) live animals, birds or live fish which are not used for human consumption while they are alive; (ii) fodder or feeding stuffs for animals, birds or fish; (iii) drugs or medicine as defined in the Pharmacy Act; and (iv) hormonal products or veterinary products for use in livestock feed" (*Food Act 1998*, 2.)
Morocco	" … any plant or animal product, raw or wholly or partly processed, intended for human consumption including beverages, gum and all products that have been used for the production and preparation or processing of food. This term does not cover plants before harvest and live animals, with the exception of those prepared for human consumption, as they are, such as shellfish, and do not cover medicines, cosmetics and tobacco" (*Loi n° 28-7*, Art. 2.1)

Table 4. *Cont.*

Country	Definition (Regulation)
Mozambique	" … any substance which is consumed in the natural state, semi-prepared or processed, intended for human consumption, including beverages, chewing gum and any other substance used in their manufacture, preparation or treatment. Cosmetics, tobacco and substances used solely as medicine are excluded." (*Decreto nº 15/2006 de 22 de Juhno*, Art. 1 g))
Namibia	" … means any article or substance (except a medicine as defined in the Medicines and Related Substances Control Act, 1965 [Act 101 of 1965]) ordinarily eaten or drunk by man, or purporting to be suitable, or manufactured or sold, for human consumption, and includes any part or ingredient of any such article or substance, or any substance used or intended or destined to be used as a part or ingredient of any such article or substance" (*Foodstuffs, Cosmetics and Disinfectants Ordinance 18 of 1979*, 1)
Niger	" … treated, partially treated or unprocessed substance intended for human consumption and including beverages, chewing gums and all substances used in the manufacture, preparation and processing of foods excluding substances used solely in the form of medicines, cosmetics or tobacco" (*Décret nº 2011-616/PRN/MEL*, Art. 7/3)
Nigeria	" … includes any article manufactured, processed, packaged, sold or advertised for use as food or drink for human consumption, chewing gum and any ingredient which may be mixed with food for any purpose whatsoever and excludes (a) live animals, birds or fish; (b) articles or substances used as drugs" (*Food and Drugs (Amendment) Decree 1999*, 7)
Rwanda	" … any items other than pharmaceutical products, cosmetics and tobacco used as food or drink for human beings and include any substance used in the manufacture or treatment of food" (*Itegeko nº 47/2012 ryo kuwa 14/01/2013*, Art. 2)
Senegal	" … any treated, partially treated or unwrought substance intended for human consumption, and includes beverages, chewing gum and all substances used in the manufacture, preparation and processing of foods, excluding substances used only in form of drugs, cosmetics or tobacco" (*Règlement nº 007/2007/CM/UEMOA*, 1)
Seychelles	" … means any substance, whether processed, semi-processed or raw, which is prepared, sold, represented or intended for human consumption, and includes drinks, bottled and packaged water, chewing gum, other products of similar nature or use and any article, substance or ingredients used in the composition, manufacture, preparation or treatment of food but does not include (a) cosmetics; (b) tobacco; (c) plants prior to harvesting; (d) live animals, birds or live fish which are not used for human consumption while they are alive, (excluding shellfish), unless they are prepared for placing on the market for human consumption; (e) fodder or feed for animals, birds or fish; (f) drugs or medicinal products; (g) hormonal products or veterinary products for use in livestock feed; and (h) residues and contaminants" (*Food Act, 2014*, Art. 2)
Sierra Leone	" … means any article used as food or drink for human consumption, other than drugs or water, and includes (a) any article which is intended for use in the composition or preparation of food; and (b) any flavouring matter or condiment and (c) any colouring matter intended for use in food" (*Act №23*, 2)
Somalia	" … animal products such as meat, milk, eggs, honey, oil, bones, skin, etc" ** (*Veterinary Law Code*, Section 1)
South Africa	" … any article or substance (except a medicine defined in Medicine and Related Substances Act (Act 101 of 1965)) ordinarily eaten or drunk by a person or purporting to be suitable, or manufactured or sold, for human consumption, and includes any part or ingredient of any such article or substance, or any substance used or intended or destined to be used as a part or ingredient of any such article or substance" (*Foodstuffs, Cosmetic, and Disinfectants Act №54 of 1972*, 1)
Sudan	" … means foods or beverages that are prepared or distributed are intended for use in human consumption and include any other substances that are part of their manufacture or part of these substances, including also dairy products." (قانون رقم ٩٦٥ لسنة ١٣٧٩م لرقابة الأطعمة, 2)
Tanzania	" … means any article other than drugs, cosmetics and tobacco used as food or drink for human consumption and includes any substance used in manufacture or treatment of food" (*The Tanzania Food, Drugs, and Cosmetics Act*, Art. 3)
Tunisia	" … all material or transformed or untransformed product that is intended for eating or chewing or adapted to eating or chewing by man" (*قانون عدد 25 لسنة* 2019, 4.1)
Uganda	" … includes drink, chewing gum and other products of a like nature and use, and articles and substances used as ingredients in the preparation of food or drink or of such products, but does not include (i) water, live animals or birds; (ii) fodder or feeding stuffs for animals, birds or fish; or (iii) articles or substances used only as drugs" (*The Food and Drugs Act*, 1)

Table 4. *Cont.*

Country	Definition (Regulation)
Zambia	" … includes any article manufactured, sold or represented for use as food or drink for human consumption, chewing gum, and any ingredient of such food, drink or chewing gum" (*The Food and Drugs Act*, 2)
Zimbabwe	" … means any substance which is, in whole or in part, intended for human consumption or which is intended for entry into, or to be used in the manufacture of, any such substance" (*Food and Food Standards Act*, 3)

* This part must have been amended since it is no longer part of this Act, ** definition of 'animal product'.

Although these definitions basically include edible insects, a further insertion into food law is not possible. This is due to the fact that food inspection framework usually includes a definition of the food category, e.g., 'meat', 'fishery product', etc. With insects being foodstuffs of animal origin, the term 'animal' is defined, if at all, in terms of livestock and species obtained via fishing, sometimes including game species. Interestingly, some regulations seem to make a difference between "animal", "bird", and "fish", suggesting that "animal" as a food provider is perceived as a mammal. In fact, *The Food and Drug Act* (Art. 1) from Uganda excludes birds and fish from its 'animal' definition. In any case, these definitions usually exclude invertebrates. Some, however, mention honeybees, e.g., *Lei n° 30/VIII/2013* of Cape Verde. One of the relatively few countries that actually mentions insects in the animal definition is the *Food Act, 2014*, of the Seychelles (Art. 2). The *Veterinary Law Code* of Somalia (Section 1) does not contain an overall definition of foodstuff, but for 'animal product', an 'animal' being "any vertebrate or invertebrate animal other than the human being".

Some nations also have addressed the idea of novel foodstuffs. To provide an example, Niger's *Décret n° 2011-616/PRN/MEL* defines them as "products or foodstuffs for which human consumption in Niger has so far been unknown or marginal, as well as foods and ingredients made from genetically modified organisms" (Art. 7/2). A definition for novel foodstuffs is also provided by *007/2007/CM/UEMOA*. It may be debated, however, if edible insects actually are a novel food in African countries, particularly if marginal consumption is used as a criterion. However, as can be seen in Table 3, the amount of edible insect species varies from country to country, as will the extent of traditional entomophagy. In this way, the issue of insects as novel food will have to be answered on the nation level, also for those cases in which edible insects are traded from one country to another.

3.2. Wildlife Resources Management

Africa is known worldwide for its spectacular wildlife, which has been exploited for centuries as game, providing meat and trophies. Fearing the extinction of these valuable resources, most countries have developed a comprehensive regulatory framework to manage these species. Laws pertaining wildlife were issued in order to regulate the usage of the large African animals, particularly mammals, birds, and reptiles such as crocodiles, be it as a trophy obtained via sports hunting, be it as a foodstuff during regular hunting. Hunting legislation basically refers to the species that can be hunted, the methods permitted, and the fee to be paid for the right to hunt under a combination of the aforementioned parameters. Inclusion of arthropods such as insects and arachnids must be seen as incidental and a consequence of the definition for 'animal', particularly if regulations date from the previous century.

Then, and in view of the many other natural resources, it became necessary to regulate the management of, e.g., wood or mineral resources, i.e., activities that compromise the integrity of certain areas. Finally, natural reserves attract tourism, and so, regulation of these also became necessary.

AU has issued the *African convention on the conservation of nature and natural resources*. It addresses the need of balancing environment protection with sustainable use of the natural resources. In Article V, some basic terms are defined, e.g., "natural resource", "species", "specimen" (which refers to any non-human life-form, dead or alive), "product" (any part or derivative of a specimen), and a series of conservation areas types are recognized (defined more precisely in Annex 2). Among others, Article

XIV addresses the sustainable use of these resources and Article XVII the traditional rights of local ethnics to make use of them. In Article VI on land and soil, the term "sustainable farming and forestry practices" is coined and considered a goal to follow. Managing these specimens is the basic objective in terms of species diversity (Article IX), be it inside conservation areas, be it outside of them. In the latter case, harvestable populations are allowed. Among other objectives, the introduction of non-native species is to be controlled strictly, and existing populations should be eradicated, pests controlled, and diseases eliminated. This goes along with the phyto-sanitary convention which basically seeks to minimize the impact of, among others, insect pests by restricting the importation of potential sources. Member states are encouraged to regulate the extraction of specimens, avoiding a series of extraction methods specified in Annex 3. Although it may be expected that this passage refers to vertebrates, some of these banned means are used for regular insect hunting, e.g., artificial light sources, target illumination devices, nets (with exceptions as agreed upon by the Conference of the Parties), and traps. It calls the member states to regulate the domestic trade with, transport of, and possession of specimens (Article XI), to ensure the sustainable management of natural resources within development plans (Article XIV), to respect traditional rights and knowledge (e.g., of farmers; Article XVII), enable research (Article XVIII), and promote capacity building and technology transfer (Articles XIX and XX). Still, this paper did not contain concise regulations, but should be seen as a general framework from which other (inter)national regulations may start from.

Below the AU, one of the larger organizations is the Commision des Forêtes d'Afrique Centrale (COMIFAC), which includes Burundi, Cameroun, Central African Republic, Chad, Congo-Brazzaville, Congo-Kinshasa, Gabon, Equatorial Guinea, Rwanda, and São Tomé and Príncipe. In their constituting treaty, they expressed their commitment to support and develop a sustainable use of the forest resources, creating income from these resources (Article 1).

Of all these bases, edible insects only play a minor role, being one of the many forest products that may, or may be not, harvested. The main focus is clearly the game species, either for sport hunting or for bushmeat. Besides, the permitted use of insects (trophy, foodstuff, feedstuff, medicine, etc.) is frequently not stated. In this way, the first approach of asserting the legal status of insects within this area was to determine if insects actually were (respectively could be) included in the 'animal' definitions these regulations provide (Table 5). In some cases (e.g., Burkina Faso), the regulation did contain a concise definition, suggesting that 'animal' was defined using common sense, i.e., a non-human member of the Animalia kingdom. If this is actually the case, then the *Décret n⁰ 96-061/PRES/PM/MEE/ MATS/MEFP/MCIA/ MTT* would in fact be applicable to insects as it refers to "animals".

In other cases, regulations exclude insects. To give an example, *Loi n⁰ 1/17 du Septembre 2011 portant Commerce de Faune et de Flore Sauvages* from Burundi refers, in terms of animals, to a list of mammal, bird, reptile, and plant species.

Table 5. Summary of the references towards edible insects within African wildlife management legislation; numbers refer to the specific article of the regulation in question; for Cape Verde, Namibia, Seychelles, Somalia, South Sudan, and Zambia, regulations' definitions were not applicable to insects; no applicable regulations were found for Libya, São Tomé and Príncipe, Senegal, Sudan, and Zimbabwe.

Country	Regulation	Insects Included	Insects a Usable Resource	Insects Used as Foodstuff	Insect Used Privately	Insects Used Commercially	Additional Information
Algeria	*Loi n⁰ 14/07*	2	2	2	2	2	permission needed (Art. 5); impact study requested (Art. 8); payment for rights

Table 5. *Cont.*

Country	Regulation	Insects Included	Insects a Usable Resource	Insects Used as Foodstuff	Insect Used Privately	Insects Used Commercially	Additional Information
Angola	*Lei n° 6/17*	4.45	4.65	12.1	65, 98	4.43	Title II: sustainable management of forests, Title III: that of animals
Benin	*Loi n° 2002-16*	4	36, 45	96	97	75, 102	Art. 34 states that non-endangered species still benefit from general animal protection measures; permit needed (Art. 73); Chapter III on wildlife rearing
Botswana	*Wildlife Conservation and National Parks Act*	2	19	19	19	39	Special game licenses for veld produce gatherers (30); wildlife management areas (Part III)
	Agricultural Resources Conservation (Utilization of Veld Products) Regulations	s *	2	2	3	4	Applies to phane caterpillars (*Gonimbrasia belina*) being a "veld product"; permit if harvest exceeds 10 kg/person/month
Burkina Faso	*Décret n° 96-061/PRES/ PM/MEE/ MATS/ MEFP/MCIA/MTT*	?	III	III	21	27, 29	No definition of "animal" provided; mammals and birds: permits depending on the level of catching and the species
	Loi n ° 003-2011/AN	107	115	115	115	115	Title III on the wildlife usage incl. farming
Burundi	*Loi n° 1/07*	4.25, 4.26	144	144	144	144	Chapter IV, Section 3 deals with "non-wood forest products"
Cameroon	*Loi n° 94/01*	3	8	8	8	101	Local populations harvesting forest products for personal use (Art. 8), permission is needed (Art. 99)
Central African Republic	*Loi n° 08.022*	65	65	65	65	66–75	Only industrial usage of "forest products others than wood" is addressed
Chad	*Loi n° 14/PR/2008*	I.2, II.2, III. 95	II.2, III. 141	III. 76	III. 72–77, III. 143	III. 78–88, III. 154, III. 179	Definition of "forest product" is prone to include insects; Extent of usage depends on the forest type; traditional usages is free and does not require a permit when used for personal needs (III.76); commercial use is subject to taxes and can be done by the owners; special section for ranched fauna (III. 180–189).
Comoros	*Loi-cadre n° 94-018*	40	40	40	40	40	All wildlife usage (41) and introduction of foreign species (44) need an official permit
Congo, Democratic Republic (Kinshasa)	*Loi n° 82-0022*	2					"game animals" are part of the term "fauna", which comprises all vertebrate and invertebrate species
	Arrêté n° 014/CAB/MIN/ENV/2004		33			38	Taxation and basic requirements for keeping wild animals
Congo, Republic (Brazzaville)	*Loi n° 48/83 du 21/04/1983*	2					Insects classified as in no need of special protection; all hunting requires a permit (Article 7); law does not refer to private or commercial uses.
Djibouti	*Décret n° 2004-0065/PR/MHUEAT*	2	2	2	2	2	Insects included in "biodiversity"; no wild animals may be managed.
Egypt	*Law No. 102 of 1983*	2	2	2	2	2	Applies to protected areas only
Equatorial Guinea	*Ley n.° 8/1.988*	2	3	3	46	49–63	Permit needed for hunting (Article 9)

Table 5. *Cont.*

Country	Regulation	Insects Included	Insects a Usable Resource	Insects Used as Foodstuff	Insect Used Privately	Insects Used Commercially	Additional Information
Eritrea	*Proclamation № 155/2006*	2	25	25	25, 29, 30	25, 30	Permit needed for hunting (Article 25); specifications in case hunting is done due to necessity (29) and for wildlife farming (30)
Eswatini	*Environment Management Act, 2002*	2(1)					Insects part of "natural resources", attended by the government
Ethiopia	*Proclamation № 541/2007*	2	2	2, 8	8	8, 12	Permit needed for hunting (Article 8) and trading (12)
Gabon	*Loi n° 016/01*	4	163	199	199, 252–261	177, 233–243	Permit needed for any kind of wildlife-related activity
	Décret n° 692/PR/MEFEPEPN	2	2	2	7	7	Subsidiary hunting is allowed and free, but products may be sold only within the community.
Gambia	*National Environment Management Act (NEMA)*	2					Insects are part of "biological diversity", but no hints to a (sustainable) use are provided
	Biodiversity and Wildlife Act, 2003	2	VI	VI	VI	VIII	Same definition as in NEMA; harvesting license mandatory; commercial use of biological resources prohibited
Ghana	*Wild Animals Preservation Act, 1961*	12					Some edible species would be in the so-called "5th schedule" which allows one to take large amounts of animals to reduce their numbers.
	Wildlife Conservation Regulations						Theoretically, insects would enter the "4th schedule" in which an application has to be written to the authorities.
Guinea	*Loi U97/038/An du 9 Decembre*	2	61	61	61	62	Insects would be wild fauna, although other (vertebrate) taxa are mentioned in particular; no commercialization allowed.
Guinea-Bissau	*Decreto-Lei n° 5/2011*	2	15	15	15	24	Insects part of "forest resources"
Ivory Coast	*Loi n° 65-255 &94-442*						Insects are beyond the animal definition
Kenya	*Wildlife Conservation and Management Act, 2013, No. 47 of 2013*	3	81	81	81	8th s *	Permits required; game farming allowed for butterflies (10th schedule),
Lesotho	*Historical Monuments, Relics, Fauna and Flora Act, Act 41 of 1967*	2					Among insects, only refers to bees; any activity requires written consent by the authorities
Liberia	*New Wildlife and National Parks Act*	1	27	27	27	28	Keeping wild animals only with a permit
Madagascar	*Ordonnance n° 60-126 du 3 Octobre 1960*	1	3	3	3	3	Insects may fit into the category "vermin birds and other animals" and may be harvested indiscreetly
Malawi	*National Parks and Wildlife Act, № 11 of 1992*	2	45	45	69	Part X	Insects seem to fall into the unprotected animal category
Mali	*Loi n° 2018-036/ du 27 Juin 2018 fixant les principes de la gestion de la faune et de son habitat*	2	78	166	143	113, 143	Insects are part of fauna; harvesting permit necessary; rearing of any kind of wild animal species in confinement is allowed (Chapter X)
Mauritania	*Loi n° 97-006 abrogeant et remplaçant la loi n° 75-003 du 15 janvier portant la chasse et de la protection de la nature*	7	7	7	7	20	Animal gathering is termed hunting and is regulated, but insects are not mentioned explicitly, and fauna is divided in completely and partially protected vertebrates

Table 5. *Cont.*

Country	Regulation	Insects Included	Insects a Usable Resource	Insects Used as Foodstuff	Insect Used Privately	Insects Used Commercially	Additional Information
Mauritius	*Native Terrestrial Biodiversity and National Parks Act 2015*	2	26	26	26	26	Insect gathering may be seen as a kind of hunting and thus requires a permit
Morocco	*Loi n° 29-05 relative à la protection des espèces de flore et de faune sauvages et au contrôle de leur commerce*	2					Definition in Art. 2 include insects, but the hunting regulation refers to endangered (vertebrate) species only; other hunting regulations do not refer to species.
Mozambique	*Lei n° 10/99*	1	20	20	20	23	Permits are required
Niger	*Loi n° 98-07 du 29 avril 1998 fixant le Régime de la Chasse et de la Protection de la Faune*	4	5	5	5	18	Permits requested for hunting and selling
Nigeria	*Forest Regulations*	2	20	20	20	20	Insects are "minor forest produce" as defined by Art. 2 Forest Law; harvesting them requires a permit
	Wild Animals Law	2	43	43	43		Killing animals in times of famine
Rwanda	*Itegeko n° 70/2013 ryokuwa 07/09/2013 rigenga urosobe rw'ibinyabuzima mu Rwanda*	2	29	29	29		Using biodiversity requires a permit
Sierra Leone	*Wildlife Conservation Act, 1972*	2	37	37	37	37	Permits required for hunting animals not contemplated in the schedules (e.g., insects)
South Africa	*National Forests Act*	2	22	22	22	28	Regulates the use and selling of "forest produces" in state forests
Tanzania	*Wildlife Conservation Act, 2008*	3	55	55	55	55, 89	Permits and licenses required for any wildlife-related activity
Togo	*Loi n° 2008-09 portant Code Forestier*	2	Title IV	Title IV	Title IV	14, Title IV	Permits and licenses required for any wildlife-related activity
Tunisia	*Loi n° 20 portant Code forestier*	3	215	215	211, 215		Animal products from forests may not be placed on the market
Uganda	*Uganda Wildlife Statute, 1996*	2	Part VI	Part VI	Part VI	Part VI, part VII	Permits and licenses required for any wildlife-related activity

* s = schedule.

Wildlife can be managed or protected. In most countries, wildlife protection is mainly focused on large animals, typically hunting species. Some nations, however, also include insects. The *Décret n° 2004-0065/PR/MHUEAT* of Djibouti puts all national animal species under protection, i.e., also the insects. In Cape Verde, the *Decreto-Regulamentar N° 7/2002* lists several beetle species as protected, i.e., the scarabids *Aphodius* spp., the hydrophilids *Berosus* spp., the buprestid *Chrysobothris dorsata*, the dytiscid *Eretes sticticus*, the cerambycids *Xystrocera* spp., and the tenebrionids *Alphitobius laevigatus*, *Tenebrio* spp., and *Zophobas* spp., which are all edible. So, local consumers have to take care not to feed on these species.

- Cape Verde. No regulation on wildlife management could be found. However, the *Decreto-Regulamentar n° 7/2002* classifies the national fauna into several types, with different levels of protection. It contains a large list of beetles, of which according to [11] some are edible, at least at the genus level. Unless bred in captivity, these species may not be extracted from the wild or managed in any commercial way (Article 9). For the commercial use of endemic species, a permit from the government is mandatory. Introducing exotic species (as farming insects would be) can only be done after authorization from the government (Article 13). As mentioned before, the list also includes the classical tenebrionid genera *Alphitobius*, *Tenebrio*, and *Zophobas*. Although the text refers to *Tenebrio guineensis*, all *Tenebrio* species are considered edible. The mentioned "*Zophobas atratus concolor*" is basically *Z. atratus*. This one and *Alphitobius laevigatus* are typically used for food. Installing tenebrionid farms on Cape Verde may therefore be a challenge because the introduction of foreign populations, certifying the captivity origin of national populations or catching the initial stock from the wild are all subject to authorization.

- Congo-Democratic. The situation appears unclear. Unlike other nations, "non-wood forestry products" as defined in the forest act (*Loi n° 11-2002*) only includes plant products. Wildlife management seems to be regulated by the national hunting law (*Loi n° 82-0022*) and derived acts. Its definition of "game" refers to all vertebrates but amphibians and fish (Article 2) in contrast to "fauna", which includes both vertebrate and invertebrate species. The rest of the law only refers to game hunting issues. However, the *Arrêté interministèriel n° 003/CAB/MIN/ECN-EF/2006* and *n° 099/CAB/MIN/FINANCES/2006*, which fix the financial framework for wildlife usage deals with "fauna" and also includes "cocoons", "nymphs", and "dead insects". They are part of the "sub-product" section of the partially protected animals' tax table. It is suggested that these sub-products refer to ornamental insects sold as dried specimens rather than collecting edible insects for consumption. Still, the implementing regulation *Arrêté n° 014/CAB/MIN/ENV/2004* includes specifications for capturing (Article 23) and keeping wild animals (33, 34, and 37–40), i.e., animals in terms of "fauna" and this would include insects. These specifications require regular veterinary controls.

- Djibouti. Not much could be found on wildlife management. *Loi n° 43/AN/83/1re L* has prohibited any wildlife management for a period of ten years, this law dating from 1983. The more recent *Décret n° 2004-0065/PR/MHUEAT* on biodiversity protection follows this idea, prohibiting any kind of wildlife management.

- Gabon. There is an interesting, publically available legislation on wildlife management. The forest code (*Loi n° 016/01*) regulates the management of forest resources of which insects are part of. Subject to several permits, the fauna can be used economically and also within traditional frames for basic necessity requirements. *Décret n° 18/PR/MEFEPEPN* regulates the establishment of wild animal breeding units to produce, among others, foodstuffs (Article 2), while *Décret n° 692/PR/MEFEPEPN* focuses on the customary usage of forest products. A position paper [16] promotes a sustainable usage of these resources for the coming years.

- Guinea. The *Loi U97/038/An* has been clearly made for vertebrates and, in particular, game species. However, insects also fit into the "wildlife fauna" definition, and Chapter 9 regulates the use of unprotected species (a concrete list of the protection status is, however, not provided, and including insects in this section is an assumption by the authors), claiming that handling up to five specimens of the given species at a time is allowed, but contacting the local forest inspector is mandatory if this number is exceeded. This would be the case in harvesting edible insects.

Finally, a modern way of handling insects is by farming them. A survey conducted via FAOLEX in search of applicable data did not yield any results.

So, edible insects can be handled in many African countries in a traditional way, but typically after an official permit was provided.

3.3. Edible Insects in Pest Legislation

Some pest insects are edible, e.g., mealworms or locusts which, when swarming, are responsible not only for marked economic losses but may also trigger famines. Thus, there are other joint ventures in which several African nations cooperate. Two of them are CLCPRO (هيئة مكافحة الجراد الصحراوي في المنطقة الغربية)—Commission de Lutte contre le Criquet Pèlerin dans la Région Occidentale,

Commission for Fighting the Desert Locust in the Western Region) and CRC (هيئة مكافحة الجراد الصحراوي في المنطقة الوسطى, Commission for Controlling the Desert Locust in the Central Region), hosted by FAO. Both commissions are dedicated to preventing damages to crops caused by swarming desert locusts (*Schistocerca gregaria*). The use of this natural resource as foodstuff has not been contemplated so far, mostly out of the sheer size of locust swarms, and so no regulation in this issue has been established.

Besides, most African nations have more or less detailed pest legislation. There is usually one basic regulation that defines pests in a way that insects are included (Table 6) and provide measures on what to do to ensure pest reduction. In addition, many countries have specific acts dealing with plant diseases and pests in which the corresponding insect species are listed (Tables 7–10). To give an example, Morocco issued an interesting regulation for the control for the African palm weevil (*Rhynchophorus ferruguineus*). This regulation makes fighting this weevil obligatory, and a Moroccan governorate was classified as a quarantine zone, forbidding all transport of palms out of this area. Palms affected by the weevil must be destroyed (*Arrêté conjoint du ministre de l'agriculture et de la pêche maritime et du ministre de l'intérieur n° 287-09 du 30 janvier 2009 édictant des mesures d'urgence destinées à la lutte contre le Charançon rouge du palmier (Rhynchophorus ferrugineus)*). However, for the scope of the present contribution, the fact that these species are not permitted is the decisive information rather than the measures taken to control these pest species. In this way, rearing and placing on the market this particular species would be against the law.

Table 6. Pest insect legislation pertaining edible insect species (contained in "all pest insect species"); No data could be obtained for Angola, Djibouti, Equatorial Guinea, Ethiopia, Gabon, Lesotho, Senegal, Sierra Leone, Somalia, South Sudan, Sudan, and Togo.

Country	Law
Algeria	*Loi n° 87-17*
Benin	*Loi n° 91-001, Décret n° 92-258*
Botswana	*Plant Protection Act 2007*
Burkina Faso	*Loi n° 025-2017/AN*
Burundi	*Décret-loi n° 1/033*
Cameroon	*Loi n° 2003/003*
Cape Verde	*Lei n° 29/VIII/2013.*
Central African Republic	*Loi n° 62-350*
Chad	*Loi n° 14/PR/95*
Comoros	*Décret du 24 juin 1903*
Congo-Kinshasa	*Décret n° 05/162*
Congo-Brazzaville	*Loi n° 52-1256*
Egypt	قانون الزراعة رقم 53 لسنة *1966*
Eritrea	*Plant Quarantine Proclamation*
Eswatini	*Plant Control Act, 1981*
Gambia	*Plant Importation and Regulation Act.*
Ghana	*Plants and Fertilizer Act, 2010*
Guinea	*Loi L/92/027/CTRN*
Guinea-Bissau	*Decreto-Lei NQ 4/99*
Ivory Coast	*Loi n° 64-490*
Kenya	*Plant Protection Act*
Liberia	*Agricultural Law*
Libya	قانون رقم 27 لسنة *1968* بشأن وقاية النباتات
Madagascar	*Ordonnance n° 86-013*
Malawi	*Plant Protection Act, 1969*
Mali	*Loi n° 02-013 du 03 juin 2002*
Mauritania	*Loi n° 2000-042*
Mauritius	*Plant Protection Act 2006*

Table 6. *Cont.*

Country	Law
Morocco	*Dahir du 20 septembre 1927*
Mozambique	*Decreto nº 5/2009 de 1 de Junho*
Namibia	*Plant Quarantine Act 2008*
Niger	*Loi nº 2015-35 du 26 mai 2015*
Nigeria	*Agriculture (Control of Importation) Act*
Rwanda	*Itegeko nº 16/2016*
São Tomé & Príncipe	*Decreto Lei nº 5/2016*
Seychelles	*Plant Protection Act*
South Africa	*Agricultural Pests Act*
Tanzania	*Plant Protection Act*
Tunisia	*Loi nº 92-72*
Uganda	*Plant Protection and Health Act*
Zambia	*Plant Pests and Diseases Act*
Zimbabwe	*Plant Pests and Diseases Act*

Swarming locusts have been a threat for the agriculture for millennia. Along with the CLCPRO and CRC initiatives, many countries have adopted national regulations. However, only some do actually include concrete measures. In other countries, the fight against locusts is documented legally by the will to build up corresponding organizations and initiatives, i.e., the national anti-locust agency (ANLA) in the Chadian *Loi nº 005/PR/2007*, IFVM in Madagascar (*Décret nº 2017-064 du 31 du janvier 2017*), the CNLP in Mali (*Loi nº 06-065 du 29 décembre 2006* and subsequent texts), the PLUCP in Niger (*Arrêté nº 13/MDA/DPV*), and the campaigns in Morocco (*Loi nº 57-02*) and Tunisia (*Décret nº 88-1751*). In those countries, locust may be combatted efficiently, but this is not reflected in the law.

Table 7. Pest insect legislation pertaining edible insect species (as presented in [11]): locusts (Orthoptera; all species are Acrididae, except for *Zonocerus variegatus* (Zygomorphidae)).

Species	Country	Law
Acrididae spp. ("locusts")	Botswana	*Locusts Act*
	Burundi	*Ordonnance nº 53*
	Chad	*Loi nº 005/PR/2007*
	Egypt	قرار وزاري رقم ١٨ لسنة ١٩٦٧
	Keyna	*Plant Protection Rules*
	Lesotho	*Locust Destruction Proclamation*
	Madagascar	*Décret nº 2017-064*
	Mauritania	*Décret nº 2002-062 du 25 juillet 2002*
	Namibia	*Regulations relating to the Destroying of Locusts*
	Sudan	قانون أبادة الجراد لسنة 1907م
Anacridium melanorhodon	Mauritius	*Plant Protection Act 2006*
	Seychelles	*Plant Protection Act*
Locusta migratoria migratorioides	Kenya	*Plant Protection Order, Plant Protection Rules*
	Zambia	*Plant Pests and Diseases (Pests and Alternate Hosts) Order*
	Zimbabwe	*Locust Control Act [Chapter 19:06].*
Locustana pardalina	Eswatini *	*Plant Control Act, 1981*
	Zimbabwe	*Locust Control Act*
Melanoplus differentialis	Mauritius	*Plant Protection Act 2006*
	Eswatini	*Plant Control Act, 1981*
Nomadacris septemfasciata	Kenya	*Plant Protection Order, Plant Protection Rules*
	Zambia	*Plant Pests and Diseases (Pests and Alternate Hosts) Order*
	Zimbabwe	*Locust Control Act [Chapter 19:06].*
Schistocerca gregaria	Egypt	قرار وزاري رقم ١٧ لسنة ١٩٦٧ بوضع نظام مكافحة الجراد الصحراوي
	Kenya	*Plant Protection Order, Plant Protection Rules*
	Zimbabwe	*Locust Control Act*
Zonocerus variegatus	Mozambique	*Decreto nº 5/2009*

* termed "Locusta pardelina" in this act.

Table 8. Pest insect legislation pertaining edible insect species (as presented in [11]: butterflies and moths (Lepidoptera).

Family	Species	Country	Law
Cossidae	*Cossus cossus*	Algeria	*Décret exécutif n° 95-387*
Crambidae	*Chilo* spp.	Kenya	*Plant Protection Order*
		Mozambique	*Decreto n° 5/2009*
	Ostrinia furnacalis	Botswana	*Plant Protection Act 2007*
Gelechiidae	*Pectinophora gossypiella*	Zambia	*Plant Pests and Diseases (Pest Control) Regulations, Plant Pests and Diseases (Pests and Alternate Hosts) Order*
	Busseola fusca	Kenya	*Plant Protection Order*
	Helicoverpa spp.	Cape Verde	*Portaria n° 37/2015*
	Heliothis spp.	Botswana	*Plant Protection Act 2007*
		Seychelles	*Plant Protection Act*
	H. zea	Tunisia	*Arrêté du Ministre de l'agriculture du 31 mai 2012*
Noctuidae		Botswana	*Plant Protection Act 2007*
	Spodoptera spp.	Mozambique	*Decreto n° 5/2009*
		Tunisia	*Arrêté du Ministre de l'agriculture du 31 mai 2012*
		Mozambique	*Decreto n° 5/2009*
	S. frugiperda	South Africa	*Control Measures Relating to Fall Armyworm*
		Tunisia	*Arrêté du Ministre de l'agriculture du 31 mai 2012*
Pieridae	*Pieris* spp.	Mozambique	*Decreto n° 5/2009*
Pyralidae	*Chilo* spp.	Benin	*Arrêté interministériel n° 128*
		Botswana	*Plant Protection Act 2007*
	Acherontia styx	Mozambique	*Decreto n° 5/2009*
Sphingidae	*Agrius convolvuli*	Botswana	*Plant Protection Act 2007*
	Clanis bilineata	Mozambique	*Decreto n° 5/2009*
	Erinnyis ello	Mozambique	*Decreto n° 5/2009*

Table 9. Pest insect legislation pertaining edible insect species (as presented in [11]): beetles (Coleoptera); Eswatini also refers to "woodborer beetles", but without any species information, a term that, in fact, can be used to several taxa.

Family	Species	Country	Law
Ceram-bycidae	*Anoplophora* spp.	Tunisia	*Arrêté du Ministre de l'agriculture du 31 mai 2012*
	A. chinensis	Mauritius	*Plant Protection Act 2006*
	A. glabripennis	Mauritius	*Plant Protection Act 2006*
	Monochamus spp.	Tunisia	*Arrêté du Ministre de l'agriculture du 31 mai 2012*
	Leptinotarsa spp.	Benin	*Arrêté interministériel n° 128*
		Malawi	*Plant Protection Act, 1969*
		Mauritania	*Arrêté n° R-0031257 du 12 novembre 2002*
		Mauritius	*Plant Protection Act 2006*
Chryso-melidae	*L. decemlineata*	Mozambique	*Decreto n° 5/2009*
		Seychelles	*Plant Protection Act*
		Tunisia	*Arrêté du Ministre de l'agriculture du 31 mai 2012*
		Zambia	*Plant Pests and Diseases (Pest Control) Regulations, Plant Pests and Diseases (Pests and Alternate Hosts) Order*
	Anthonomus spp.	Mozambique	*Decreto n° 5/2009*
	Cosmopolites sordidus	Kenya	*Plant Protection Order*
		Mauritania	*Arrêté n° R-0031257 du 12 novembre 2002*
	Rhynchophorus spp.	Benin	*Arrêté interministériel n° 128*
		Mauritania	*Arrêté n° R-0031257 du 12 novembre 2002*
		Mauritius	*Plant Protection Act 2006*
Curculi-onidae	*R. ferrugineus*	Morocco	*Arrêté conjoint (…) n° 287-09*
		Tunisia	*Arrêté du Ministre de l'agriculture du 31 mai 2012*
	R. phoenicis	Mauritius	*Plant Protection Act 2006*
	Scyphophorus acupunctatus	Kenya	*Plant Protection Order*
		Mozambique	*Decreto n° 5/2009*
	Sitophilus oryzae	Kenya	*Plant Protection Order*
		Zambia	*Plant Pests and Diseases (Pests and Alternate Hosts) Order*

Table 9. *Cont.*

Family	Species	Country	Law
Scarabeidae	*Adoretus* spp.	Botswana	*Plant Protection Act 2007*
		Benin	*Arrêté interministériel n⁰ 128*
	Amphamalion spp.	Botswana	*Plant Protection Act 2007*
	Anomala spp.	Botswana	*Plant Protection Act 2007*
	Leucophora spp.	Mozambique	*Decreto n⁰ 5/2009*
	Oryctes spp.	Benin	*Arrêté interministériel n⁰ 128*
	Popilia spp.	Benin	*Arrêté interministériel n⁰ 128*
		Malawi	*Plant Protection Act, 1969*
	P. japonica	Tunisia	*Arrêté du Ministre de l'agriculture du 31 mai 2012*
		Zambia	*Plant Pests and Diseases (Pest Control) Regulations*
	Xylotrupes gideon	Botswana	*Plant Protection Act 2007*
Tenebrionidae	*Alphitobius* spp.	Kenya	*Plant Protection Order*
	A. diaperinus	Zambia	*Plant Pests and Diseases (Pests and Alternate Hosts) Order*
	A. levigatus	Mozambique	*Decreto n⁰ 5/2009*
		Zambia	*Plant Pests and Diseases (Pests and Alternate Hosts) Order*
	Tenebrio spp.	Kenya	*Plant Protection Order*
		Zambia	*Plant Pests and Diseases (Pests and Alternate Hosts) Order*
	T. molitor	Kenya	*Plant Protection Order*
		Zambia	*Plant Pests and Diseases (Pests and Alternate Hosts) Order*
	Tribolium castaneum	Kenya	*Plant Protection Order*
		Zambia	*Plant Pests and Diseases (Pests and Alternate Hosts) Order*
	T. confusum	Kenya	*Plant Protection Order*
		Zambia	*Plant Pests and Diseases (Pests and Alternate Hosts) Order*

While locusts, butterfly caterpillars, and beetle grubs and larvae are relative well-known food insects, some of the species contained in Table 10 are not, particularly the cockroach and fly species. It should be stressed that the species listed are known to be edible, at least in some parts of their ranges, which may not necessarily be Africa. The record of edibility of *Periplaneta americana* comes from Brazil, that of *Anastrepha ludens* from Mexico.

Table 10. Pest insect legislation pertaining edible insect species (as presented in [11]): other orders.

Order	Family	Species	Country	Law
Blattodea	Blattidae	*Blatta orientalis*	Zambia	*Plant Pests and Diseases (Pests and Alternate Hosts) Order*
		Periplaneta americana	Zambia	*Plant Pests and Diseases (Pests and Alternate Hosts) Order*
	Ectobiidae	*Blatella germanica*	Zambia	*Plant Pests and Diseases (Pests and Alternate Hosts) Order*
Diptera	Tephritidae	*Anastrepha* spp.	Benin	*Arrêté interministériel n⁰ 128*
			Botswana	*Plant Protection Act 2007*
			Mauritania	*Arrêté n⁰ R-0031257 du 12 novembre 2002*
			Mozambique	*Decreto n⁰ 5/2009*
		A. ludens	Mauritius	*Plant Protection Act 2006*
			Tunisia	*Arrêté du Ministre de l'agriculture du 31 mai 2012*
	Tipulidae	*Tipula paludosa*	Botswana	*Plant Protection Act 2007*
Hemiptera	Alydidae	*Leptocorisa acuta*	Mozambique	*Decreto n⁰ 5/2009*
		L. oratorius	Mozambique	*Decreto n⁰ 5/2009*
	Psyllidae	*Psylla* spp.	Mozambique	*Decreto n⁰ 5/2009*
	Pyrrhocoridae	*Dysdercus* spp.	Botswana	*Plant Protection Act 2007*
	Tessaratomidae	*Tessaratoma papillosa*	Mozambique	*Decreto n⁰ 5/2009*
Hymenoptera	Formicidae	*Atta cephalotes*	Botswana	*Plant Protection Act 2007*

As a summary of the previous tables, Table 11 provides an overview of the current legal situation of edible in the different African nations.

Table 11. Summary of the legal context of edible insects in Africa.

Country	Native Edible Species	Edible Insects Tolerated	Suitable Foodstuff Definition	Wildlife Management —Private	Wildlife Management —Commercial	General Pest Management	Edible Specified as Pests
Algeria	(−)	+	+	+	+	+	+
Angola	+		+	+	+		
Benin	+	+	+	+	+	+	+
Botswana	+	+		+	+	+	+
Burkina Faso	+	+	+	−	−	+	
Burundi	+	+		+	+	+	+
Cameroon	+	+		+	−	+	
Cape Verde	−	(−)	+	−	−	+	+
Central African Republic	+	+	+	+	+	+	
Chad	+			+	+	+	+
Comoros	−			+	+	+	
Congo, Democratic Republic (Kinshasa)	+	+		+	+	+	
Congo, Republic (Brazzaville)	+			+	+	+	
Djibouti	−			−	−		
Egypt	+		+	+	+	+	+
Equatorial Guinea	+			+	+		
Eritrea	−			+	+	+	
Eswatini	+		+	+	+	+	+
Ethiopia	+		+	+	+		
Gabon	+			+	+		
Gambia	+			+	−	+	
Ghana	+		+	+		+	
Guinea	+	+		+	−	+	
Guinea-Bissau	+		+	+	+	+	
Ivory Coast	+		+	−	−	+	
Kenya	+		+	+	+	+	+
Lesotho	+			+	+		+
Liberia	+		+	+	+	+	
Libya	+					+	
Madagascar	+	+	+	+	+	+	+
Malawi	+		+	+	+	+	+
Mali	+		+	+	+	+	
Mauritania	(−)		+	+	+	+	+
Mauritius	+		+	+	+	+	
Morocco	+	+	+			+	+
Mozambique	+		+	+	+	+	+
Namibia	+		+	−	−	+	+
Niger	+	+	+	+	+	+	
Nigeria	+		+	+	+	+	
Rwanda	(−)		+	+	+	+	
São Tomé and Príncipe	+					+	
Senegal	+		+				
Seychelles	−		+	−	−	+	+
Sierra Leone	+		+	+	+		
Somalia	(−)		+	−	−		
South Africa	+	+	+	+	+	+	+
South Sudan	+			−	−		
Sudan	+		+				+
Tanzania	+	+	+	+	+	+	
Togo	+	+		+	+	.	
Tunisia	(−)	+	+	+	−	+	+
Uganda	+		+	+	+	+	
Zambia	+		+	−	−	+	+
Zimbabwe	+		+			+	+

4. Discussion

4.1. Data Availability

One of the main difficulties in carrying out the present survey was data gathering. Getting first-hand information from the corresponding authorities was challenging. However, as the group of authors comprised African and European scientists, the African colleagues had better chances in getting in contact with these authorities. A clear cut was seen between the responses of countries where French and Arabic are spoken (attended by the African colleagues) and those in which English or Portuguese are the official languages (attended by the European colleagues), with the latter responding very late, if at all (despite several attempts made to get an answer), while the African colleagues were able to provide results within two weeks. Differences between the authorities and citizens asking for their advice is a global phenomenon, but also African researchers make special emphasis on the local conditions [17].

Data availability is also an important issue on the internet; some countries did not provide their current legislation on the corresponding government portals. Instead, data had to be traced via other sites, especially FAOLEX. Still, gaps remained, either due to technical or internet safety reasons, or simply because some governments do not put these sources online.

In this context and beyond the current survey, it was noted that most regulatory texts were written either in French, English, or Portuguese, i.e., the administrative languages from colonial times. The only exceptions were the regulations presented (sometimes exclusively) in Arabic of the North African countries, and the full set of regulations from Ethiopia and Rwanda, which were either bilingual (Amharic and English) or even trilingual (Kinyarwanda, English, and French). While this was rather advantageous for the researchers, this may pose a problem for the local African citizens. In Africa, there are more than 2000 languages spoken, and in many countries, the percentage of the population that understands these official languages completely may be relatively low. This, of course, varies strongly among countries, ranging between 35% (Mali) and 88% (Eswatini) in 2018 (http://data.uis.unesco.org/Index.aspx?queryid=166#). So, there is a risk that a certain portion of a country's population may not be able to understand the laws of their own country. However, the situation is changing in some countries. The web presence of the government of South Africa (https://www.gov.za/) has been presenting regulations in several of its eleven official languages, usually English plus another one.

4.2. Edible Insects in African Food Legislations

With the exception of phane in Botswana, edible insects have not been considered in the African food law, neither on the national nor on international level. This may be surprising in view of the rich entomophagy tradition that exists in Africa. Van Huis [6] stressed the importance of edible insects in the different regions of the continent, which range from using them in those seasons of the year in which other natural foodstuffs are rarer to the simple reason that insects are consumed because of their taste. This is a completely different scenario from, e.g., the EU where insect-consuming tradition is almost inexistent, and insects are treated as truly novel foods that enter the European food market. In this way, the EU novel food regulation opens this market, and future legal acts will incorporate insects into the public health surveillance, adopting as much as possible from existing regulation, but attending, at the same time, the idiosyncrasies of these foodstuffs, e.g., microbiological criteria. This may be flanked by national European guidelines [7]. However, food law in Africa by itself is the key to this finding.

It should be stressed that, on one hand, African food legislations are merely starting to adapt to the current standards experienced in other parts of the world, and that, on the other hand, the progress in that varies markedly among countries. Some laws cited in the present contribution date down to the 1900s and are still in power. Yet, these are exceptions, and many laws date from the 1990s to the 2010s. In some countries, no food law could be encountered, and if available, many laws seem to lack a concrete depth. In some way, many of these texts express expectations about the issues the address, leaving a very large range for interpretations, making law less transparent. However, this phenomenon is far from being exclusively African. Many EU laws also contain these kinds of expectations, e.g., *REG (EC) 178/2002*. However, subsequent acts provide the details of the bases set in these general laws, e.g., *REG (EC) 2073/2005* on the microbiological criteria for a series of foodstuffs. Still, the current EU novel food regulation (*REG (EC) 2015/2283*) also sets the bases for novel food certifications, but does not contain specific indication on production, processing, or quality control. A corresponding amendment of *REG (EC) 2073/2005* for the EU is dearly awaited.

So, African food laws are sometimes outdated and lack depth, which would increase their applicability. However, food safety in Africa is a basic problem, regardless of how efficient the legislation is. Kussaga et al. (2013) [18] present a large list of microbiological and chemical risks detected in African foodstuffs, ranging from pesticides, mycotoxins, heavy metals, over many bacteria to a series of vermiform parasites. Apart from that, Africa is experiencing a serious problem with antibiotics residues in its foodstuffs. An extensive review concluded that in some African countries, the prevalence of veterinary drugs contained in foodstuffs amounted to up to 94% in certain countries. This may affect both the food-processing industry (e.g., fermented dairy products) and the consumer alike [19]. This shows that food safety in Africa is a very immanent problem.

This condition becomes even more complicated. On one hand, traditional food habits have changed in Africa, creating products not clearly addressed in the current legislation. As an example, for Western Africa, not only the total amount of food required has increased, but consumers' preferences have shifted to convenience products, consuming more perishable products, and a growing awareness of food quality issues [20]. Improving policy coordination and policy implementation are, in fact, two of the key recommendations the aforementioned authors give to modernize the Western African food sector.

On the other hand, the "modernization" of traditional foods does not always improve their quality. As traditional food production has experienced labor ease by means of mechanization of e.g., cutting and grating processes, the production increase has not lead to a growing awareness of food safety, good manufacturing practices, and poor hygiene. Selling food that requires a certain degree of air circulations in tight-sealing plastic bags may also affect the quality by favoring the growth of microaerophilic or even anaerobial flora [17]. Since insects are traditional foodstuffs, these risks may also apply to them.

The degree of entomophagy also varies across the continent. Table 3 shows the different amounts of REIS per country, and while insect consumption may be rather unimportant in e.g., Djibouti or São Tomé and Príncipe, in Botswana, Benin, or Madagascar it plays a larger role. Besides and strictly speaking, introducing an edible species from one African nation to another where the consumption is unknown would make this species a novel food as coined in UEMOA, Niger, or EU regulation.

Another important issue is to tell food safety from food security. Strictly speaking, food security refers to the supply with (any kind of) food (in any quality), while food safety, in contrast, means to provide a foodstuff in a quality that does not compromise the consumer's health, i.e., free of pathogens, toxic substances, foreign bodies, radiation, etc. In many countries, food security is considered more important, particularly in those in that climate and other factors may lead to food shortages. However, AU documents include food safety in food security. Given the importance of food insects in times of the year when food is scarce (the rainy season, time before harvest, etc. [21]), the lacking of the consideration of insects still remains interesting. A possible reason is the fact that a certain portion of the current food legislation originates from colonial times, and the omission of edible insects in the first place may be part of the entomophobia of the colonial rulers. In fact, many people in Europe still consider eating insects either disgusting or a staple for the poor and primitive that cannot afford any "true" food" [22], although these concepts do not, by any chance, reflect the reality. In fact, this seems to be one of the major reasons why entomophagy is still not more popular in Europe, despite corresponding awareness campaigns.

All in all and in comparison to other areas of the world, food safety regulation in many African countries appears scarce or even not existing. One possible explanation may be the fact that governments were forced to focus on food security rather than food safety in response to natural catastrophes and warfare. However, as international African documents exposed here showed, the idea that food security mandatorily has to include food safety is becoming increasingly attended. Besides, the AU has underlined that poor food safety is not only a problem for the population (with 91 million cases of food-borne diseases in Africa per year and 137,000 casualties related to them) but also a problem decreasing the competitiveness of African agriculture inside and outside the continent [23]. The goals formulated in the Malabo Declaration to end hunger in Africa by 2025 will not be reached if food safety is left unattended. In response to that, the first FAO/WHO/AU International Food Safety Conference was held in Addis Ababa in February of 2019, and edible insects were addressed in one session [24], underlining the need to establish concrete policies also with regard to farming insects since the absence of specific regulations may trigger the establishment of insect farms that use—in terms of food safety—unsafe substrates, e.g., slaughterhouse wastes. The ideas raised in Addis Ababa were presented again at a subsequent meeting in Geneva in April of 2019, where the need of harmonizing food safety regulation was stressed [25].

Taking up the misleading conceptions that appear in Western societies on entomophagy, the reality in Africa is, at least locally and seasonally, a huge variety of vivid entomophagous traditions, far from being an "emergency foodstuff" (Figures 1 and 2). Most government bodies interviewed directly consider this tradition, tolerating the consumption and the trade with these animals, even though no concise framework exists. Table 4 showed that most African countries work with foodstuff definitions that permit the inclusion of edible insects, although they may not have been considered at the moment of publishing the corresponding regulations. This is a positive fact, and future framework can be based on these definitions. The will of some of the interview partners to fill this legal gap is also a promising perspective.

Figure 1. Fried locusts sold on a Western African market (image by S. Tchibozo).

This was also confirmed by the minutes of a meeting of several governmental and non-governmental organisations (GO and NGO) from in and outside Africa, held in Kenya [26]. One of the meeting's goals was to create awareness of the importance of the insect sector and the need to establish a corresponding legislation, which ideally could be harmonized among countries to ensure a comparable quality. These regulations should include both food safety standards and best practice codes. The future will show what will become of this initiative.

Figure 2. Living caterpillars sold on a Central African Market (image by S. Tchibozo).

4.3. Food Insects Obtained from the Wild

Following the tradition, insects have been caught from the wild in order to be processed and consumed [5]. African wildlife management legislations referring to hunting and/or natural resource management (as that of forests) usually base themselves on "wildlife" definitions that do include insects. In this context, and strictly speaking of law, this can result as a disadvantage, particularly in the moment of hunting respectively gathering them. As in the food law, the application of wildlife management laws to insects may result by mere extrapolation as the regulations were written while having traditional livestock or game species in mind. In those, using some trapping and hunting techniques is clearly banned. However, they do apply to what the definitions understand as "wildlife", and so, light traps, which are typically used to attract insects in the night, may be in fact illegal. The same is true for the permits many regulations request. It may be expected that, except phane in Botswana, gathering edible insects from the wild will be done without asking for these permits, perhaps because it is a traditional way of eating and earning money, perhaps because insects do not have the perceived relevance as cattle or high-priced game species.

Still, forest and wildlife management offer an interesting way to use the insect resource in a sustainable way. These species may be managed in their corresponding biotopes. This, however, requires the awareness that only sustainable use of these insects will ensure a longer-lasting income, rather than exterminating a species by over-harvesting. For this, reliable data on the degree to which a population may be extracted without endangering its survival are necessary, and this data used to be inexistent for insects. Only local gatherers may have this knowledge out of their experience.

In any way, Table 5 shows a large degree of variation among countries. Some allow wildlife management, some do not, and some regulations only refer to specific animal or product groups. Due to the complexity of this issue, it is highly recommended to study the actual situation in a given country intensively before any major uses of insects are made. Tchibozo et al. [27] summarize the necessary requirements to use wild populations thoroughly:

- Good knowledge of the edible species of a given region;
- Thorough species inventory of the area in question;
- Identification of the edible species;
- Good knowledge of the species' biology including their host plants;
- Detection of breeding sites;
- Detection of places where they are/can be modified.

With knowledge, the particular breeding behavior and the feasibility for production can be assessed. This model is laid out for satisfying a local market and is the base of obtaining a solid colony for breeding in captivity, with the option of selling the offspring also abroad [27].

Besides, food safety concerns remain, since forest-derived foodstuffs are even less subjected to quality controls than livestock-derived ones are. Although tradition is prone to cope with traditional food safety problems like spoilage or pathogens, more modern risks as contaminants like insecticides are likely to pass undetected. Besides, modern storage methods like plastic bags also may impair the quality of insect products sold on the market.

A special situation is those species of insects that are edible and a pest at the same time. The idea of consuming edible pest insects is more than tempting, particularly since many of these species have a better nutritional profile that than the crops they feed on. To give an example, teff (*Eragrostis tef*) is a typical East African crop, and contains 12 g of protein/100 g dry matter (value calculated from Baye [28], while orthopterans range between 41 and 91 g/100 g dry matter [6]. Using pest insects may be beneficial from the nutritional point of view opening up a new food resource; it reduces the damage to the crops and is, by itself, part of the environmental management of a given area. Legally, this may be difficult as many countries have established corresponding legislations to combat them. This applies mainly to locusts, but also to other species, as can be seen in Tables 7–10. From the food safety point of view however, the proper use of this food resource will depend fundamentally on the toxicological status of these animals (pesticides) and the possibility to process these animals right after harvesting into a storable product with a stable quality. In fact, orthopterans killed by insecticides are sometimes gathered and sold on West African markets [5].

Finally, some countries protect some native insect species from usage, e.g., Cape Verde and Djibouti, including some that are edible. This is a regular proceeding to protect the national fauna, particularly if the territory is relatively small and, in the case of Cape Verde, a set of island relatively far off the continental coast.

Thus, managing edible insects in the wild is an ancient tradition. However, a sound legal system to ensure sustainability will be necessary to make a larger use of edible insects.

4.4. Insect Farming

When wildlife management may be difficult because of legal uncertainties and food safety concerns, farming them may be an option [4]. Southeast Asian countries like Thailand, Cambodia, Laos, and Myanmar have started a larger attempt to rear insects and produce them as "mini-livestock" [29]. Two so-called Thai Agricultural standards that deal with crickets (*TAS 8202-2017*) and silkworms (*TAS 8201/2012*) are examples on how insect farming is regulated by the government. This option is basically also open to African countries.

From the legal point of view, livestock regulations in Africa are even scarcer than food-related ones and refer, if available at all, exclusively to classical and local domestic species.

Some countries allow wildlife respectively game farms, and if definitions fit, they could be nominally also applied to insects (Table 5). However, the choice of indigenous farmed species must be done carefully in order not to infringe any species protection, wildlife management, or pest control regulation (see above). Bringing foreign productive species (certain cricket and mealworm species or the black soldier fly) into the country may also be hazardous and may infringe on other nature protection or trading laws.

Raheem et al. [4] mention first insect farms for the black soldier fly (*Hermetia illucens*) in South Africa (which in fact is the largest in the world) and Nigeria, but for feed production. Edible crickets are reared in Kenya and Uganda. This implies that at least in those two countries, the setup of these farms was legal.

5. Conclusions

In this way, asking for food quality standards for edible insects may be challenging if some countries have not addressed food quality standards for meat, fish, or dairy products so far. However, the world progresses, and Africa is a part of the global community just like any other region of the world is, and if foodstuffs' quality will be addressed in the future, those for edible insects should be included. The regulatory framework should address the entire production chain, starting from two different points of primary production, i.e., wildlife resource and farmed insects.

Regarding gathering from the wild, many African countries already have a solid legal base that could be applied to insects. However, population control is vital in order to protect existing natural populations. For food safety, the quality of the gathered animals is critical, particularly in terms of residues and contaminants, and harvested animals should be tested on a regular base, e.g., using Codex Alimentarius recommendations.

The possibility of farming insects in Africa must be assessed thoroughly, taking care not to interfere with land ownership, limited possibilities of using a given area, and, if applicable, farming regulations for livestock and/or wildlife. Animal species' protection, animal imports, and pest legislation must also enter this evaluation. The farmer must be able to control the life cycle of the farmed species. Special care must be taken to avoid the escape of the farmed insects respectively the entry of insect predators into the farm. Once established, a farm ensures food quality by providing a controlled environment to the farmed animals, from the materials used for construction and bedding to the feedstuffs. This is where residues and contaminants can also enter the production chain of a farm. Farms should also be evaluated on a regular base, e.g., to obtain a certificate that it works according to the national requirements. During these evaluations, the inspectors should pay special attention to animal welfare, meaning that the five freedoms also apply to farming insects. Whenever possible, farming must be done according to national law.

Regardless the origin of the harvested animals, killing must be done according to the animal welfare state-of-the-art procedures, which currently is freezing them or mushing them quickly in a blender. As soon as possible, insects must be heated thoroughly (e.g., 100 °C for 10 min) to reduce bacterial counts, and rinsed. Storage of merely cooked insects must be in the refrigerator or an ice box, or, better, frozen. Insects preserved by other methods, e.g., drying, smoking, or fermenting can be stored traditionally. However, microbial counts should be assessed for these products in order to provide best-before dates for the consumer. Besides, general hygiene rules also apply to insects just as any other foodstuff.

In terms of public health surveillance, insects sold publicly, regardless if gathered or farmed, should be included in regular monitoring programs. There is no need to treat insect-based products any differently as more established foodstuffs are. Appropriate labeling follows national guidelines and should, by any chance, show the name of the insect (ideally including the scientific name), a best-before date and the information about allergens; all insects contain tropomyosin, which may lead to cross reactions of patients that are allergic to dust (mites) and crustaceans. Laboratory analyses should focus, on one hand, on the gross chemical composition (this may change from batch to batch if feeding regime was changed between batches), antibiotics (using substance specific tests rather than generic inhibitory ones as many insects contain innate inhibitors, which could lead to false-positive results), residues, and contaminants. The other focus is microbiology, with salmonellae, listeriae, total bacterial counts, coagulase-positive staphylococci, *Escherichia coli* and other Enterobacteriaceae, presumptive *Bacillus cereus*, yeasts, and fungi as the most important parameters, plus the ones established in national food law.

Foods **2020**, *9*, 502

Author Contributions: Conceptualization, N.T.G. and S.T.; Methodology, N.T.G. and S.T.; Software, N.T.G., S.T., A.A., F.A., M.M.G., and W.A.A.S.; Validation, S.T., A.A., F.A., M.M.G., W.A.A.S., and M.P.; Formal Analysis, N.T.G.; Investigation, N.T.G., S.T., A.A., F.A., M.M.G., and W.A.A.S.; Resources, N.T.G., S.T., M.P.; Data Curation, N.T.G.; Writing—Original Draft Preparation, N.T.G.; Writing—Review and Editing, S.T., A.A., F.A., M.M.G., W.A.A.S., and M.P.; Visualization, S.T.; Supervision, N.T.G., S.T., and M.P.; Project Administration, N.T.G.; Funding Acquisition, M.P. and N.T.G. All authors have read and agreed to the published version of the manuscript.

Funding: This research was funded by Deutsche Forschungsgemeinschaft and University of Veterinary Medicine Hannover, Foundation within the funding programme Open Access Publishing.

Acknowledgments: The authors wish to express their gratitude to Bernard Cole (Australia, for South Africa) and Jessy Kamwi (Namibia), Embassy of the Central African Republic in Paris, and all contributors from the authors' networks.

Conflicts of Interest: The authors declare no conflict of interest.

Appendix A

Regulation Documents:

(Note: in bi or trilingual texts, the order is from left to right; enumeration is strictly alphabetical and numerical)

Algeria

- Décret exécutif n° 91-53 relatif aux conditions d'hygiène lors du processus de la mise à la consommation des denrées alimentaires. http://www.fao.org/faolex/results/details/en/c/LEX-FAOC004085.
- Décret exécutif n° 95-387 du 5 Rajab 1416 correspondant au 28 novembre 1995 fixant la liste des ennemis des végétaux et les mesures de surveillance et de lutte qui leur sont applicables. (JORA N° 73 du 29-11-1995). http://www.fao.org/faolex/results/details/en/c/LEX-FAOC043461.
- Loi n° 14-07 du 13 Chaoual 1435 correspondant au 9 aout 2014 relative aux ressources biologiques. http://www.fao.org/faolex/results/details/en/c/LEX-FAOC171802.
- Loi n° 87-17 du Ier août 1987 relative à la protection phytosanitaire. http://www.fao.org/faolex/results/details/en/c/LEX-FAOC003579.

Angola

- Decreto Presidencial 179/18 - Aprova o Regulamento sobre a Sujeição a Análises Laboratoriais dos Produtos Destinados ao Consumo Humano e Animal. http://www.fao.org/faolex/results/details/en/c/LEX-FAOC178359.
- Lei n° 6/17 – Lei de bases de florestas e fauna salvagem. http://www.fao.org/faolex/results/details/en/c/LEX-FAOC162520.

Benin

- Arrêté interministériel n° 128 MDR/MF/DC/CC/CP relatif au contrôle phytosanitaire des végétaux et des produits végétaux à l'importation et à l'exportation. http://www.fao.org/faolex/results/details/en/c/LEX-FAOC007526.
- Décret n° 92-258 fixant les modalités d'application de la Loi n° 91-004 du 11 février 1991 portant réglementation phytosanitaire. http://www.fao.org/faolex/results/details/en/c/LEX-FAOC002350
- Loi n° 84-009 du 15 mars 1984 sur le contrôle des denrées. http://www.fao.org/faolex/results/details/en/c/LEX-FAOC086098.
- Loi n° 91-004 portant réglementation phytosanitaire. http://www.fao.org/faolex/results/details/en/c/LEX-FAOC002349.
- Loi n° 2002-16 du 18 octobre 2004 portant régime de la faune enRépublique du Bénin. http://www.fao.org/faolex/results/details/en/c/LEX-FAOC166784.

AU

- African convention on the conservation of nature and natural resources. http://www.fao.org/faolex/results/details/en/c/LEX-FAOC045449.l

Botswana

- Agricultural Resources Conservation (Utilization of Veld Products) Regulations. http://www.fao.org/faolex/results/details/en/c/LEX-FAOC091414.
- Food Control Act. http://www.fao.org/faolex/results/details/en/c/LEX-FAOC066060.
- Locusts Act (Chapter 35:03). http://www.fao.org/faolex/results/details/en/c/LEX-FAOC006483.
- Plant Protection Act 2007 (Chapter 35:02). http://www.fao.org/faolex/results/details/en/c/LEX-FAOC126405.
- Wildlife Conservation and National Parks Act. http://www.fao.org/faolex/results/details/en/c/LEX-FAOC004728.

Burkina Faso

- Décret n° 96-061/PRES/PM/MEE/MATS/MEFP/MCIA/MTT portant réglementation de l'exploitation de la faune au Burkina Faso. http://www.fao.org/faolex/results/details/en/c/LEX-FAOC004885.
- Loi n° 003-2011/AN portant Code forestier au Burkina Faso. http://www.fao.org/faolex/results/details/en/c/LEX-FAOC106703.
- Loi n° 025-2017/AN du 15 mai 2017 portant protection des végétaux au Burkina Faso. http://www.fao.org/faolex/results/details/en/c/LEX-FAOC171520.

Burundi

- Décret-loi n° 1/033 portant protection des végétaux au Burundi. http://www.fao.org/faolex/results/details/en/c/LEX-FAOC004668.
- Loi N° 1/7 du 15 juillet portant Révision du Code Forestier. http://www.fao.org/faolex/results/details/en/c/LEX-FAOC163753.
- Loi N° 1/17 du 10 septembre 2011 portant Commerce de Faune et de Flore Sauvages. http://www.fao.org/faolex/results/details/en/c/LEX-FAOC162986.
- Ordonnance n° 53/agri - assistance dans la lutte contre les invasions de sauterelles et de criquets. http://www.fao.org/faolex/results/details/en/c/LEX-FAOC039378.

Cameroon

- Loi n° 2003/03 portant protection phytosanitaire. http://www.fao.org/faolex/results/details/en/c/LEX-FAOC050036.

Cape Verde

- Decreto Legislativo n° 3/2009. http://www.fao.org/faolex/results/details/en/c/LEX-FAOC097520.
- Decreto-Regulamentar n° 7/2002. http://www.fao.org/faolex/results/details/en/c/LEX-FAOC051798.
- Lei n° 29/VIII/2013. http://www.fao.org/faolex/results/details/en/c/LEX-FAOC123988.
- Portaria n° 37/2015 de 13 de Agosto. http://www.fao.org/faolex/results/details/en/c/LEX-FAOC148050.

Central African Republic

- Loi n° 03-04 du 20 janvier 2003 portant Code d'hygiène en République centrafricaine. http://www.fao.org/faolex/results/details/en/c/LEX-FAOC176540.
- Loi n° 08/022 portant Code forestier de la République Centrafricaine. http://www.fao.org/faolex/results/details/en/c/LEX-FAOC107432.

- Loi n° 62-350 relative à l'organisation de la protection des végétaux en République centrafricaine. http://www.fao.org/faolex/results/details/en/c/LEX-FAOC007180.

Chad

- Décret n° 10/PR/MA/99 fixant les modalités d' application de la loi n° 14/PR/95 relative à la protection des végétaux. http://www.fao.org/faolex/results/details/en/c/LEX-FAOC126372.
- Loi n° 005/PR/2007 du 16 avril 2007 portant création d'une Agence Nationale de Lutte Antiacridienne (ANLA). http://www.fao.org/faolex/results/details/en/c/LEX-FAOC180033.
- Loi n° 14/PR/95 relative à la protection des végétaux. http://www.fao.org/faolex/results/details/en/c/LEX-FAOC004040.
- Loi n° 14/PR/2008 portant régime de forêts, de la faune et de ressource halieutiques. http://www.fao.org/faolex/results/details/en/c/LEX-FAOC117920.

COMESA

- COMESA treaty amended by council meeting 2009. https://www.comesa.int/comesa-treaty/.

COMIFAC

- Traité relatif à la conservation et à la gestion durable des ecosystème forestiers d'Afrique Centrale et instituant la Commission de Forêts d'Afrique Centrale (COMIFAC). http://www.fao.org/faolex/results/details/en/c/LEX-FAOC071928.

Comoros

- Code de la santé publique et de l'action sociale pour le bien être de la population - Loi n° 95-013 du 24 juin 1995. http://droit-afrique.com/upload/doc/comores/Comores-Code-1995-sante-publique.pdf.
- Décret du 24 juin 1903 relatif aux mesures à prendre en cas de maladie contagieuse et parasitaire des plantations. *http://droit-afrique.com/upload/doc/comores/Comores-Decret-1903-maladie-des-plantations.pdf*.
- Loi-cadre n° 94-018 relative à l'environnement. http://www.fao.org/faolex/results/details/en/c/LEX-FAOC011718.

Congo-Brazzaville

- Loi n° 48/83 du 21/04/1983 définissant les conditions de la conservation de et de l'exploitation de la faune sauvage. http://www.fao.org/faolex/results/details/en/c/LEX-FAOC004177.
- Loi n° 52-1256 du 26 novembre 1952 relative à l'organisation de la protection des végétaux dans les territoires relevant du Ministère de la France d'outre-mer. http://www.fao.org/faolex/results/details/en/c/LEX-FAOC161165.

Congo-Kinshasa

- Arrêté interministériel n° 003/CAB/MIN/ECN-EF/2006 et n° 099/CAB/MIN/FINANCES/2006 du 13 juin 2006 portant fixation des taux des droits, taxes et redevances à percevoir, en matière de faune et de flore, à l'initiative du ministère de l'environnement, conservation de la nature, eaux et forêts. http://www.fao.org/faolex/results/details/en/c/LEX-FAOC071929.
- Arrêté n° 014/CAB/MIN/ENV/2004 du 29 avril 2004 relatif aux mesures d'exécution de la loi n° 82-002 du 28 mai 1982 portant réglementation de la chasse. http://www.fao.org/faolex/results/details/en/c/LEX-FAOC061848.
- Décret n° 05/162 du 18 novembre 2005 portant réglementation phytosanitaire en République Démocratique du Congo.
- Loi n° 82-002 portant réglementation de la chasse. http://www.fao.org/faolex/results/details/en/c/LEX-FAOC004275.

- Loi n° 11-2002 portant Code forestier. http://www.fao.org/faolex/results/details/en/c/LEX-FAOC034383.

Djibouti

- Décret n° 2004-0065/PR/MHUEAT portant protection de la biodiversité. http://www.fao.org/faolex/results/details/en/c/LEX-FAOC046297.
- Loi n° 43/AN/83/1re L portant modification de l'article 1er de la délibération n° 268/7 L du 3 avril 1971 portant interdiction totale de la chasse sur toute l'étendue du territoire. http://www.fao.org/faolex/results/details/en/c/LEX-FAOC004183.

EAC

- EAC food security action plan (2011–2015). https://www.eac.int/index.php?option=com_documentmananger&task=download.document&file=bWFpbl9kb2N1bWVudHNfcGRmX2pLRWtST2pEalpSeE1qQ2pzeWlkVkRNRUFDDIEZvb2QgU2VjdXJpdXpdHkgQWN0aW9uIFBsYW4=&counter=109.

ECCAS

- Treaty establishing the Economic Community of Central African States. https://www.wipo.int/edocs/lexdocs/treaties/en/eccas/trt_eccas.pdf.

ECOWAS

- Revised treaty. Abuja, Nigeria, http://www.ecowas.int/wp-content/uploads/2015/01/Revised-treaty.pdf.

Egypt

- قانون الزراعة رقم 53 لسنة 1966 (Law Nr. 53 of 1966: Promulgation of the Agriculture Law), extwprlegs1.fao.org › docs › pdf › egy153081.
- قرار وزاري رقم ١٧ لسنة ١٩٦٧ بوضع نظام مكافحة الجراد الصحراوي (Ministerial Decree No. 17 of 1967 on establishing a system to control the Desert Locust). http://www.fao.org/faolex/results/details/en/c/LEX-FAOC122033.
- قرار وزاري رقم ١٨ لسنة ١٩٦٧ بشأن مقاومة حشرات الجراد المصري والمستوطن وأنواع النطاط (Ministerial Decree No. 18 of 1967 on the Egyptian locust resistance and the types of hoppers). http://www.fao.org/faolex/results/details/en/c/LEX-FAOC122031.
- Law No. 1/2017 on Promulgating National Food Safety Authority Law. http://nfsa.gov.eg/Images/App_PP/DeskTop/App_Web/1/MyWebMedia/PDF/NFSA%20Law-English.pdf.
- Law No. 102 of 1983 on Natural Protected Areas. http://www.fao.org/faolex/results/details/en/c/LEX-FAOC020777.

Equatorial Guinea

- Ley n.° 8/1.988, de fecha 31 de diciembre, Reguladora de la Fauna Silvestre, Caza y Áreas Protegidas. http://www.endangeredspecieslaws.com/endangered-species-laws-of-equatorial-guinea/.

Eritrea

- Forestry and Wildlife Conservation and Development Proclamation (No. 155/2006). http://www.fao.org/faolex/results/details/en/c/LEX-FAOC068045.
- Plant Quarantine Proclamation (No. 156/2006). http://www.fao.org/faolex/results/details/en/c/LEX-FAOC068047.

Eswatini

- Environmental Management Act 2002 (Act No. 5 of 2002). http://www.fao.org/faolex/results/details/en/c/LEX-FAOC047833.
- Plant Control Act, 1981. http://www.fao.org/faolex/results/details/en/c/LEX-FAOC078838.
- Veterinary Public Health Act. http://www.gov.sz/images/veterinary%20public%20health%20act_%202013.pdf.

Ethiopia

- አዋጅ ቁጥር 541/2007 - የዱር አንላት ልማት፣ ጥበቃና አጠቃቀም አዋጅ /Proclamation No. 541/2007, a Proclamation to Provide for the Development [sic!] Conservation and Utilization of Wildlife. http://www.fao.org/faolex/results/details/en/c/LEX-FAOC095249.
- አዋጅ ቁጥር 661/2009 - ስለምግብ፣ መድኃኒትና ጤናአብካቤ አስተዳደርና ቁጥጥር የመጣአዋጅ/ Proclamation No. 661/2009, a proclamation to provide for food, medicine, and health care administration and control. http://www.fao.org/faolex/results/details/en/c/LEX-FAOC094419.

EU

- Commission Regulation (EC) No 2073/2005 of 15 November 2005 on microbiological criteria for foodstuffs (Text with EEA relevance). https://eur-lex.europa.eu/legal-content/EN/TXT/?uri=celex%3A32005R2073.
- Regulation (EC) No 178/2002 of the European Parliament and of the Council of 28 January 2002 laying down the general principles and requirements of food law, establishing the European Food Safety Authority and laying down procedures in matters of food safety. https://eur-lex.europa.eu/legal-content/EN/TXT/HTML/?uri=CELEX:32002R0178&from=DE.
- Regulation (EU) 2015/2283 of the European Parliament and of the Council of 25 November 2015 on novel foods, amending Regulation (EU) No 1169/2011 of the European Parliament and of the Council and repealing Regulation (EC) No 258/97 of the European Parliament and of the Council and Commission Regulation (EC) No 1852/2001 (Text with EEA relevance). https://eur-lex.europa.eu/eli/reg/2015/2283/oj.
- Risk profile related to production and consumption of insects as food and feed https://www.efsa.europa.eu/en/efsajournal/pub/4257.
- Ref. Ares(2019)382900 - 23/01/2019: Annex to the Commission Regulation (EU) … / … amending Annex III to Regulation (EC) No 853/2004 of the European Parliament and of the Council as regards specific hygiene requirements for insects intended for human consumption [sic]. https://ec.europa.eu/info/law/better-regulation/have-your-say/initiatives/2079-Specific-hygiene-rules-for-insects-intended-for-human-consumption.

Gabon

- Décret n° 18/PR/MEFEPEPN du 6 janvier 2005 fixant les conditions de création d'unités d'élevage d'espèces animales sauvages. http://www.fao.org/faolex/results/details/en/c/LEX-FAOC143387/.
- Décret n° 692/PR/MEFEPEPN du 24 août 2004 fixant les conditions d'exercice des droits d'usage coutumiers en matière de forêt, de faune, de chasse et de pêche. http://www.fao.org/faolex/results/details/en/c/LEX-FAOC174469/.
- Loi n° 016-01 portant code forestier en République gabonaise. http://www.fao.org/faolex/results/details/en/c/LEX-FAOC029255/.

GAFTA

- Economic and Social Council (Ed., 1997):, قرار المجلس الاقتصادي‌إعلان منطقة التجارة الحرة العربية الكبرى والاجتماعى رقم 1317 د.ع 59 بتاريخ 1997/2/19. http://www.aproarab.com/Down/Etfaqiat/Ar/EtfaqiatAr4.doc. 3 pp.

Gambia

- Biodiversity and Wildlife Act, 2003. http://www.fao.org/faolex/results/details/en/c/LEX-FAOC158129.
- National Environment Management Act, 1994 (NEMA). http://www.fao.org/faolex/results/details/en/c/LEX-FAOC006275.
- Plant Importation and Regulation Act. http://www.fao.org/faolex/results/details/en/c/LEX-FAOC041167.

Ghana

- Plants and Fertilizer Act, 2010 (No. 803). http://www.fao.org/faolex/results/details/en/c/LEX-FAOC168842.
- Public Health Act, 2012 (Act No. 851 of 2012). http://www.fao.org/faolex/results/details/en/c/LEX-FAOC136559.
- Wild Animal Preservation Act, 1961. http://www.fao.org/faolex/results/details/en/c/LEX-FAOC040827.
- Wildlife Conservation Regulations, 1971. http://www.fao.org/faolex/results/details/en/c/LEX-FAOC040817.

Guinea

- Loi L/92/027/CTRN instituant le contrôle phytosanitaire des végétaux à l'importation et à l'exportation. http://www.fao.org/faolex/results/details/en/c/LEX-FAOC004005.
- Loi U97/038/An du 9 Decembre Adoptant et promulguant Ie Code de Protection de la faune sauvage et Réglementation de la Chasse. http://www.fao.org/faolex/results/details/en/c/LEX-FAOC01273.

Guinea-Bissau

- Decreto n° 62-E/92 do 30 Dezembro: Regime do Controlo Sanitário dos Géneros Alimentícios. http://www.fao.org/faolex/results/details/en/c/LEX-FAOC016119.
- Decreto-Lei n° 2/2004. http://www.fao.org/faolex/results/details/en/c/LEX-FAOC012733.
- Decreto-Lei n° 5/2011. http://www.fao.org/faolex/results/details/en/c/LEX-FAOC118220.
- Decreto-Lei n° 4/99 Define as medidas de protecção fitossanitárias destinadas a evitar a introdução no Pais de organismos prejudiciais aos vegetais ou produtos vegetais. http://www.fao.org/faolex/results/details/en/c/LEX-FAOC023411.

IGAD

- IGAD Regional Strategy, Volume 1: the framework. https://igad.int/documents, 84 pp.
- IGAD Regional Strategy, Volume 2: implementation plan 2016–2020. https://igad.int/documents, 124 pp.

IvoryCoast

- Décret n° 92-487 du 26 août 1992 portant étiquetage et présentation des denrées alimentaires. http://www.fao.org/faolex/results/details/en/c/LEX-FAOC178128.
- Loi n° 64-490 du 21 décembre 1964 relative à la protection des végétaux. http://www.fao.org/faolex/results/details/en/c/LEX-FAOC175761.
- Loi n° 65-225 relative à la protection de la faune et à l'exercice de la chasse. http://www.fao.org/faolex/results/details/en/c/LEX-FAOC089113.
- Loi n° 94-442 portant modification de la loi n° 65-225 relative à la protection de la faune et à l'exercice de la chasse. http://www.fao.org/faolex/results/details/en/c/LEX-FAOC089114.

Kenya

- Food, Drugs and Chemical Substances Act. http://www.fao.org/faolex/results/details/en/c/LEX-FAOC106440.
- Plant Protection Act (Cap. 324). http://www.fao.org/faolex/results/details/en/c/LEX-FAOC018403.
- Plant Protection Order (Cap. 324). http://www.fao.org/faolex/results/details/en/c/LEX-FAOC018406.
- Plant Protection Rules (Cap. 324). http://www.fao.org/faolex/results/details/en/c/LEX-FAOC018416.
- Wildlife Conservation and Management Act, 2013, No. 47 of 2013. *http://www.fao.org/faolex/results/details/en/c/LEX-FAOC134375*.

Lesotho

- Historical Monuments, Relics, Fauna and Flora Act, Act 41 of 1967. http://www.fao.org/faolex/results/details/en/c/LEX-FAOC041401.
- Locust Destruction Proclamation (No. 3 of 1925). http://www.fao.org/faolex/results/details/en/c/LEX-FAOC128717.

Liberia

- Agricultural Law (Title 3 of the Revised Liberian Code of Laws). http://www.fao.org/faolex/results/details/en/c/LEX-FAOC005338.
- New Wildlife and National Parks Act. http://www.fao.org/faolex/results/details/en/c/LEX-FAOC003548.
- Public Health Law – Title 33 – Liberian Code of Laws Revised. http://www.fao.org/faolex/results/details/en/c/LEX-FAOC174510.

Libya

- قانون رقم 27 لسنة 1968 بشأن وقاية النباتات. Law No. 27 of 1968 on the protection of plants. http://www.fao.org/faolex/results/details/en/c/LEX-FAOC046426.

Madagascar

- Décret n° 2017-064 du 31 du janvier 2017 Portant création et organisation du Centre de Lutte Antiacridienne de Madagascar (IFVM). http://www.fao.org/faolex/results/details/en/c/LEX-FAOC171287.
- Ordonnance n° 86-013 relative à la législation phytosanitaire à Madagascar. http://www.fao.org/faolex/results/details/en/c/LEX-FAOC007459.
- Ordonnance n° 60-126 du 3 Octobre1960. http://www.fao.org/faolex/results/details/en/c/LEX-FAOC004210.

Malawi

- National Parks and Wildlife Act (Act No. 11 of 1992). http://www.fao.org/faolex/results/details/en/c/LEX-FAOC004733.
- Plant Protection Act, 1969 (No. 11 of 1969). http://www.fao.org/faolex/results/details/en/c/LEX-FAOC063795.

Mali

- Décret n° 06-259/P-RM du 3 juin 2006 instituant l'autorisation de mise sur le marché des denrées alimentaires, des aliments pour animaux et des additifs alimentaires. http://www.fao.org/faolex/results/details/en/c/LEX-FAOC153390.

- Loi n° 02-013 du 03 juin 2002 Instituant le contrôle phytosanitaire en République du Mali. http://www.fao.org/faolex/results/details/en/c/LEX-FAOC153268.
- Loi n° 06-065 du 29 décembre 2006 portant création du Centre National de Lutte Contre le Criquet Pèlerin. http://www.fao.org/faolex/results/details/en/c/LEX-FAOC158047.
- Loi n° 2018-036/ du 27 juin 2018 fixant les principes de la gestion de la faune et de son habitat. http://www.fao.org/faolex/results/details/en/c/LEX-FAOC180235.

Mauritania

- Arrêté n° R-0031257 du 12 novembre 2002 fixant la liste des organismes de quarantaine. http://www.fao.org/faolex/results/details/en/c/LEX-FAOC138454.
- Décret n° 2002-062 du 25 juillet 2002 portant application de la loi 042-2000 du 26 juillet 2000 relative à la protection des végétaux. http://www.fao.org/faolex/results/details/en/c/LEX-FAOC086852.
- Loi n° 97-006 abrogeant et remplaçant la loi n° 75-003 du 15 janvier portant la chasse et de la protection de la nature. http://www.fao.org/faolex/results/details/en/c/LEX-FAOC010273.
- Loi n° 2000-042 relative à la protection des végétaux. http://www.fao.org/faolex/results/details/en/c/LEX-FAOC061959.

Mauritius

- Food Act 1998 (Act No. 1 of 1998). http://www.fao.org/faolex/results/details/en/c/LEX-FAOC028943.
- Native Terrestrial Biodiversity and National Parks Act 2015 (No. 14 of 2015). http://www.fao.org/faolex/results/details/en/c/LEX-FAOC161113.
- Plant Protection Act (Act No. 10 of 2006). http://www.fao.org/faolex/results/details/en/c/LEX-FAOC074277.

Morocco

- Arrêté conjoint du ministre de l'agriculture et de la pêche maritime et du ministre de l'intérieur n° 287-09 du 30 janvier 2009 édictant des mesures d'urgence destinées à la lutte contre le Charançon rouge du palmier (*Rhynchophorusferrugineus*). http://www.fao.org/faolex/results/details/en/c/LEX-FAOC092510.
- Dahir du 20 septembre 1927 (23 rebia I 1346) portant règlement de police sanitaire des végétaux en zone française de l'Empire Chérifien. http://www.fao.org/faolex/results/details/en/c/LEX-FAOC178301.
- Loi n° 28-07 relative à la sécurité sanitaire des produits alimentaires. http://www.fao.org/faolex/results/details/en/c/LEX-FAOC096767.
- Loi n° 29-05 relative à la protection des espèces de flore et de faune sauvageset au contrôle de leur commerce. http://www.fao.org/faolex/results/details/en/c/LEX-FAOC106726.
- Loi n° 57-02 portant approbation, quant au principe, de la ratification du Royaume du Maroc de l' Accord portant création d'une Commission de lutte contre le criquet pèlerin dans la région occidentale, fait à Rome en novembre 2000. http://www.fao.org/faolex/results/details/en/c/LEX-FAOC171836.

Mozambique

- Decreto n° 5/2009 de 29 de Dezembro. http://www.fao.org/faolex/results/details/en/c/LEX-FAOC112022.
- Decreto n° 5/2009 de 1 de Junho. http://www.fao.org/faolex/results/details/en/c/LEX-FAOC112020.
- Decreto n° 15/2006 – Regulamento sobre os Requisitos Higiénico-Sanitários de Produção, Transporte, Comercialização e Inspecção e Fiscalização dos Géneros Alimentícios, http://www.fao.org/faolex/results/details/en/c/LEX-FAOC110946.

- Lei nº 10/99 – estabelece os princípios e normas básicos sobre a protecção, conservação e utilização sustentável dos recursos florestais e faunísticos. http://www.fao.org/faolex/results/details/en/c/LEX-FAOC020106.

Namibia

- Foodstuffs, Cosmetics and Disinfectants Ordinance 18 of 1979 – Ordinance to control the sale, manufacture and importation of foodstuffs, cosmetics and disinfectants; and to provide for incidental matters. https://laws.parliament.na/cms_documents/foodstuffs,-cosmetics-and-disinfectants-a17dca6f1d.pdf.
- Nature Conservation Ordinance, 1975. http://www.fao.org/faolex/results/details/en/c/LEX-FAOC018007.
- Plant Quarantine Act, 2008 (No. 7 of 2008). http://www.fao.org/faolex/results/details/en/c/LEX-FAOC094398.
- Regulations relating to the Destroying of Locusts (GN no. 30 of 1989). http://www.fao.org/faolex/results/details/en/c/LEX-FAOC188741.

Niger

- Arrêté nº 13/MDA/DPV du 11 février 2005, portant création et organisation du projet de lutte d'urgence contre le criquet pèlerin. http://www.fao.org/faolex/results/details/en/c/LEX-FAOC065246.
- Décret nº 2011-616/PRN/MEL du 25 novembre 2011 réglementant l'inspection d'hygiène des denrées animales et des denrées alimentaires d'origine animale. http://www.fao.org/faolex/results/details/en/c/LEX-FAOC176645.
- Loi nº 98-07 fixant le régime de la chasse et de la protection de la faune. http://www.fao.org/faolex/results/details/en/c/LEX-FAOC080736.
- Loi nº 2015-35 du 26 mai 2015 relative à la protection des végétaux. http://www.fao.org/faolex/results/details/en/c/LEX-FAOC175085.

Nigeria

- Agriculture (Control of Importation) Act. http://www.fao.org/faolex/results/details/en/c/LEX-FAOC120040.
- Forest Law. http://www.fao.org/faolex/results/details/en/c/LEX-FAOC003330.
- Forest Regulations. http://www.fao.org/faolex/results/details/en/c/LEX-FAOC003331.
- Food and Drugs (Amendment) Decree 1999, Decree No. 21, http://www.fao.org/faolex/results/details/en/c/LEX-FAOC034264.
- Wild Animals Law. http://www.fao.org/faolex/results/details/en/c/LEX-FAOC041660.

Rwanda

- Itegeko nº 16/2016 ryo ku wa 10/05/2016 rigena uburyo bwo kurengera ubuzima bw'ibimera mu Rwanda/Law N° 16/2016 of 10/05/2016 on plant health protection in Rwanda/Loi N°16/2016 du 10/05/2016 portant protection de la santé des végétaux au Rwanda. http://www.fao.org/faolex/results/details/en/c/LEX-FAOC188020.
- Itegeko nº 47/2012 ryo ku wa 14/01/2013 rigenga imicungire n'igenzura ry'ibiribwa n'imiti/Law No 47/2012 of 14/01/2013 relating to the regulation and inspection of food and pharmaceutical products/Loi No 47/2012 du 14/01/2013 portant règlementation et inspection des produits alimentaires et pharmaceutiques. http://www.fao.org/faolex/results/details/en/c/LEX-FAOC131821.

- Itegeko n° 70/2013 ryo ku wa 07/09/2013 rigenga urosobe rw'ibinyabuzima mu Rwanda/Law n° 70/2013 of 07/09/2013 governing biodiversity in Rwanda/Loi n° 70/2013 du 07/09/2013 régissant la biodiversité au Rwanda. http://www.fao.org/faolex/results/details/en/c/LEX-FAOC131764.

SADC

- Dar-es-Salam declaration on agriculture and food security in the SADC region. https://www.sadc.int/documents-publications/show/813.
- Protocol on wildlife conservation and law enforcement. https://www.sadc.int/documents-publications/show/824.
- Regional guidelines for the regulation of food safety in SADC member states. https://www.sadc.int/documents-publications/show/4087.

São Tomé & Príncipe

- Decreto Lei n° 5/2016 Aprova a Lei-quadro de Sanidade Vegetal. http://www.fao.org/faolex/results/details/en/c/LEX-FAOC160888.

Seychelles

- Food Act, 2014 (No. 8 of 2014). http://www.fao.org/faolex/results/details/en/c/LEX-FAOC143993.
- Plant Protection Act 1996 (Cap. 171A). http://www.fao.org/faolex/results/details/en/c/LEX-FAOC007782.

Sierra Leone

- Act No. 23, 1960, Public Health Ordinance. http://www.fao.org/faolex/results/details/en/c/LEX-FAOC181412.
- Wildlife Conservation Act, 1972. http://www.fao.org/faolex/results/details/en/c/LEX-FAOC041659.

Somalia

- Veterinary Law Code – 2016. http://www.fao.org/faolex/results/details/en/c/LEX-FAOC171698.

South Africa

- Agricultural Pests Act (No. 36). http://www.fao.org/faolex/results/details/en/c/LEX-FAOC107871.
- Control Measures Relating to Fall Armyworm (G.N. No. 449 of 2017). http://www.fao.org/faolex/results/details/en/c/LEX-FAOC167448.
- Foodstuffs, Cosmetic, and Disinfectants Act No. 54 of 1972. http://www.fao.org/faolex/results/details/en/c/LEX-FAOC085626.
- National Forests Act. http://www.fao.org/faolex/results/details/en/c/LEX-FAOC123699.

Sudan

- قانون رقم ٥٦٩ لسنة ١٩٧٣م لرقابة الأطعمة. http://extwprlegs1.fao.org/docs/pdf/sud68386.pdf.
- قانون أبادة الجراد لسنة 1907م (Locust Destruction Act of 1907). http://www.fao.org/faolex/results/details/en/c/LEX-FAOC027490.

Tanzania

- Plant Protection Act (No. 13 of 1997). http://www.fao.org/faolex/results/details/en/c/LEX-FAOC019688.

- Tanzania Food, Drugs, and Cosmetics Act. http://www.fao.org/faolex/results/details/en/c/LEX-FAOC053027.
- Wildlife Conservation Act, 2008. http://www.fao.org/faolex/results/details/en/c/LEX-FAOC097858.

Thailand

- Thai Agricultural Standard TAS 8201/2012 – Good practices for silk cocoon production. http://web.acfs.go.th/eng/system_standard.php?pageid=6
- Thai Agricultural Standard TAS 8202/2017 – Good agricultural practices for cricket farm. http://web.acfs.go.th/eng/system_standard.php?pageid=9

Togo

- Loi n° 2008-09 portant Code Forestier. http://www.fao.org/faolex/results/details/en/c/LEX-FAOC085011.

Tunisia

- قانون عدد 25 لسنة 2019 مؤرخ في 26 فيفري 2019 يتعلق بالسلامة الصحية للمواد الغذائية وأغذية الحيوانات. http://www.legislation.tn/detailtexte/Loi-num-2019-25-du----jort-2019-024__2019024000251.
- Arrêté du Ministre de l'agriculture du 31 mai 2012, fixant la liste des organismes de quarantaine. http://www.fao.org/faolex/results/details/en/c/LEX-FAOC113265.
- Décret n° 88-1751 fixant l'organisation et les modalités de fonctionnement de la campagne de lutte anti-acridienne.
- Loi n° 20 portant Code forestier. http://www.fao.org/faolex/results/details/en/c/LEX-FAOC002805.
- Loi n° 92-72 portant refonte de la législation relative à la protection des végétaux. http://www.fao.org/faolex/results/details/en/c/LEX-FAOC011737.

UEMOA

- Règlement n° 007/2007/CM/UEMOA relatif à la sécurité sanitaire des végétaux, des animaux et des aliments dans l'UMEOA. http://www.droit-afrique.com/upload/doc/uemoa/UEMOA-Reglement-2007-07-securite-sanitaire.pdf.

Uganda

- Plant Protection and Health Act (Cap. 31). http://www.fao.org/faolex/results/details/en/c/LEX-FAOC158615.
- The Food and Drugs Act, Chapter 27. http://www.fao.org/faolex/results/details/en/c/LEX-FAOC096144.
- Uganda Wildlife Statute, 1996. http://www.fao.org/faolex/results/details/en/c/LEX-FAOC009000.

Zambia

- Plant Pests and Diseases Act (Cap. 233). http://www.fao.org/faolex/results/details/en/c/LEX-FAOC046759.
- Plant Pests and Diseases (Pests and Alternate Hosts) Order (Cap. 233). http://www.fao.org/faolex/results/details/en/c/LEX-FAOC046785.
- Plant Pests and Diseases (Pest Control) Regulations (Cap. 233). http://www.fao.org/faolex/results/details/en/c/LEX-FAOC046784.
- The Food and Drugs Act, Chapter 303. http://www.fao.org/faolex/results/details/en/c/LEX-FAOC048635.

Zimbabwe

- Food and Food Standards Act, Chapter 15:04. http://www.fao.org/faolex/results/details/en/c/LEX-FAOC024975.
- Locust Control Act [Chapter 19:06]. http://www.fao.org/faolex/results/details/en/c/LEX-FAOC060738.
- Plant Pests and Diseases Act [Chapter 19:08]. http://www.fao.org/faolex/results/details/en/c/LEX-FAOC060741.

References

1. Amadi, E.N.; Kiin-Kabari, D.B. Nutritional composition and microbiology of some edible insects commonly eaten in Africa, hurdles and future prospects: A critical review. *J. Food Microbiol. Saf. Hyg.* **2016**, *1*, 107. [CrossRef]
2. Jideani, A.I.O.; Nethsiheni, R.K. Selected edible insects and their products in traditional medicine, food and pharmaceutical industries in Africa: Utilization and prospects. In *Future Foods*; Mikkola, H., Ed.; IntechOpen: London, UK, 2017; pp. 55–69. Available online: https://www.intechopen.com/books/future-foods/selected-edible-insects-and-their-products-in-traditional-medicine-food-and-pharmaceutical-industrie (accessed on 15 April 2020).
3. Kelemu, S.; Niassy, S.; Torto, B.; Fiaboe, K.; Affognon, H.; Tonnag, H.; Maniania, N.K.; Ekesi, S. African edible insects for food and feed: Inventory, diversity, commonalities and contribution to food safety. *J. Insects Food Feed* **2015**, *1*, 103–119. [CrossRef]
4. Raheem, D.; Carrascosa, C.; Oluwole, O.B.; Nieuwland, M.; Saraiva, A.; Millán, R.; Raposo, A. Traditional consumption of and rearing edible insects in Africa, Asia and Europe. *Crit. Rev. Food Sci. Nutr.* **2018**, *59*, 2169–2188. [CrossRef] [PubMed]
5. Tchibozo, S.; Lecoq, M. Edible Orthoptera from Africa: Preservation and promotion of traditional knowledge. *Metaleptea* **2017**, *37*, 24–29.
6. Huis, A. Insects as food in Sub-Saharan Africa. *Int. J. Trop. Insect Sci.* **2003**, *23*, 163–185. [CrossRef]
7. Grabowski, N.T.; Ahlfeld, B.; Lis, K.A.; Jansen, W.; Kehrenberg, C. The current legal status of edible in Europe. *BMTW* **2019**, *132*, 295–311.
8. De-Magistris, T.; Pascucci, S.; Mitsopoulos, D. Paying to see a bug on my food: How regulations and information can hamper radical innovations in the European Union. *Br Food J.* **2015**, *117*, 1777–1792. [CrossRef]
9. Halloran, A.; Vantomme, P.; Hanboonsong, Y.; Ekesi, S. Regulating edible insects: The challenge of addressing food security, nature conservation, and the erosion of traditional food culture. *Food Sec.* **2015**, *7*, 739–746. [CrossRef]
10. Lähteenmäki, A.; Grmelová, N.; Hénoult-Ethier, L.; Deschamps, M.-H.; Vandenberg, G.W.; Zhao, A.; Zhang, Y.; Yang, B.; Nemane, V. Insects as food and feed: Laws of the European Union, United States, Canada, Mexico, Australia, and China. *EFFL* **2017**, *12*, 22–36.
11. Jongema, Y. *List of Edible Insects of the World*; Wageningen University and Research: Wageningen, The Netherlands, 2017; Available online: https://www.wur.nl/en/Research-Results/Chair-groups/Plant-Sciences/Laboratory-of-Entomology/Edible-insects/Worldwide-species-list.htm (accessed on 15 April 2020).
12. Commission de l'Ocean Indien (COI). Conférence des Bailleurs Sur la Sécurité Alimentaire Dans L'indianocéanie. Available online: https://www.commissionoceanindien.org/conference-des-bailleurs-la-communaute-internationale-sengage-aux-cotes-de-la-commission-de-locean-indien-coi-pour-la-securite-alimentaire-dans-lindianoceanie/ (accessed on 15 April 2020).
13. Vink, N.; Sandrey, R. Chapter 8: Regional integration in SACU's agricultural sector. In *Monitoring Regional Integration of Southern Africa Yearbook*; Trade Law Centre for Southern Africa: Western Cape, South Africa, 2009; pp. 165–189. Available online: https://www.kas.de/wf/doc/kas_20303-544-2-30.pdf (accessed on 15 April 2020).
14. Agricultural Trade Forum (ATF). Namibia Food Safety Policy. Available online: https://www.atf.org.na/issues/article.php?blogID=27 (accessed on 15 April 2020).
15. Djankla, M.T. Analyse de la Législation Vétérinaire Togolaise Relative à la Santé Publique au Regard des Lignes Directrices de l'OIE. Master's Thesis, Université Cheik Anta Diop de Dakar, Dakar, Senegal, 2011; 32p.

16. Direction Générale de la Faune et des Aires Protegées. Plan d' Actions National sur l'Utilisation Durable de la Faune Sauvage par les Populations Locales au Gabon 2017–2019. Available online: http://www.fao.org/faolex/results/details/en/c/LEX-FAOC178881/ (accessed on 15 April 2020).

17. Oguntoyinbo, F.A. Safety challenges associated with traditional foods in West Africa. *Food Rev. Int.* **2014**, *30*, 338–358. [CrossRef]

18. Kussaga, J.B.; Jacxses, L.; Tiisekwa, B.P.M.; Luning, P.A. Food safety management systems performance in African food processing companies: A review of deficiencies and possible improvement strategies. *J. Sci. Food Agric.* **2013**, *94*, 2154–2169. [CrossRef] [PubMed]

19. Mensah, S.E.P.; Koudané, O.D.; Sanders, P.; Laurentie, M.; Mensah, G.A.; Abiola, F.A. Résidus d'antibiotiques et denrées d'origine animale en Afrique: Risques de santé publique. *Rev. Sci. Tech. Off. Int. Epiz* **2014**, *33*, 975–986. [CrossRef]

20. Staatz, J.; Hollinger, F. *West African Food Systems and Changing Consumer Demands*; West African Papers 4; OECD Publishing: Paris, France, 2016; 23p.

21. Huis, A. Edible insects contributing to food security? *Agric. Food Secur.* **2015**, *4*. [CrossRef]

22. Grabowski, N.T. *Speiseinsekten*; Behr's Verlag: Hamburg, Germany, 2017; 77p.

23. AU. Food Safety coordination and tracking on the agenda at a side meeting of the 14th CAADP Partnership Platform. 2018. Available online: https://au.int/en/pressreleases/20180425/food-safety-coordination-and-tracking-agenda-side-meeting-14th-caadp (accessed on 15 April 2020).

24. Binder, E.V. Alternative food and feed products. In Proceedings of the First FAO/WHO/AU International Food Safety Conference, Addis Ababa, Ethiopia, 12–13 February 2019; IFSC-1/19/TS2.5. Available online: https://www.who.int/docs/default-source/resources/bp-alternative-food-and-feed-products.pdf (accessed on 15 April 2020).

25. Anon. Thematic briefing: Promoting harmonized food safety regulation in a time of change, innovation, and globalized trade. In Proceedings of the Future of Food Safety—International Conference on Food Safety, Geneva, Switzerland, 23–24 April 2019; Available online: https://www.who.int/docs/default-source/resources/promoting-harmonized-food-safety-regulation-en.pdf?sfvrsn=6ce81d24_2 (accessed on 15 April 2020).

26. GREEiNSECT. Report on International Conference on Legislation and Policy on the Use of Insects as Food and Feed in East Africa, the Vic Hotel, Kisumu, Kenya, 2–3 March 2016. Available online: https://www.google.com/url?sa=t&rct=j&q=&esrc=s&source=web&cd=1&cad=rja&uact=8&ved=2ahUKEwijhrDY0cHlAhXHUxUIHVZRBVkQFjAAegQIAhAC&url=https%3A%2F%2Fgreeinsect.ku.dk%2Fevents%2FConference_Report.pdf&usg=AOvVaw1o7CZ840MQpbsf1yo2WIaG (accessed on 15 April 2020).

27. Tchibozo, S.; Malaisse, F.; Mergen, P. Insectes consommés par l'homme en Afrique occidentale francophone. *Geo-Eco-Trop* **2017**, *40*, 105–114.

28. Nutrient Composition and Health Benefits. Ethiopia Strategy Support Program Working Paper # 67, Addis Abeba/ETH. Available online: http://ebrary.ifpri.org/cdm/ref/collection/p15738coll2/id/128334 (accessed on 15 April 2020).

29. Hanboonsong, Y.; Jamjanya, T.; Durst, P. *Six-Legged Livestock—Edible Insect Farming, Collecting and Marketing in Thailand*; FAO—Regional Office for Asia and the Pacific: Bangkok, Thailand, 2013; 57p.

Article

Anti-Thrombotic, Anti-Oxidant and Haemolysis Activities of Six Edible Insect Species

Su-Jin Pyo [1], Deok-Gyeong Kang [1], Chuleui Jung [2] and Ho-Yong Sohn [1,*]

[1] Department of Food and Nutrition, Andong National University, Andong 36729, Korea; vytn0608@naver.com (S.-J.P.); kangdk107@naver.com (D.-G.K.)
[2] Department of Plant Medicals, Andong National University, Andong 36729, Korea; cjung@andong.ac.kr
* Correspondence: hysohn@anu.ac.kr; Tel.: +82-54-820-5491

Received: 28 February 2020; Accepted: 17 March 2020; Published: 1 April 2020

Abstract: In Korea, various insect species such as crickets and grasshoppers, as well as honey bee and silkworm pupae, have been consumed as food and used in oriental medicine. In this study to evaluate useful the bioactivities and potentially adverse effects of edible insects, ethanol extracts of *Allomyrina dichotoma* (AD), *Tenebrio molitor* (TM), *Protaetia brevitarsis* (PB), *Gryllus bimaculatus* (GB), *Teleogryllus emma* (TE), and *Apis mellifera* (AM) were prepared and evaluated with regard to their anti-thrombosis, anti-oxidant and haemolysis activities against human red blood cells. AD and TE extracts showed strong anti-oxidant activities, which were not related to polyphenol content. All ethanol extracts, except AM extract, showed strong platelet aggregation activities. The platelet aggregation ratios of the extracts were 194%–246% of those of the solvent controls. The effects of the AD, TM, PB, GM, and AM extracts on thrombin, prothrombin and various coagulation factors were negligible. Only the extract of TM showed concentration-dependent anti-coagulation activities, with a 1.75-fold aPTT (activated Partial Thromboplastin Time) extension at 5 mg/mL. Of the six insect extracts, TM and AM extracts exhibited potent haemolytic activity. Our results on the insect extracts' functional properties suggest that edible insects have considerable potential not just as a food source but as a novel bio-resource as well.

Keywords: edible insect; blood coagulation; platelet aggregation; haemolysis; *Teleogryllus emma*

---

1. Introduction

As the imbalance between food production and consumption increases, the environmental burden to produce sufficient food creates diverse socio-economic concerns, compounded by climate change. To identify alternative food sources has thus become an important issue [1,2]. Insects are receiving particular attention as a sustainable alternative to meet future food demands and satisfy nutritional requirements [3–7]. It was estimated that at least 2000 species have been considered as edible by perhaps 2 billion people of various ethnic groups, representing approximately 30% of the global population [5].

In Korea, the use of insects in the traditional Korean medicine is an age-old practice [8,9] and certain species like silkworms (*Bombyx mori*) and grasshoppers (*Oxyoa sinuosa*) have been accepted as regular food items throughout the country [9–12]. Several studies have demonstrated the potential and commercial value of insects in terms of their protein content, composition of essential amino acids, fatty acids and mineral content [12,13]. The results came from analyses of three species of beetles (*Allomyrina dichotoma*: Dynastidae; *Protaetia brevitarsis*: Cetoniidae; *Tenebrio molitor*: Tenebrionidae), two species of crickets (*Teleogryllus emma* and *Gryllus bimaculatus*: Gryllidae) and the honey bee *Apis mellifera ligustica* [13,14]. Recently, the useful bioactivities of certain insects, such as the anti-adipogenic effect of *T. molitor* larvae in 3T3-L1 adipocytes and anti-obesity effects of *T. molitor* larvae in high-fat diet-induced obese mice [15], as well as the cytotoxic effect of *T. molitor* larvae extract against prostate (PC3 and 22Rv1), cervix (HeLa), liver (PLC/PRF5, HepG2, Hep3B, and SK-HEP-1), colon (HCT116), lung (NCI-H460), breast (MDA-MB231), and ovary (SKOV3) cancer cells [16], and lipid and homocysteine-lowering effects of *T. molitor* larvae in the plasma and liver of obese Zucker rats [17], were reported. In addition, the anti-microbial peptides from *A. mellifera* (royalisin), *Sacophaga peregrine* (sapecin), *Anopheles gambiae* and *B. mori* (defensin A, B, C) [18], the haemolysis activity of crude venom and mellitin from *A. mellifera* and wasp [19,20] and anti-exercise-fatigue activity of drone pupae extract of wasp in mice [21] were reported. However, limited information is available on their role as anti-oxidants and their bio-medical functional impact in connection with thrombosis-related issues and haemolysis remains unknown.

In this study, the effects of the extracts of six edible insect species on blood coagulation, platelet-aggregation, and anti-oxidant activities were investigated to develop functional food ingredients using edible insects. Since adverse effects of insect extracts, such as anemia due to haemolysis, are well reported [19,20], haemolysis involving red blood cells of the insect extracts was also investigated. Our results regarding the insect extracts' functional properties suggest that edible insects have considerable potential not just as a food source, but as a novel bio-resource as well.

2. Materials and Methods

2.1. Materials

Samples of *Allomyrina dichotoma* (AD), *Tenebrio molitor* (TM), *Protaetia brevitarsis* (PB), *Gryllus bimaculatus* (GB), and *Teleogryllus emma* (TE) were obtained from commercial insect farms in Korea during May 2016, and sample of *Apis mellifera* (AM) was collected by A.B. Jessen, University of Copenhagen, Demark during the late summer in 2016 (Table 1). Details of the farming methods and culturing environments for the six edible insect species have previously been reported [13,14]. Insect samples were previously washed off by room-temperature water and dried. Not all insects were starved, except AD and PB, which were starved at the last instar for days to clean out the digestive tracks. The wings of honey bee drone adults were removed. All the samples were freeze-dried for at least 72 h at or below −50 °C, since killing insects with chemical or other physical means would have an influence on their bio-functionality. The dried samples were ground into powder and stored at −20 °C until further analyses. Reagent-grade chemicals were purchased from Fisher Scientific (Hampton, NH, USA). The blood samples were provided from Daegu-Gyeungbook Blood Center in Korea as Research blood. The study was conducted in accordance with the Declaration of Helsinki, and the protocol was approved by the Ethics Committee of Andong National University in Korea (IRB-Andong National Univ-482: 1040191-201602-BR-001-01, and IRB-Andong National Univ-1312: 1040191-202002-BR-001-01). Blood tests were conducted as a received state on the received day (same-day experiments), as recommended by suppliers.

Table 1. List of preferred insects used as food, feed, and oriental medicine in Korea.

Sample	Scientific Name	Classification	Feed	Life Stage
AD	*Allomyrina dichotoma*	Coleoptera Dynastidae	Fermented sawdust	Larvae
TM	*Tenebrio molitor*	Coleoptera Tenebrionidae	Rice, bran	Larvae
PB	*Protaetia brevitarsis*	Coleoptera Cetoniidae	Rice, bran, Organic soil	Larvae
GB	*Gryllus bimaculatus*	Orthoptera Gryllidae	Rice, bran, Vegetables	Late nymph
TE	*Teleogryllus emma*	Orthoptera Gryllidae	Rice, bran, Vegetables	Late nymph
AM	*Apis mellifera*	Hymenoptera Apidae	Honey, pollen	Adult

2.2. Sample Preparation

To prepare ethanol extracts, 5 g of the lyophilized powders of the six edible insect species were dissolved with 100 mL of ethanol (95%, Daejung Chemicals & Metals Co., Ltd. Siheung, Korea) and extracted three times at room temperature. After that, the extract was filtered by paper (Whatsman No. 2) and concentrated under reduced pressure (Eyela Rotary evaporator, N-1000, Tokyo Rikakikai Co., Ltd., Tokyo, Japan). The powders of insects extracts were prepared using freeze dryer equipment (FD5508, Ilshin Lab Co. Ltd., Korea), and the insect samples were dissolved in dimethylsulfoxide (DMSO) at suitable concentrations for functional component analyses.

2.3. In-vitro Anti-Coagulation Activity

The insect extracts were dissolved in DMSO, and anti-coagulation activity was measured by determining the clotting times required for the activation of thrombin (TT: Thrombin Time), prothrombin (PT: Prothrombin Time) and coagulation factors (aPTT: activated Partial Thromboplastin Time). TT, PT and aPTT were determined following the supplier's instructions (Ameulung coagulometer, Lemgo, Germany) [22]. For the TT assay, 50 μL thrombin (0.5U, Sigma Co., St Louis, MO, USA), 50 μL CaCl$_2$ (50 mM) and 10 μL insect extract were added to a preheated (37 °C) test tube and incubated at 37 °C for 3 min. Then, 100 μL preheated (37 °C) control plasma (MD Pacific Technology Co., Ltd., Huayuan Industrial Area, Tianjin, China) was added; the time (in seconds) from the addition of the control plasma to the detection of clot formation was measured and defined as the TT. For the PT assay, 70 μL control plasma and 10 μL insect extract were added to a preheated (37 °C) test tube and incubated at 37 °C for 3 min. Then, 130 μL preheated (37 °C) PT reagent (MD Pacific Technology Co., Ltd., Huayuan Industrial Area, Tianjin, China) was added; the time (in seconds) from the addition of PT reagent to the detection of clot formation was measured and defined as the PT. For the aPTT assay 70 μL control plasma and 10 μL insect extract were mixed in a preheated test tube at 37 °C for 3 min, and 65 μL preheated (37 °C) aPTT reagent (MD Pacific Technology Co., Ltd., Huayuan Industrial Area, Tianjin, China) was added and incubated at 37 °C for 3 min. Then, 65 μL CaCl$_2$ (35 mM) preheated to 37 °C was added. The time (in seconds) from the addition of CaCl$_2$ to the detection of clot formation was measured and defined as the aPTT [22]. Aspirin (1.5 mg/mL of acetylsalicylic acid, Sigma, St. Louis, MO, USA), which inhibits thrombin generation and clot formation at high doses (300–500 mg/day) [23], was used as positive control. All data are presented as the mean ± SD values of triplicates.

2.4. In Vitro Platelet Aggregation Activity

The effects of the insect extracts on platelet aggregation were measured using Whole Blood Aggregometer (Chrono-log, Havertown, PA, USA) [24]. The PRP obtained from human blood was washed with washing buffer, and the washed platelets (5 × 10^8 cells/mL) were suspended using buffer containing 138 mM NaCl, 2.7 mM KCl, 12 mM NaHCO$_3$, 0.36 mM NaH$_2$PO$_4$, 5.5 mM glucose, 0.49 mM MgCl$_2$, and 0.25% gelatin (pH 7.4). The platelets were then incubated with the insect extracts, followed by stimulation with 2.5 μL of collagen (1 mg/mL) at 37 °C for 12 min. Impedance changes were monitored and amplitude, slope and area under the curve (AUC) were calculated using Aggrolink program (Aggrolink 5.2.3, Chrono-log, Havertown, PA, USA). Amplitude was expressed as ohms by the maximum extent of platelet aggregation and slope (rate of reaction) was determined by drawing a tangent through the steepest part of the curve. AUC was calculated from the platelet aggregation curve [22]. Platelet aggregation was calculated according to the following equation: (AUC of samples/AUC of DMSO) × 100 [24]. Aspirin, which inhibits the COX-1 and reduce thromboxane A2 synthesis at low concentration (100 mg/day) [25], was used as positive control of anti-platelet aggregation.

2.5. Haemolysis Activity

A human blood sample was diluted in phosphate-buffered saline (PBS; pH 7.4) and centrifuged at 200× g for 10 min. After three washes, the final concentration of the erythrocytes was adjusted to 4%. The erythrocyte suspension was transferred into 96-well plates and incubated with the insect extract at 37 °C for 1 h. The plate was centrifuged at 200× g for 10 min. The supernatant was collected and aliquoted and haemolysis was evaluated by determining the release of haemoglobin from the 4%

human erythrocyte suspension at 414 nm with an ELISA reader. Zero percent and 100% haemolysis activities were evaluated with PBS alone and 0.1% Triton X-100, respectively. Haemolysis percentage was calculated using the following equation: haemolysis (%) = (($A_{414\,nm}$ in the bee samples − $A_{414\,nm}$ in PBS)/($A_{414\,nm}$ in 0.1% Triton X-100 − $A_{414\,nm}$ in PBS)) × 100 [26].

2.6. In-Vitro Anti-oxidant Activity

DPPH (1,1-diphenyl-2-picryl hydrazyl) anion scavenging ability, ABTS (2,2-azobis(3-ethylbenzothiazoline-6-sulfonate)) cation scavenging activity, nitrite scavenging activity and reducing power were evaluated following previously reported methods [27,28]. Ascorbic acid (Sigma Co., St. Louis, MO, USA) was used as positive control. Each activity rating was expressed as the mean and the deviation of each of three replicates.

DPPH radical scavenging activity of insect extracts was assessed according to the method described by Suzuki et al. [27]. The ability of insect extracts to scavenge DPPH free radicals was evaluated with reference to the results obtained for the ascorbic acid standard curve (0–1 mg 100 mL^{-1}). The absorbance at 516 nm was measure using an Ultraviolet-visible spectrophotometer (Epouch Microplate reader, Biotek Instrument Inc., Winooski, VT, USA).

ABTS•$^+$ radical scavenging activity was measured using the ABTS radical cation decolorization assay, as described by Suzuki et al. [27]. The ability of insect extracts to scavenge ABTS•$^+$ radicals was evaluated with reference to the results obtained for the ascorbic acid (Sigma Co., St. Louis, MO, USA) standard curve (0–1 mg 100 mL^{-1}). The absorbance at 734 nm was measured using a UV-visible spectrophotometer (Epouch Microplate reader, Biotek Instrument Inc., Winooski, VT, USA).

The nitrite scavenging activity of the insect extracts was assessed according to the method described by Kim et al. [29] with some modification. A total of 5.0 mL insect extracts were mixed with sodium nitrite (1 mL, 0.01%) and citrate–phosphate buffer (pH 1.2) and adjusted to a volume of 10 mL with distilled water. The reaction solution was incubated at 37 °C for 1 h. Then, 3 mL of the reaction solution were mixed with 3 mL Griess reagent. After vigorous mixing with a vortex, the mixture was placed at room temperature for 15 min and the absorbance at 520 nm was recorded as A; B was the absorbance of the control group (water instead of sample solutions); C was the absorbance of the insect extract. Nitrite scavenging activity was calculated by the following equation: (1 − (A − C)/B) × 100 [27].

The reducing power of the insect extracts was assayed according to the method used by Jayaprakasha et al. [30] with some modifications. Briefly, 2.5 mL of insect extracts in ethanol were mixed with 2.5 mL of phosphate buffer (0.2 mol/L, pH 6.6) and 2.5 mL of potassium ferricyanide (10%). The mixtures were incubated for 20 min at 50 °C. Then, 2.5 mL of trichloroacetic acid (10%) and 0.5 mL FeCl$_3$ (0.1%) were added to the mixture and incubated for 10 min; absorbance was measured at 700 nm against the buffer. Ascorbic acid was used as the standard.

2.7. Component Analysis

The total polyphenol content of the six insect extracts was measured colorimetrically by using Folin-Ciocalteu's phenol reagent [31]. Distilled water of 2.6 mL and 200 µL of Folin-Ciocalteu's phenol reagent were added to 200 µL of the sample and mixed together; the mixture was allowed to react for 6 min at room temperature and then 2 mL of 7% (w/v) Na$_2$CO$_3$ solution was added. The mixture was then allowed to react for 30 min at 30 °C, and the absorbance was measured using a spectrophotometer (Epouch Microplate reader, Biotek Instrument Inc., Winooski, VT, USA) at 700 nm. A standard curve was constructed using rutin as the standard.

The total flavonoid content in the six insect extracts was measured by the aluminium colorimetic method [32]. Distilled water of 320 µL and 15 µL of 5% (w/v) NaNO$_2$ added to 100 µL of the sample were mixed together and allowed to react for 5 min. Then, 10% AlCl$_3$ solution was added and the mixture was reacted for 1 min before 1M NaOH was added; subsequently, the absorbance was measured at 510 nm. A standard curve was constructed using tannic acid as a standard substance. Total sugar content was measured by the phenol–sulfuric acid method in microplate format [33], and the reducing sugar content was measured by the DNS method [34]. A standard curve was constructed using sucrose and glucose, respectively.

2.8. Statistical Analysis

Each activity value was expressed as the mean and the standard deviation of each of three replicates. Data were analyzed using SPSS packages (version 25). One-way analysis of variance (ANOVA) and Duncan Multiple Range Test at the level ($p < 0.05$) were used for the overall analysis of variance and mean separation, respectively.

3. Results

3.1. Component Anlysis of Ethanol Extracts

The ethanol extracts for AD, TM. PB, GB, TE and AM were prepared and their extraction yields are shown in Table 2. The highest yield was observed in TM (40.5%) and followed by TE, PB, AD, GB and AM yields. The ethanol extraction yields were closely related to the content of crude lipid and oil of the insect samples [13,14]. Total polyphenol content showed that the GB and TE extracts possess 15.5–15.6 mg/g, whereas the AD extract has only 1.3 mg/g. Analysis of total flavonoid content showed that the AD extract possesses 5.7 mg/g, whereas the TM extract contains only 0.1 mg/g. Total sugar (89.3 ± 0.8 mg/g) and reducing sugar content (56.7 ± 1.5 mg/g) of the extract of AM were found to be the highest of the six insect extracts.

Table 2. Yields and component assay of the ethanol extracts of 6 different insect species.

Extract	Yields (%)	Component (mg/g)			
		Total Polyphenol	Total Flavonoid	Total Sugar	Reducing Sugar
AD	23.1	1.3 ± 0.1 [a]	5.7 ± 0.1 [d]	13.0 ± 0.1 [b]	1.8 ± 0.0 [a]
TM	40.5	2.6 ± 0.2 [b]	0.1 ± 0.1 [a]	28.7 ± 1.0 [c]	8.5 ± 0.2 [c]
PB	24.9	11.8 ± 0.3 [c]	3.5 ± 0.3 [b]	31.5 ± 0.4 [d]	13.0 ± 0.3 [d]
GB	19.7	15.6 ± 0.3 [d]	3.9 ± 0.3 [b]	7.9 ± 0.3 [a]	7.5 ± 0.3 [c]
TE	30.1	15.5 ± 0.9 [d]	4.7 ± 0.2 [c]	8.5 ± 0.2 [a]	6.1 ± 0.4 [b]
AM	14.2	12.4 ± 0.2 [c]	4.5 ± 0.3 [c]	89.3 ± 0.8 [e]	56.7 ± 1.5 [e]

Different superscripts within a column indicate statistically significant differences ($p < 0.05$).

3.2. In Vitro Anti-Coagulation Activity

The anti-coagulation activities of the ethanol extracts of the six insect species were determined by measuring their TT, PT and aPTT. Treatment of aspirin (1.5 mg/mL) as positive control extended the clotting time to 1.64-fold of TT, 1.39-folds of PT, and 1.37-folds of aPTT compared with non-treatment (DMSO as solvent control) (Table 3). The treatment of high concentrations of aspirin (5 mg/mL) extended the clotting time to above 15-fold of TT, PT and aPTT compared with non-treatment. Treatment of the insect extracts (5.0 mg/mL) of AD, TM. PB, GB, and AM did not show significant changes in the clotting times of TT, PT and aPTT. However, the TE extract (5.0 mg/mL) was found to significantly extend TT (1.31-fold) and aPTT (1.75-fold) (Table 3), and the anti-coagulation effects of TE extract could be shown to be concentration-dependent (Figure 1).

Table 3. Effect of the ethanol extracts of 6 different insect species on blood coagulation.

Extract/Chemicals	Concentration (mg/mL)	Anti-Coagulation Activity (Multiplication of Control)		
		TT [1]	PT [2]	aPTT [3]
DMSO	-	1.00 ± 0.07 [bc]	1.00 ± 0.00 [a]	1.00 ± 0.01 [b]
Aspirin	1.5	1.64 ± 0.03 [f]	1.39 ± 0.02 [e]	1.37 ± 0.03 [d]
	5.0	>15.0	>15.0	>15.0
AD	5.0	1.10 ± 0.01 [d]	1.02 ± 0.02 [ab]	1.11 ± 0.10 [c]
TM	5.0	0.90 ± 0.00 [a]	1.13 ± 0.03 [c]	0.93 ± 0.01 [ab]
PB	5.0	1.04 ± 0.03 [c]	1.20 ± 0.01 [d]	0.95 ± 0.01 [ab]
GB	5.0	0.96 ± 0.00 [b]	1.18 ± 0.00 [d]	0.90 ± 0.02 [a]
TE	5.0	1.31 ± 0.01 [e]	1.05 ± 0.02 [b]	1.75 ± 0.02 [e]
AM	5.0	1.04 ± 0.04 [c]	1.03 ± 0.04 [ab]	0.96 ± 0.00 [ab]

Anti-coagulation activity was calculated on the clotting time of a given sample divided by the clotting time of the solvent control in blood coagulation assays. The thrombin (TT[1]), prothrombin (PT[2]) and activated partial thromboplastin times (aPTT[3]) of the solvent control (DMSO) were 31.5 s, 20.5 s and 80.9 s, respectively. Data are means $\pm$ SD of triplicate determinations. Different superscripts within a column indicate statistically significant differences ($p < 0.05$).

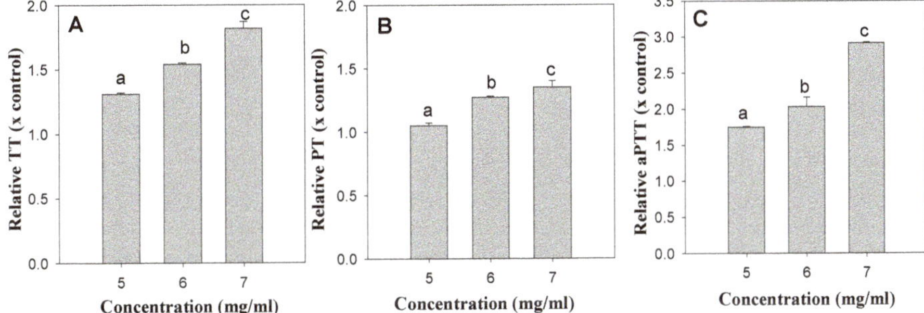

Figure 1. The concentration-dependent anti-coagulation activities of TE extract. (**A**): TT changes by TE extract, (**B**): PT changes by TE extract, (**C**): aPTT changes by TE extract. Different superscripts within a panel indicate statistically significant differences ($p < 0.05$).

3.3. In-vitro Platelet Aggregation Activities

Since the platelet aggregation is strongly inhibited by aspirin [25], the changes in PAR (platelet aggregation ratio) by aspirin treatments were determined. At 0.25 and 0.125 mg/mL of aspirin, the PARs were 38.9% and 54.3% compared with non-treatment (Table 4, Figure 2). Strong and rapid activations of platelet aggregation were identified by treatment of the insect extract (0.25 mg/mL) of AD, TM, PB, GB and TE. The PARs of these insect extracts were 194.5%–246.1% compared with non-treatment. Only AM treatment (0.25 mg/mL) did not alter the platelet aggregation

Table 4. Effect of the ethanol extracts of 6 different insect species on platelet aggregation.

Extract/Chemicals	Concentration (mg/mL)	Amplitude (ohm)	Slope	Lag Time (s)	Area Under Curve	PAR [1] (%)
DMSO	-	20	3	51	130.6	102.7
	-	20	3	59	123.7	97.3
Aspirin	0.25	9	1	60	49.4	38.9
	0.125	13	2	67	69.1	54.3
AD	0.25	27	4	17	215.3	194.5
TM	0.25	27	6	13	249.1	195.9
PB	0.25	26	8	8	262.5	206.4
GB	0.25	28	6	8	248.9	195.8
TE	0.25	25	12	1	272.4	246.1
AM	0.25	18	2	28	113.9	102.5

[1] PAR: Platelet Aggregation Ratio. Data are presented as representative results relative to three independent determinations. Amplitude is expressed as ohms by maximum extent of platelet aggregation, and slope (rate of reaction) is determined by drawing a tangent through the steepest part of curve. Area under the curve (AUC) was calculated from the platelet aggregation curve.

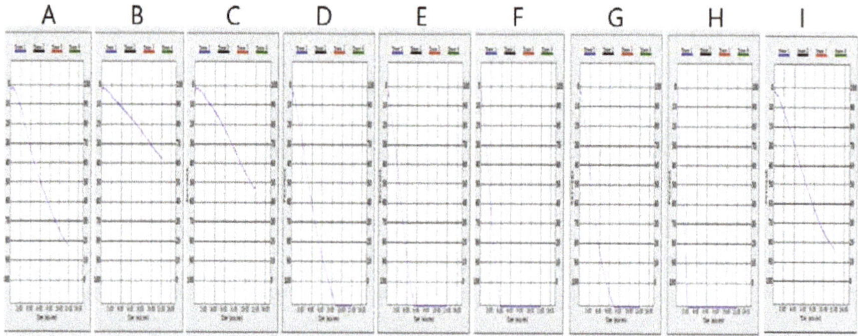

Figure 2. Impedance changes during platelet aggregation after addition of aspirin and the insect extracts in whole blood aggregometer. (**A**) DMSO, (**B**) aspirin (0.25 mg/mL), (**C**) aspirin (0.125 mg/mL), (**D**) AD extract (0.25 mg/mL), (**E**) TM extract (0.25 mg/mL), (**F**) PB (0.25 mg/mL), (**G**) GB extract (0.25 mg/mL), (**H**) TE extract (0.25 mg/mL) and (**I**) AM extract (0.25 mg/mL), respectively.

3.4. Haemolytic Activities

Haemolytic activities of the insect extracts were evaluated. Treatment with amphotericin B, known to possess potent haemolytic activity [26], and used as antifungal and anticancer agent, resulted in 53.4% haemolysis at 0.0125 mg/mL and 89.0% haemolysis at 0.025 mg/mL (Table 5). Among the insect extracts, AD, TM, PB and GB did not produce haemolysis up to 1.0 mg/mL (0.0–7.6% of haemolysis). However, the TE and AM extracts revealed potent haemolytic activity (94.5–97.5%) at 1.0 mg/mL. The calculated HC_{50}s (the concentration for 50% haemolysis) for amphotericin B, TE and AM extracts were 0.015, 0.236 and 0.396 mg/mL, respectively.

Table 5. Haemolytic activities of the ethanol extracts of 6 different insect extracts.

Extract/Chemicals	Concentration (mg/mL)	Haemolysis (%)
DMSO	-	0.0 ± 1.4 [a]
Triton X-100	1.0	100.0 ± 0.4 [e]
Amphotericin B	0.05	99.1 ± 0.4 [e]
	0.025	89.0 ± 0.6 [d]
	0.0125	53.4 ± 0.2 [c]
	0.0063	1.1 ± 0.8 [a]
	0.0032	0.0 ± 0.4 [a]
AD	1.0	2.4 ± 0.5 [a]
TM	1.0	7.6 ± 1.3 [b]
PB	1.0	0.0 ± 3.5 [a]
GB	1.0	1.1 ± 0.5 [a]
TE	1.0	97.5 ± 2.2 [e]
AM	1.0	94.5 ± 8.7 [e]

Different superscripts within a column indicate statistically significant differences ($p < 0.05$).

3.5. In-Vitro Anti-Oxdant Activities

Oxidative stress is closely related to fibrin clot formation. Inappropriate posttranslational modification of fibrin and RBC haemolysis play roles in thrombotic disorders [35–37]. The assay of anti-oxidant activities showed that the AD and TE extracts (0.5 mg/mL) have potent DPPH anion scavenging power (59.9% for AD extract and 68.3% for TE extract), and ABTS cation scavenging activities (64.3% for AD extract and 62.1% for TE extract). The reducing powers of the AD, TE and AM extracts were found to be stronger than the extracts of TM, PB, GB and AM. Potent nitrite scavenging activity was observed in the PB and AM extracts.

4. Discussion

Insects and their solvent extractions have been used as medicinal resources and to promote health in Asian countries including China, Japan, and Korea and for a long time [38]. In Korea, the extract of PB has been used to treat inflammatory disease [39] and the extract of GB and *Aspongopus chinensis* has widely been used to treat inflammation [40,41]. Recent research demonstrated that various invertebrates have potent anti-thrombotic activities. The anti-thrombotic and fibrinolytic activities of *Lumbricus rubellus* [42] and anticoagulants from *Holotrichia diomphalia* larvae [43] are well reported. Furthermore, several anti-thrombotic compounds were identified, for example, the N-acetyltyramine, hydroxytyrosol and several diketopiperazines, which have anti-FXa and anti-platelet activities as in the TM [44], anti-platelet alkaloids from *Scolopendra subspinipes* [45], anti-platelet simplagrin from *Simulium nigrimanum* salivary glands [46], anticoagulant peptides from the mosquito *Culex pipiens* [47] and horsefly salivary glands [48], as well as *Bombyx batryticatus* [49]. However, assessments of the anti-thrombosis activities along with haemolytic and anti-oxidant activities of various edible insects are scarce.

In this study, the high polyphenol contents (12.4–15.6 mg/g) and flavonoid contents (3.5–4.7 mg/g) were found in PB, GB, TE and AM extract (Table 2). However, the TM extract has 2.6 mg/g of polyphenol and 0.1 mg/g of flavonoid. The TM extract showed the lowest anti-oxidant activities among the insect extracts used. The contents of total sugar and reducing sugar in AM extract showed 3–10 folds higher than the other insect extracts. High level of sugars in AM would have been influenced by the feeding honey in the hive. However, the bee venom would not influence the results because we used only drones which lack the venom gland. Some insects also produce toxic substances such as benzoquinones from *Tenebrio molitor* under stress [50]. However, we tried to standardize the sampling and preparation process generic to all species we studied.

We have determined the effects of the ethanol extracts of six edible insects (AD, TM, PB, GB, TE and AM) on blood coagulation and platelet aggregation. Extracts of these insect species have traditionally been considered as a part of anti-thrombotic oriental medicines in Korea. Only the TE extract showed significant and concentration-dependent inhibition against thrombin, prothrombin and blood coagulation factors (Table 3, Figure 1). However, the TE extract at 0.25 mg/mL showed the most potent platelet aggregation among the six insect extracts (Table 4, Figure 2). Considering the importance of platelet aggregation in thrombus formation as a primary haemostasis plug [24], the TE extract may result in accelerated thrombus formation.

The TE and AM extracts had strong haemolytic activities against red blood cells (Table 5). Red blood cells (RBC) are the most abundant cell type in the blood and play an essential role in oxygen transfer. Several natural plant extracts and compounds of plants and/fungi, such as saponins, phenylhydrazine, penicillin, and cephalosporin induce haemolysis and result in erythrokatalysis [51–54]. Haemolysis of the RBC results in an exposure of intracellular phosphatidylserine, which accelerates blood coagulation and hyper-coagulable states [55]. Therefore, strong haemolysis activities of the TE extract may also contribute to the formation of fibrin clots.

The extracts of AD and TE exhibit strong radical scavenging activities along with reducing powers (Table 6). Previous reports [35–37] have suggested that oxidative stress increases the modification of fibrin and RBC haemolysis. Thus, the strong anti-oxidant activities of these insect extracts may contribute to delayed blood clot formation, and ROS-induced cellular damage. However, our in vitro results on the coagulation and platelet aggregation assay did not suggest an anti-thrombotic effect of the insect extracts. Our results suggest that the ethanol extracts have haemostatic rather than anti-thrombotic activities. Further research on the purification of the thrombosis-related compounds and their underlying mechanism in vivo is required.

Table 6. Anti-oxidant activities of the ethanol extracts of 6 different insect species.

Extract	Scavenging Activity [SA] (%)			Reducing Power Abs. 700
	DPPH SA	ABTS SA	Nitrite SA	
AD	59.9 ± 0.8 [e]	64.3 ± 2.5 [d]	24.5 ± 0.7 [ab]	0.269 ± 0.007 [d]
TM	4.8 ± 1.7 [a]	18.4 ± 1.5 [a]	16.6 ± 0.2 [a]	0.034 ± 0.008 [a]
PB	15.9 ± 1.7 [b]	32.8 ± 1.8 [b]	46.8 ± 1.0 [d]	0.151 ± 0.007 [b]
GB	25.5 ± 0.6 [d]	39.0 ± 0.9 [c]	26.8 ± 1.4 [b]	0.143 ± 0.003 [b]
TE	68.3 ± 0.4 [f]	62.1 ± 0.5 [d]	20.6 ± 0.8 [ab]	0.337 ± 0.009 [e]
AM	18.5 ± 1.4 [c]	40.1 ± 2.3 [c]	40.4 ± 6.3 [c]	0.230 ± 0.001 [c]

The concentrations used for DPPH, ABTS, and reducing power assays were 500 µg/mL; for the nitrite scavenging activity assay 200 µg/mL were used. Different superscripts within a column indicate statistically significant differences ($p < 0.05$).

The potent nitrite scavenging activities of the PB and AM extracts are interesting (Table 6). Nitrite, as an important food additive in the meat processing industry, reacts with secondary amines to form nitrosoamine, which has emerged as a putative candidate responsible for colorectal cancer [56]. The consumption of excess amounts of nitrite could oxidize haemoglobin. The negative effects of nitrite on human health, including possible allergenic and vasodilator effects, were also reported [57]. Therefore, a reduction in nitrite exposure in protein-rich foods is helpful to reduce the development of colorectal cancer [56], and the idea of replacing nitrite with phytochemicals extracted from oregano or thyme provides an idea for the production of innovative processed meat [58]. However, there are no reports for the production of processed meat with PB and AM extracts to date Our results suggest the possible use of PB and AM extracts as anti-nitrite food ingredients in processed meat products, such as hams, sausages, bacons or crushed meats (Table 6). Furthermore, the extracts of PB did not show significant haeamolytic activity against RBC up to 1 mg/mL. Therefore, the evaluation of nitrite reduction and increased antioxidant of the PB extract-containing processed meat is worth further study.

The antioxidant activities of the insect extracts are not related to the total polyphenol content (Tables 2 and 6). The correlation coefficients of total polyphenol to DPPH scavenging, ABTS scavenging, nitrite scavenging, and reducing power were 0.003, 0.009, 0.125, and 0.094, respectively. The correlation coefficients of total polyphenol in respect to anti-coagulation activities (0.121 for TT, 0.387 for PT, and 0.085 for aPTT), and platelet aggregation rate (0.002) were not significant. These results suggest that phenolic compounds are not connected with anti-oxidant, anti-coagulation and platelet aggregation. It could be argued that edible insects like those examined in this paper could serve as a potential source of anti-oxidant ingredients. However, since the anti-oxidant activity was not related to the total polyphenol content, this suggests that edible insects are endowed with redox ingredients, ranging from phenolics and proteins to unidentified compounds, able to counteract oxidative stress from water and a lipophilic environment [59]. The purification of active compounds is in progress, and non-phenolic, lipophilic molecules may be related to the anti-thrombosis and anti-oxidant effects.

5. Conclusions

Analyses of anti-coagulation, platelet aggregation, anti-oxidant and haemolytic activities of the six edible insects (*Allomyrina dichotoma, Tenebrio molitor, Protaetia brevitarsis, Gryllus bimaculatus, Teleogryllus emma* and *Apis mellifera*) have revealed that these insects have their own biological functionalities. Useful bioactivities of the six edible insects, such as the TE extract as an anti-coagulation, AD and TE extracts as anti-oxidation, PB and AM extracts as nitrite scavenging and TE and AM as hemolysis, were considered. The contents of polyphenol and flavonoids were not related to the antioxidant activity, which is dissimilar to the plant antioxidant system. All insect extracts except AM promoted platelet aggregation.

Author Contributions: Conceptualization, H.-Y.S.; methodology, S.-J.P., D.-G.K.; software, H.-Y.S., and S.-J.P.; validation, H.-Y.S. and S.-J.P.; formal analysis, S.-J.P. and D.-G.K.; investigation S.-J.P.; resources, C.J.; data curation, H.-Y.S. and C.J.; writing—original draft preparation, H.-Y.S.; writing—review and editing, C.J.; visualization, H.-Y.S.; supervision, H.-Y.S.; project administration, C.J.; funding acquisition, C.J. All authors have read and agreed to the published version of the manuscript.

Funding: This work was funded by the Ministry of Education, Science, and Technology (NRF-2018R1A6A1A03024862).

Acknowledgments: This work was supported by the Basic Research Program through the National Research Foundation of Korea (NRF) funded by the Ministry of Education, Science, and Technology (NRF-2018R1A6A1A03024862).

Conflicts of Interest: The authors declare no conflict of interest.

References

1. FAO. The State of Food Insecurity in the World. Available online: http://www.fao.org/hunger/keymessages/en/ (accessed on 15 January 2020).
2. Dossey, A.T.; Morales-Ramos, J.A.; Rojas, M.G. *Insects as Sustainable Food Ingredients: Production, Processing and Food Applications*; Elsevier: Amsterdam, The Netherlands, 2016.
3. Van Huis, A.; Van Itterbeeck, J.; Klunder, H.; Mertens, E.; Halloran, A.; Muir, G.; Vantomme, P. *Edible Insects: Future Prospects for Food and Feed Security*; Food and Agriculture Organization of the United Nations: Rome, Italy, 2013.
4. Cortes Ortiz, J.; Ruiz, A.T.; Morales-Ramos, J.; Thomas, M.; Rojas, M.; Tomberlin, J.; Yi, L.; Han, R.; Giroud, L.; Jullien, R. Insect mass production technologies. In *Insects as Sustainable Food Ingredients*; Elsevier: Amsterdam, The Netherlands, 2016.
5. Van Huis, A. Potential of insects as food and feed in assuring food security. *Annu. Rev. Entomol.* **2013**, *58*, 563–583. [CrossRef] [PubMed]

50. Blankespoor, C.L.; Pappas, P.W.; Eisner, T. Impairment of the chemical defense of the beetle, *Tenebrio molitor*, by metacestodes (cysticeroides) of the tapeworm, *Hymenolepis diminuta*. *Parasitology* **1997**, *115*, 105–110. [CrossRef] [PubMed]

51. Shander, A.; Javidroozi, M.; Ashton, M.E. Drug-induced anemia and other red cell disorders: A guide in the age of polypharmacy. *Curr. Clin. Pharmacol.* **2011**, *6*, 295–303. [CrossRef]

52. Matarrese, P.; Straface, E.; Pietraforte, D.; Gambardella, L.; Vona, R.; Maccaglia, A.; Minetti, M.; Malorni, W. Peroxynitrite induces senescence and apoptosis of red blood cells through the activation of aspartyl and cysteinyl proteases. *FASEB J.* **2005**, *19*, 416–418. [CrossRef]

53. Petz, L.D.; Fudenberg, H.H. Coombs-positive hemolytic anemia caused by penicillin administration. *N. Engl. J. Med.* **1966**, *274*, 171–178. [CrossRef]

54. Ehmann, W.C. Cephalosporin-induced hemolysis: A case report and review of the literature. *Am. J. Hematol.* **1992**, *40*, 121–125. [CrossRef]

55. Franck, P.; Bevers, E.M.; Lubin, B.H.; Comfurius, P.; Chiu, D.T.; Op den Kamp, J.A.; Zwaal, R.F.; van Deenen, L.L.; Roelofsen, B. Uncoupling of the membrane skeleton from the lipid bilayer. The cause of accelerated phospholipid flip-flop leading to an enhanced procoagulant activity of sickled cells. *J. Clin. Investig.* **1985**, *75*, 183–190. [CrossRef] [PubMed]

56. Crowe, W.; Elliott, C.T.; Green, B.D. A review of the In Vivo evidence investigating the role of nitrite exposure from processed meat consumption in the development of colorectal cancer. *Nutrients* **2019**, *11*, 2673. [CrossRef] [PubMed]

57. Rivera, N.; Bunning, M.; Martin, J. Uncured-labeled meat products produced using plant-derived nitrates and nitrites: Chemistry, safety, and regulatory considerations. *J. Agric. Food Chem.* **2019**, *67*, 8074–8084. [CrossRef] [PubMed]

58. Hung, Y.; Verbeke, W.; de Kok, T.M. Stakeholder and consumer reactions towards innovative processed meat products: Insights from a qualitative study about nitrite reduction and phytochemical addition. *Food Control* **2016**, *60*, 690–698. [CrossRef]

59. Mattia, C.D.; Battista, N.; Sacchetti, G.; Serafini, M. Antioxidant activities in vitro of water and liposoluble extracts obtained by different species of edible insects and Invertebrates. *Front. Nutr.* **2019**, *6*, 106. [CrossRef] [PubMed]

 *foods*

Article

Nutritional Composition of *Apis mellifera* Drones from Korea and Denmark as a Potential Sustainable Alternative Food Source: Comparison Between Developmental Stages

Sampat Ghosh [1], Ho-Yong Sohn [2], Su-Jin Pyo [2], Annette Bruun Jensen [3], Victor Benno Meyer-Rochow [1,4] and Chuleui Jung [1,5,*]

[1] Agricultural Science and Technology Research Institute, Andong National University, Andong 36729, Korea; sampatghosh.bee@gmail.com (S.G.); meyrow@gmail.com (V.B.M.-R.)

[2] Department of Food and Nutrition, Andong National University, Andong 36729, Korea; hysohn@anu.ac.kr (H.-Y.S.); vytn0608@naver.com (S.-J.P.)

[3] Department of Plant and Environmental Science, University of Copenhagen, 1871 Frederiksberg, Denmark; abj@plen.ku.dk

[4] Department of Genetics and Ecology, Oulu University, SF-90140 Oulu, Finland

[5] Department of Plant Medicals, Andong National University, Andong 36729, Korea

* Correspondence: cjung@andong.ac.kr; Tel.: +82-54-820-6229

Received: 28 February 2020; Accepted: 26 March 2020; Published: 27 March 2020

Abstract: We compared nutrient compositions of honey bee (*Apis mellifera*) drones of different developmental stages from two different populations—the Italian honey bee reared in Korea and Buckfast bees from Denmark. Analyses included amino acid, fatty acid, and mineral content as well as evaluations of antioxidant properties and haemolysis activities. The compositions of total amino acids, and thus protein content of the insects, increased with development. A similar trend was observed for minerals presumably due to the consumption of food in the adult stage. In contrast, total fatty acid amounts decreased with development. Altogether, seventeen amino acids, including all the essential ones, except tryptophan, were determined. Saturated fatty acids dominated over monounsaturated fatty acids in the pupae, but the reverse held true for the adults. Drones were found to be rich in minerals and the particularly high iron as well as K/Na ratio was indicative of the nutritional value of these insects. Among the three developmental stages, adult Buckfast drones exhibited the highest antioxidant activity. Bearing in mind the overall high nutritional value, i.e., high amino acids, minerals and less fatty acids, late pupae and adult drones can be useful for human consumption while the larvae or early pupal stage can be recommended as feed. However, owing to their relatively high haemolysis activity, we advocate processing prior to the consumption of these insects.

Keywords: Amino acids; fatty acids; minerals; antioxidant; antimicrobial; supplement; sustainable food; food safety

---

1. Introduction

In the past few decades, following an earlier suggestion by Meyer-Rochow [1] to consider insects in combatting global nutritional impasses, insects have received significant attention as a potential sustainable food item for humans. Entomophagy, i.e., the practice of eating insects [2], has existed and still exists as a cultural attribute in many different traditional communities around the globe [3–11].

Scientific reports, possibly exaggerated, claim that almost 2 billion people eat insects regularly as a planned part of their diet [12], but the total number of insects considered edible around the world is almost certainly more than the 2000 mentioned by Jongema [13]. Priority has been given to insects primarily because of their rich nutrient composition and sustainability—insects require fewer resources to rear and have lower environmental impacts than conventional livestock [12,14]. However, the domestication of insect species appears not to have been easy or much needed, judging by the small number of truly domesticated species like silkworms and honey bees. Honey bee domestication started 3–5000 years ago in the civilizations of the Middle East. The nutritional as well as therapeutic value of honey has been appreciated and mentioned in ancient texts like Ayurveda, the Bible and the Quran. Beeswax, royal jelly, bee pollen, propolis and bee venom are further significant products of bees and even the word "medicine" has its origin from the fermented honey known to Anglo-Saxon tribals as 'mead'.

What is, however, less appreciated is that the bees themselves and not just their honey is nutritious. To cite some examples, Australian Aboriginal people consume the honey plus specimens of native bees as food [15]; the Hazda people of Tanzania do not remove bee larvae from the combs that they eat [16] and in Thailand more than 10,000 colonies of *Apis florea* are annually harvested by 'bee hunters' as people consume larvae, pupae and the honey together [17]. Use of honey bee brood as food is also common in Mexico [18], Ecuador [19], Zambia [20], Senegal [21,22], and China [23].

A sensory analysis of drone brood found developmental stages, worker larvae and pupae to possess differences in their sensory profiles [24]. Nutrient composition and values of different developmental stages and honey bee species like *Apis mellifera, A. cerana, A. dorsata* have been studied [25–27], but these studies focused on worker brood, whereas the drones remained unobserved. In contrast to female queen and worker honey bees, drones are produced parthenogenetically principally by queen bees and to a minor extent by workers, and they are thus haploid [28].

The major task of the drones is to mate with virgin queens and to assure genetic mixing and natural reproduction. The bigger bodies of drone brood in comparison with worker bees likely makes them more susceptible to the attack by *Varroa* mites and removing drone broods from the hive as a way to control *Varroa* mite infections is a common practice in many countries [29]. Thus, the drone brood remains unused (despite a call to use them as food by Ambühl [30], who also provided recipes to prepare them). It makes sense, both ecologically and economically, to use drone brood, a by-product of the hive [22,31], as human food or animal feed. The usage of drone bees could help enhance the economic condition especially of the small- to medium-scale beekeepers. However, reports of the nutrient composition of honey bee drones are limited. It therefore seemed important to examine and put on record the chemical composition, including amino and fatty acids as well as mineral content of drones, their larvae and pupae to gauge the nutritional value of the different developmental stages. The data would then allow us to draw conclusions on likely functional consequences of drone consumption. The study's aim was further to clarify if there were any differences in the body composition of honey bee drones from two locations of the northern hemisphere separated by 8500 km and 20 degrees of latitude.

2. Materials and Methods

2.1. Nutritional Composition of Drones of Italian Bees and Buckfast Bees

2.1.1. Sample Preparation

Drone larvae, pupae and adults of the honey bee *Apis mellifera* were obtained from two different countries in two different continents separated by 8500 km and 20 degrees of latitude ($n = 20$ for each studied developmental stage of the drones)—Buckfast bees from Europe (Denmark) and Italian bees from Asia (South Korea). The Buckfast bee is a hybrid bee bred according to the Buckfast system, which is a mixture of many *A. mellifera* subspecies, primarily *Apis mellifera ligustica* and *A. m. mellifera*. In Denmark, drones of the Buckfast bee were collected from a healthy colony in the apiary located at the University of Copenhagen, Frederiksberg Campus, during the late summer of 2016. Drones of the Italian bees were obtained from a healthy colony of the experimental apiary of Andong

National University, South Korea during the autumn of 2019. The samples were categorized based on developmental stage—larvae (fifth larval instars from the open-cell phase), early pupae (when eye colour was not yet present), late pupae (when dark eye colour was apparent), young adults (just emerging) and old adults >7 days. Wings of the adult drones were removed. All the samples were freeze-dried for at least 72 h at −50 °C (Alpha 1-4 LD Plus, Christ, Osterode am Harz, Niedersachsen, Germany). The dried samples were ground into powder and stored at −20 °C until further analyses. All glassware used for the chemical analyses was meticulously cleaned and the chemicals used for the analytical purposes were of pure HPLC grade. Weights of the different developmental stages of the Italian bee drones were taken with the help of Acculab weighing equipment (ALC310.3, Kern, Kingswinford, West Midlands, UK) with a precision of 0.001 g. However, Buckfast drones were received in a freeze-dried condition from Denmark, which is why their fresh weights could not be recorded.

2.1.2. Amino Acid Analysis

The amino acid composition was estimated using a Sykam Amino Acid analyzer S433 (Sykam GmbH, Eresing, Bavaria, Germany) equipped with Sykam LCA L-07 column following the standard method [32]. The samples (20 mg) were hydrolyzed in 6 N HCl for 24 h at 110 °C under a nitrogen atmosphere and then concentrated in a rotary evaporator. The concentrated samples were reconstituted with sample dilution buffer supplied by the manufacturer (physiological buffer 0.12N citrate buffer, pH 2.20). The hydrolyzed samples were analyzed for amino acid composition.

2.1.3. Fatty Acid Analysis

Fatty acid composition was analyzed following the standard method of the Korean Food Standard Codex [33] using gas chromatography-flame ionization detection (GC-14B, Shimadzu, Tokyo, Japan) equipped with an SP-2560 column. The samples were derivatized into fatty acid methyl esters (FAMEs). Identification and quantification of FAMEs were accomplished by comparing the retention times of peaks with those of pure standards purchased from Sigma (Yongin, Republic of Korea) and analyzed under the same conditions.

2.1.4. Mineral Analysis

Minerals were analyzed following standard procedures according to the Korean Food Standard Codex [33]. Samples were digested with nitric and hydrochloric acid (1:3) at 200 °C for 30 min. Each sample was then filtered using Whatman filter paper (0.45 micron) and stored in washed glass vials before analyses could commence. The mineral contents were analyzed using an inductively coupled plasma-optical emission spectrophotometer (ICP-OES 720 series; Agilent; Santa Clara, CA, USA).

2.2. Functional Properties of Buckfast Drone Bee Ethanol Extracts

2.2.1. Sample Preparation

To prepare ethanol extracts, 10% (w/v) of bee larva, pupa and adult lyophilized powders was dissolved in 95% ethanol by powder weight (Daejung Chemicals & Metals Co., Ltd., Siheung, Korea) and extracted three times at room temperature. After that, the extract was paper-filtered (Whatman No. 2) and concentrated under reduced pressure (Eyela Rotary evaporator, N-1000, Tokyo Rikakikai Co., Ltd, Tokyo, Japan). A freeze dryer (FD5508, Ilshin Lab Co. Ltd, Dongduchun, Korea) was used to prepare the powder. Extracted samples were dissolved in DMSO at a suitable concentration, for functional component analysis. Functional analyses were restricted only to Buckfast drone because of the limited accessibility. We therefore carried out the examination of functional properties.

2.2.2. Total Polyphenol, Flavonoid, Reducing Sugar

Total polyphenol content in bee larvae, pupae and adults was measured colorimetrically by using Folin–Ciocalteu's phenol reagent [34]. Distilled water of 2.6 mL and 200 µL of Folin-Ciocalteu's phenol reagent added to 200 µL of the sample were mixed together; the mixture was allowed to react for 6 min at room temperature and then 2 mL of 7% (w/v) Na_2CO_3 solution were added. The mixture was subsequently allowed to react for 30 min at 30 °C and the absorbance was measured using a spectrophotometer (Epouch Microplate reader, Biotek Instrument Inc., Winooski, VT, USA) at 700 nm. A standard curve was constructed using rutin as the standard.

The total flavonoid content of bee larvae, pupae and adults was measured by the aluminium colorimetric method [35]. Distilled water of 320 µL and 15 µL of 5% (w/v) $NaNO_2$ to 100 µL of the sample were mixed together and allowed to react for 5 min. Then, 10% $AlCl_3$ solution was added and the mixture was reacted for 1 min, before 1M NaOH was added. Subsequently, the absorbance was measured at 510 nm. A standard curve was constructed using tannic acid as a standard substance. Total sugar content was determined by the phenol-sulphuric acid method in microplate format [36] and the reducing sugar content was measured by the DNS method [37]. Standard curves were constructed using sucrose and glucose, respectively.

2.2.3. Antioxidant Activity

DPPH (1,1-diphenyl-2-picryl hydrazyl) anion scavenging ability, ABTS (2,2-azobis(3-ethylbenzothiazoline-6-sulfonate)) cation scavenging activity, nitrite scavenging activity and reducing power were evaluated following previously reported methods [38]. Vitamin C (Sigma Co., St. Louis, MO, USA) was used as a positive control.

2.2.4. Antimicrobial Activity

The bacteria used were Listeria monocytogenes KACC 10550, Staphylococcus epidermidis ATCC 12228, Staphylococcus aureus KCTC 1916, Bacillus subtilis KCTC 1924 for Gram-positive, and Escherichia coli KCTC 1682, Pseudomonas aeruginosa KACC 10186, Salmonella typhimurium KCTC 1926 and Proteus vulgaris KCTC 2433 for Gram-negative bacteria) cultured in nutrient broth (Difco Co., Sparks, MD, USA) at 37 °C for 24 h. Yeasts like Candida albicans KCTC 1940 and Saccharomyces cerevisiae IF0 0233 were cultured in Sabouraud dextrose agar (Difco Co. Sparks, MD, USA) at 30 °C for 24 h. Antimicrobial activity was determined by the disc diffusion method [39]. A volume of 100 µL of each preculture was spread on agar plates; filter paper (Whatman No.2) soaked with 5 µL of bee extracts was loaded on the agar plate. After 24 h of culture, the size of the growth inhibition zone was measured. Ampicillin and miconazole were used as positive controls to be compared with the ethanol extracts of the bee samples.

2.2.5. Haemolysis Activity

A human blood sample was diluted in phosphate-buffered saline (PBS; pH 7.4) and centrifuged at 200× g for 10 min. After three washes, the final concentration of the erythrocytes was adjusted to 4%. The erythrocyte suspension was transferred into 96-well plates and incubated with the extracts of the bee larvae, pupae and adults at 37 °C for 1 h. The plate was centrifuged at 200× g for 10 min. The supernatant was collected and aliquoted, and haemolysis was evaluated by determining the release of haemoglobin from the 4% human erythrocyte suspension at 414 nm with an ELISA reader. The 0% and 100% haemolysis activities were evaluated with PBS alone and 0.1% Triton X-100, respectively. Haemolysis percentage was calculated using the following equation haemolysis (%) = (($A_{414\,nm}$ in the bee samples − $A_{414\,nm}$ in PBS)/($A_{414\,nm}$ in 0.1% Triton X-100 − $A_{414\,nm}$ in PBS)) × 100 [40].

2.3. Statistical Analysis

In order to increase reliability, all the chemical analyses were performed in triplicate and expressed as mean ± SD except for the case of fatty acid and mineral analysis. For body weight, $n = 9$ individuals were taken for late pupae, early adult and adult Italian bee drones, and $n = 5$ individuals were taken for early pupae. The composite sampling technique has been followed for sampling for each experiment. To test the differences, in case of the drone's weight and the functional properties of Buckfast drones, one-way analysis of variance (ANOVA) followed by Posthoc test (LSD) were carried out using SPSS 16.0. If the *p* value e was found ≤0.05 (CI = 95%), the null hypothesis was rejected.

3. Results and Discussion

3.1. Body Weight of Different Developmental Stages of Drone

Body weights of Italian honey bee drones were found to differ significant between pupae and adults, ranging from 0.34 to 0.27 g (ANOVA, $p < 0.05$) (Figure 1). Early pupae (0.34 ± 0.02 g) were found to have a higher body weight than late pupae (0.30 ± 0.02 g) and early adults (0.27 ± 0.01 g). However, the body weight then increased again as the adults aged. After emergence, drones are initially fed by workers. The former then commence reproductive preparations for their mating flights, which would take place after 7–12 days following eclosion [41].

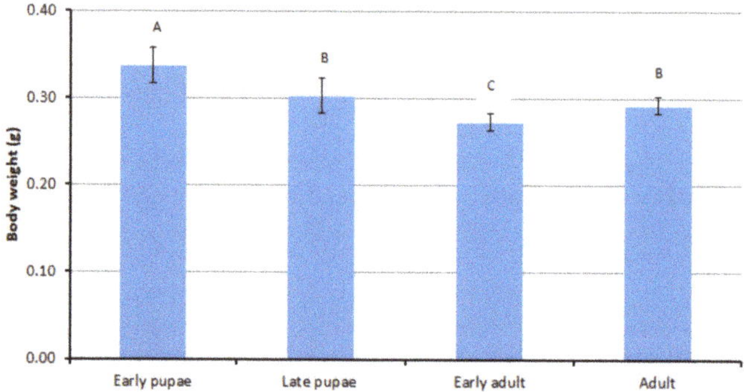

Figure 1. Body weights of different developmental stages of Italian honey bee drones from Korea. Different superscripts indicate statistically significant differences ($p < 0.05$).

3.2. Nutritional Composition of Drone Bees

3.2.1. Amino Acid Composition

The amino acid composition of different developmental stages of drones is shown in Table 1. Altogether, seventeen amino acids were found. A significantly increasing amount of amino acid content (and therefore protein) was found to be associated with the developmental stage of a drone (ANOVA, $p < 0.05$ for Italian bees and Buckfast bees). This was in agreement with previous reports on worker honey bees [26,27]. Assuming the total amino acid content reflects protein content of a drone, the value was higher or comparable to the protein content of conventional foods of animal origin (pork: 27.7%, beef: 40.5%, chicken: 54.7%, egg: 52.7%), i.e., values for conventional food obtained from the USDA database and calculated on dry matter basis. Glutamic acid was the most abundant and very important amino acid as it is the precursor of γ-amino butyric acid (GABA), an inhibitory neurotransmitter. Leucine followed by lysine was dominant among the essential amino acids. A similar pattern of amino acid distribution was reported by Kim et al. [42], but methodological differences to our study existed. Lysine receives attention as it is a limiting amino acid in many cereal-based diets [43,44], which are most prevalent in developing countries. Lysine is the precursor of carnitine, which plays a critical role in the carrying process of fatty acids [45].

Phenylalanine and tyrosine together were present in high amounts. Phenylalanine is the precursor of tyrosine, which synthesizes dopamine, adrenaline (epinephrine) and noradrenaline (norepinephrine) [46]. Tryptophan was not determined and cysteine and methionine were not recovered entirely, presumably because of the incomplete acid hydrolysis of the sample and the relatively small amounts of these amino acids [47]. However, the presence of sizable amounts of essential amino acids increases the food value of drone brood. Valine, leucine and isoleucine are three branched-chain amino acids playing essential roles in the maintenance of muscle tissue. Threonine is an important component of structural proteins. Moreover, it plays a critical role in the maintenance of the intestinal mucosal integrity and barrier function [48]. Histidine is the precursor of histamine, which plays a vital role in immune responses [49]. Methionine plays important roles in metabolic affairs and detoxification.

Table 1. Amino acid composition (g/100 g dry matter basis) of different developmental stages of Italian honey bee drones from Korea (K-) and Buckfast honey bee drones from Denmark (D-).

Amino Acid	D-Larvae	D-Late Pupae	D-Adult	K-Early pupae	K-Late Pupae	K-Early Adult	K-Adult
Valine *	2.87 ± 0.099	2.97 ± 0.198	3.79 ± 0.480	2.56 ± 0.016	2.97 ± 0.030	4.07 ± 0.043	4.22 ± 0.022
Isoleucine *	2.43 ± 0.021	2.56 ± 0.014	3.28 ± 0.340	2.13 ± 0.004	2.44 ± 0.014	3.16 ± 0.025	3.27 ± 0.007
Leucine *	3.96 ± 0.014	4.26 ± 0.092	5.51 ± 0.062	3.54 ± 0.004	4.14 ± 0.036	5.53 ± 0.043	5.65 ± 0.010
Lysine *	3.52 ± 0.035	3.68 ± 0.021	4.35 ± 0.521	3.00 ± 0.007	3.51 ± 0.023	4.43 ± 0.004	4.56 ± 0.009
Tyrosine **	2.55 ± 0.042	2.76 ± 0.007	2.77 ± 0.219	2.20 ± 0.003	2.77 ± 0.038	3.04 ± 0.019	2.87 ± 0.053
Threonine *	1.86 ± 0.035	1.57 ± 0.134	1.95 ± 0.084	1.89 ± 0.303	1.93 ± 0.005	3.23 ± 0.006	2.66 ± 0.001
Phenylalanine *	2.08 ± 0.028	2.15 ± 0.049	2.35 ± 0.189	1.83 ± 0.009	2.00 ± 0.003	2.29 ± 0.009	2.38 ± 0.006
Histidine *	1.21 ± 0.042	1.27 ± 0.000	1.55 ± 0.439	0.94 ± 0.001	1.12 ± 0.016	1.42 ± 0.000	1.41 ± 0.000
Methionine *	1.15 ± 0.007	1.16 ± 0.000	1.44 ± 0.004	0.17 ± 0.031	0.44 ± 0.047	1.91 ± 0.581	2.28 ± 0.076
Arginine ***	2.18 ± 0.099	2.45 ± 0.042	3.67 ± 0.063	2.20 ± 0.018	2.55 ± 0.016	3.35 ± 0.001	3.55 ± 0.000
Aspartic acid	3.23 ± 0.028	3.22 ± 0.028	3.68 ± 0.180	2.50 ± 0.012	2.72 ± 0.020	3.16 ± 0.007	3.40 ± 0.005
Glutamic acid	7.94 ± 0.262	8.78 ± 0.014	8.74 ± 0.863	10.01 ± 0.044	10.55 ± 0.036	12.16 ± 0.065	12.39 ± 0.050
Serine	2.03 ± 0.092	2.40 ± 0.141	2.91 ± 0.112	1.75 ± 0.111	2.09 ± 0.006	3.19 ± 0.021	2.93 ± 0.023
Glycine	2.29 ± 0.014	2.65 ± 0.007	4.19 ± 0.832	2.10 ± 0.004	2.84 ± 0.039	4.58 ± 0.042	4.40 ± 0.003
Alanine	2.36 ± 0.014	2.87 ± 0.000	5.28 ± 0.055	2.56 ± 0.009	3.44 ± 0.027	5.82 ± 0.069	5.97 ± 0.001
Cysteine	0.25 ± 0.014	0.35 ± 0.014	1.93 ± 0.957	0.19 ± 0.001	0.28 ± 0.032	0.39 ± 0.077	0.38 ± 0.003
Proline	1.58 ± 0.000	1.52 ± 0.028	2.33 ± 0.124	2.99 ± 0.026	3.60 ± 0.035	4.61 ± 0.044	4.70 ± 0.010
Total	**43.49**	**46.62**	**59.72**	**42.56**	**49.39**	**66.34**	**67.02**

* Essential amino acid; ** Conditional essential amino acid; *** Essential amino acid for children.

The amino acid contents measured were within the range of previously reported values for edible insects belonging to different orders [50–55]. Adult Buckfast honey bee drones contained less total amino acid than adult *A. mellifera ligustica* drones did. The differences between the amino acid content are likely to be due to the different geographical locations and environmental conditions as well as the locally available pollen resources for the bees' nutrition. The ages of the adults and their physiologic adaptations are additional factors to be considered. Younger drones are capable of synthesizing some amino acids like proline more successfully than the older adults do [56]. However, it was not possible to categorize the adult drone population samples according to age while estimating their nutritional composition.

3.2.2. Fatty Acid Composition

Table 2 represents the fatty acid compositions of different developmental stages of drone bees. A gradual decreasing trend, although not statistically examined, with developmental stage was found for total fatty acid content (Table 2); the observation being relevant with regard to previous reports on worker bees of *A. mellifera* and *A. cerana* [26,27]. SFA (saturated fatty acid) and MUFA (monounsaturated fatty acid) proportions sharply decreased with the progression of the developmental stages, while in contrast, the proportion of PUFAs (polyunsaturated fatty acids) increased. In all cases the proportion of MUFAs dominated. Palmitic acid followed by stearic acid was the most abundant saturated fatty acids. On the other hand, oleic and α-linolenic acid dominated among the MUFAs and PUFAs, respectively.

From a nutritional point of view, SFAs, particularly lauric, myristic and palmitic acid, but not stearic acid, are deemed to be undesirable as they are thought to increase body cholesterol [57]. All of the three undesirable fatty acids were decreasing as the drones got older. In almost all of the cases, only very small amounts of PUFAs were found, a finding that agrees with results obtained earlier from worker honey bees like *A. mellifera*, *A. cerana* and *A. dorsata* workers [26,27]. Linoleic and α-linolenic acids are essential for most insects including honey bees, although some insect species are capable of synthesizing linoleic acid [58]. Being the major source of ω-3 fatty acids, α-linolenic acid plays important roles in the context of reproduction. Studies by Ma et al. [59] have revealed that α-linolenic acid provided with a 2% to 4% pollen substitute was found to be optimal for the highest reproductive demands while deficiencies of ω-3 fatty acids were found to impair learning in honey bees [60]. The higher MUFA content of drones rather than worker bees is advantageous from the standpoint of human nutrition as it reduces low density lipoprotein (LDL) [61,62].

Table 2. Fatty acid composition (mg/100 g dry matter basis) of different developmental stages of Italian honey bee drones from Korea (K-) and Buckfast honey bee drones from Denmark (D-).

Fatty Acid	D-Larvae	D-Late Pupae	D-Adult	K-Early Pupae	K-Late Pupae	K-Early Adult	K-Adult
Lauric acid (C12:0)	25.95	31.37	4.08	32.48	33.41	14.17	6.14
Myristic acid (C14:0)	359.51	365.50	15.97	333.07	258.05	48.35	18.31
Palmitic acid (C16:0)	4809.97	4879.12	294.67	4517.45	3570.83	802.94	384.12
Stearic acid (C18:0)	1110.26	1302.45	257.09	1356.94	1267.04	592.54	341.51
Arachidic acid (C20:0)	ND	56.17	35.92	120.62	145.82	157.05	104.24
Behenic acid (C22:0)	ND	ND	62.86	14.38	23.34	51.35	46.46
Lignoceric acid (C24:0)	ND	ND	ND	39.17	42.64	39.99	34.95
Subtotal (SFA)	**6305.69**	**6634.61**	**670.59**	**6414.11**	**5341.13**	**1706.39**	**935.73**
Palmitoleic acid (C16:1)	56.35	51.92	166.58	47.65	48.33	74.29	92.91
Elaidic acid (C18:1n9t)	ND	ND	ND	6.75	0.00	0.00	0.00
Oleic acid (C18:1n9c)	4720.25	5104.52	1783.36	4902.83	4412.01	2545.19	1900.32
cis11-Eicosenic acid (C20:1n9)	ND	ND	127.09	8.69	10.38	14.01	9.57
Subtotal (MUFA)	**4776.60**	**5156.44**	**2077.03**	**4965.92**	**4470.72**	**2633.49**	**2002.80**
Linoleic acid (C18:2n6c)	ND	67.87	61.8	22.76	30.69	37.43	35.92
Linolenic acid (C18:3n3)	ND	ND	ND	61.24	83.23	108.50	104.85
cis-13,16-Docosadienoic acid (C22:2)	ND	ND	ND	15.20	17.23	21.54	17.56
Subtotal (PUFA)	**ND**	**67.87**	**61.8**	**99.20**	**131.15**	**167.47**	**158.33**
Total	**11,082.29**	**11,858.92**	**2809.42**	**11,479.23**	**9943.00**	**4507.35**	**3096.86**

3.2.3. Mineral Content

Table 3 represents the content of nutritionally important minerals of the different developmental stages of drone bees. In general, adult drones contained quantitatively higher amounts of minerals than those reported for the brood, although not statistically examined. The most abundant mineral was phosphorus followed by potassium (Table 3). Phosphorus is essential for ATP and energy production. Phosphorus is also required to form mineral salts with calcium that play critical roles in the formation of bones and teeth. Calcium plays important roles in several other physiological functions including muscle contraction and relaxation, nerve functioning, blood clotting, etc. The most common food sources of calcium are dairy products and meat, items which are often unaffordable for people in developing and underdeveloped regions of the world. High potassium and a high K/Na ratio are advantageous for human nutrition particularly for those sections of the population suffering from hypertension [63,64]. Sodium and potassium are required for the appropriate maintenance of electrolyte and proper fluid balance, muscle contraction and nerve transmission. The sodium content of Buckfast honey bee drones from Denmark was much higher than that of Italian honey bee drones from Korea. In contrast, the copper content of Buckfast honey bee drones from Denmark was less than that of Italian honey bee drones from Korea. Differences in the mineral contents are presumably due to different resources available for the bees' nourishment in the two different geographical locations. Iron is a mineral of interest as its deficiency often leads to anaemia and because the most vulnerable populations for iron deficiencies are underprivileged women and children in the poorest regions of the world [65,66]. Adult drones contained higher amounts of iron than their pupae. Assuming good bioavailability, it can be expected that a diet supplemented with drones can help ameliorate iron deficiencies. Zinc, another essential mineral, plays critical roles in many metabolic pathways including DNA replication, transcription and protein synthesis [67], and the relatively high amounts of this element in drones has to be seen as a bonus in connection with drones as part of the human nutrition.

Table 3. Mineral contents (mg/100 g dry matter basis) of different developmental stages of Italian honey bee drones from Korea (K-) and Buckfast honey bee drones from Denmark (D-).

Minerals	D-Larvae	D-Late Pupae	D-Adult	K-Early Pupae	K-Late Pupae	K-Adult
Calcium	34.21	38.7	60.72	43.72	49.29	66.19
Magnesium	68.06	81.86	121.45	82.89	95.03	123.18
Sodium	30.08	38.02	79.45	7.29	8.52	11.33
Potassium	891.08	1101.98	1465.23	544.55	643.06	784.04
Phosphorus	686.88	802.61	1166.06	774.03	892.41	1132.35
Iron	5.62	5.99	12.23	4.86	5.67	10.58
Zinc	5.10	6.04	15.86	5.25	5.88	8.40
Copper	0.11	0.37	1.39	1.82	1.94	2.59
Manganese	0.87	ND	1.71	0.28	0.29	0.52

3.3. Functional Properties of Buckfast Honey Bee Drone Bee Ethanol Extract

3.3.1. Total Polyphenol, Flavonoids, Reducing Sugar Content

Figure 2 represents total polyphenol, total flavonoid, total sugar and reducing sugar content of different developmental stages of drones. The total polyphenol content of adult drones (12.4 ± 0.2 mg/g) was found to be higher than that of the brood stages (6.8 ± 0.9 mg/g for larvae, and 5.6 ± 0.2 mg/g for pupae). The polyphenol content of the brood stage of the drone was comparable to the value reported for tannin content of edible insects like *Oecophylla smaragdina* (4.97 ± 0.5 mg/g) and *Odontotermes* sp. (6.15 ± 0.6 mg/g) [52]. Adeduntan reported tannin content of edible herbivorous insects from Nigeria within the range of 2.5 to 11.5 mg/g [68]. No significant differences were found in regard to flavonoid content of different developmental stages (5.8 ± 1.4, 4.7 ± 0.9 and 4.5 ± 0.3 mg/g for larvae, pupae, and adults, respectively). Total sugar and reducing sugars were found to be more abundant in the adults (56.7 ± 1.5 mg/g) than other developmental stages (2.7 ± 0.1 mg/g for larvae and 2.1 ± 0.2 mg/g for pupae). Total sugar and reducing sugar content of adult drones was found to be much higher than that reported for five edible insects like *Allomyrina dichotoma*, *Tenebrio molitor*, *Protaetia brevitarsis*, *Gryllus bimaculatus* and *Teleogyllus emma* [69]. Reactive oxygen species (ROS), including hydrogen peroxide, hypochlorous acid and free radicals such as hydroxyl radical and superoxide anions, are readily produced by several oxidation processes in the human body as part of the metabolism. These ROS are often potentially damaging to cells and prolonged oxidative stress has even been associated with the risk of developing diabetes, heart disease, cancer and aging processes [70]. To balance the oxidative stress, animals including humans have developed a complex system of antioxidants such as glutathione and enzymes like catalase and superoxide dismutase. However, dietary supplements of antioxidants are also immensely important to balance oxidative stresses. Polyphenols are an important group of compounds showing antioxidant activity, which is due primarily to resonance stabilization of the phenoxyl radical after oxidation [71].

Flavonoids are phenolic compounds, also exhibiting antioxidant activity [72]. To cite but a few examples, flavonoids inhibit enzymes responsible for superoxide anion production such as xanthin oxidase, protein kinase, etc. [73,74], and a number of flavonoids have been shown chelating trace metals, which play critical roles in enhancing ROS formation [72]. Flavonoids also act as antimicrobials, photoreceptors, visual attractors and feeding repellents of herbivores in plants [72]. Reducing sugars are capable of acting as reducing agents since they contain free aldehyde or ketone groups. Excessive sugar consumption is one factor promoting an overweight status and obesity leading to diabetes [75]. In contrast, an intake of reducing sugars lowers the risk of becoming overweight or obese. Thus, the consumption of drones, including the adults, should not increase the risk of obesity and associated complications.

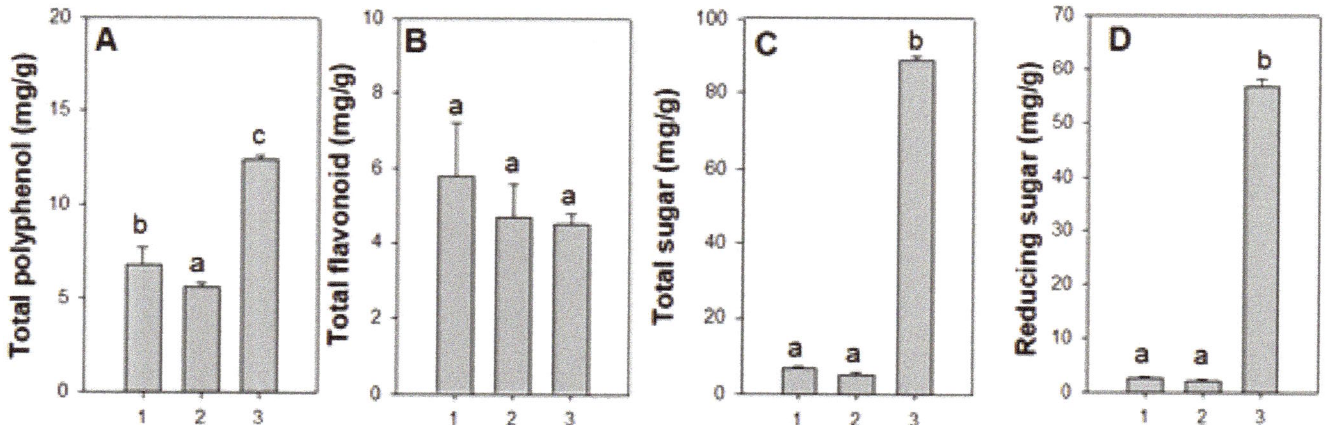

Figure 2. Functional component analyses of ethanol extracts of Buckfast honey bee drone larvae, pupae and adults from Denmark; **A**: Total polyphenol (mg/g), **B**: Total flovaonoid (mg/g), **C**: Total sugar (mg/g), and **D**: Reducing sugar (mg/g). Numbers 1, 2, 3 represent larvae, late pupae and adults, respectively. Different superscripts indicate statistically significant differences ($p < 0.05$).

3.3.2. Antioxidant and Antimicrobial Profiles

Table 4 represents antioxidant activity, including DPPH, ABTS, nitrite and reduction potential of drone larvae, pupae and adults. Adult drones exhibited the highest antioxidant activity followed by pupae and larvae, which is presumably due to the presence of high amounts of polyphenols. Furthermore, high amounts of amino acids like aspartic acid, glutamic acid, cysteine and lysine can improve the consequences of oxidative stress [76,77]. The gradual increase in their amount along with the developmental stages of the drones can justify the higher antioxidant activity of adults. Table 5 represents the antimicrobial activities of the different developmental stages. No antibacterial and antifungal activities were noticed in the present study.

Table 4. Antioxidant activities of ethanol extracts of Buckfast honey bee drone larvae, pupae and adults from Denmark. The concentration used for 1,1-diphenyl-2-picryl hydrazyl (DPPH), 2,2-azobis (3-ethylbenzothiazoline-6-sulfonate (ABTS), and the reducing power assay was 500 µg/mL and for the nitrite scavenging assay the concentration was 200 µg/mL. Vitamin C (ascorbic acid) was used as a standard. Different superscripts within a column indicate statistically significant differences ($p < 0.05$).

Extract (mg/mL)	Antioxidant Activity (%)			Reducing Power (700 nm)
	DPPH	ABTS	Nitrite	
Larvae (0.5)	0.3 ± 0.4 [a]	10.4 ± 0.2 [a]	25.6 ± 4.5 [a]	0.018 ± 0.001 [b]
Late pupae (0.5)	1.3 ± 0.4 [a]	10.5 ± 0.5 [a]	20.9 ± 4.1 [a]	0.008 ± 0.002 [a]
Adult (0.5)	18.5 ± 1.4 [b]	40.1 ± 2.3 [b]	40.4 ± 6.3 [b]	0.230 ± 0.001 [c]
Vitamin C (0.1)	92.5 ± 0.6 [c]	95.2 ± 0.3 [c]	85.6 ± 2.6 [c]	1.545 ± 0.064 [d]

Table 5. Antimicrobial activities of hot water and ethanol extracts of Buckfast honey bee drone larvae, pupae and adults from Denmark against pathogenic and food spoilage microorganisms. The concentration of extracts and standard chemicals (Ampicillin and Miconazole) used were 500 µg/disc and 1.0 µg/disc, respectively.

Extract	Antimicrobial Activity (Clear Zone: mm)									
	Gram Positive Bacteria				Gram Negative Bacteria				Fungi	
	LM	SE	SA	BS	EC	PA	ST	PV	CA	SC
Larvae	–	–	–	–	–	–	–	–	–	–
Late pupae	–	–	–	–	–	–	–	–	–	–
Adult	–	–	–	–	–	–	–	–	–	–
Ampicillin	13 ± 0.1	21 ± 0.2	15 ± 0.1	12 ± 0.2	6 ± 0.1	8 ± 0.2	11 ± 0.1	18 ± 0.2	–	–
Miconazole	–	–	–	–	–	–	–	–	8 ± 0.1	13 ± 0.2

LM: *Listeria monocytogenes*, SE: *Staphylococcus epidermidis*, SA: *Staphylococcus aureus*, BS: *Bacillus subtilis*, EC: *Escherichia coli*, PA: *Pseudomonas aeruginosa*, ST: *Salmonella typhimurium*, PV: *Proteus vulgaris*, CA: *Candida albicans*, and SC: *Saccharomyces cerevisiae*. '—'indicates no activity.

3.3.3. Haemolysis Activity

Figure 3 represents the haemolytic activity of drone larvae, pupae and adults. Ethanolic extraction of drone powder exhibited dose-dependent haemolysis activity. All three different stages showed high haemolytic activity at a concentration of 1000 µg/mL, which is comparable to the antifungal drug Amphotericin B at a concentration of 100 µg/mL. The haemolytic activity shown by the drone brood and adults was found to be much higher than that reported for other studied edible insects (communicated 69) as well as many medicinal plants [78,79]. Out of 222 plants, Hossaini, for example, discovered 20 species that exhibit haemolytic activities [80]. However, a detailed study on the mechanism of the observed haemolysis and the active compounds responsible for the haemolytic activity is yet to be carried out. Therefore, based on the present results we do not advocate the raw consumption of drone bees, but recommend that the drone brood and adults be processed. This could involve blanching, drying, boiling, etc., and should lead to increased safety [81].

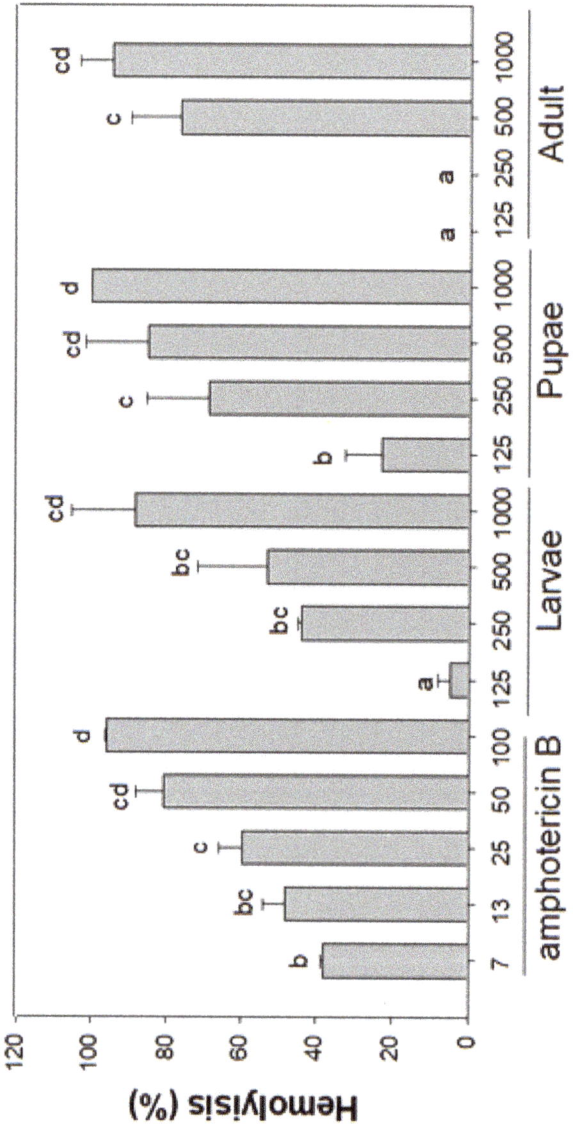

Figure 3. Haemolytic activities of the ethanol extracts of Buckfast honey bee drone; larvae, late pupae and adults from Denmark. Amphotericin B was used as positive control. Different superscripts indicate statistically significant differences among treatment concentrations ($p < 0.05$).

4. Conclusions

Drone brood (especially at the late pupal stage and as adults) can be recommended as an alternative and nutritious food item for humans as it contains almost all the essential amino acids needed by humans, as also proposed by Ulmer et al. [82]. Similarities in the nutrient content between Italian and Buckfast drones especially with regard to amino and fatty acids were apparent and most likely represent a reflection of the physiological processes shared by the two breeds. On the other hand, discrepancies in the values of some minerals like sodium and potassium in adult populations of both drones very likely stem from distinct environments and their ecological characteristics as well as the different food plants used by the bees. From a nutritional standpoint, the relatively small amounts of fatty acids and the richness in minerals as well as antioxidant activities would provide an extra advantage.

On the other hand, the larvae or the early pupal stage, with less total amino acid and, thus, protein, than further developed stages, can be recommended as animal feed. However, the examination of the functional properties of Italian drones remains a task for future investigation. Scientific studies showed the efficacy of drone brood, although not at the exact stage mentioned, in terms of improvement of reproductive quality of pigs [83]. However, marketing of drone brood as food per se or a food supplement requires legislation from governmental agencies and permits for the commercialization of this insect. Obviously, a standard dossier of beekeeping practices is essential to ensure the hygiene of the honey bee drone production and to address potential safety issues. Once these obstacles are overcome, the use of drones as food or feed can be a win-win solution for bee keepers and as well as the consumers.

Author Contributions: Conceptualization, C.J., A.B.J., S.G.; methodology, S.G., H.-Y.S., S.-J.P.; software, S.G., H.-Y.S.; validation, S.G., H.-Y.S., C.J.; formal analysis, S.G., H.-Y.S., S.-J.P.; investigation, C.J., A.B.J., S.G., H.-Y.S.; resources, C.J.; data curation, H.-Y.S. and C.J.; writing—original draft preparation, S.G.; writing—review and editing, C.J., A.B.J., V.B.M.-R., S.G.; visualization, C.J.; supervision, C.J.; project administration, C.J.; funding acquisition, C.J. All authors have read and agreed to the published version of the manuscript.

Funding: This work was funded by the Ministry of Education (NRF-2018R1A6A1A03024862).

Acknowledgments: This research was funded by the BSRP through the National Research Foundation of Korea (NRF), Ministry of Education, grant number NRF-2018R1A6A1A03024862 and by the Velux Foundation; Denmark, grant number 32726.

Conflicts of Interest: The authors declare no conflict of interest.

References

1. Meyer-Rochow, V.B. Can insects help to ease the problem of world food shortage? *Search* **1975**, *6*, 261–262.
2. Evans, J.; Alemu, M.H.; Flore, R.; Frøst, M.B.; Halloran, A.; Jensen, A.B.; Maciel-Vergara, G.; Meyer-Rochow, V.B.; Münke-Svendsen, C.; Olsen, S.B.; et al. 'Entomophagy': An evolving terminology in need of review. *J. Insects Food Feed* **2015**, *1*, 293–305. [CrossRef]
3. Bequaert, J. Insects as food: How they have augmented the food supply of mankind in early and recent years. *Nat. Hist. J.* **1921**, *21*, 191–200.
4. Bergier, E. *Peuples Entomophages et Insectes Comestibles: Étude sur les Moeurs de L'homme et de L'insecte*; Imprimérie Rullière Frères: Avignon, France, 1941.
5. Bodenheimer, F.S. *Insects as Human Food*; W. Junk Publishers: The Hague, The Netherlands, 1951.
6. DeFoliart, G.R. Insects as food: Why the western attitude is important. *Ann. Rev. Entomol.* **1999**, *44*, 21–50. [CrossRef] [PubMed]
7. Van Huis, A. Insects as food in Sub-Saharan Africa. *Insect Sci. Appl.* **2003**, *23*, 163–185. [CrossRef]
8. Chakravorty, J.; Ghosh, S.; Meyer-Rochow, V.B. Comparative survey of entomophagy and entomotherapeutic practices in six tribes of eastern Arunachal Pradesh (India). *J. Ethnobiol. Ethnomed.* **2013**, *9*, 50. [CrossRef]
9. Costa-Neto, E.M.; Dunkel, F.V. Insects as Food: History, culture, and modern use around the world. In *Insects as Sustainable Food Ingredients*; Dossey, A.T., Morales-Ramos, J.A., Rojas, M.G., Eds.; Elsevier: Amsterdam, The Netherlands, 2016; pp. 29–58. [CrossRef]

10. Mitsuhashi, J. *Edible Insects of the World*; CRC Press: Boca Raton, FL, USA, 2017.

11. Van Itterbeeck, J.; Andrianavalona, I.N.R.; Rajemison, F.I.; Rakotondrasoa, J.F.; Ralantoarinaivo, V.R.; Hugel, S.; Fisher, B.L. Diversity and use of edible grasshoppers, locusts, crickets, and katydids (Orthoptera) in Madagascar. *Foods* **2019**, *8*, 666. [CrossRef]

12. Van Huis, A.; van Itterbeeck, J.V.; Klunder, H.; Mertens, E.; Halloran, A.; Muir, G.; Vantomme, P. Edible insects: Future prospects for food and feed security. In *FAO Forestry Paper 171*; Food and Agriculture Organization of the United Nations: Rome, Italy, 2013.

13. Jongema, Y. Worldwide List of Edible Insects 2017. Available online: https://www.wur.nl/en/Research-Results/Chair-groups/Plant-Sciences/Laboratory-of-Entomology/Edible-insects/Worldwide-species-list.htm (accessed on 15 January 2020).

14. Oonincx, D.G.A.B.; Itterbeck, J.V.; Heetkamp, M.J.W.; Brand, H.V.D.; Loon, J.J.A.V.; van Huis, A. An exploration on greenhouse gas and ammonia production by insect species suitable for animal or human consumption. *PLoS ONE* **2010**, *5*, e14445. [CrossRef]

15. Cherry, R.H. Use of insects by Australian Aborigines. *Am. Entomol.* **1991**, *37*, 8–13. [CrossRef]

16. Murray, S.S.; Schoeninger, M.J.; Bunn, H.T.; Pickering, T.R.; Marlett, J.A. Nutritional composition of some wild plant foods and honey used by Hazda foragers of Tanzania. *J. Food Comp. Anal.* **2001**, *13*, 3–13. [CrossRef]

17. Wongsiri, S.; Lekprayoon, C.; Thapa, R.; Thirakupt, K.; Rinderer, T.E.; Sylvester, H.A.; Oldroyd, B.P.; Booncham, U. Comparative biology of *Apis andreniformis* and *Apis florea* in Thailand. *Bee World* **1997**, *78*, 23–35. [CrossRef]

18. Ramos-Elorduy, J.; Moreno, J.M.P.; Prado, E.E.; Perez, M.A.; Otero, J.L.; de Guavara, O.L. Nutritional value of edible insects from the state of Oaxaca, Mexico. *J. Food Comp. Anal.* **1997**, *10*, 142–157. [CrossRef]

19. Onore, G. A brief note on edible insects in Ecuador. *Ecol. Food Nutr.* **1997**, *36*, 277–285. [CrossRef]

20. Mbata, K.J. Traditional use of arthropods in Zambia. I. The food insects. *Food Insects Newsl.* **1995**, *8*, 5–7.

21. Gessain, M.; Kinzler, T. Miel et Insectes à miel chez les Bassari et D'autres Populations du Sénégal Oriental [Honey and Honey Making Insects in the Bassari and Other Populations of Eastern Senegal]. In *L'homme et L'annimal*; Pujol, R., Ed.; Premier Colloque d'Ethnozoologie: Paris, France, 1975; pp. 247–254.

22. Jensen, A.B.; Evans, J.; Jonas-Levi, A.; Benjamin, O.; Martinez, I.; Dahle, B.; Roos, N.; Lecocq, A.; Foley, K. Standard methods for *Apis mellifera* brood as human food. *J. Apic. Res.* **2019**, *58*, 1–28. [CrossRef]

23. Lou, Z.-Y. Insects as food in China. *Ecol. Food Nutr.* **1997**, *36*, 201–207. [CrossRef]

24. Evans, J.; Müller, A.; Jensen, A.B.; Dahle, B.; Flore, R.; Eilenberg, J.; Frøst, M.B. A descriptive sensory analysis of honeybee drone brood from Denmark and Norway. *J. Insects Food Feed* **2016**, *2*, 277–283. [CrossRef]

25. Finke, M.D. Nutrient composition of bee brood and its potential as human food. *Ecol. Food Nutr.* **2005**, *44*, 257–270. [CrossRef]

26. Ghosh, S.; Jung, C.; Meyer-Rochow, V.B. Nutritional value and chemical composition of larvae, pupae, and adults of worker honey bee, *Apis mellifera ligustica* as a sustainable food source. *J. Asia-Pac. Entomol.* **2016**, *19*, 487–495. [CrossRef]

27. Ghosh, S.; Chuttong, B.; Burgett, M.; Meyer-Rochow, V.B.; Jung, C. Nutritional value of brood and adult workers of the Asia honeybee species *Apis cerana* and *Apis dorsata*. In *African Edible Insects as Alternative Source of Food, Oil, Protein and Bioactive Components*; Mariod, A.A., Ed.; Springer: Basel, Switzerland, 2020; pp. 265–273. [CrossRef]

28. Winston, M.L. *The Biology of the Honey Bee*; Harvard University Press: Cambridge, MA, USA, 1987.

29. Charrière, J.-D.; Imdorf, A.; Bachofen, B.; Tschan, A. The removal of capped drone brood: An effective means of reducing the infestation of Varroa in honey bee colonies. *Bee World* **2003**, *84*, 117–124. [CrossRef]

30. Ambüehl, D. *Beezza! The kingbee Cook Book. Skyfood*; Selbstverlag: Unterterzen, Switzerland, 2016.

31. Lecocq, A.; Foley, K.; Jensen, A.B. Drone brood production in Danish apiaries and its potential for human consumption. *J. Apic. Res.* **2018**, *57*, 331–336. [CrossRef]

32. AOAC. *Official Methods of Analysis*, 15th ed.; Association of Official Analytical Chemists: Washinton, DC, USA, 1990.

33. *Korean Food Standard Codex*; Ministry of Food and Drug Safety: Cheongju, Korea, 2010.

34. Singleton, V.L.; Orthofer, R.; Lamuela-Raventos, R.M. Analysis of total phenols and other oxidation substrates and antioxidants by means of folin-ciocalteu reagent. *Methods Enzymol.* **1999**, *299*, 152–178.

35. Jing, L.; Ma, H.; Fan, P.; Gao, R.; Jia, Z. Antioxidant potential, total phenolic and total flavonoid contents of Rhododendron anthopogonoides and its protective effect on hypoxia-induced injury in PC12 cells. BMC Complement. *Altern. Med.* **2015**, *15*, 287. [CrossRef]

36. Masuko, T.; Minami, A.; Iwasaki, N.; Majima, T.; Nishimura, S.; Lee, Y.C. Carbohydrate analysis by a phenol-sulfuric acid method in microplate format. *Anal. Biochem.* **2005**, *339*, 69–82. [CrossRef] [PubMed]

37. Silveira, M.H.; Aquiar, R.S.; Siika-aho, M.; Ramos, L.P. Assessment of the enzymatic hydrolysis profile of cellulose substrates based on reducing sugar release. *Bioresur. Technolol.* **2014**, *151*, 392–396. [CrossRef]

38. Taipong, K.; Boonprakob, U.; Crosby, K.; Cisneros-Zevallos, L.; Byrne, D.H. Comparison of ABTS, DPPH, FRAP, and ORAC assays for estimating antioxidant activity from guava fruit extracts. *J. Food Comp. Anal.* **2006**, *19*, 669–675. [CrossRef]

39. Jumina, J.; Mutmainah, M.; Purwono, B.; Kurniawan, Y.S.; Syah, Y.M. Antibacterial and antifungal activity of three monosaccharide monomyristate derivatives. *Molecules* **2019**, *24*, 3692. [CrossRef] [PubMed]

40. Lee, H.; Woo, E.; Lee, D.G. (-)-Nortrachelogenin from Partrinia scabiosaefolia elicits an apoptotoc response in Candida albicans. *FEMS Yeast Res.* **2016**, *16*, fow013. [CrossRef]

41. Rowell, G.A.; Taylor, O.R., Jr.; Locke, S.J. Variation in drone mating flight times among commercial honey bee stockes. *Apidologie* **1986**, *17*, 137–158. [CrossRef]

42. Kim, S.G.; Woo, S.O.; Bang, K.W.; Jang, H.R.; Han, S.M. Chemical composition of drone pupa of *Apis mellifera* and its nutritional evaluation. *J. Apic.* **2018**, *33*, 17–23. [CrossRef]

43. Kent, N.L.; Evers, A.D. *Technology of Cereals*, 4th ed.; Pergamon Press: Oxford, UK, 1994.

44. WHO/FAO/UNU Expert Consultation. *Protein and Amino Acid Requirements in Human Nutrition*; WHO Technical Report Series 935; World Health Organization: Geneva, Switzerland, 2002.

45. Longo, N.; Frigeni, M.; Pasquali, M. Carnitine transport and fatty acid oxidation. *Biochim. Biophys. Acta Mol. Cell Res.* **2016**, *1863*, 2422–2435. [CrossRef] [PubMed]

46. Metzler, D.E. *Biochmeistry: TheChemical Reactions of Living Cells, Vol. 1 and 2*, 2nd ed.; Elsevier Academic Press: Cambridge, MA, USA, 2001.

47. Pickering, M.V.; Newton, P. Amino acid hydrolysis: Old problems, new solutions. *Lc Gc* **1990**, *8*, 778–781.

48. Mao, X.; Zeng, X.; Qiao, S.; Wu, G.; Li, D. Specific roles of threonine in intestinal mucosal integrity and barrier function. *Front. Biosci.* **2011**, *3*, 1192–1200. [CrossRef]

49. Schneider, E.; Rolli-Derkinderen, M.; Arock, M.; Dy, M. Trends in histamine research: New functions during immune responses and hematopoiesis. *Trends Immunol.* **2002**, *23*, 255–263. [CrossRef]

50. Rumpold, B.A.; Schlüter, O.K. Nutritional composition and safety aspects of edible insects. *Mol. Nutr. Food Res.* **2013**, *53*, 802–823. [CrossRef]

51. Chakravorty, J.; Ghosh, S.; Jung, C.; Meyer-Rochow, V.B. Nutritional composition of *Chondacris rosea* and *Brachutrupes orientalis*: Two common insects used as food by tribes of Arunachal Pradesh, India. *J. Asia-Pac. Entomol.* **2014**, *17*, 407–415. [CrossRef]

52. Chakravorty, J.; Ghosh, S.; Megu, K.; Jung, C.; Meyer-Rochow, V.B. Nutritional and anti-nutritional composition of *Oecophylla smaragdina* (Hymenoptera: Formicidae) and *Odototermes* sp. (Isoptera: Termitidae): Two preferred edible insects of Arunachal Pradesh, India. *J. Asia-Pac. Entomol.* **2016**, *19*, 711–720. [CrossRef]

53. Ghosh, S.; Lee, S.-M.; Jung, C.; Meyer-Rochow, V.B. Nutritional composition of five commercial edible insects in South Korea. *J. Asia-Pac. Entomol.* **2017**, *20*, 686–694. [CrossRef]

54. Ghosh, S.; Jung, C.; Choi, K.; Choi, H.Y.; Kim, H.W.; Kim, S. Body fatty and amino acid composition of a native bumblebee, *Bombus ignitus* relative to *B. terrestris* of foreign origin in Korea. *J. Apic.* **2017**, *32*, 111–117. [CrossRef]

55. Ghosh, S.; Debelo, D.G.; Lee, W.; Meyer-Rochow, V.B.; Jung, C.; Dekebo, A. Termites in human diet: An investigation into their nutritional profile. In *African Edible Insects as Alternative Source of Food, Oil, Protein and Bioactive Components*; Mariod, A.A., Ed.; Springer: Basel, Switzerland, 2020; pp. 293–306. [CrossRef]

56. Berger, B.; Crailsheim, K.; Leonhard, B. Proline, leucine and phynylalanine metabolism in adult honeybee drones (*Apis mellifera carnica* Pollm.). *Insect Biochem. Mol. Biol.* **1997**, *27*, 587–593. [CrossRef]

57. Grundy, S.M. What is the desirable ratio of saturated, polyunsaturated, and monounsaturated fatty acids in the diet? *Am. J. Clin. Nutr.* **1997**, *66*, 988S–990S. [CrossRef] [PubMed]

58. De Renobales, M.; Cripps, C.; Stanley-Samuelson, D.W.; Jurenka, R.A.; Blomquist, G.B. Biosynthesis of linoleic acid in insects. *TIBS* **1987**, *12*, 364–366. [CrossRef]

59. Ma, L.; Wang, Y.; Hang, X.; Wang, H.; Yang, W.; Xu, B. Nutritional effect of alpha-linolenic acid on honey bee colony development (*Apis mellifera* L.). *J. Apic. Sci.* **2015**, *59*, 63–72. [CrossRef]

60. Arien, Y.; Dag, A.; Zarchin, S.; Masci, T.; Shafir, S. Omega-3 deficeincy impairs honey bee learning. *Proc. Natl. Acad. Sci. USA* **2015**, *112*, 15761–15766. [CrossRef]

61. Mensink, R.P.; Katan, M.B. Effect of monounsaturated fatty acids versus complex carbohydrates on high-density lipoproteins in healthy men and women. *Lancet* **1987**, *329*, 122–125. [CrossRef]

62. Kris-Etherton, P.M. AHA Science Advisory: Monounsaturated fatty acids and risk of cardiovascular disease. *J. Nutr.* **1999**, *129*, 2280–2284.

63. Treasure, J.; Ploth, D. Role of potassium in the treatment of hypertension. *Hypertension* **1983**, *5*, 864–872. [CrossRef]

64. Houston, M.C. The importance of potassium in managing hypertension. *Curr. Hypertens. Rep.* **2011**, *13*, 309–317. [CrossRef]

65. Milman, N. Anemia-still a major health problem in many parts of the world! *Ann. Hematol.* **2011**, *90*, 369–377. [CrossRef]

66. Miller, J.L. Iron deficiency anemia: A common and curable disease. *Cold Spring Harb. Perspect. Med.* **2013**, *3*, a011866. [CrossRef]

67. Wu, F.Y.-H.; Wu, C.-W. Zinc in DNA replication and transcription. *Ann. Rev. Nutr.* **1987**, *7*, 251–272. [CrossRef] [PubMed]

68. Adeduntan, S.A. Nutritional and antinutritional characteristics of some insects foraging in Akure reserve Ondo state, Nigeria. *J. Food Technol.* **2005**, *3*, 563–567.

69. Pyo, S.-J.; Kang, D.-G.; Jung, C.; Sohn, H.-Y. Anti-thrombotic, antioxidant and haemolysis activities of six edible insect species. *Commun. Foods* **2020**. Accepted.

70. Liguori, I.; Russo, G.; Curcio, F.; Bulli, G.; Aran, L.; Della-Morte, D.; Gargiulo, G.; Testa, G. Oxidative stress, aging and disease. *Clin. Interv. Aging* **2018**, *13*, 757–772. [CrossRef]

71. Zhou, L.; Elias, R.J. Chapter 9—Understanding antioxidant and prooxidant mechanisms of phenolics in food lipids. In *Lipid Oxidation Challenges in Food Systems*; Logan, A., Nienaber, U., Pan, X., Eds.; Elsevier; Academic Press; AOCS Press: Amsterdam, The Netherlands, 2013; pp. 297–321.

72. Pietta, P.-G. Flavonoids as antioxidants. *J. Nat. Prod.* **2000**, *63*, 1035–1042. [CrossRef]

73. Hanasaki, Y.; Ogawa, S.; Fukui, S. The correlation between active oxygens scavenging and antioxidative effects of flavonoids. *Free Radic. Biol. Med.* **1994**, *16*, 845–850. [CrossRef]

74. Ursini, F.; Maiorino, M.; Morazzoni, P.; Roveri, A.; Pifferi, G. A novel antioxidant flavonoid (IdB 1031) affecting molecular mechanisms of cellular activation. *Free Radic. Biol. Med.* **1994**, *16*, 547–553. [CrossRef]

75. WHO-EM/NUT/270/E. In *Summary Report on the Technical Consultation on Reducing Sugar Intake in the Eastern Mediterranean Region*; World Health Organization Regional Office for the Eastern Mediterranean: Amman, Jordan, 2015.

76. Duan, J.; Yin, J.; Ren, W.; Liu, T.; Cui, Z.; Huang, X.; Wu, L.; Kim, S.W.; Liu, G.; Wu, X.; et al. Dietary supplementation with L-glutamate and L-aspartate alleviates oxidative stress in weaned piglets challenged with hydrogen peroxide. *Amino Acids* **2016**, *48*, 53–64. [CrossRef]

77. Anraku, M.; Shintomo, R.; Taguchi, K.; Kragh-Hansen, U.; Kai, T.; Maruyama, T.; Otagiri, M. Amino acids of importance for the antioxidant activity of human serum albumin as revealed by recombinant mutants and genetic variants. *Life Sci.* **2015**, *134*, 36–41. [CrossRef]

78. Kumar, G.; Karthik, L.; Rao, K.V.B. Haemolytic activity of Indian medicinal plants toward human erythrocytes: An *in vitro* study. *Appl. Bot.* **2011**, *40*, 5534–5537.

79. Zohra, M.; Fawzia, A. Hemolytic activity of different herbal extracts used in Algeria. *Int. J. Pharm. Sci. Res.* **2014**, *5*, 495–500.

80. Hossaini, A.A. Hemolytic and hemagglutinating activities of 222 plants. *Vox Sang.* **1968**, *15*, 410–417. [CrossRef] [PubMed]

81. Melgar-Lalanne, G.; Hernández-Áivarez, A.-J.; Salinas-Castro, A. Edible insects processing: Traditional and innovative technologies. *Compr. Rev. Food Sci. Food Saf.* **2019**, *18*, 1166–1190. [CrossRef]

82. Ulmer, M.; Smetana, S.; Heinz, V. Utilizing honeybee drone brood as a protein source for food products: Life cycle assessment of apiculture in Germany. *Resour. Conserv. Recycl.* **2020**, *154*, 104576. [CrossRef]
83. Bolatovna, K.S.; Rustenov, A.; Eleuqalieva, N.; Omirzak, T.; Akhanov, U.K. Improving reproductive qualities of pigs using drone brood homogenate. *Biol. Med.* **2015**, *7*, 2.

 *foods*

Article

Insects as Novel Food: A Consumer Attitude Analysis through the Dominance-Based Rough Set Approach

Rocco Roma [1], Giovanni Ottomano Palmisano [2] and Annalisa De Boni [1,*

[1] Department of Agricultural and Environmental Sciences, University of Bari Aldo Moro, 70126 Bari, Italy; rocco.roma@uniba.it
[2] International Center for Advanced Mediterranean Agronomic Studies (CIHEAM Bari), 70010 Valenzano, Italy; ottomano@iamb.it
* Correspondence: annalisa.deboni@uniba.it

Received: 28 February 2020; Accepted: 17 March 2020; Published: 27 March 2020

Abstract: In Western societies, the unfamiliarity with insect-based food is a hindrance for consumption and market development. This may depend on neophobia and reactions of disgust, individual characteristics and socio-cultural background, and risk-perceptions for health and production technologies. In addition, in many European countries, the sale of insects for human consumption is still illegal, although European Union (EU) and the European Food Safety Authority (EFSA) are developing regulatory frameworks and environmental and quality standards. This research aims to advance the knowledge on entomophagy, providing insights to improve consumer acceptance in Italy. This is done by carrying out the characterization of a sample of consumers according to their willingness to taste several types of insect-based food and taking into account the connections among the consumers' features. Thus, the dominance-based rough set approach is applied using the data collected from 310 Italian consumers. This approach provided 206 certain decision rules characterizing the consumers into five groups, showing the consumers' features determining their specific classification. Although many Italian consumers are willing to accept only insects in the form of feed stuffs or supplements, this choice is a first step towards entomophagy. Conversely, young Italian people are a niche market, but they can play a role in changing trends.

Keywords: entomophagy; consumer analysis; DRSA

---

1. Introduction

In Western societies the practice of eating insects, also known as entomophagy, is not usual in traditional diets, so that insects are rarely considered as edible [1]. However, insects can become a possible alternative to animal protein source thanks to their richness in protein, fat, minerals and vitamins [2], lower request of land and water [3], lower environmental impacts in terms of fewer greenhouse gases emissions and ammonia production [4], and also due to their more efficient feed conversion rate with respect to conventional meats [5,6]. In spite of the growing interest towards these benefits and the subsequent debate around the theme of insects as food, most of Western consumers still have reactions of disgust and rejection against them [7]. Generally, the main obstacles for consumers' acceptance of novel food (defined by EU Commission as "food that had not been consumed to a significant degree by humans in the EU before 15 May 1997") are food taboos and socio-cultural and psychological barriers, so that the aspect of a food can cause a disgust-based food rejection [8,9]. Indeed, the evident references of a food's origin to an animal (i.e., its "animalness") are strong determiners of a disgust response [10,11]. Moreover, food neophobia, defined as aversion to eating new and unfamiliar food, plays a key role in the acceptance of novel food [12,13].

As a consequence, the unfamiliarity with insects as food may represent also a hindrance for consumption and market development, especially in cultures where insects' consumption is not usual [11]. In particular, neophobia and organoleptic features of edible insects in comparison with the features of other well known food (e.g., meat and legumes) is seen as a decisive obstacle to consumers' acceptance [9,14–17]. In order to tackle these issues, previous studies suggested to integrate invisible insects in food preparation and/or to associate them with attractive flavors [7,18]. Furthermore, many authors underlined how the food product preparation affects the willingness to eat insects [9,11,19]. For instance, adding insects to familiar preparations (e.g., bread or pasta) or incorporating minced or powdered insects into ready-to-eat preparations, seemed to effectively increase the liking and willingness to try this kind of food in comparison to adding visible insects to meals or proposing them in their "whole form" [20–23]. Other authors highlight how consumers may show different behaviors towards the quality and presentation of insect-based food according to their own individual features and socio-cultural background [11,20,24,25], and also in relation to their risk-perceptions in terms of worries for health and production technologies [26,27]. Therefore, it is clear that consumer acceptance of insect-based food may depend on the amount, quality and source of information they receive and provide [26,28,29].

Moving the focus from research to policymaking, in many European countries, the sale of insects for human consumption is still illegal, even though the EU Commission is developing regulatory frameworks, environmental and quality standards to prevent risks for consumers from the consumption of novel food [30]. Specifically, the EU "Novel Food Regulation no. 2015/2283", in effect since 1st January 2018, allows to request the authorization for the commercialization of novel food [6] and the European Food Safety Authority (EFSA) is working on an evaluation of the risk profile related to the production and consumption of insects as food and feed [31]. However, only some EU countries (e.g., Belgium, Netherlands, Denmark, Finland and Germany) have adopted their own internal regulations for the trading of insect-based food, and this affects the spread of retailers selling insect products and their availability on the market, and at the same time, may cause concerns in consumers about safety and healthiness of these food products.

In the light of this complex scenario, the use of a comprehensive consumer-oriented approach is crucial to simultaneously analyze the factors influencing entomophagy and thus to provide overall insights for its diffusion [6,32]. Therefore, this research aims to advance the knowledge on entomophagy by supplying information to improve consumer acceptance in Italy. This is done by carrying out the characterization of a sample of consumers according to their willingness to taste several types of insect-based food and by taking into account the direct connections among the consumers' features.

Thus, a multiple criteria decision aiding (MCDA) approach is applied here, starting from the data gathered from direct consumer questionnaires. MCDA is an umbrella term describing a collection of formal approaches, which take into account multiple criteria in helping individuals or groups to explore decisions that matter [33]. A decision can be tackled through MCDA when there are different choices or alternatives to be judged as more desirable than others by means of criteria that may be in conflict to a substantial extent [34]. Specifically, the applied MCDA approach is called the dominance-based rough set approach (DRSA) because it is based on the rough sets theory and seeks to characterize the groups of consumers by means of simple "*If ... then ...* " decision rules [35].

Indeed, the decision rules inform about the relationships between conditions and decisions; in this way, the rules enable traceability of the decision support process and give understandable justifications for the decision to be made, so that the resulting preference model constitutes a 'glass box' [36]. Then, DRSA has been successfully applied in a variety of fields such as medical diagnosis, engineering reliability, empirical studies of material data, airline market and evaluation of bankruptcy risk [36,37]. On the other hand, applications on food science with focus on consumer analysis are still scarce [38,39]. Moreover, despite a growing interest towards entomophagy both by civil society and scholars, this topic is rather unexplored, showing a knowledge gap between curiosity-driven tasting and actual acceptance [20], which should be filled by applying discovering approaches [40]. For all these reasons, DRSA is considered suitable to explore the topic of entomophagy and to provide in depth analysis of consumers' attitude towards insect-based food.

The paper is organized as follows. After describing the methodology for the data collection, the DRSA is illustrated both from the theoretical and the empirical perspectives (Section 2). Then, Section 3 shows the results of the descriptive statistics of the sample and the DRSA application. Section 4 provides a detailed discussion of the results with a focus on those obtained from DRSA. Finally, the concluding remarks are reported in Section 5.

2. Materials and Methods

2.1. Data Collection

A questionnaire for the evaluation of the willingness to taste insect-based food was managed via "Google forms" in September 2019. The sample inclusion criteria were: being Italian, ≥18 years old and not being vegetarian or vegan; the sample consisted of 310 Italian consumers (61.6% female and 38.4% male) with age ranging from 18 to 81 years old. The questionnaire was supplied together with the definition of entomophagy, a brief overview of the environmental and nutritional benefits of edible insects, and a summary of the EU "Novel Food Regulation no. 2015/2283". Moreover, the questionnaire consisted of 18 questions and was structured into the following 4 sections.

- **Section 1—Willingness to taste insect-based food.** This section aimed to investigate the attitude towards the consumption of insects by humans and by cattle, pigs and chickens as feed. Therefore, four different kinds of food, in terms of the degree of processing and perception of the insects as an ingredient, were presented as pictures in the online survey: (1) meat, fish, eggs and milk obtained from animals raised with insect-based feed; (2) protein food supplements based on insect flour (e.g., cricket flour); (3) cookies made from wheat and insect flour; (4) cookies containing visible insects. Consumers were asked to choose which product they were more willing to taste, or else to declare that they were not willing to taste insect-based food in any form or preparation.
- **Section 2—Socio-demographic information and consumers' habits.** The second section looked to obtain specific consumer information regarding gender, age, monthly income, habitual travel outside Europe [16,41], sports activity [2], raw seafood consumption [42] and insect-based food knowledge [16,18]. Moreover, consumers' care for food nutritional and environmental aspects were investigated, asking them to assign a score (from 1 = irrelevant to 3 = determinant) to assess how nutritional and environmental features determine their food choices [43,44].
- **Section 3—Consumers' attitude towards novel food and innovative technologies for food preparation.** In this section, consumers' neophobia and trust towards innovative technologies for food preparation were assessed, because these individual features have been indicated as important predictors of the acceptance of insects as food [1,16,45,46]. Hence, consumers were asked to assign a score from 1 (I do not trust novel food not even if I know what it contains/The innovative technologies for food preparation are useless and can be harmful) to 4 (I am always looking for novel and different food, I taste everything /The innovative technologies can be fundamental to produce nourishing and sustainable food) in order to investigate their approach towards the consumption of novel food as well as with respect to the proposed innovative technologies.
- **Section 4—Factors affecting the willingness to taste insect-based food.** This last section aimed to evaluate the role of specific insect-based food characteristics in determining consumers' acceptance. Therefore, consumers were asked to assign the importance to a food safety certification and to a pleasant taste, smell and consistency by expressing a score from 1 (Irrelevant) to 4 (Crucial). In addition, the role of attending a "cooking show" or a "bug banquet" held by a well-known chef [47] was also considered an important factor, so the consumers were asked to assign a score from 1 (No, I would not taste anyway) to 3 (A presentation held by a well-known chef would be relevant for my acceptance).

2.2. The Dominance-Based Rough Set Approach

DRSA is a multiple criteria decision aiding (MCDA) method developed by Greco et al. [48,49] and it is an extension of the classical rough set approach (CRSA) [50,51], because it takes into account the preferences of a decision-maker and the typical inconsistency of decision problems [52]. Hence, DRSA substitutes the indiscernibility relation with a dominance relation in the rough approximation of decision classes, making it possible to discover the inconsistencies with respect to the dominance principle [53]. In other words, DRSA is a conceptual framework for the discovery of decision rules that have a syntax concordant with the dominance principle [48,54].

DRSA is applied by following the stepwise procedure described by Błaszczyński et al. [55,56] and reported hereafter.

- **Decision table**: Let us consider a decision table including a finite universe of objects U evaluated on a finite set of condition attributes $F = \{f_1, \ldots, f_n\}$ and on a single decision attribute d. The set of the indices of attributes is denoted by $I = \{1, \ldots, n\}$. Without loss of generality, $f_i : U \rightarrow \Re$ for each $i = 1, \ldots, n$, and, for all objects $x, y \in U$, $f_i(x) \geq f_i(y)$ means that "x is at least as good as y with respect to attribute i", that is denoted by $x \succeq_i y$. Therefore, it is supposed that $\succeq_i y$ is a complete preorder defined on U based on the quantitative and qualitative evaluations $f_1(\cdot)$. Moreover, the decision attribute d makes a partition of U into a finite number of decision classes $Cl = \{Cl_1, \ldots, Cl_m\}$ such that each $x \in U$ belongs to one and only one class Cl_t, $t = 1, \ldots, m$. It is assumed that the classes are preference ordered, e.g., for all $r, s = 1, \ldots, m$, such that $r > s$, the objects from Cl_r are preferred respect to the objects from Cl_s. More precisely, if $\geq$ is a comprehensive weak preference relation on U, e.g., if for all $x, y \in U$, $x \geq y$ means that x is at least as good as y, then it is supposed that $[x \in Cl_r, y \in Cl_s, r > s] \Rightarrow x > y$, where $x > y$ means $x \geq y$ and not $y \geq x$. These assumptions are typical of an ordinal classification problem with monotonicity constraints, where the decision table includes examples of ordinal classification that represent an input preference information to be analyzed by using DRSA. The sets to be approximated are defined as upward union and downward union of decision classes, respectively:

$$Cl_t^{\geq} = \bigcup_{s \geq t} Cl_s, \ Cl_t^{\leq} = \bigcup_{s \leq t} Cl_s, \ t = 1, \ldots, m.$$

  The statement $x \in Cl_t^{\geq}$ means that x belongs to at least class Cl_t, while $x \in Cl_t^{\leq}$ means that x belongs to at most class Cl_t. Let us highlight that $Cl_1^{\geq} = Cl_m^{\leq} = U$, $Cl_m^{\geq} = Cl_m$ and $Cl_1^{\leq} = Cl_1$. In addition, for $t = 2, \ldots, m$,

$$Cl_{t-1}^{\leq} = U - Cl_t^{\geq} \text{ and } Cl_t^{\geq} = U - Cl_{t-1}^{\leq}.$$

- **Dominance cones**: The approximation of upward and downward unions of decision classes is represented by granules of knowledge that are generated by attributes (i.e., criteria). These granules are also defined as dominance cones in the attribute values space. Specifically, x dominates y with respect to the set of attributes $P \subseteq F$ (that is x P – dominates y), denoted by xD_Py, if for every attribute $f_i \in P$, $f_i(x) \geq f_i(y)$. The relation of P-dominance is reflexive and transitive, namely it is a partial preorder. Given a set of attributes $P \subseteq I$ and $x \in U$, the granules of knowledge used for approximation in DRSA are as follows: a set of objects dominating x, called P-dominating set, given by $D_P^+(x) = \{y \in U : yD_Px\}$; a set of objects dominated by x, called P-dominated set, given by $D_P^-(x) = \{y \in U : xD_Py\}$. In other words, object x dominating object y on all the considered attributes also dominates y on the decision (i.e., the object should be assigned to at least as good decision class as y). The objects that satisfy the dominance principle are consistent, while those violating the dominance principle are called inconsistent.

- **Approximation of ordered decision classes:** The P-lower approximation of $Cl_t^{\geq}$ denoted by $\underline{P}\big(Cl_t^{\geq}\big)$, and the P-upper approximation of $Cl_t^{\geq}$, denoted by $\overline{P}\big(Cl_t^{\geq}\big)$, are defined as follows ($t = 2, \ldots, m$):

$$\underline{P}\big(Cl_t^{\geq}\big) = \big\{x \in U : D_P^+(x) \subseteq Cl_t^{\geq}\big\}, \overline{P}\big(Cl_t^{\geq}\big) = \big\{x \in U : D_P^-(x) \cap Cl_t^{\geq} \neq \varnothing\big\}.$$

In the same way, one can define the P-lower approximation and P-upper approximation of $Cl_t^{\leq}$ as follows ($t = 1, \ldots, m - 1$):

$$\underline{P}\big(Cl_t^{\leq}\big) = \big\{x \in U : D_P^-(x) \subseteq Cl_t^{\leq}\big\}, \overline{P}\big(Cl_t^{\leq}\big) = \big\{x \in U : D_P^+(x) \cap Cl_t^{\leq} \neq \varnothing\big\}.$$

The P-lower approximation and P-upper approximation satisfy the following inclusion property, for all $P \subseteq F$:

$$\underline{P}\big(Cl_t^{\geq}\big) \subseteq Cl_t^{\geq} \subseteq \overline{P}\big(Cl_t^{\geq}\big), \, t = 2, \ldots, m, \underline{P}\big(Cl_t^{\leq}\big) \subseteq Cl_t^{\leq} \subseteq \overline{P}\big(Cl_t^{\leq}\big), \, t = 1, \ldots, m - 1.$$

The P-lower approximation and P-upper approximation of $Cl_t^{\geq}$ and $Cl_t^{\leq}$ hold the complementarity property, according to which:

$$\underline{P}\big(Cl_t^{\geq}\big) = U - \overline{P}\big(Cl_{t-1}^{\leq}\big) \text{ and } \overline{P}\big(Cl_t^{\geq}\big) = U - \underline{P}\big(Cl_{t-1}^{\leq}\big), \, t = 2, \ldots, m, \underline{P}\big(Cl_t^{\leq}\big) = U - \overline{P}\big(Cl_{t+1}^{\geq}\big) \text{ and } \overline{P}\big(Cl_t^{\leq}\big) = U - \underline{P}\big(C$$

The P-boundary of $Cl_t^{\geq}$ and $Cl_t^{\leq}$ denoted by $Bn_P\big(Cl_t^{\geq}\big)$ and $Bn_P\big(Cl_t^{\leq}\big)$, respectively, are defined as follows:

$$Bn_P\big(Cl_t^{\geq}\big) = \overline{P}\big(Cl_t^{\geq}\big) - \underline{P}\big(Cl_t^{\geq}\big), \, t = 2, \ldots, m, Bn_P\big(Cl_t^{\leq}\big) = \overline{P}\big(Cl_t^{\leq}\big) - \underline{P}\big(Cl_t^{\leq}\big), \, t = 1, \ldots, m - 1.$$

Due to this complementarity property, $Bn_P\big(Cl_t^{\geq}\big) = Bn_P\big(Cl_{t-1}^{\leq}\big)$, for $t = 2, \ldots, m$.

- **Quality of approximation:** For every $P \subseteq F$, the quality of approximation of the ordinal classification Cl by a set of attributes P is defined as the ratio of the number of objects P-consistent with the dominance principle and the number of all the objects in U. Since the P-consistent objects are those which do not belong to any P-boundary $Bn_P\big(Cl_t^{\geq}\big)$, $t = 2, \ldots, m$, or $Bn_P\big(Cl_t^{\leq}\big)$, $t = 1, \ldots, m - 1$, the quality of approximation of the ordinal classification Cl by a set of attributes P, can be written as follows:

$$\gamma_P(Cl) = \frac{\left|U - \left(\bigcup_{t=2, \ldots, m} Bn_P\big(Cl_t^{\geq}\big)\right)\right|}{|U|} = \frac{\left|U - \left(\bigcup_{t=1, \ldots, m-1} Bn_P\big(Cl_t^{\leq}\big)\right)\right|}{|U|},$$

$\gamma_P(Cl)$ is seen as a degree of consistency of the objects from U, where P is the set of attributes (i.e., criteria) and Cl is the considered ordinal classification. Furthermore, for every $P \subseteq F$, the accuracy of approximation of union of ordered classes $Cl_t^{\geq}$, $Cl_t^{\leq}$ by a set of attributes P is defined as the ratio of the number of objects belonging to P-lower approximation and to P-upper approximation of the union. The accuracy of approximation $\alpha_P\big(Cl_t^{\geq}\big)$, $\alpha_P\big(Cl_t^{\leq}\big)$ can be written as follows:

$$\alpha_P\big(Cl_t^{\geq}\big) = \frac{\big|\underline{P}\big(Cl_t^{\geq}\big)\big|}{\big|\overline{P}\big(Cl_t^{\geq}\big)\big|}, \, \alpha_P\big(Cl_t^{\leq}\big) = \frac{\big|\underline{P}\big(Cl_t^{\leq}\big)\big|}{\big|\overline{P}\big(Cl_t^{\leq}\big)\big|}.$$

- **Reduction of attributes:** Each minimal subset $P \subseteq F$ such that $\gamma_P(Cl) = \gamma_F(Cl)$ is called reduct of Cl, and it is denoted by RED_{Cl}. There are more than one reduct for a given set of U.

Indeed, the intersection of all reducts is defined the core and it is denoted by $CORE_{Cl}$. The attributes in this core cannot be removed without affecting the quality of approximation. Therefore, in set F, there are the following three categories of attributes: (i) indispensable attributes included in the core; (ii) exchangeable attributes included in some reducts but not in the core; (iii) redundant attributes, neither indispensable neither exchangeable, which are not included in any reduct.

- **Decision rules**: The approximations of upward and downward unions lead to a generalized description of objects in terms of *"if ... , then ... "* decision rules. For a given upward or downward union of classes, $Cl_t^{\geq}$ or $Cl_s^{\leq}$, the decision rules induced under a hypothesis that objects belonging to $\underline{P}(Cl_t^{\geq})$ or $\underline{P}(Cl_s^{\leq})$ are positive examples, and all the others are negative, suggest a certain assignment to class Cl_t or better, or to class Cl_s or worse, respectively. Nevertheless, the decision rules induced under a hypothesis that objects belonging to $\overline{P}(Cl_t^{\geq})$ or $\overline{P}(Cl_s^{\leq})$ are positive examples, and all the others are negative, suggest a possible assignment respectively to class Cl_t or better, or to class Cl_s or worse. Finally, the decision rules induced under a hypothesis that objects belonging to the intersection $\overline{P}(Cl_s^{\leq}) \cap \overline{P}(Cl_t^{\geq})$ are positive examples, and all the others are negative, suggest an approximate assignment to some classes between Cl_s and Cl_t ($s < t$). In the case of preference ordered description of objects, the set U is composed of examples of ordinal classification. Then, the following five types of decision rules are considered:

(1) Certain $D_{\geq}$ decision rules, which provide the lower profile descriptions for objects belonging to $\underline{P}(Cl_t^{\geq})$: if $f_{i1}(x) \geq r_{i1}$ and ... and $f_{ip}(x) \geq r_{ip}$, then $x \in Cl_t^{\geq}$, $\{i_1, ..., i_p\} \subseteq I$, $t = 2, ..., m$, $r_{i1}, ..., r_{ip} \in \mathfrak{R}$;

(2) Possible $D_{\geq}$ decision rules, which provide the lower profile descriptions for objects belonging to $\overline{P}(Cl_t^{\geq})$: if $f_{i1}(x) \geq r_{i1}$ and ... and $f_{ip}(x) \geq r_{ip}$, then x *possibly belongs to* $Cl_t^{\geq}$, $\{i_1, ..., i_p\} \subseteq I$, $t = 2, ..., m$, $r_{i1}, ..., r_{ip} \in \mathfrak{R}$;

(3) Certain $D_{\leq}$ decision rules, which provide the upper profile descriptions for objects belonging to $\underline{P}(Cl_t^{\leq})$: if $f_{i1}(x) \leq r_{i1}$ and ... and $f_{ip}(x) \leq r_{ip}$, then $x \in Cl_t^{\leq}$, $\{i_1, ..., i_p\} \subseteq I$, $t = 1, ..., m-1$, $r_{i1}, ..., r_{ip} \in \mathfrak{R}$;

(4) Possible $D_{\leq}$ decision rules, which provide the upper profile descriptions for objects belonging to $\overline{P}(Cl_t^{\leq})$: if $f_{i1}(x) \leq r_{i1}$ and ... and $f_{ip}(x) \leq r_{ip}$, then x *possibly belongs to* $Cl_t^{\leq}$, $\{i_1, ..., i_p\} \subseteq I$, $t = 1, ..., m-1$, $r_{i1}, ..., r_{ip} \in \mathfrak{R}$;

(5) Approximate $D_{\geq \leq}$ decision rules, which provide simultaneously lower and upper profile descriptions for objects belonging to $Cl_s \cup Cl_{s+1} \cup ... \cup Cl_t$, without possibility of discerning to which class: if $f_{i1}(x) \geq r_{i1}$ and ... and $f_{ik}(x) \geq r_{ik}$ and $f_{ik+1}(x) \leq r_{ik+1}$ and ... and $f_{ip}(x) \leq r_{ip}$, then $x \in Cl_s \cup Cl_{s+1} \cup ... \cup Cl_t$, $\{i_1, ..., i_p\} \subseteq I$, $s, t \in \{1, ..., m\}$ $s < t$, $r_{i1}, ..., r_{ip} \in \mathfrak{R}$.

The first and the third types of rule represent the certain knowledge extracted from data, while the second and the fourth types of rule represent the possible knowledge. The fifth type of rule represents doubtful knowledge, because it is supported only by inconsistent objects [52,57]. Furthermore, a set of decision rules is complete when it covers all the considered objects (i.e., the examples of ordinal classification) in such a way that the consistent objects are re-assigned to their original classes, while the inconsistent objects are assigned to clusters of classes referring to this inconsistency. A set of decision rules is minimal when it is complete and non-redundant [58].

2.3. DRSA Empirical Model

DRSA is applied in this research, because it shows the following advantages [36,59]: (i) it connects directly the choice to the condition attributes that determine it (i.e., the type of insect-based food as a decision attribute with the consumers' features); (ii) it links the condition attributes to the decision attribute through a GAIN-type preference information and a COST-type preference information (i.e., the higher the education level the higher the willingness to taste insect-based food, or the lower the age the higher the willingness to taste insect-based food); (iii) it analyzes the decision rules to identify the object (i.e., consumer) supporting the choice and that justify it; (iv) it deals with both quantitative and qualitative data, and also with the inconsistencies that do not need to be removed before the analysis; (v) it acquires a posteriori information (i.e., consumer groups) about the most relevant attributes that delineate the objects in the form *"if ..., then ..."* decision rules, which are easy to interpret.

Table 1 shows the DRSA input information. More specifically, the names and codes of the criteria as well as the scale of measurement are based on the questionnaire, while the preference for each criterion (GAIN or COST) was identified in accordance with the relevant scientific literature. It should be recalled that GAIN means that the greater the value of a criterion, the greater the preference, while COST means that the lower the value of a criterion, the greater the preference [48,49].

This input information, together with the data gathered from the questionnaires, was elaborated using the software "jMAF" developed by the Laboratory of Intelligent Decision Support Systems (Poznań University of Technology, Poland) [55,56]. The extraction of the minimal set of certain decision rules (i.e., the first and the third types of rule) was carried out by applying the Dominance-based Learning from Examples Module (DOMLEM) algorithm [48,60] already implemented in the software. The subsequent characterization of the consumer groups was carried by interpreting the decision rules after the evaluation of their performance through the reclassification of the inconsistent objects [36].

Table 1. List of criteria with associated codes, scale of measurement and preference information according to the relevant scientific literature.

Criterion and Code	Scale of Measurement	Preference	Reference
SECTION 1: Willingness to taste insect-based food			
Food (FOOD)	(1, 2, 3, 4, 5)	GAIN (decision field)	[2,20,29]
SECTION 2: Socio-demographic information and consumers' habits			
Gender (GENDER)	(0, 1)	COST	[16,18]
Age (AGE)	(Continuous)	COST	[16,61]
Education (EDU)	(1, 2, 3)	GAIN	[61,62]
Income (INC)	(1, 2, 3, 4)	COST	[5,63]
Sport (SPORT)	(1, 2, 3, 4)	GAIN	[2]
Travel (TRAV)	(1, 2, 3)	GAIN	[11]
Knowledge (KNOW)	(1, 2, 3)	GAIN	[16,18]
Raw seafood (SEAF)	(1, 2, 3)	GAIN	[42]
Nutrition (NUT)	(1, 2, 3)	GAIN	[2,16]
Environment (ENV)	(1, 2, 3)	GAIN	[6,16]
SECTION 3: Consumers' attitude towards novel food and innovative technologies for food preparation			
Novel food (NEO)	(1, 2, 3, 4)	GAIN	[16,46]
Technology (TECH)	(1, 2, 3, 4)	GAIN	[16]
SECTION 4: Intrinsic and extrinsic factors affecting the willingness to taste insect-based food			
Chef (CHEF)	(1, 2, 3)	GAIN	[11]
Taste (TASTE)	(1, 2, 3, 4)	GAIN	[18,40]
Smell (SMELL)	(1, 2, 3, 4)	GAIN	[20,40]
Consistency (CONS)	(1, 2, 3, 4)	GAIN	[20]
Certification (CERT)	(1, 2, 3)	GAIN	[64]

3. Results

3.1. Descriptive Statistics of the Sample

The mean age of respondents was 33 years old (S.D. = 10.2 years). More than a half of them hold a high school degree and a third hold a university degree, showing a slight bias towards higher education levels. This bias may be attributed to the use of the electronic form to carry out the survey. The descriptive statistics of the sample are summarized in Table 2.

Table 2. Sample statistics on socio-demographic information and consumers' habits.

Criterion	Scale of Measurement	Frequency (%) or Mean ± SD
Age	Continuous	32.5 ± 10.2
Gender	0 = Male	38.4%
	1 = Female	61.6%
Education	1 = Compulsory school	14.8%
	2 = High school	54.8%
	3 = University degree or postgraduate	30.4%
Income	1 = Up to 1000 €	20.3%
	2 = From 1100 to 2000 €	45.8%
	3 = From 2100 to 3000 €	18.7%
	4 = More than 3000 €	15.2%
Sport	1 = Never	18.1%
	2 = Occasionally	45.5%
	3 = Regularly (at least twice per week)	30.9%
	4 = Competitive level	5.5%
Travel	1 = Never	32.2%
	2 = Occasionally (less than once per year)	42.6%
	3 = Regularly (at least once per year)	25.2%
Knowledge	1 = I have never heard about insect-based food for human consumption	12.2%
	2 = I have heard about insect-based food, but I do not know what it means	49.4%
	3 = I have heard about insect-based food and I know what it means	38.4%
Raw seafood consumption	1 = Never	32.9%
	2 = Occasionally	57.7%
	3 = Regularly (at least once per week)	9.4%
Care of nutritional aspects in food choice	1 = Irrelevant	5.8%
	2 = Relevant but not determining	51.3%
	3 = Absolutely fundamental	42.9%
Care of environmental aspects in food choice	1 = Irrelevant	10.7%
	2 = Relevant but not determining	59.0%
	3 = Absolutely fundamental	30.3%

Table 3 summarizes the frequencies concerning the consumers' attitude towards novel food and innovative technologies for food preparation, and frequencies of other factors affecting the willingness to taste insect-based food. Regarding the consumers' food choices, the answers showed that most of the participants in the survey (50.6%) were willing to taste meat, fish, eggs or milk made from animals raised with insect-based feed. Furthermore, 22.9% of consumers were not willing to taste insect-based food in any form of preparation; 16.8% declared a willingness to taste cookies made from wheat and insect flour with invisible insects; 7.4% of consumers were willing to take protein food supplements based on insect flour, such as cricket flour; and only 2.3% agreed to taste cookies containing visible insects.

Table 3. Sample statistics on acceptance of novel food, innovative technologies for food preparation and other factors affecting the willingness to taste insect-based food.

Criterion	Scale of Measurement	Frequency (%)
Type of insect-based food	1 = I am not willing to eat insect-based food in any form or preparation	22.9%
	2 = I am willing to eat meat, fish, eggs or milk made from animals raised with insect-based feed	50.6%
	3 = I am willing to eat protein food supplements based on insect flour	7.4%
	4 = I am willing to eat cookies made from wheat and insect flour	16.8%
	5 = I am willing to eat cookies with visible insects	2.3%
Novel food acceptance	1 = I do not trust novel food, not even if I know what it contains	7.8%
	2 = I sometimes taste novel food, but only if I am well informed about its characteristics	37.7%
	3 = I am glad to taste novel food if its appearance and smell are attractive	44.5%
	4 = I am always looking for novel and different food, I taste everything	10%
Innovative technologies acceptance	1 = The innovative technologies for food preparation are useless and can be harmful	9%
	2 = There is plenty tasty and nourishing food available on the market, so there is no need to use innovative technologies to produce more food	23%
	3 = The benefits of innovative food technologies are often overrated and can reduce the natural quality of food	21.9%
	4 = The innovative technologies can be fundamental to produce nourishing and sustainable food	46.1%
Presence of a food safety certification	1 = No, I would not taste anyway	26.5%
	2 = It would not change so much	30.6%
	3 = Yes, I would taste	42.9%
Participation to a tasting session leaded by a well-known chef	1 = No, I would not taste anyway	35.2%
	2 = It would not change so much	41.6%
	3 = Yes, I would taste	23.2%

On the other hand, the results of the survey showed a general positive attitude towards novel food. Indeed, almost half of consumers (44.5%) declared to be glad to taste novel food if its appearance and smell are attractive. Moreover, less than half of them (37.7%) declared to sometimes taste novel food, but only if they are well informed about the food's characteristics, while 10% of the sample is always looking for novel and different food and hence, taste everything. A limited share of respondents (7.8%) do not trust novel food, and was not willing to taste even if well informed about its characteristics or ingredients.

Innovative technologies for food preparation appeared to be fundamental in obtaining nourishing and sustainable food according to 46.1% of consumers, while 44.9% of respondents believed that the benefits of these technologies are often overrated, can reduce the natural quality of food and there is already plenty tasty and nourishing food on the market. A limited share of consumers (9%) were against innovative food technologies, because they are considered useless and harmful.

The presence of a food safety certification was perceived as reassuring for 42.9% of consumers, while 30.6% of them declared that certification does not improve their current acceptance of insect-based food and 26.5% of consumers would not taste anyway.

The majority of respondents considered intrinsic food features (taste, smell, consistency) crucial in their willingness to taste insect-based food (Table 4). Moreover, about 30% of the sample declared not to be willing to taste any kind of this food although it can be organoleptically pleasant. In addition, smell was rated as important but not fundamental according to almost a quarter of respondents, while less than 10% of respondents deemed that consistency is not very important.

Table 4. Sample statistics about the importance of organoleptic features of insect-based food.

Criterion	Scale of Measurement	Frequency (%)
Importance of taste when evaluating insect-based food	1 = I would not taste any kind of insect-based food	28.4%
	2 = Not very important	5.5%
	3 = Important but not fundamental	16.4%
	4 = Crucial	49.7%
Importance of smell when evaluating insect-based food	1 = I would not taste any kind of insect-based food	29.4%
	2 = Not very important	4.2%
	3 = Important but not fundamental	25.8%
	4 = Crucial	40.6%
Importance of consistency when evaluating insect-based food	1 = I would not taste any kind of insect-based food	30.0%
	2 = Not very important	7.4%
	3 = Important but not fundamental	22.6%
	4 = Crucial	40.0%

Finally, attending an insect-based food event (e.g., cooking show or bug banquet) held by a famous chef was considered a boost for tasting this food by 23.2% of those interviewed, while 35.2% of them would not taste it anyway and 41.6% of them thought that these events are irrelevant in improving insect-based foods acceptance.

3.2. Characterization of the Consumer Groups

The application of DRSA enabled to obtain the minimal set of 206 certain decision rules that are organized as follows according to the four insect-based foods plus the not-eat option. In particular, 25 decision rules classified the consumers as not willing to taste insect-based food in any form or preparation (i.e., *"at most FOOD 1"* rules); 81 rules classified the consumers willing to taste meat, fish, eggs or milk from animals raised with insect-based feed (i.e., *"at least FOOD 2"* and *"at most FOOD 2"* rules); 57 rules classified the consumers willing to taste protein food supplements based on insect flour (i.e., *"at least FOOD 3"* and *"at most FOOD 3"* rules); 38 rules classified the consumers willing to taste cookies made from wheat and insect flour (i.e., *"at least FOOD 4"* and *"at most FOOD 4"*); and 5 rules classified the consumers willing to taste cookies with visible insects (i.e., *"at least FOOD 5"*).

Table 5 shows, as an example, 8 certain decision rules classifying the consumers into the 5 food classes. The interpretation of the rules follows a general structure where the criteria and their values (i.e., the consumers' features) are in the first part of the rule (*If . . .*), while a certain insect-based food class is reported in the second part of the rule (*then . . .*). For instance, rule no. 82 regarding the class "Food 1" should be interpreted as follows: *"IF consumers are at least 36 years old with an education level between compulsory and high school, they have never heard of insect-based food and are not willing to taste this kind of food although it can be organoleptically pleasant, THEN they are not willing to taste insect-based food in any form or preparation"*. In the same way, rule no. 4 concerning the class "Food 5" should be interpreted as follows: *"IF the consumers are male between 18 and 22 years old with an income up to 1000 € per month, they think that taking part in a tasting session held by a well-known chef may strongly improve their willingness to eat insect-based food, and they also give great importance to taste in their food choices, THEN they would be willing to taste cookies with visible insects"*.

Table 5. Examples of certain decision rules for each food class.

Rule no.	Decision Rule	Food Class
82	If (Age ≥ 36) & (Edu ≤ 2) & (Know ≤ 1) & (Smell ≤ 1) then (Food ≤ 1) |CERTAIN, AT_MOST, 1|	Food 1
64	If (Edu ≥ 2) & (Neo ≥ 3) & (Cert ≥ 2) & (Chef ≥ 2) & (Taste ≥ 2) & (Smell ≥ 2) then (Food ≥ 2) |CERTAIN, AT_LEAST, 2|	Food 2
105	If (Seaf ≤ 2) & (Cert ≤ 1) & (Taste ≤ 1) then (Food ≤ 2) |CERTAIN, AT_MOST, 2|	Food 2
29	If (Age ≤ 37) & (Sport ≥ 3) & (Trav ≥ 3) & (Chef ≥ 3) then (Food ≥ 3) |CERTAIN, AT_LEAST, 3|	Food 3
152	If (Cert ≤ 1) & (Taste ≤ 2) then (Food ≤ 3) |CERTAIN, AT_MOST, 3|	Food 3
9	If (Edu ≥ 3) & (Sport ≥ 2) & (Neo ≥ 4) & (Cons ≥ 3) then (Food ≥ 4) |CERTAIN, AT_LEAST, 4|	Food 4
193	If (Neo ≤ 3) & (Taste ≤ 3) then (Food ≤ 4) |CERTAIN, AT_MOST, 4|	Food 4
4	If (Gender ≤ 0) & (Age ≤ 22) & (Inc ≤ 1) & (Chef ≥ 3) & (Taste ≥ 4) then (Food ≥ 5) |CERTAIN, AT_LEAST, 5|	Food 5

In order to know if the rules can be used to characterize the groups of consumers, the analysis of the rules' performances was carried out through the reclassification of the consumers for which these rules were induced [56]. The results of this reclassification are summarized as follows. Firstly, the confusion (or misclassification) matrix (Table 6) defines both the consistent objects that were reassigned to their original decision classes and the inconsistent objects that were reclassified into the decision classes referring to this inconsistency. In other words, the confusion matrix identifies the consumers who gave consistent answers and thus actually belong to the declared food class, and the consumers that need to be reclassified in a different food class since they gave incorrect answers with respect to the declared food class.

Table 6. The confusion matrix. Each column represents the objects in a predicted class, while each row represents the objects in an actual class.

Food Class	PREDICTED				
ACTUAL	1	2	3	4	5
1	52	18	1	0	0
2	0	129	0	0	0
3	0	0	13	0	0
4	0	0	0	33	0
5	0	2	0	0	5

This matrix shows that 232 consumers (74.8% of respondents) provided consistent answers and thus are reassigned to the food class declared in the questionnaire; these consumers can be found along the diagonal of the matrix. On the other hand, 21 consumers (6.8% of respondents) represent the incorrect cases that are reassigned to different decision classes. In particular, 19 consumers who declared that they are not willing to taste insect-based food in any form or preparation (class "Food 1") have the same characteristics as consumers belonging to other food classes. Indeed, 18 consumers are actually willing to taste meat, fish, eggs or milk from animals raised with insect-based feed (class "Food 2"), while one consumer might be willing to taste protein food supplements based on insect flour (class "Food 3"). In addition, two consumers who were willing to taste cookies with visible insects (class "Food 5") have the same features as those consumers willing to taste meat, fish, eggs or milk from animals raised with insect-based feed (class "Food 2").

The confusion matrix does not include the ambiguous cases, which represent the inconsistent consumers that can neither be reassigned to their original decision class nor reclassified. When DRSA is applied to large real-life data sets, ambiguous cases may occur due to significant differences between lower and upper approximations of the unions of decision classes and also to weak decision rules (i.e., rules supported by few objects from lower approximations) [36,56]. These cases are shown in a separated table providing their distribution within the different food classes (Table 7).

Table 7. The ambiguous cases for each food class and their distribution. Each number in brackets identifies an unclassified consumer.

Food Class	No. of Ambiguous Cases	Distribution of Ambiguous Cases in Food Classes		
		<2, 3>	<2, 4>	<3, 4>
2	28	6 (30, 89, 136, 138, 210, 278)	22 (5, 21, 27, 56, 68, 70, 79, 80, 84, 86, 108, 123, 129, 163, 169, 190, 200, 213, 235, 252, 270, 282)	
3	10	8 (40, 45, 49, 55, 174, 201, 211, 259)	1 (305)	1 (155)
4	19		18 (2, 33, 61, 88, 142, 146, 184, 215, 249, 262, 264, 274, 276, 284, 285, 287, 307, 309)	1 (106)

Table 7 shows that there are 57 unclassified consumers (18.4% of respondents). In particular, 28 consumers initially declared to be willing to taste meat, fish, eggs or milk from animals raised with insect-based feed, but 6 of them would also actually taste protein food supplements based on insect flour, and likewise 22 consumers would also taste cookies made from wheat and insect flour. Moreover, among the 10 consumers at first willing to taste protein food supplements based on insect flour, there are 8 consumers that would also taste eggs or milk from animals raised with insect-based feed, while one consumer is willing to taste both this last type of food and also cookies made from wheat and insect flour. Within this third food class, there is another consumer that is actually willing to taste the cookies together with the food declared in the questionnaire. Finally, 19 consumers were initially willing to taste cookies made from wheat and insect flour, but actually 18 of them would also taste meat, fish, eggs or milk from animals raised with insect-based feed, while one consumer is also willing to taste protein food supplements based on insect flour.

The reclassification of the consumers based on the extracted decision rules confirmed their validity to characterize the groups of consumers with the exclusion of the ambiguous cases reported in Table 7. Therefore, the interpretation of the minimal set of certain decision rules enabled to characterize 5 groups of consumers according to the most relevant features displayed by the rules and described hereafter.

- **Group 1—Consumers not willing to eat insect-based food in any form or preparation.** The first consumer group consists of 17 males and 35 females between 35 and 50 years old, their education is mostly at the secondary high school level and the monthly income is over 2100 €. These consumers practice sports occasionally, they have partial knowledge of insect-based food and they never eat raw seafood. Moreover, they are willing to taste new types of food only if detailed information is supplied, and they also think that innovative technologies are not useful for preparing food, because they can be harmful and useless. Finally, this group of consumers is not very interested in the food nutritional aspects and a good smell does not encourage tasting.

- **Group 2—Consumers willing to eat meat, fish, eggs or milk from animals raised with insect-based feed.** The second consumer group consists of 42 males and 107 females, between 26 and 32 years old with a medium-high education level (high school or university degree), and their income per month is between 1100 and 3000 €. About half of them have heard of insect-based food although they do not know what it means, while the remaining consumers have sufficient knowledge and they occasionally eat raw seafood and are thus glad to taste or eat novel food if the smell and consistency are attractive. Moreover, a certification, a chef's presentation (e.g., cooking show and bug banquet) and a good smell may encourage tasting.

- **Group 3—Consumers willing to eat protein food supplements based on insect flour (i.e., cricket flour).** Consumers of the third group are 7 males and 7 females between 20 and 34 years old, their monthly income is between 2100 and 3000 €. They travel less than once a year, and they are willing to taste novel food only if well informed about its features. These consumers do not trust new technologies for food preparation and they are also doubtful about certification as a tool to encourage tasting and eating insect-based food. In addition, a good taste is considered important but not fundamental in their choices.

- **Group 4—Consumers willing to eat cookies made from wheat and insect flour.** There are 17 males and 16 females between 21 and 25 years old, their monthly income is between 1100 and 2000 €; they travel less than once a year, and occasionally practice sports. Moreover, these consumers like to taste novel food if its appearance and smell are attractive, and they think that innovative technologies for food preparation are overrated and reduce food quality. Finally, they are not encouraged by a chef's presentation of insect-based food (e.g., cooking show and bug banquet), even though they think that taste is important in determining their choice.

- **Group 5—Consumers willing to eat cookies with visible insects.** There are 3 males and 2 females under 23 years old, their hold a university degree, their income is mostly under 2000 € per month, and they regularly practice a sport. Furthermore, these consumers are always looking for novel and different food and thus are willing to taste everything. Consequently, they believe that innovative technologies are fundamental to produce nourishing and sustainable food. In addition, they think that taking part in presentations held by a well-known chef (e.g., cooking show and bug banquet) strongly improves their own willingness to taste insect-based food.

4. Discussion

The main finding of the survey still confirms the low level of acceptance of insect-based food in Italy [13,61,65], because about 23% of respondents declared to be not willing to taste any kind of this food. Moreover, most of the participants in the survey expressed their acceptance of the introduction of insects into their diet only as feed or food supplement, in a form as different as possible from visible insects. In addition, less than 20% of interviewed consumers declared to be available to taste familiar food containing processed insects like cookies made with insect flour or with visible insects.

The analysis of the consumer groups obtained from the decision rules shows that the consumers strongly rejecting insect-based food in any form or preparation (i.e., Group 1) are the oldest, in accordance with other authors highlighting that an increase in age is associated with a decrease in the probability of accepting insects as a foodstuff [5,16]. These consumers hold an intermediate education level and have partial knowledge of insect-based food. In this sense, some studies [6,7,16,66] reported how complete information about the sustainability and environmental perspectives of edible insects, can positively influence the consumers' attitude towards tasting and consuming insect-based products. In addition, providing consumers with comprehensive information on edible insects may reduce their fear and increase their purchase probability [32]. Moreover, this group of consumers seems more interested in organoleptic rather than health features of food and are not willing to compromise taste for health and environmental benefits [62,67]. These consumers are also very careful in tasting novel food, as they declare to be unfamiliar with eating raw seafood. This may be considered a further cause of scepticism towards similar products in terms of appearance, like insect-based food. Hence, food neophobia has a strong role in influencing the consumers' attitude towards edible insects [7,9,15,16,19]. On the contrary, the familiarity with raw seafood consumption may represent an encouragement to taste insect-based food, due to the similarity of grasshoppers and cicadas to crustaceans and the resemblance of aquatic insect larvae's taste to fish served as '*ceviche*' [68].

The largest group of consumers (i.e., Group 2) showed a good level of acceptance of insects as feed for cattle and fish. Although insect-based feed does not yet exist on the EU market due to the lack of a defined regulatory framework, currently these consumers' positive attitude may bode good market perspectives in compliance with the findings of other scholars [67,69,70]. However, this also confirms that the majority of consumers are clearly not ready to incorporate insects as such into their diets [13]. In this sense, the acceptance of edible insects may be improved by a proper knowledge of this food, as well as by a higher inclination to taste novel food, especially if it is attractive in terms of consistency and smell. Likewise, these consumers are familiar with raw seafood consumption, hence the availability of further information regarding the similarity between insects and crustaceans (e.g., exoskeletons and antennae, their presence both in the aquatic and terrestrial systems), could improve the acceptability of entomophagy in their food habits [19]. Finally, being involved in thematic cooking shows and bug banquets may also strengthen their positive attitude.

The third group of consumers declared to be ready to accept protein food supplements based on insect flour. It should be recalled that, also in this case, the original product is not directly recognizable due to the high level of processing. The consumers belonging to this group have a medium-high monthly income, and this finding is in line with some studies [42,63] highlighting that unprocessed insects are usually consumed by people with medium-low income especially in developing countries, because their production process is cheap and they can afford a food with a high amount of protein, especially in a period of food shortage. However, recent research [27,32] found that the consumption of unprocessed insects is increasing also among high-income consumers, due to their better awareness about the nutritional benefits of this food. Moreover, the consumers in this group are still sceptic about tasting novel food, which explains why they prefer a food in which the insect is completely processed. In addition, they travel more than the consumers of other groups, but this feature still does not positively affect the tasting and consumption of unprocessed insects. Even though a regular tourist may have a strong propensity to look for unknown food and to appreciate the novelty of gastronomic experiences in a foreign environment, sometimes local food consumption is not a priority when the consumption of food is not the main goal of a trip and it is treated as a daily task, where the experience of meals does not increase the curiosity to try new food [71,72]. These consumers do not trust innovative technologies for food preparation, and they believe that a certification cannot encourage the tasting of insect-based food. This scepticism may be due to food safety concerns such as microbiological and chemical hazards especially in some developed countries, where potential consumers are deterred from incorporating or even thinking of including insects in their diets [26,27].

The fourth group is willing to eat cookies containing insect flour and includes consumers between 21 and 25 years old with a medium income. These findings are in line with studies showing that young and medium-income consumers are readier to taste and eat insect-based food as they have a better awareness of their nutritional benefits and entomophagy is seen as a thrilling experience [27,32,68]. It should be also noted that this group shows a gender balance, differently from other studies reporting that women are more disgusted by products of animal origin and more reluctant to accept insect-based food [15,61,73]. In this sense, a favorable behavior towards insect-based food both by males and females may be related to the curiosity about the taste and consistency of such a new food, which is typical of young people [19,40,66]. Hence, curiosity and attention towards taste and smell is the most important factor in their acceptance of this specific type of insect-based food [40], as they think that innovative technologies for food production may decrease food quality and the cooking shows or bug banquets do not influence their choices.

Consumers included in the fifth group are the youngest and hold a university degree. Indeed, a high education level plays a key role in the acceptance of edible insects [61]. Moreover, they practice sports regularly and this finding is in compliance with a study reporting that people who exercise regularly take care of the protein profile of their food, thus they are more positive towards unprocessed insects or insects as a visible ingredient [2].

In addition, these consumers believe that innovative food technologies are of fundamental importance in food production, although their novel food knowledge and involvement in sustainability issue is limited and not directly related to their choices. They are always seeking for new food experiences such as cooking shows and bug banquets, confirming the results of other studies highlighting that entomophagy may take on, especially in young people, the symbolic role of toughness or bravery, hence more related to a thrilling experience rather than to a conscious food choice [16,68].

The overall findings of this research clearly show the low level of acceptance of insect-based food, a very low knowledge of the features of this new food, and in particular, a scarce expertise in certain subjects (e.g., health and environmental benefits) that can encourage its tasting and consumption. Therefore, providing targeted information about entomophagy's benefits [65] may increase the consumers' acceptance of insect-based food. Specifically, both the similarity of insects with raw seafood and crustaceans and the general positive nutritional and environmental aspects should be better pointed out. At the same time, supplying people with opportunities to try insects during tasting sessions, special events, food fairs and "bug banquets" held by experts could reassure consumers about the organoleptic characteristics of these food and the social acceptability of entomophagy. In this sense, some studies dealing with the sensory acceptance of insect-enriched food confirmed that if insect-based food is properly presented, consumers may show a surprisingly high acceptance [74–76].

Young people currently represent only a niche market, but they can play a role in changing trends as future buyers and consumers. Conversely, for the majority of consumers, the willingness to accept insects in the form of feed stuffs or supplements can represent a first step towards entomophagy, if the positive characteristics of the products are properly communicated. It should be also highlighted that accepting to taste invisible insects may represent a first step towards overcoming the rejection of unfamiliar food and can further increase peoples' willingness to eat unprocessed insects [9]. As other authors point out [1,64,77], previous positive experiences and repeated consumption facilitate food preferences and willingness to introduce novel items in food habits [78]. With a view on legalizing the trade of these products, it is reasonable to think that food products containing processed insects as ingredients can be more promising than marketing products containing entire insects [1,9]. However, it is also essential to put in place the appropriate strategies to communicate the various positive characteristics of these products.

5. Conclusions

This research provides a useful contribution to understand how consumers' features may affect different behaviors towards entomophagy, since it advances the knowledge on this topic through new and detailed information. The DRSA enabled to inquire the consumers' attitudes dealing with uncertain data and taking into account the interactions among the individuals' features, thus providing a clear link between the consumers' choice and the factors that determine it. Although the decision rules cannot be used to characterize the entire Italian population due to the limited sample, the results can be considered a first step for future broader investigations and can suggest possible educational and communication strategies to improve consumer acceptance. According to the results obtained, future research will be carried out by implementing direct tasting sessions to reduce the uncertainty of data because of the illustration of insect-based food with pictures, and also by applying the variable-consistency dominance-based rough set approach (VC-DRSA) that will be more suitable to deal with a large real-life data set and thus to reduce the number of inconsistent cases.

Author Contributions: Conceptualization: A.D.B.; R.R. and G.O.P.; methodology: A.D.B.; R.R. and G.O.P.; software use: G.O.P.; validation: A.D.B.; R.R. and G.O.P.; formal analysis: A.D.B.; R.R. and G.O.P.; investigation: A.D.B.; R.R. and G.O.P.; data curation: A.D.B.; R.R. and G.O.P.; writing—original draft preparation: A.D.B.; R.R. and G.O.P.; writing—review and editing: A.D.B.; R.R. and G.O.P. All authors have read and agreed to the published version of the manuscript.

Funding: This research received no external funding.

Acknowledgments: Authors would like to thank Angela Natilla for providing support in data collection.

Conflicts of Interest: The authors declare no conflict of interest.

References

1. Hartmann, C.; Siegrist, M. Becoming an insectivore: Results of an experiment. *Food Qual. Prefer.* **2016**, *51*, 118–122. [CrossRef]
2. Collins, C.M.; Vaskou, P.; Kountouris, Y. Insect Food Products in the Western World: Assessing the Potential of a New 'Green' Market. *Ann. Entomol. Soc. Am.* **2019**, *112*, 518–528. [CrossRef] [PubMed]
3. Van Huis, A. Potential of insects as food and feed in assuring food security. *Annu. Rev. Entomol.* **2013**, *58*, 563–583. [CrossRef] [PubMed]
4. Oonincx, D.G.A.B.; van Itterbeeck, J.; Heetkamp, M.J.W.; van den Brand, H.; van Loon, J.J.A.; van Huis, A. An exploration on greenhouse gas and ammonia production by insect species suitable for animal or human consumption. *PLoS ONE* **2010**, *5*, 1–7. [CrossRef] [PubMed]
5. Schlup, Y.; Brunner, T. Prospects for insects as food in Switzerland: A tobit regression. *Food Qual. Prefer.* **2018**, *64*, 37–46. [CrossRef]
6. Mancini, S.; Sogari, G.; Menozzi, D.; Nuvoloni, R.; Torracca, B.; Moruzzo, R.; Paci, G. Factors predicting the intention of eating an insect-based product. *Foods* **2019**, *8*, 270. [CrossRef] [PubMed]
7. Lombardi, A.; Vecchio, R.; Borrello, M.; Caracciolo, F.; Cembalo, L. Willingness to pay for insect-based food: The role of information and carrier. *Food Qual. Prefer.* **2019**, *72*, 177–187. [CrossRef]
8. Meyer-Rochow, V.B. Food taboos: Their origins and purposes. *J. Ethnobiol. Ethnomed.* **2009**, *5*, 1–10. [CrossRef]
9. Hartmann, C.; Shi, J.; Giusto, A.; Siegrist, M. The psychology of eating insects: A cross-cultural comparison between Germany and China. *Food Qual. Prefer.* **2015**, *44*, 148–156. [CrossRef]
10. Rozin, P.; Fallon, A.E. A Perspective on Disgust. *Psychol. Rev.* **1987**, *94*, 23–41. [CrossRef]
11. Tan, H.S.G.; Fischer, A.R.H.; Tinchan, P.; Stieger, M.; Steenbekkers, L.P.A.; van Trijp, H.C.M. Insects as food: Exploring cultural exposure and individual experience as determinants of acceptance. *Food Qual. Prefer.* **2015**, *42*, 78–89. [CrossRef]
12. Pliner, P.; Hobden, K. Development of a scale to measure the trait of food neophobia in humans. *Appetite* **1992**, *19*, 105–120. [CrossRef]
13. Laureati, M.; Proserpio, C.; Jucker, C.; Savoldelli, S. New sustainable protein sources: Consumers' willingness to adopt insects as feed and food. *Ital. J. Food Sci.* **2016**, *28*, 652–668.
14. Raudenbush, B.; Frank, R.A. Assessing Food Neophobia: The Role of Stimulus Familiarity Department of Psychology, Wheeling Jesuit University. *Appetite* **1999**, *32*, 261–271. [CrossRef] [PubMed]
15. Tuorila, H.; Lähteenmäki, L.; Pohjalainen, L.; Lotti, L. Food neophobia among the Finns and related responses to familiar and unfamiliar foods. *Food Qual. Prefer.* **2001**, *12*, 29–37. [CrossRef]
16. Verbeke, W. Profiling consumers who are ready to adopt insects as a meat substitute in a Western society. *Food Qual. Prefer.* **2015**, *39*, 147–155. [CrossRef]
17. Fenko, A.; Backhaus, B.W.; van Hoof, J.J. The influence of product- and person-related factors on consumer hedonic responses to soy products. *Food Qual. Prefer.* **2015**, *41*, 30–40. [CrossRef]
18. Caparros Megido, R.; Gierts, C.; Blecker, C.; Brostaux, Y.; Haubruge, É. Consumer acceptance of insect-based alternative meat products in Western countries. *Food Qual. Prefer.* **2016**, *52*, 237–243. [CrossRef]
19. Caparros Megido, R.; Sablon, L.; Geuens, M.; Brostaux, Y.; Alabi, T.; Bleker, C.; Drugmand, D.; ÉRIC Haubrugei, E.; Francis, F. Edible insects acceptance by belgian consumers: Promising attitude for entomophagy development. *J. Sens. Stud.* **2014**, *29*, 14–20. [CrossRef]
20. Tan, H.S.G.; van den Berg, E.; Stieger, M. The influence of product preparation, familiarity and individual traits on the consumer acceptance of insects as food. *Food Qual. Prefer.* **2016**, *52*, 222–231. [CrossRef]
21. Pelchat, M.L.; Pliner, P. Try it. You'll like it'. Effects of information on willingness to try novel foods. *Appetite* **1995**, *24*, 153–165. [CrossRef]
22. Tuorila, H.; Andersson, Å.; Martikainen, A.; Salovaara, H. Effect of product formula, information and consumer characteristics on the acceptance of a new snack food. *Food Qual. Prefer.* **1998**, *9*, 313–320. [CrossRef]
23. Wansink, B. Changing eating habits on the home front: Lost lessons from World War II research. *J. Public Policy Mark.* **2002**, *21*, 90–99. [CrossRef]

24. Chakravorty, J.; Ghosh, S.; Meyer-Rochow, V.B. Practices of entomophagy and entomotherapy by members of the Nyishi and Galo tribes, two ethnic groups of the state of Arunachal Pradesh (North-East India). *J. Ethnobiol. Ethnomed.* **2011**, *7*, 1–14. [CrossRef] [PubMed]

25. Obopile, M.; Seeletso, T. Eat or not eat: An analysis of the status of entomophagy in Botswana. *Food Secur.* **2013**, *5*, 817–824. [CrossRef]

26. Belluco, S.; Losasso, C.; Maggioletti, M.; Alonzi, C.C.; Paoletti, M.G.; Ricci, A. Edible insects in a food safety and nutritional perspective: A critical review. *Compr. Rev. Food Sci. Food Saf.* **2013**, *12*, 296–313. [CrossRef]

27. Imathiu, S. Benefits and food safety concerns associated with consumption of edible insects. *NFS J.* **2020**, *18*, 1–11. [CrossRef]

28. Rollin, F.; Kennedy, J.; Wills, J. Consumers and new food technologies. *Trends Food Sci. Technol.* **2011**, *22*, 99–111. [CrossRef]

29. Pascucci, S.; De-Magistris, T. Information bias condemning radical food innovators? The case of insect-based products in the Netherlands. *Int. Food Agribus. Manag. Rev.* **2013**, *16*, 1–16.

30. Frewer, L.J.; Fischer, A.R.H.; Brennan, M.; Bánáti, D.; Lion, R.; Meertens, R.M.; Rowe, G.; Siegrist, M.; Verbeke, W.; Vereijken, C.M.J.L. Risk/Benefit Communication about Food—A Systematic Review of the Literature. *Crit. Rev. Food Sci. Nutr.* **2016**, *56*, 1728–1745. [CrossRef]

31. EFSA Scientific Committee. Risk profile related to production and consumption of insects as food and feed. *EFSA J.* **2015**, *13*, 1–60.

32. Liu, A.J.; Li, J.; Gómez, M.I. Factors Influencing Consumption of Edible Insects for Chinese Consumers. *Insects* **2020**, *11*, 10. [CrossRef] [PubMed]

33. Belton, V.; Stewart, T. *Multiple Criteria Decision Analysis. An Integrated Approach*; Springer: Berlin, Germany, 2002.

34. Fagioli, F.F.; Rocchi, L.; Paolotti, L.; Słowiński, R.; Boggia, A. From the farm to the agri-food system: A multiple criteria framework to evaluate extended multi-functional value. *Ecol. Indic.* **2017**, *79*, 91–102. [CrossRef]

35. Greco, S.; Matarazzo, B.; Slowinski, R. Dominance-based on Rough Set Approach. *Knowl. Creat. Diffus. Util.* **2007**, 325–339.

36. Greco, S.; Matarazzo, B.; Slowinski, R. Decision Rule Approach. In *Multiple Criteria Decision Analysis: State of the Art Surveys*, 2nd ed.; Figueira, J., Greco, S., Ehrgott, M., Eds.; Springer Science + Business Media: New York, NY, USA, 2016; pp. 497–552.

37. Liou, J.J.H.; Tzeng, G.H.A. Dominance-based Rough Set Approach to customer behavior in the airline market. *Inf. Sci.* **2010**, *180*, 2230–2238. [CrossRef]

38. Kashima, T.; Matsumoto, S.; Ishii, H. Decision support system for menu recommendation using rough sets. *Int. J. Innov. Comput. Inf. Control* **2011**, *7*, 2799–2808.

39. Celotto, E.; Ellero, A.; Ferretti, P. Conveying Tourist Ratings into an Overall Destination Evaluation. *Procedia—Soc. Behav. Sci.* **2015**, *188*, 35–41. [CrossRef]

40. Sogari, G. Entomophagy and Italian consumers: An exploratory analysis. *Prog. Nutr.* **2015**, *17*, 311–316.

41. Woolf, E.; Zhu, Y.; Emory, K.; Zhao, J.; Liu, C. Willingness to consume insect-containing foods: A survey in the United States. *Lwt* **2019**, *102*, 100–105. [CrossRef]

42. Van Huis, A.; Van Itterbeeck, J.; Klunder, H.; Mertens, E.; Halloran, A.; Muir, G.; Vantomme, P. *Edible Insects—Future Prospects for Food and Feed Security*; FAO Forestry Paper 171; Food and Agriculture Organization of the United Nations: Rome, Italy, 2013; Available online: http://www.fao.org/3/i3253e/i3253e.pdf (accessed on 10 February 2020).

43. Roininen, K.; Lähteenmäki, L.; Tuorila, H. Quantification of consumer attitudes to health and hedonic characteristics of foods. *Appetite* **1999**, *33*, 71–88. [CrossRef]

44. House, J. Consumer acceptance of insect-based foods in the Netherlands: Academic and commercial implications. *Appetite* **2016**, *107*, 47–58. [CrossRef] [PubMed]

45. Gmuer, A.; Nuessli Guth, J.; Hartmann, C.; Siegrist, M. Effects of the degree of processing of insect ingredients in snacks on expected emotional experiences and willingness to eat. *Food Qual. Prefer.* **2016**, *54*, 117–127. [CrossRef]

46. Menozzi, D.; Sogari, G.; Veneziani, M.; Simoni, E.; Mora, C. Eating novel foods: An application of the Theory of Planned Behaviour to predict the consumption of an insect-based product. *Food Qual. Prefer.* **2017**, *59*, 27–34. [CrossRef]

47. Hamerman, E.J. Cooking and disgust sensitivity influence preference for attending insect-based food events. *Appetite* **2016**, *96*, 319–326. [CrossRef]

48. Greco, S.; Matarazzo, B.; Slowinski, R. Rough sets theory for multi-criteria decision analysis. *Eur. J. Oper. Res.* **2001**, *129*, 1–47. [CrossRef]

49. Greco, S.; Matarazzo, B.; Slowinski, R. Rough approximation by dominance relations. *Int. J. Intell. Syst.* **2002**, *17*, 153–171. [CrossRef]

50. Pawlak, Z. Rough sets. *Int. J. Inf. Comput. Sci.* **1982**, *11*, 341–356. [CrossRef]

51. Pawlak, Z. *Rough Sets. Theoretical Aspects of Reasoning about Data*; Kluwer Academic Publishers: Dordrecht, The Netherlands, 1991.

52. Kadziński, M.; Greco, S.; Słowiński, R. Robust Ordinal Regression for Dominance-based Rough Set Approach to multiple criteria sorting. *Inf. Sci.* **2014**, *283*, 211–228. [CrossRef]

53. Sawicki, P.; Żak, J. The application of dominance-based rough sets theory for the evaluation of transportation systems. *Procedia Soc. Behav. Sci.* **2014**, *111*, 1238–1248. [CrossRef]

54. Boggia, A.; Rocchi, L.; Paolotti, L.; Musotti, F.; Greco, S. Assessing Rural Sustainable Development potentialities using a Dominance-based Rough Set Approach. *J. Environ. Manage.* **2014**, *144*, 160–167. [CrossRef]

55. Błaszczyński, J.; Greco, S.; Matarazzo, B.; Słowiński, R.; Szelag, M. jMAF—Dominance-Based Rough Set Data Analysis Framework. Available online: http://idss.cs.put.poznan.pl/site/uploads/media/jMAFmanual.pdf (accessed on 10 February 2020).

56. Błaszczyński, J.; Greco, S.; Matarazzo, B.; Słowiński, R.; Szelag, M. jMAF—Dominance-based rough set data analysis framework. In *Rough Sets and Intelligent Systems—Professor Zdzisław Pawlak in Memoriam*; Suraj, Z., Skowron, A., Eds.; Springer: Berlin, Germany, 2013; pp. 185–209.

57. Ferretti, P.; Zolin, M.B.; Ferraro, G. Relationships among sustainability dimensions: Evidence from an Alpine area case study using Dominance-based Rough Set Approach. *Land Use Policy* **2020**, *92*, 1–10. [CrossRef]

58. Chakhar, S.; Ishizaka, A.; Labib, A.; Saad, I. Dominance-based rough set approach for group decisions. *Eur. J. Oper. Res.* **2016**, *251*, 206–224. [CrossRef]

59. Abastante, F.; Asunis, S.; Bottero, M.; Greco, S.; Lami, I.M. The Dominance-based Rough Set Approach for Sustainability Assessment: The Evaluation of Alternative Transport Links between Two Italian Cities. In Proceedings of the Sustainable Architecture and Urban Development SB10, Amman, Jordan, 12–14 July 2010; Volume II, pp. 345–360.

60. Boggia, A.; Massei, G.; Pace, E.; Rocchi, L.; Paolotti, L.; Attard, M. Spatial multicriteria analysis for sustainability assessment: A new model for decision making. *Land Use Policy* **2018**, *71*, 281–292. [CrossRef]

61. Cicatiello, C.; Rosa, B.; Franco, S.; Lacetera, N. Consumer approach to insects as food: Barriers and potential for consumption in Italy. *Br. Food J.* **2016**, *118*, 2271–2286. [CrossRef]

62. Verbeke, W. Consumer acceptance of functional foods: Socio-demographic, cognitive and attitudinal determinants. *Food Qual. Prefer.* **2005**, *16*, 45–57. [CrossRef]

63. Gahukar, R.T. Entomophagy and human food security. *Int. J. Trop. Insect Sci.* **2016**, *31*, 129–144. [CrossRef]

64. De Magistris, T.; Pascucci, S.; Mitsopoulos, D. Paying to see a bug on my food. *BFJ* **2015**, *117*, 1777–1792. [CrossRef]

65. Sogari, G.; Menozzi, D.; Mora, C. Exploring young foodies' knowledge and attitude regarding entomophagy: A qualitative study in Italy. *Int. J. Gastron. Food Sci.* **2017**, *7*, 16–19. [CrossRef]

66. Pambo, K.O.; Okello, J.J.; Mbeche, R.M.; Kinyuru, J.N.; Alemu, M.H. The role of product information on consumer sensory evaluation, expectations, experiences and emotions of cricket-flour-containing buns. *Food Res. Int.* **2018**, *106*, 532–541. [CrossRef]

67. Bigliardi, B.; Galati, F. Innovation trends in the food industry: The case of functional foods. *Trends Food Sci. Technol.* **2013**, *31*, 118–129. [CrossRef]

68. Shelomi, M. Why we still don't eat insects: Assessing entomophagy promotion through a diffusion of innovations framework. *Trends Food Sci. Technol.* **2015**, *45*, 311–318. [CrossRef]

69. Mancuso, T.; Baldi, L.; Gasco, L. An empirical study on consumer acceptance of farmed fish fed on insect meals: The Italian case. *Aquac. Int.* **2016**, *24*, 1489–1507. [CrossRef]

70. Arru, B.; Furesi, R.; Gasco, L.; Madau, F.A.; Pulina, P. The introduction of insect meal into fish diet: The first economic analysis on European sea bass farming. *Sustainability* **2019**, *11*, 1697. [CrossRef]

71. Ji, M.; Wong, I.K.A.; Eves, A.; Scarles, C. Food-related personality traits and the moderating role of novelty-seeking in food satisfaction and travel outcomes. *Tour. Manag.* **2016**, *57*, 387–396. [CrossRef]
72. Özdemir, B.; Seyitoğlu, F. A conceptual study of gastronomical quests of tourists: Authenticity or safety and comfort? *Tour. Manag. Perspect.* **2017**, *23*, 1–7. [CrossRef]
73. Gere, A.; Székely, G.; Kovács, S.; Kókai, Z.; Sipos, L. Readiness to adopt insects in Hungary: A case study. *Food Qual. Prefer.* **2017**, *59*, 81–86. [CrossRef]
74. Biró, B.; Fodor, R.; Szedljak, I.; Pásztor-Huszár, K.; Gere, A. Buckwheat-pasta enriched with silkworm powder: Technological analysis and sensory evaluation. *LWT* **2019**, *116*, 108542. [CrossRef]
75. Haber, M.; Mishyna, M.; Itzhak Martinez, J.J.; Benjamin, O. The influence of grasshopper (Schistocerca gregaria) powder enrichment on bread nutritional and sensorial properties. *LWT* **2019**, *115*, 108395. [CrossRef]
76. Schouteten, J.J.; De Steur, H.; De Pelsmaeker, S.; Lagast, S.; Juvinal, J.G.; De Bourdeaudhuij, I.; Verbeke, W.; Gellynck, X. Emotional and sensory profiling of insect-, plant- and meat-based burgers under blind, expected and informed conditions. *Food Qual. Prefer.* **2016**, *52*, 27–31. [CrossRef]
77. Looy, H.; Dunkel, F.V.; Wood, J.R. How then shall we eat? Insect-eating attitudes and sustainable foodways. *Agric. Hum. Values* **2014**, *31*, 131–141. [CrossRef]
78. Loewen, R.; Pliner, P. Effects of prior exposure to palatable and unpalatable novel foods on children's willingness to taste other novel foods. *Appetite* **1999**, *32*, 351–366. [CrossRef] [PubMed]

Article

Degree of Hydrolysis Affects the Techno-Functional Properties of Lesser Mealworm Protein Hydrolysates

Giulia Leni [1], Lise Soetemans [1,2], Augusta Caligiani [1], Stefano Sforza [1] and Leen Bastiaens [2,*]

[1]   Department of Food and Drug, University of Parma, 43124 Parma, Italy; giulia.leni@studenti.unipr.it (G.L.);
       lise.soetemans@vito.be (L.S.); augusta.caligiani@unipr.it (A.C.); stefano.sforza@unipr.it (S.S.)
[2]   VITO, Flemish Insititute for Technolgical Research, 2400 Mol, Belgium
*   Correspondence: leen.bastiaens@vito.be; Tel.: +32-143-356-34

Received: 28 February 2020; Accepted: 22 March 2020; Published: 25 March 2020

Abstract: Protein hydrolysates from lesser mealworm (*Alphitobius diaperinus*, LM) were obtained by enzymatic hydrolysis with protease from *Bacillus licheniformis*. A preliminary test performed for five hours of hydrolysis generated an insect protein hydrolysate with 15% of degree of hydrolysis (DH), optimum solubility property and oil holding capacity, but emulsifying and foaming ability were completely impaired. In order to investigate the potential implication of DH on techno-functional properties, a set of protein hydrolysates with a different DH was obtained by sub-sampling at different time points during three hours of enzymatic hydrolysis process. An increase in DH% had positive effects on the solubility property and oil holding ability, while a reduced emulsifying ability was observed up to five hours of hydrolysis. These results demonstrated that the enzymatic hydrolysis, if performed under controlled conditions and not for a long period, represents a valid method to extract high quality protein from insects with tailored techno-functionality, in order to produce tailored ingredients for feed and food purpose.

Keywords: edible insect; protein hydrolysate; enzymatic hydrolysis; degree of hydrolysis; techno-functional properties; novel proteins

---

1. Introduction

According to the last Food and Agriculture Organization (FAO) reports, a huge effort must be deployed to meet the future food demand connected to the increasing world population. The world in 2050 will host about 9 billion people, 30% higher than nowadays, and a strong lack of farmland, water, nutrients, and non-renewable energy is expected [1]. Along the food chain, the meat production represents the most impacting field and for these reasons novel food protein sources have been envisioned to meet the future demand [2].

Insects represent a good source of proteins for feed and food applications. In fact, the amount of proteins in insects can range between 30% and 60% on dry matter basis, with a high-quality profile rich in essential amino acids [3,4]. Furthermore, insects, in comparison to the common livestock, are characterized by many environmental advantages, such as less land use, feed, and water requirement, and a high food conversion ratio [5]. In the European Union, the legalization on insect proteins in food and feed applications has been granted by including them in the novel food category and by permitting their use for pet, fur and aquaculture [6–8], respectively.

As future protein source, insects have been studied and different extraction protocols for proteins were explored. For examples, Yi et al. performed a chemical extraction on *Tenebrio molitor* by combining concentrated salt and alkali pH for an overnight extraction [9]. Bußler et al. and Zhao et al. applied an alkali extraction only after defatting the insects with hexane or ethanol [10,11]. Soetemans et al. reported the use of organic acids to obtain protein and lipid enriched fractions from black soldier fly larvae, after a mechanical removal of chitin [12]. Caligiani et al. compared the ability of three different protocols to fractionate and separate the main components of insect: Chitin, lipid, and protein. The protein fraction of black soldier fly was recovered with chemicals, by following both an alkali extraction and the milder Osborne fractionation, and enzymatic processes [4]. Indeed, the enzymatic method constitutes an essential part of the processes used by modern companies to produce, from complex matrices, a large and diversified range of products for human and animal consumption [13]. The use of exogenous enzymes, instead of chemicals, allows not only to control the process, but also to prevent protein degradation due to milder reaction condition. Proteases are efficient to separate proteins from lipids and insoluble compounds (e.g., fibers) by hydrolyzing peptide bonds and releasing peptides and free amino acids in solution with high efficiency [14]. Peptides, compared to the parental proteins, are characterized by an enhanced in gastro-intestinal digestibility and bio-accessibility [15]. In literature many studies have investigated the bioactivity of protein hydrolysates obtained from insects, such as antioxidant, angiotensin-converting enzyme-inhibitory, antidiabetic, and antihypertensive activity [16]. On the other side, limited publications are available regarding the techno-functional properties of insect protein hydrolysates. Purschke et al. [17] demonstrated the ability of targeted enzymatic hydrolysis to produce protein hydrolysates from *Locusta migratoria* protein flour with tailored techno-functional properties. They observed an increase in protein solubility, emulsifying activity, foam ability, and oil binding capacity in a broad spectrum of pH. Hall et al. [18] proved that Alcalase hydrolysis could represent an efficient biotechnology tool to improve the techno-functionality of cricket proteins. In particular, they demonstrated the ability of this enzyme to enhance the solubility, emulsion and foam capacity of insect protein hydrolysates obtained at different enzyme concentrations and time of hydrolysis.

Enzymatic assisted extraction has been demonstrated to be a valid method to extract proteins from insects in form of peptides. In our previous work, we have tested on lesser mealworm (*Alphitobius diaperinus*, LM) seven different enzymes from microbial, vegetable and animal origins, by performing at a laboratory scale an end-point hydrolysis [14]. The hydrolysates were characterized for the degree of hydrolysis (DH), the yield of extraction and the presence of free amino acids, but for their future involvement as insect-based protein ingredients in food or feed formulations, it is necessary to assess also their techno-functional properties. For the first time, the present work investigates the techno-functional properties of protein hydrolysates isolated from LM, focusing on the effect of the DH.

2. Materials and Methods

2.1. Insect Samples

LM, provided by Protifarm (Ermelo, The Netherlands), were reared as described in our previous work [14]. Larvae were killed by liquid nitrogen, packed under vacuum sealed and frozen at −20 °C. After one week, samples were freeze-dried for 36 h (Christ, gamma 1–16 LSC, Osterode am Harz, Germany) and stored at −20 °C for the future analysis. Samples were grinded for 2 min with a laboratory grinder (Microtron MB 550, Kinematica, Luzern, Switzerland) at maximum speed before each analysis.

2.2. Enzymatic Hydrolysis for Protein Extraction

2.2.1. Preliminary Enzymatic Assisted Extraction

The enzymatic assisted extraction of proteins was carried out by commercial protease from *Bacillus licheniformis* (≥2.4 U/g; EC Number 3.4.21.62) at optimal conditions for hydrolysis as suggested by the supplier (Sigma-Aldrich, St. Louis, MO, USA). More specifically, 200 g of dried ground larvae were mixed with 1 L of a buffer solution (Na_2HPO_4 10 mM) and 0.25% of enzyme in a 2 L reactor, combined with the pH-STAT system (Metrohm, Varese, Italy) to control the pH during the reactions by the addition of NaOH 1 M. By knowing the amount of NaOH added, it was possible to back-calculate the degree of hydrolysis as described by Butrè et al. [19]. The hydrolysis reaction was performed for five hours and after this time the solution was heated at 90 °C for 5 min for enzyme inactivation. The hydrolysates were then centrifuged (Eppendorf, 5810/5810 R, Milano, Italy) at 2683 g at 4 °C for 30 min. The supernatant was separated from the pellet and lyophilized with LIO-5PDGT freeze-dryer (5pascal, Milano, Italy). The freeze-dried protein hydrolysate was then defatted with diethyl ether, followed by quantification for protein content with Kjeldhal analysis, as described by Leni et al. [14].

2.2.2. Set of Protein Hydrolysates Collected at Different Time-Points

The enzymatic hydrolysis as previously described, was performed again in order to obtain a set of hydrolysates collected at different time-points. More specifically, 1.5 kg of dried ground larvae were mixed with 7.5 L of a buffer solution (Na_2HPO_4 10 mM) in a 10 L flask at pH 7.5 and at 60 °C. Before addition of enzyme, the mixture was homogenized in an incubator (New Brunswick, Innova 42 shaker, Eppendorf, Milano, Italy) at 130 rpm for 30 min, while checking the pH (ProfiLine pH 3310, WTW, Xylem Analytics LLC, Weilheim, Germany). The pH fluctuations were adjusted by adding NaOH 50%. After this period, 1.5 L of solution was subsampled and collected as control (time 0). Next, 0.25% of enzyme was added and the hydrolysis was performed for 180 min, with sub-sampling of 1.5 L aliquots of hydrolysate at different time points, being after 30 min, 60 min, 120 min, and 180 min. Each sub-sample, control included, was heated at 90 °C for 5 min to inactive the endogenous and exogenous enzymes and subsequently centrifuged at 4 °C for 30 min at 3220 g. The supernatant was separated from the pellet and the lipid upper layer using a 500 μm sieve and were freeze-dried to generate hydrolysates for the further analysis. Each hydrolysis reaction was performed in duplicate.

2.2.3. Bulk Composition

The hydrolysates and the intact larvae were characterized in terms of humidity, lipids, total N and ash. The dry matter content was determined after drying of the samples in oven at 105 °C for 24 h. Total ash was determined after mineralization at 550 °C for two times 5 h. For crude lipid quantification an automatized Soxhlet extractor (SER 148/3 VELP SCIENTIFICA, Monza e Brianza, Italy) was used with diethyl ether. The total N content was measured by Vario EL Cube (Elementar, Langenselbold, Germany) instrument by the supplier. Briefly, the sample was burned in an oxygen rich environment at 1150 °C in the combustion tube. All the burning gasses flowed through the reduction tube (helium as support gas) and were reduced to N_2, CO_2, H_2O, and SO_2. These different components were adsorbed at Selective Trap Columns and separated liberated (purge and trap technique). The detection of the components was performed with a thermo conductivity detector (TCD) cell. The proteinaceous N contribution was separated from the chitinous one, assuming that 87% of total N in LM was from protein origin, and then multiplied for the N to protein conversion factor 5.67, as we described in our previous work [14]. For each sub-sample, the protein content (g/100 g DM) was calculated, as well as the protein concentration as described in Equation (1):

$$\text{Protein concentration} \left(\frac{g}{L}\right) = \frac{g \text{ of protein in the sub} - \text{sample after freeze dry expressed on DM}}{L \text{ of sub} - \text{sample collected}} \quad (1)$$

2.2.4. Degree of Hydrolysis

The DH, defined as the percentage of cleaved peptide bonds in a protein hydrolysate, was calculated using o-phtaldialdehyde (OPA) as described by Leni et al. [14] with some modification. In particular, the supernatants were diluted in a 2% (*w/v*) sodium dodecyl sulphate, stirred for 20 min and stored at 4 °C overnight before the assay. The OPA/NAC (N-acetyl-cysteine) reagent (100 mL) was prepared by combining 10 mL of 50 mM OPA (in methanol) and 10 mL of NAC 50 mM, 5 mL of 20% (*w/v*) SDS, and 75 mL of borate buffer (0.1 M, pH 9.5). The reagent was covered with aluminum foil to protect it from light and allowed to stir for at least 1 h before use. The OPA assay was carried out (in triplicates) by the addition of 5 μL of sample (or standard) to 215 μL of OPA/NAC reagent in microplates. The absorbance of this solution was measured, after 10 min of shacking, at 340 nm with Tecan Infinite® 200 PRO spectrophotometer (Tecan, Männedorf, Switzerland) against a control cell containing the reagent and 5 µL of the buffer used for the sample. The intrinsic absorbance of the samples was measured before OPA addition and subtracted. The standard curve was prepared using L-isoleucine (0–2 mg/mL). The DH was calculated as the ratio between the free nitrogen groups after hydrolysis and the total nitrogen groups: DH% = (N free/N total) × 100. The first value was calculated by the OPA reactivity. The total moles of nitrogen atoms involved in peptide bonds before hydrolysis were calculated by considering that the 87% of total N [14] in insects is from protein origin and is involved in peptide bonds.

2.3. Techno Functional Properties

2.3.1. Solubility

The protein solubility was determined at pH 3, pH 5, and pH 7. The hydrolysates collected at different time points, were solubilized in demineralized water till a final concentration of 1% (*w/w*). The pH was adjusted with NaOH and HCl and the solution was mixed with an overhead shaker (Trayster digital, IKA, Königswinter, Germany) for 30 min. The pH was checked and adjusted if needed and the samples were placed on an overhead shaker for another 30 min. After that, the tubes were weighted and centrifuged at 5910 g for 20 min at 4 °C. The supernatant was separated and collected for the analysis in order to quantify the soluble N and the ionic strength. The total N was determined by a chemiluminescence detector (Multi N/C 3100 Analytik Jena, Jena, Germany). Briefly, the sample was oxidized by catalytic combustion in an oxygen atmosphere at 800 °C, to N oxides. The formed measuring gas was dried and entered in the reaction chamber of the chemiluminescence detector. There, the N monoxide present in the measuring gas, was oxidized with ozone into activated N dioxide. By emitting light photons (luminescence) the molecules of the N dioxide returned to their original state. The luminescence, proportional with the N monoxide concentration, was detected using a photomultiplier. The ionic strength was measured with a conductivity meter (ProfiLine Cond 3310, WTW, Weilheim, Germany).

The protein solubility was calculated as described in Equation (2):

$$\text{Protein solubility } (\%) = \frac{\text{g } N \text{ in the supernatant}}{\text{g proteinaceous } N \text{ in the sample}} \times 100 \tag{2}$$

The N-content in the supernatant measured by chemiluminescence analysis was assumed to be only from protein origin, whereas the amount of proteinaceous N in the sample was determined as described in 2.2.3. The analysis was performed in triplicate. In order to compare the results obtained with food standard proteins, the procedure was repeated with egg white (Sigma-Aldrich, St. Louis, MO, USA).

2.3.2. Emulsification Properties

The emulsification property was determined following the method proposed by Purschke et al. [17] with some modification. Briefly, the hydrolysates were diluted with demineralized water at a concentration of 0.1% (*w/w*) and mixed with an overhead shaker for 30 min. The ionic strength was measured with a conductivity meter (ProfiLine Cond 3310, WTW, Weilheim, Germany). The solution was centrifuged at 3220 g for 15 min and the supernatant mixed with commercial corn oil (Vandemoortele, Ghent, Belgium) (1:1 *v/v*) and emulsified at 11.000 rpm for 30 s using a homogenizer (ULTRA-TURRAX® T18, IKA, Königswinter, Germany). An aliquot of the emulsion was immediately transferred into scaled tubes and centrifuged at 3220 g for 15 min at 20 °C (fixed angle centrifuge). The height of the resulting emulsified layer (Hel), even if not homogeneous in the scaled tube, and the total height of solution (Hs) were used to calculate the emulsification ability as described below (3). The analyses were performed in triplicate.

$$\text{Emulsifying activity (\%)} = \frac{\text{Hel}}{\text{Hs}} \times 100 \tag{3}$$

Casein from bovine milk and egg white (Sigma-Aldrich, St. Louis, MO, USA) were subjected to the same procedure in order to evaluate and compare their emulsification properties with insect protein hydrolysates.

2.3.3. Oil Holding Capacity

For the oil holding capacity (OHC), 1 g of hydrolysate was transferred to a falcon tube and 10 g of commercial corn oil (Vandemoortele, Ghent, Belgium) were added. The solution was mixed with on overhead shaker for 5 min at 55 rpm. After 30 min, the tube was centrifuged at 3000 g for 30 min at 20 °C. The sample was re-weighed after 10 min of decantation upside-down (45° angle) and the holding capacity calculated as described in the equation below (4):

$$\text{OHC}_{\text{(g oil/g sample)}} = \frac{\text{W2} - \text{W1}}{\text{W0}} \tag{4}$$

where W0 was the weight of the sample, W1 was the weight of the tube and the sample, W2 was the weight of the tube after decantation. The analyses were performed in triplicate. In order to compare insect protein hydrolysate with standard food proteins, egg white, and casein from bovine milk (Sigma-Aldrich, St. Louis, MO, USA) were subjected to the same procedure and their OHC determined.

2.3.4. Foaming Capacity

The foaming capacity were measured with a homemade apparatus, composed by a graduated glass cylinder, in which was placed the solution, and a pump which fluxed air inside the mixture. The hydrolysate was suspended in demineralized water at a final concentration of 1% and mixed on an overhead shaker for 30 min. The ionic strength was measured with a conductivity meter (ProfiLine Cond 3310, WTW, Weilheim, Germany) and then the solution was transferred in the foam tube and the starting volume (in cm) was noted. The air flow was bubbled through the sample at a flow rate of 2 L/h for 1 min and the final volume reached by the foam was red after pump stopping and reported. Casein from bovine milk and egg white (Sigma-Aldrich, St. Louis, MO, USA) were subjected to the same procedure for evaluating and comparing their foaming capacity with insect protein hydrolysates. The analysis was performed in triplicate. The foam capacity was measured with the following Equation (5):

$$\text{Foam capacity (\%)} = \frac{\text{Volume foam}}{\text{Volume start}} \times 100 \tag{5}$$

2.4. Statistical Analysis

Data are expressed as the mean ± standard deviation. Statistical analysis was performed using SPSS version 21.0 (SPSS Inc., Chicago, IL, USA). The data were subjected to one-way analysis of variance (ANOVA) to determine the differences between samples. Significant differences were compared at a level of $p < 0.05$.

3. Results

3.1. Techno Functional Assay of Protein Hydrolysate

Insect protein hydrolysate was obtained from LM with the protease from *Bacillus licheniformis*. The whole insect starting material contained 52 ± 0.2% of proteins on dry matter basis. After five hours of hydrolysis, a hydrolysate with a DH of 14.9 ± 0.2% and protein concentration of 51.39 ± 0.02 g/L was obtained. The final protein hydrolysate, after the defatting step, contained 58.2 ± 1.3% of proteins on dry matter basis (Table 1). The protein hydrolysate presented a high solubility property, with 95 ± 4% of total proteins soluble at pH 3, 5, and 7, explicable by the presence of peptides and free amino acids. Further, the protein hydrolysate presented also good capacity to hold oil, with 6.7 ± 0.6 g oil per g of sample. This value was 5 times more than the ability evaluated for casein and egg white (Table 2). On the contrary, the protein hydrolysate did not display foaming ability and only a slight capacity to form emulsions. In order to better explore if the high DH% reached could have affected foaming and emulsifying property, additional enzymatic hydrolysis were performed aiming at generating hydrolysates with different DHs.

3.2. Generation of Protein Hydrolysates with Different Degrees of Hydrolysis

Via subsampling during enzymatic assisted hydrolyses of LM starting material, water soluble protein hydrolysates with a DH ranging between 2.9% and 9.8% were obtained. The water-soluble fraction of the control sample, even if obtained without the enzyme addition, presented itself a small DH%, which could be related to a mild denaturation and hydrolysis occurred during the heat inactivation. This is in accordance with the findings of Purschke et al. [17] and Hall et al. [18], who both demonstrated an initial DH of about 5%. No significant differences were determined on DH% from 30 min to 60 min of hydrolysis, while a significant increase was monitored from 60 min till 180 min.

The proximate composition of the soluble fraction of the control and the hydrolysates was determined in terms of humidity, protein, lipid, and ash content and is reported in Table 1. All samples presented a similar compositional profile, rich in protein (66 ± 4% on DM on average) and with a slight amount of lipid (on average 15 ± 4% on DM basis).

Table 1. Bulk composition of hydrolysates and freeze-dried control samples collected at different time points during the enzymatic hydrolysis with information about DH% and protein concentration.

Hydrolysis Time	Protein% on DM	Lipid% on DM	Ash% on DM	DH%°	Protein Concentration (g/L)
		Preliminary test			
300 min *	58.2 ± 1.3 [a]	nd	nd	14.9 ± 0.2 [e]	51.39 ± 0.02 [b]
		Detailed test			
Control (0 min)	68.8 ± 4.9 [b]	16.3 ± 0.9 [ab]	10.7 ± 0.1 [a]	2.9 ± 0.5 [a]	34.2 ± 4.3 [a]
30 min	69.2 ± 4.5 [bc]	10.9 ± 6.3 [a]	9.3 ± 1.3 [a]	4.8 ± 0.5 [b]	38.9 ± 4.1 [a]
60 min	65.41 ± 1.05 [bc]	12.7 ± 2.3 [ab]	9.6 ± 0.8 [a]	5.7 ± 0.5 [b]	46.72 ± 1.01 [ab]
120 min	63.5 ± 1.7 [cd]	16.04 ± 2.36 [ab]	8.66 ± 1.98 [a]	8.3 ± 0.5 [c]	51.4 ± 1.9 [ab]
180 min	62.1 ± 0.3 [d]	17.3 ± 2.6 [b]	9.1 ± 0.6 [a]	9.8 ± 0.7 [d]	50.78 ± 10.96 [ab]

Results are expressed as the mean ± standard deviation ($n = 6$). Values followed by different letters within one column are significantly different ($p < 0.05$). Abbreviation: degree of hydrolysis, DH; dry matter, DM; lesser mealworm, LM; nd, not determined; ° For samples collected from 0 to 180 min DH% was calculated by OPA assay, for sample obtained after 300 min of hydrolysis DH% was calculated from pH-STAT method; * hydrolysate produced in a separated enzymatic hydrolysis and subjected to a defatting step with diethyl-ether, as described in 1.2.1.

The addition of exogenous enzyme increased the protein concentration in the supernatant of each sub-sample, even if, after 120 min, it was determined a flatting of rise near to a plateau, as reported in the enzymatic kinetic curve in Figure 1. Nevertheless, the protein hydrolysate collected after 180 min of hydrolysis had a protein concentration 48% higher than the one of control sample, thus collected before enzyme addition.

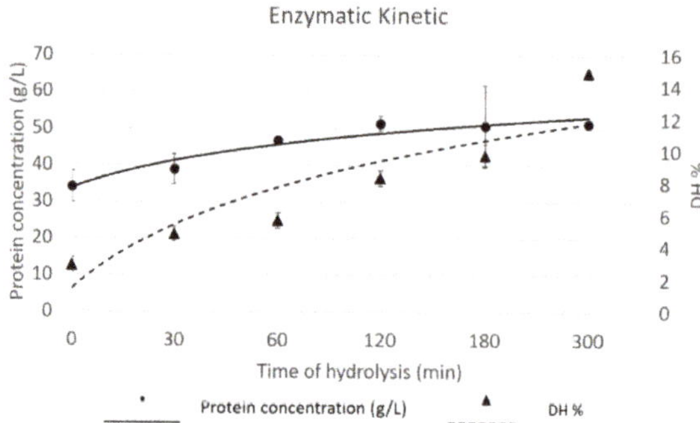

Figure 1. Enzymatic kinetic described by protein concentration (g/L) and DH% of hydrolysates collected at different time points.

The enzymatic kinetic is also here described by evaluating the increase in DH% during the hydrolysis (Figure 1). By combining the slopes of the two curves, it is possible to assume that enzyme, during the hydrolysis, preferred to act on protein material, which was already extracted and in solution, rather than extract more protein material from the residual insect biomass. In fact, if the protein concentration did not show significative differences between the different hydrolysates, the results from DH% underlined a significative increase of free amino groups, which can be described by the hydrolysis of protein which are already in solution.

3.3. Functional Properties

3.3.1. Solubility

The solubility property in function of pH is an important parameter of the protein hydrolysates in view of their potential industrial application. Solubility of proteins relates to surface hydrophobic (protein–protein) and hydrophilic (protein–solvent) interaction; in food case, such solvent is the water, and therefore the protein solubility is classified as a hydrophilic property [20]. In the present work, all hydrolysates were more soluble than native proteins at both pH 3 and pH 7 with values higher than 82%, even if only after 60 min of hydrolysis the difference became significant. In general, all samples, control included, presented better solubility at these pH than at pH 5 (Table 2). The lower solubility at pH 5 is due to the proximity to the isoelectric point. This is in accordance with what reported by Bußler et al. [10], who discovered that *Tenebrio molitor* intact proteins presented lowest solubility round pH 4. Purschke et al. identified at pH 5 the isoelectric point of protein hydrolysate obtained from *Locusta migratoria*, belonging to a different insect species [17].

The proximity to the isoelectric point could also justify the high standard deviation (15% on average) of trials performed at pH 5. In fact, at isoelectric pH, the zero net charge reduces the repulsive electrostatic forces, whereas the attraction forces predominate and cause the protein aggregation and precipitation. In the hydrolysate produced after 300 min of hydrolysis no significant differences were identified at pH 3, 5, and 7. This is in agreement with wat reported by Hall et al. [18] where cricket hydrolysate produced after 90 min with 1.5% of Alcalase enzyme did not present big differences at pH 3, 7, 8, and 10.

Table 2. Techno-functional properties of freeze-dried samples collected at different time: protein solubility (pH 3, 5, 7), emulsification ability, oil holding and foaming capacity. Information about casein and egg white functionality are also reported.

Sample	Protein Solubility%			Emulsification Activity%	Oil Holding Capacity g oil/g Sample	Foaming Capacity%
	pH 3	pH 5	pH 7			
Preliminary test						
300 min[*]	94.8 ± 4.8 [ab]	96.1 ± 4.2 [b]	94.3 ± 0.6 [d]	1.2 ± 0.3 [a] **	6.7 ± 0.6 [d]	0
Detailed test						
Control	86.7 ± 0.3 [a]	60.3 ± 28.6 [a]	81.04 ± 0.95 [a]	20.99 ± 4.06 [e] **	1.4 ± 0.1 [a]	0
30 min	91.1 ± 2.1 [ab]	67.8 ± 22.4 [a]	82.1 ± 1.6 [ab]	11.8 ± 3.1 [d] **	1.4 ± 0.1 [a]	0
60 min	92.7 ± 1.8 [b]	58.1 ± 12.9 [a]	85.8 ± 1.7 [bc]	6.7 ± 2.7 [c] **	1.7 ± 0.3 [ab]	5.3 ± 4.2 [a]
120 min	93.6 ± 1.6 [b]	67.1 ± 17.4 [a]	90.4 ± 2.7 [cd]	3.1 ± 1.1 [bc] **	1.99 ± 0.25 [bc]	12.59 ± 7.02 [a]
180 min	94.9 ± 2.6 [b]	69.9 ± 2.6 [a]	90.9 ± 2.5 [d]	2.03 ± 0.36 [b] **	2.2 ± 0.2 [c]	73.6 ± 16.1 [c]
Standard protein						
Casein	nd	nd	nd	34.3 ± 2.3 [f]	1.52 ± 0.03 [a]	43 ± 5 [b]
Egg white	85.9 ± 2.6 [a]	80.4 ± 4.3 [a]	86.4 ± 1.5 [bc]	52.5 ± 3.5 [g]	1.1 ± 0.2 [a]	123 ± 24 [d]

Results are expressed as the mean the mean ± standard deviation (n = 6). Values followed by different letters within one column are significantly different ($p < 0.05$). nd: not determined. * hydrolysate produced in a separated enzymatic hydrolysis as described in 1.2.1. ** Incomplete emulsion layers were observed—value also given as indication of trends.

The % of solubilized protein in the current study was higher compared to Purschke et al., [17] where the solubility of *Locusta migratoria* protein hydrolysates obtained from Neutrase and Flavourzyme did not exceed the 6%. The differences could be mainly related to the different enzymes used for the hydrolysis, which are characterized by a diverse cleavage specificity. Instead, Hall et al. obtained protein hydrolysates from Alcalase activity on cricket proteins with a solubility profile similar to our results [18].

In Figure 2 the % of solubilized protein at different pH was plotted against the DH% and a positive correlation was observed for solubility at pH 3 and 7 (respectively, r = 0.842 and r = 0.876), whereas close to the isoelectric point, protein solubility stays mostly unchanged. An increase in solubility with increasing DH-values may be associated with an increase of small peptides, that exposes more ionizable amino and carboxylic groups. These groups promote electrostatic repulsion and enhances the formation of hydrogen bonds with water molecules and, as such, solubility improvement [21]. This positive correlation was also determined by Purschke et al. when Alcalase, Neutrase and Papain were used to produce protein hydrolysates from locust, while Flavourzyme did not show any correlation at pH 3, 5, 7, and 9. The latter maybe due to the different proteolytic activity of this enzyme [17]. The same correlation was also determined for other animal and vegetable matrices, such as proteins originating from salmon and egg hydrolyzed with Alcalase [22,23], rice endosperm and chickpea hydrolyzed respectively with endoprotease and Alcalase [21–24]. At pH 5 we did not determine any correlation, due to the instable peptide solubility near the isoelectric point. The same phenomenon was underlined for sardinella hydrolyzed with Alcalase, where no correlation between DH and solubility was determined near the isoelectric point determined at pH 3 and 4 [25].

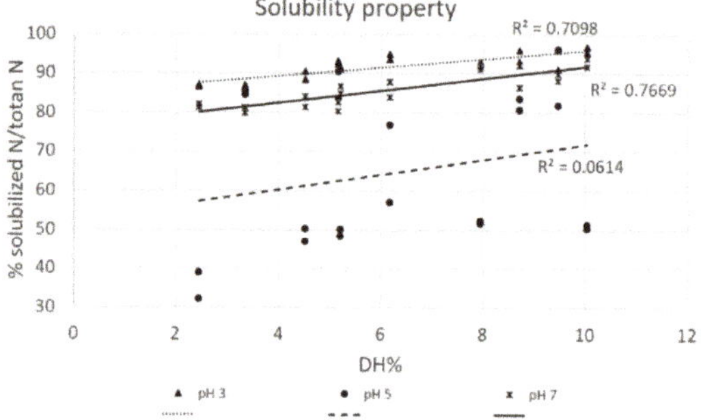

Figure 2. Solubility properties of 1% of protein hydrolysates (average ionic strength 2.8 ± 0.2 mS/cm) reported as % of solubilized *N* in function of DH%.

3.3.2. Emulsifying Activity

Emulsions consist of two immiscible phases, and the oil in water emulsions are the most common in food products. In the present work, insect protein hydrolysates were characterized by an incomplete emulsion layers, which, nevertheless, have been considered in order to give an indication of trends. The control sample presented the highest emulsifying ability (EA), which significantly decreased immediately after the enzyme addition. In particular, as illustrated in Figure 3, a negative correlation ($r = 0.903$) was recorded between emulsifying activity and DH% underlining the loss of this property during the hydrolysis, till manifesting the complete coalescence of oil droplets and phases separation.

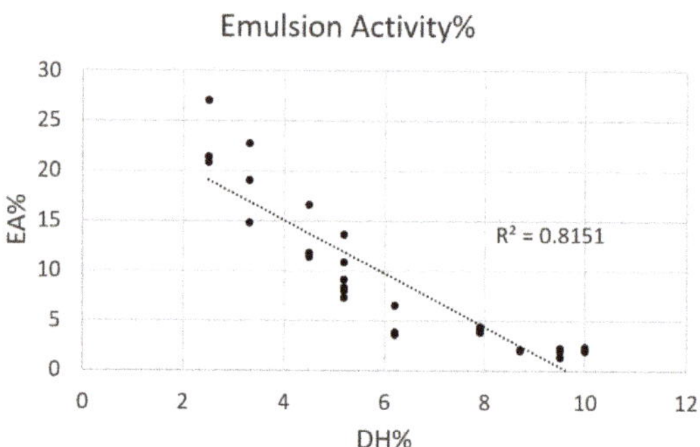

Figure 3. Emulsifying activity of 0.1% of protein samples (average pH 7.5 ± 0.2 and ionic strength of 1.3 ± 0.1 mS/cm) reported in function of DH%. The analysis was done in triplicate. Note: emulsification layers were incomplete—values only given as indication of a trend, not for direct comparison with literature data.

The protein emulsification mechanism is attributed to their migration to the surface of freshly formed oil droplets during homogenization. Proteins are able to form a protective film promoting oil-in-water emulsion due to their duality for the presence of hydrophilic and hydrophobic groups [26]. The shorter peptides, released from proteins after hydrolysis, may migrate to the interface oil/water more rapidly than proteins, but they are less efficient to reduce the interfacial tension between the two phases and to create a strong interfacial film round oil droplets [26]. Similar results were found by Quaglia and Orban [27], Kristinsson and Rasco [28], and Purschke et al. [17] working with sardine, salmon, and cricket, respectively. In this last case, a significant loss in emulsifying activity (mainly at acidic pH) in comparison to the control sample was observed when cricket protein flour was hydrolyzed with Neutrase and Flavourzyme. At neutral pH the emulsifying activity of protein hydrolysates obtained from the two enzyme activities displayed an opposite behavior. In fact, Neutrase negatively affected the emulsify ability, while Flavourzyme improved it, maybe due to the different specificity of proteolytic cleavage. Hall et al. [18] identified the same correlation when 0.5% of Alcalase/substrate concentrations were used, while at higher enzyme concentrations (3%) no correlation was identified.

3.3.3. Oil Holding Capacity

The oil holding capacity (OHC) of control and hydrolysates was correlated to the different DH% and reported in Figure 4. The lowest value, 1.4 ± 0.1 g oil/g sample, was determined for the control sample. This amount increased with an increasing DH% till the 2.2 ± 0.2 g oil/g sample only after 180 min of reaction and has expected to continuously increase. In fact, as determined in the preliminary test, after 300 min of reaction the capacity to hold oil increased till 6.7 ± 0.6 g oil/g sample.

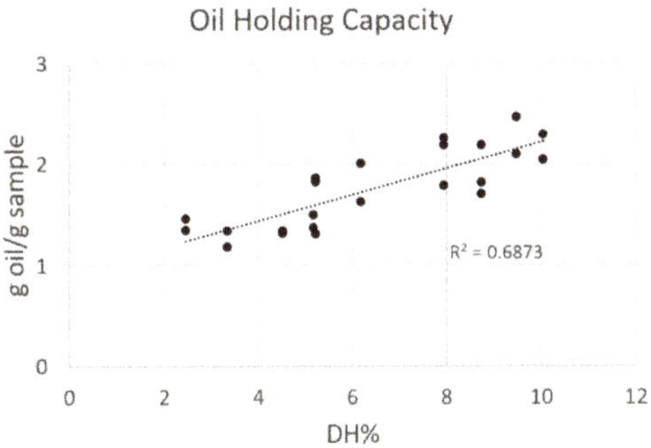

Figure 4. Oil holding capacity of the different hydrolysates and controls produced expressed as g oil/g sample in function of DH%. The analysis was done in triplicate.

This raise could be explained by the modification of protein structure and the exposure of more hydrophobic side chain of amino acids, which, before the hydrolysis, were trapped in the protein folding, promoting the physical entrapment of oil [29]. Purschke et al. demonstrated the improvement in OHC of protein hydrolysates when compared to the unhydrolyzed samples even if no correlations with DH were calculated [17]. Souissi et al. obtained results in agreement with Purschke, demonstrating the increased ability of sardinella protein hydrolysate to hold oil, if compared to intact proteins [25]. Several authors observed in plant protein hydrolysates an initial increase in OHC upon hydrolysis [28]. However, a critical point exists at which the liberation of polar ionizable groups has a larger impact on OHC, than the increased availability of hydrophobic regions [30].

3.3.4. Foaming Capacity

Foam is a colloidal system comprising a continuous aqueous phase with dispersed gas. Proteins, due to their amphiphilic nature, represent a good surfactant with the hydrophobic portion oriented to the air bubbles and the hydrophilic part to the watery phase [24]. In Figure 5, the foaming capacity of the different hydrolysates, with a comparable average pH and ionic strength, was plotted to their DH%, along with the control and an exponential correlation was determined. The control sample did not present foaming capacity, which on the contrary seemed to appear in the hydrolysates collected after 60 min of hydrolysis, reaching their maximum after 180 min of hydrolysis. Nevertheless after 300 min of hydrolysis no foam capacity was determined, defining this exponential trend only up to limit value of DH, which was between 10% to 15%.

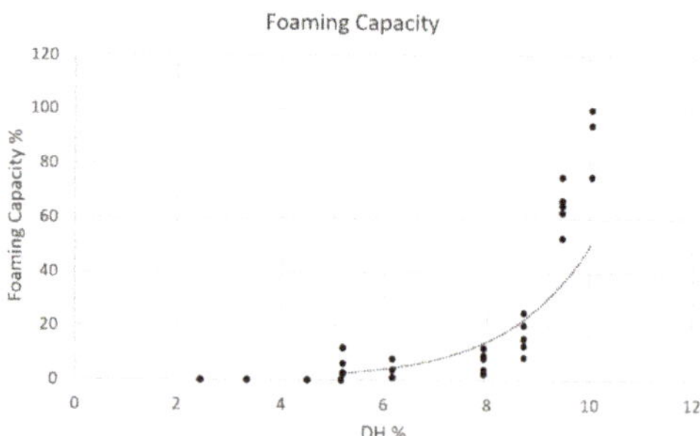

Figure 5. Foaming capacity of 1% of protein samples (average pH 7.9 ± 0.3 and ionic strength of 2.8 ± 0.2 mS/cm) reported in function of DH% with the trendline. The analysis was done in triplicate.

An increase in DH% probably led to a more pronounced amphiphilicity, which may enhance the interfacial interaction with air bubbles until a DH of 10%. However, even if a foaming capacity was observed, all formed foams did not display any stability property and after 1 min they started to collapse. These results were in line with what reported for other protein hydrolysates obtained from edible insects [17,18].

3.4. Potential Application

The enzymatic hydrolysis led to an improvement of solubility and oil holding ability, while it reduced the emulsifying property. Furthermore, it was determined that functional properties could be tailored according to their DH value. Insect protein hydrolysates, compared to casein and egg white, were characterized by higher OHC. This is not of a secondary importance since the oil holding capacity is a property appreciated especially for the meat industry. In fact, the higher the OHC, the higher the ability of a food or feed formula to retain flavors and improve the palatability [31]. The high solubility property at pH 3 and 7, and the increase in oil holding capacity demonstrated the potential for using LM hydrolysates in acidic food systems, such as sports beverages and acidified sauces, and in feed system as replacement of milk for weanling animals. The enzymatic assisted extraction affected the ability of LM proteins to act as surfactants for marinating oil and air droplets dispersed in an aqueous solution. In fact, immediately after 30 min from the enzyme addition the emulsifying ability of LM protein hydrolysates started to significantly decrease. Furthermore, the emulsify property evaluated for all the insect samples were far to the ones calculated for egg white and casein, which are known to be good emulsifiers. For this reason, for feed and food emulsions, which will be prepared with insect hydrolysates, it could be necessary to add emulsifiers to stabilize the formulation. LM proteins displayed the ability to foam after 60 min of enzymatic hydrolysis, but after 300 min of hydrolysis no foaming ability was determined. The increasing in hydrolysis time, and so DH%, could have impaired the ability of peptides to arrange round air bubbles, due to the shorter length. The maximum foam ability was determined after 180 min of hydrolysis, overcoming the foam ability of casein. Nevertheless, the formed foams did not show stability and, immediately after 1 min from their constitution, they started to collapse. For this reason, insect protein hydrolysate could not be used as foaming agents but, due to the absence of foam stability, could be included in food or feed beverages.

4. Conclusions

In conclusion, this work provides for the first time information about the influence of DH% on techno-functional properties of protein hydrolysates produced from LM under process condition that are scalable for the industrial production. These results demonstrated that the functional properties of LM could be tailored by enzymatic assisted extraction. Nevertheless, deep investigations are needed

Foods **2020**, *9*, 381

in order to evaluate how these properties could be affected when insect protein-based ingredients will be included in food/feed complex matrices.

Author Contributions: The authors contributed to the article as follows: G.L.: conceptualization, methodology, formal analysis, writing original draft preparation and editing; L.S.: methodology, review; A.C.: methodology, supervision, review; S.S.: data curation, supervision, writing—review; L.B.: data curation, supervision, writing—review, project administration. All authors have read and agreed to the published version of the manuscript.

Funding: This project received funding from the Bio Based Industries Joint Undertaking under the European Union's Horizon 2020 research and innovation program under grant agreement No. 720715 (InDIRECT project).

Acknowledgments: Natasja Gianotten (Protifarm) is gratefully acknowledged for having provided the lesser mealworm larvae, while Simons Queenie and Bert Van den Bosch (VITO) for their help with some experimental aspects.

Conflicts of Interest: The authors declare no conflict of interest.

References

1. FAO. How to Feed the World in 2050. *Insights Expert Meet FAO* **2009**, *1*, 1–35. [CrossRef]

2. Boland, M.J.; Rae, A.N.; Vereijken, J.M.; Meuwissen, M.P.M.; Fischer, A.R.H.; van Boekel, M.A.J.S.; Hendriks, W.H. The future supply of animal-derived protein for human consumption. *Trends Food Sci. Technol.* **2013**, *29*, 62–73. [CrossRef]

3. Yi, L.; Lakemond, C.M.M.; Sagis, L.M.C.; Eisner-Schadler, V.; van Huis, A.; Boekel, M.A.J.S.V. Extraction and characterisation of protein fractions from five insect species. *Food Chem.* **2013**, *141*, 3341–3348. [CrossRef] [PubMed]

4. Caligiani, A.; Marseglia, A.; Leni, G.; Baldassarre, S.; Maistrello, L.; Dossena, A.; Sforza, S. Composition of black soldier fly prepupae and systematic approaches for extraction and fractionation of proteins, lipids and chitin. *Food Res. Int.* **2018**, *105*, 812–820. [CrossRef]

5. Rumpold, B.A.; Schlüter, O.K. Potential and challenges of insects as an innovative source for food and feed production. *Innov. Food Sci. Emerg. Technol.* **2013**, *17*, 1–11. [CrossRef]

6. Regulation (EU) 2015/2283 of the European Parliament and of the Council of 25 November 2015 on novel foods, amending Regulation (EU) No 1169/2011 of the European Parliament and of the Council and repealing Regulation (EC) No 258/97 of the European Parliament and of the Council and Commission Regulation (EC) No 1852/2001. *Off. J. Eur. Union* **2015**, *L 327*, 1–22. Available online: http://data.europa.eu/eli/reg/2015/2283/oj (accessed on 20 August 2019).

7. Commission Regulation (EU) 2017/893 of 24 May 2017 amending Annexes I and IV to Regulation (EC) No 999/2001 of the European Parliament and of the Council and Annexes X, XIV and XV to Commission Regulation (EU) No 142/2011 as regards the provisions on processed animal protein. *Off. J. Eur. Union* **2017**, *L 138*, 92–115.

8. Regulation (EU) No 56/2013 of 16 January 2013 Amending Annexes I and IV to Regulation (EC) No 999/2001 of the European Parliament and of the Council Laying Down Rules for the Prevention, Control and Eradication of Certain Transmissible Spongiform Encephalopathies. *Off. J. Eur. Union* **2013**, *L 21*, 3–16. Available online: http://data.europa.eu/eli/reg/2013/56/oj (accessed on 20 August 2019).

9. Yi, L.; Van Boekel, M.A.J.S.; Lakemond, C.M.M. Extracting Tenebrio molitor protein while preventing browning: Effect of pH and NaCl on protein yield. *J. Insects Food Feed* **2017**, *3*, 21–31. [CrossRef]

10. Bußler, S.; Rumpold, B.A.; Jander, E.; Rawel, H.M.; Schlüter, O.K. Recovery and techno-functionality of flours and proteins from two edible insect species: Mealworm (Tenebrio molitor) and black soldier fly (Hermetia illucens) larvae. *Heliyon* **2016**, *2*, e00218. [CrossRef]

11. Zhao, X.; Vázquez-gutiérrez, J.L.; Johansson, D.P.; Landberg, R.; Langton, M. Yellow Mealworm Protein for Food Purposes—Extraction and Functional Properties. *PLoS ONE* **2016**, *11*, e0147791. [CrossRef] [PubMed]

12. Soetemans, L.; Uyttebroek, M.; D'Hondt, E.; Bastiaens, L. Use of Organic Acids to Improve Fractionation of the Black Soldier Fly larvae juice into Lipid and Protein Enriched Fractions. *Eur. Food Res. Technol.* **2019**, *245*, 2257–2267. [CrossRef]

13. Nielsen, P.N.; Olsen, H.S. Enzymic modification of food protein. In *Enzymes in Food Technol.*; Whitehurst, R.J., van Oort, M., Eds.; Blackwell Publishing Ltd.: Hoboken, NJ, USA, 2002.

14. Leni, G.; Soetemans, L.; Jacobs, J.; Depraetere, S.; Gianotten, N.; Bastiaens, L.; Caligiani, A.; Sforza, S. Protein hydrolysates from Alphitobius diaperinus and Hermetia illucens larvae treated with commercial proteases. *J. Insects Food Feed.* Accepted (2 March 2020).

15. Tavano, O.L. Protein hydrolysis using proteases: An important tool for food biotechnology. *J. Mol. Catal. B Enzym* **2013**, *90*, 1–11. [CrossRef]

16. Nongonierma, A.B.; FitzGerald, R.J. Unlocking the biological potential of proteins from edible insects through enzymatic hydrolysis: A review. *Innov. Food Sci. Emerg. Technol.* **2017**, *43*, 239–252. [CrossRef]

17. Purschke, B.; Meinlschmidt, P.; Horn, C.; Rieder, O.; Jäger, H. Improvement of techno-functional properties of edible insect protein from migratory locust by enzymatic hydrolysis. *Eur. Food Res. Technol.* **2018**, *244*, 999–1013. [CrossRef]

18. Hall, F.G.; Jones, O.G.; O'Haire, M.E.; Liceaga, A.M. Functional properties of tropical banded cricket (Gryllodes sigillatus) protein hydrolysates. *Food Chem.* **2017**, *224*, 414–422. [CrossRef]

19. Butré, C.I.; Sforza, S.; Gruppen, H.; Wierenga, P.A. Introducing enzyme selectivity: A quantitative parameter to describe enzymatic protein hydrolysis. *Anal. Bioanal. Chem.* **2014**, *406*, 5827–5841. [CrossRef]

20. Zayas, J.F. Solubility of Proteins. In *Functionality of Proteins in Food*; Springer: Berlin/Heidelberg, Germany, 1997.

21. Ghribi, A.M.; Gafsi, I.M.; Sila, A.; Blecker, C.; Danthine, S.; Attia, H.; Bougatef, A.; Besbes, S. Effects of enzymatic hydrolysis on conformational and functional properties of chickpea protein isolate. *Food Chem.* **2015**, *187*, 322–330. [CrossRef]

22. Gbogouri, G.A.; Linder, M.; Fanni, J.; Parmenter, M.C. Food Chemistry and Toxicology Influence of Hydrolysis Degree. *J. Food Sci.* **2004**, *69*, 615–622. [CrossRef]

23. Bao, Z.; Zhao, Y.; Wang, X.; Chi, Y.J. Effects of degree of hydrolysis (DH) on the functional properties of egg yolk hydrolysate with alcalase. *J. Food Sci. Technol.* **2017**, *54*, 669–678. [CrossRef] [PubMed]

24. Nisov, A.; Ercili-Cura, D.; Nordlund, E. Limited hydrolysis of rice endosperm protein for improved techno-functional properties. *Food Chem.* **2020**, *302*, 125274. [CrossRef] [PubMed]

25. Souissi, N.; Bougatef, A.; Triki-Ellouz, Y.; Nasri, M. Biochemical and functional properties of sardinella (Sardinetta aurita) by-product hydrolysates. *Food Technol. Biotechnol.* **2007**, *45*, 187–194.

26. Van der Ven, C.; Gruppen, H.; De Bont, D.B.A.; Voragen, A.G.J. Emulsion properties of casein and whey protein hydrolysates and the relation with other hydrolysate characteristics. *J. Agric. Food Chem.* **2001**, *49*, 5005–5012. [CrossRef] [PubMed]

27. Quaglia, G.B.; Orban, E. Influence of Enzymatic Hydrolysis on Structure and Emulsifying Properties of Sardine (Sardina pilchardus) Protein Hydrolysates. *J. Food Sci.* **1990**, *55*, 1571–1573. [CrossRef]

28. Kristinsson, H.G.; Rasco, B.A. Fish protein hydrolysates: Production, biochemical and functional prop-erties. *Crit. Rev. Food Sci. Nutr.* **2000**, *40*, 43–81. [CrossRef] [PubMed]

29. Mune, M.A. Functional Properties of Hydrolysates. *J. Food Process. Preserv.* **2015**, *39*, 2386–2392. [CrossRef]

30. Wouters, A.G.B.; Rombouts, I.; Fierens, E.; Brijs, K.; Delcour, J.A. Relevance of the Functional Properties of Enzymatic Plant Protein Hydrolysates in Food Systems. *Compr. Rev. Food Sci. Food Saf.* **2016**, *15*, 786–800. [CrossRef]

31. Kinsella, J.E.; Melachouris, N. Functional properties of proteins in foods: A survey. *Crit. Rev. Food Sci. Nutr.* **1976**, *7*, 219–280. [CrossRef]

Article

Hepatoprotective Effects of Steamed and Freeze-Dried Mature Silkworm Larval Powder against Ethanol-Induced Fatty Liver Disease in Rats

Da-Young Lee [1], Kyung-Sook Hong [1], Moon-Young Song [1], Sun-Mi Yun [1], Sang-Deok Ji [2], Jong-Gon Son [2] and Eun-Hee Kim [1,*]

[1] College of Pharmacy and Institute of Pharmaceutical Sciences, CHA University, Seongnam 13488, Korea; angela8804@naver.com (D.-Y.L.); sukihong83@gmail.com (K.-S.H.); wso219@naver.com (M.-Y.S.); sun21mi@naver.com (S.-M.Y.)

[2] Department of Agricultural Biology, National Institute of Agricultural Science, Rural Development Administration, Wanju 55365, Korea; ji35879@daum.net (S.-D.J.); sonjg@korea.kr (J.-G.S.)

* Correspondence: ehkim@cha.ac.kr; Tel.: +82-31-881-7179

Received: 12 February 2020; Accepted: 29 February 2020; Published: 4 March 2020

Abstract: Silkworm, *Bombyx mori*, contains high amounts of beneficial nutrients, including amino acids, proteins, essential minerals, and omega-3 fatty acids. We have previously reported a technique for producing steamed and freeze-dried mature silkworm larval powder (SMSP), which makes it easier to digest mature silkworm. In this study, we investigated the preventive effects of SMSP on alcoholic fatty liver disease and elucidated its mechanism of action. Male Sprague-Dawley rats treated with SMSP (50 mg/kg) or normal diet (AIN-76A) were administered 25% ethanol (3 g/kg body weight) by oral gavage for 4 weeks. SMSP administration for 4 weeks significantly decreased hepatic fat accumulation in ethanol-treated rats by modulating lipogenesis and fatty acid oxidation-related molecules such as sirtuin 1, AMP-activated protein kinase, and acetyl-CoA carboxylase 1. Moreover, SMSP administration significantly diminished the levels of triglyceride in liver tissues by as much as 35%, as well as lowering the serum levels of triglyceride, gamma glutamyl transpeptidase, alanine transaminase, and aspartate aminotransferase in ethanol-treated rats. SMSP supplementation also decreased the pro-inflammatory tumor necrosis factor-alpha and interleukin 1 beta levels and cytochrome P450 2E1 generating oxidative stress. These results suggest that SMSP administration may be possible for the prevention of alcoholic liver disease.

Keywords: steamed and freeze-dried mature silkworm larval powder; alcoholic fatty liver; ethanol; lipogenesis; fatty acid oxidation; Sprague-Dawley rats

---

1. Introduction

Alcohol-related liver disorder is one of the most common diseases in the world [1]. Chronic alcohol consumption leads to liver diseases including hepatic steatosis, hepatocellular injury, and inflammation, and can cause liver fibrosis, cirrhosis, and hepatocellular carcinoma [2,3]. In particular, as one of the most common forms of alcoholic liver disease, hepatic steatosis is generally considered to be a benign and reversible process that can progress to a more serious condition [4,5]. AMP-activated protein kinase (AMPK) is known to be a major factor of energy metabolism and is involved in lipid metabolism [6]. When AMPK activity is inhibited by several causes, lipogenic transcription factors increase, resulting in fat generation [7]. In addition, AMPK phosphorylates the serine residues of acetyl CoA carboxylase 1 (ACC1), activating catabolic pathways, including fatty acid oxidation [8]. Thus, the regulation of lipogenesis and fatty acid oxidation through AMPK signaling is a therapeutic mechanism for alcoholic steatosis.

After drinking, alcohol is metabolized by cytochrome P450 2E1 (CYP2E1), an enzyme mainly expressed in the liver [3]. Increased CYP2E1 by alcohol consumption causes oxidative stress and inflammation in the kupffer cells, resulting in liver damage [9,10]. In the liver, kupffer cells play a key role in an innate immune system and they accelerate the secretion of pro-inflammatory cytokines, such as tumor necrosis factor-alpha (TNF-α), interleukin (IL)-1β, and IL-6 [11]. Recently, it has been found that these cytokines induce the imbalance of lipid metabolism in the liver and cause the more serious liver diseases [12]. Limuro et al. reported that inhibition of TNF-α inhibited fatty liver in alcohol-treated rats [13]. Moreover, hepatic injury and steatosis induced by alcohol were alleviated in TNF-α knockout mice [13]. Other cytokines, such as IL-1β and IL-6, have been also reported to interfere with lipid metabolism in the liver [14]. Therefore, suppression of the pro-inflammatory cytokines production might have a therapeutic effect against hepatic inflammation and steatosis induced by alcohol consumption.

Recently, there has been interest in edible insects, and they have been highlighted as a future food [15]. The silkworm, *Bombyx mori*, contains high amounts of protein, omega-3 fatty acids, vitamins, and minerals [16]. Silkworms are traditionally used for the treatment of spasms, phlegm, and flatulence [15]. In addition, the silkworm has recently received scientific attention and several studies have reported its health advantages against liver injury [16], Parkinson's disease [17], and diabetic hyperglycemia [18]. However, after the third day of the fifth instar, the protein composition of silkworm glands rapidly increases, hence, there is a problem insofar as the biological activity is lowered [19]. In addition, dried or refrigerated silkworms are difficult to consume as food because the silk protein is denatured to a high strength [19]. Therefore, Ji et al. have developed a new method of processing silkworms into an easy-to-eat form, in which the mature silkworms are steamed for 130 min at 100 °C before lyophilizing and grinding [19]. We previously reported that steamed and freeze-dried mature silkworm larval powder (SMSP, 0.1, 1, and 10 g/kg) alleviated fatty liver disease in rats treated with alcohol [20]. However, in terms of human dosage, the dose of SMSP reported in the previous studies is burdensome to consume large amounts of silkworm powder per day. In the present study, we aimed to determine whether low dosage of SMSP attenuates hepatic steatosis in rats with ethanol-induced fatty liver and to explore the potential molecular mechanisms of the health benefits of SMSP, focusing on the gene expression involved in lipid metabolism and inflammation in liver.

2. Materials and Methods

2.1. Preparation of Steamed Mature Silkworm Larval Powder

SMSP was made as previously described [19]. Briefly, live mature larvae of *Bombyx mori*, 7th day of 5th instar silkworms, were immediately steamed for 130 min at 100 °C using an electric pressure-free cooking machine (KumSeong Ltd., Boocheon, Korea) and freeze-dried using freeze-drier (FDT-8612, Operon Ltd., Kimpo, Korea) for 24 h. Silkworms were then grinded using a disk mill (Disk Mill01, Korean Pulverizing Machinery Co. Ltd., Incheon, Korea) and a hammer mill (HM001, Korean Pulverizing Machinery Co. Ltd., Incheon, Korea). And the SMSP was stored at −50 °C and then used for oral gavage for rats.

2.2. Animal and Experimental Design

All animal experiment procedures were conducted in accordance with the guidelines and approval of the Institutional Animal Care and Use Committees (IACUC) of the CHA University (reference number: 180002). Three-week-old male Sprague-Dawley (SD) rats weighing 120–130 g were purchased from Orient bio (Seoul, Korea). All animals were housed in a standardized laboratory environment with a 12 h light/dark cycle at constant temperature of 24 °C. After a week of acclimatization, all animals randomly divided into 3 groups (*n* = 8). The normal group was fed with rodent chow diet (Haran 2018s) and orally received 0.2 mL of distilled water. The ethanol-treated group was fed with rodent chow diet and orally received 25% ethanol (3 g/kg body weight) once a day for four weeks.

The SMSP-treated group was also fed rodent chow diet and orally received 25% ethanol and SMSP (50 mg/kg body weight) once a day for four weeks. Rats were monitored daily for body weight change for the experimental period. At the end of the experiment, all rats were sacrificed at 4 weeks by carbon dioxide anesthesia. Blood was drawn from the abdominal aorta into a heparin tube. Serum was subsequently obtained by centrifuging the blood at 3000 rpm for 15 min at 4 °C. The livers were excised, rinsed with PBS, and weighed, and the portion of liver was fixed in 10% formalin. Serum and liver samples were stored at −80 °C until analysis.

2.3. Serum Analysis

The serum levels of triglyceride, gamma glutamyl transpeptidase (GGT), alanine aminotransferase (ALT), and aspartate aminotransferase (AST) were analyzed using Hitachi automatic analyzer 7600-210 (Hitachi High-Technologies Corporation, Tokyo, Japan). The serum concentration of TNF-α and IL-1β were measured by using a commercial rat ELISA kit (R&D Systems, Minneapolis, MN, USA) following manufacturer's instruction.

2.4. Hepatic Triglyceride Analysis

For the measurement of triglyceride levels in liver tissues, we used the commercial triglyceride assay kit (AB 65336, Abcam, Cambridge, UK). The liver triglyceride was determined as following the manufacturer's instruction.

2.5. Histological Analysis in the Liver

Formalin-fixed liver samples were embedded in paraffin, sliced at 5 μm, followed by sectioning and hematoxylin and eosin (H&E) staining by standard procedures. Histopathological scoring was assessed by an experienced pathologist, who was blinded to the treatment groups. Levels of fatty infiltration and steatosis were graded as 0 point for no hepatocytes affected, 0.5 point for slightly affected (0–5%), 1 point for mildly (5–20%), 2 points for moderately (20–50%), and 3 points for severely (>50%) [21].

2.6. RNA Isolation and Gene Expression Analysis

Total mRNA was isolated from the rat livers using Trizol reagent (Invitrogen, Carlsbad, CA, USA) and cDNA was synthesized using Labopass cDNA synthesis kit (Cosmogenetech Co., Ltd., Seoul, Korea) according to the manufacturer's instructions. The mRNA levels were analyzed by quantitative real-time PCR (qRT-PCR). The qRT-PCR was assessed as previously reported [22] and was performed on a ViiA™ 7 real-time PCR system (Life Technologies Corporation, Carlsbad, CA, USA) using Luna universal qPCR master mix (New England Biolabs, Beverly, MA, USA). 18S ribosomal RNA (18s rRNA) were used as an internal control. The primer sets for qRT-PCR and RT-PCR are listed in Table 1.

Table 1. List of primers.

Gene	Forward	Reverse	Size (bp)
CYP2E1	AAA CAG GGT AAT GAG GCC CG	AGG CTG GCC TTT GGT CTT TT	75
IL-1β	CAC CTT CTT TTC CTT CAT CTT TG	GTC GTT GCT TGT CTC TCC TTG TA	240
AdipoR1	AGG TCA AAC GTG ACG GCT C	TTA GGC CTG TCG ACT CTC CA	71
SIRT1	CCC AGA TCC TCA AGC CAT GTT	TTG TGT GTG TGT TTT TCC CCC	120
SREBP1c	CCA TGG ACG AGC TAC CCT TC	GGC ACT GGC TCC TCT TTG AT	382
PPARγ	AGA AGG CTG CAG CGC TAA AT	GGC CTG TTG TAG AGT TGG GT	405
FAS	TCG ACT TCA AAG GAC CCA GC	GGC AAT ACC CGT TCC CTG AA	228
ACC1	CTT GGG GTG ATG CTC CCA TT	GCT GGG CTT AAA CCC CTC AT	116
PGC-1α	TAA ACT GAG CTA CCC TTG GG	CTC GAC ACG GAG AGT TAA AGG AA	89
18s rRNA	GCA ATT ATT CCC CAT GAA CG	GGC CTC ACT AAA CCA TCC AA	111

2.7. Western Blotting

Liver tissues were grinded with cell lysis buffer containing protease inhibitor (Roche Applied Science, Mannheim, Germany) and samples were incubated on ice with frequent vortexing for 10 min and centrifuged for 15 min at 13,000 rpm. The protein concentration of each supernatant was quantified using Pierce^{TM} BCA protein assay kit (Thermo Fisher Scientific, Waltham, MA, USA) in accordance with the manufacturer's instructions. The proteins were loaded onto sodium dodecyl sulfate polyacrylamide gel electrophoresis (SDS-PAGE), and transferred to polyvinylidene fluoride membranes (Millipore, Burlington, MA, USA). After transfer, membranes were blocked with bovine serum albumin (BSA) solution and probed with the specified primary antibodies (diluted 1:1000) overnight at 4 °C. The membranes were washed and incubated with the appropriate secondary antibodies for 40 min. The blots were then developed using an enhanced chemiluminescence system (Thermo Fisher Scientific, Waltham, MA, USA). Antibodies for p-AMPK, AMPK, peroxisome proliferator-activated receptor gamma (PPARγ) and β-actin were purchased from Santa Cruz Biotechnology (Santa Cruz, CA, USA). Antibodies for ACC1 and p-ACC1 were purchased from Cell Signaling Technology (Danvers, MA, USA).

2.8. Statistical Analysis

All data are expressed as means ± standard deviation (SD). Statistical analysis was performed using one-way analysis of variance (ANOVA). Statistical significance was accepted at $p < 0.05$.

3. Results

3.1. SMSP Supplementation Alleviates Hepatic Steatosis in Ethanol-Treated Rats

To examine the hepatoprotective role of SMSP, we used an ethanol-induced hepatic steatosis rat model. The ethanol (3 g/kg) and SMSP (50 mg/kg) were orally injected into SD rats for 4 weeks. SMSP administration for 4 weeks significantly reduced the total liver weight compared with ethanol-treated rats without affecting body weight change (Figure 1A,B). The SMSP group, as compared with the ethanol group, had a reduction in liver weight to body weight ratio (Figure 1C). Ethanol treatment for 4 weeks successfully induced fatty liver and liver injury in rats, which were manifested by significant increases in serum triglyceride, GGT, ALT, and AST activities compared with those of normal diet-fed rats (Figure 2B–E). In the meantime, SMSP significantly reversed the ethanol-treated hepatic accumulation of triglyceride by as much as 35% (Figure 2A), as well as lowering the serum triglyceride, GGT, ALT, and AST activities by 15%, 41%, 8.3%, and 9.4%, respectively (Figure 2B–E). In addition, steatosis scores were evaluated in H&E staining images of liver tissues from all groups. As a result, hepatic lipid accumulation was remarkably increased in ethanol-treated rats (Figure 3A). As shown in Figure 3B, the ethanol-induced elevation of the steatosis score was significantly normalized in the SMSP-treated rats.

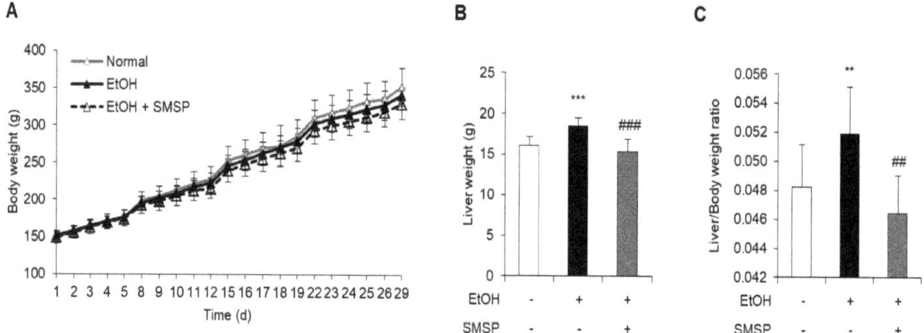

Figure 1. Effect of steamed and freeze-dried mature silkworm larval powder (SMSP) on body weight and liver weight in rats treated with ethanol (EtOH) for 4 weeks. (**A**) Body weight changes, (**B**) liver weight at the end of experiment, and (**C**) liver/body weight ratio were measured. The data represent mean ± SD ($n = 8$); ** $p < 0.01$ and *** $p < 0.001$ vs. normal group; ## $p < 0.01$ and ### $p < 0.001$ vs. EtOH group.

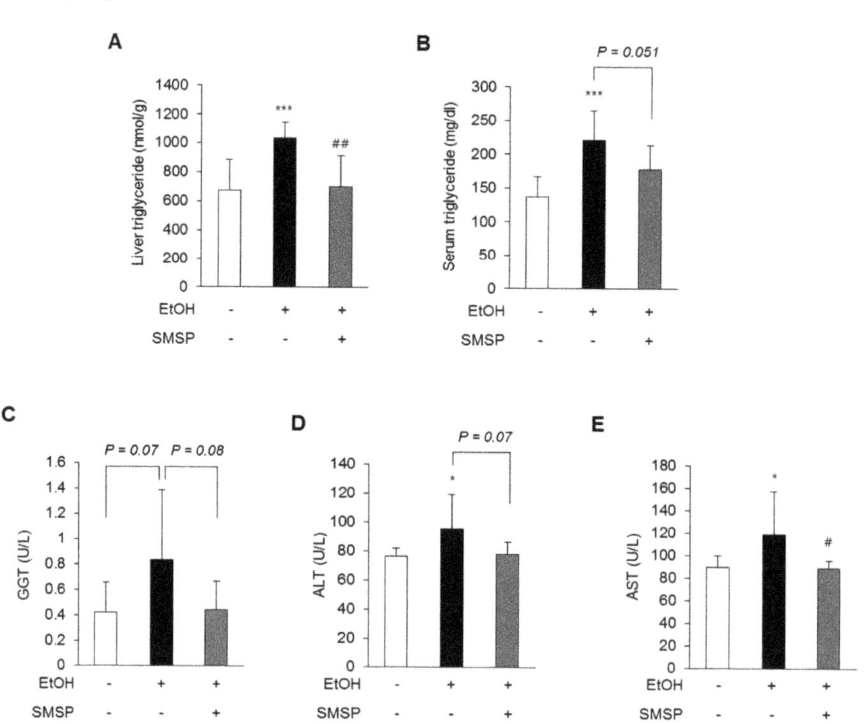

Figure 2. SMSP administration alleviates hepatic steatosis in EtOH-treated rats. (**A**) Hepatic triglyceride and serum levels of (**B**) triglyceride, (**C**) gamma-glutamyl transpeptidase (GGT), (**D**) alanine aminotransferase (ALT), (**E**) aspartate aminotransferase (AST) were measured. The data represent mean ± SD ($n = 8$); * $p < 0.05$ and *** $p < 0.001$ vs. normal group; # $p < 0.05$ and ## $p < 0.01$ vs. EtOH group.

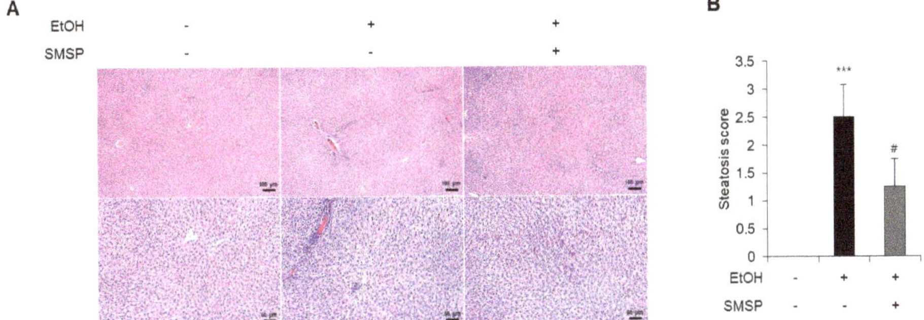

Figure 3. SMSP administration attenuates hepatic lipid accumulation in EtOH-treated rats. (**A**) Histopathological sections of liver were stained with H&E (magnification, ×40 and 100). (**B**) Hepatic steatosis scores were quantified from H&E-stained sections. The data represent mean ± SD ($n = 4$); *** $p < 0.001$ vs. normal group; # $p < 0.05$ vs. EtOH group.

3.2. SMSP Supplementation Attenuates Inflammation in Ethanol-Treated Rats

Chronic alcohol consumption leads to liver injury, fat metabolism, inflammation, and hepatocellular carcinogenesis [3]. In addition, chronic alcohol consumption induces the production of reactive oxygen species (ROS), regulated by cytochrome P450 2E1 (CYP2E1), that lead to the production of reactive aldehydes with potent pro-inflammatory properties [11]. Based on this rationale, we analyzed the serum levels of TNF-α and IL-1β. As shown in Figure 4A,B, the TNF-α and IL-1β were significantly increased in ethanol-treated rats, while SMSP treatment attenuated the serum levels of TNF-α and IL-1β. Moreover, the hepatic mRNA expression of the CYP2E1 and IL-1β were significantly decreased by SMSP administration (Figure 4C,D). These results indicate that SMSP supplementation ameliorates liver inflammation induced by ethanol.

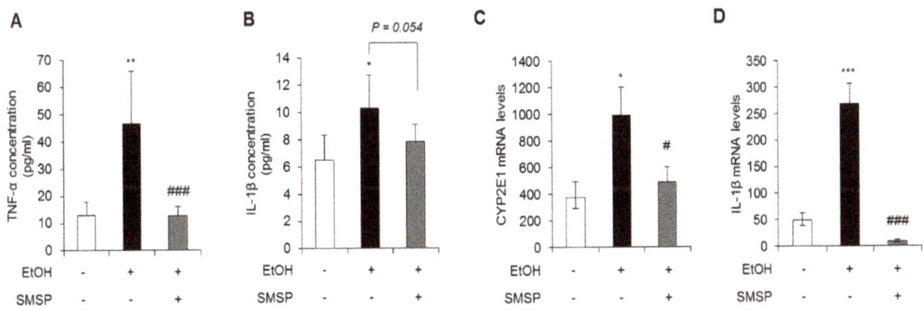

Figure 4. SMSP administration attenuates inflammation in EtOH-treated rats. Serum levels of (**A**) tumor necrosis factor-alpha (TNF-α) and (**B**) interleukin 1 beta (IL-1β) were measured by ELISA. The mRNA expressions of (**C**) cytochrome P450 2E1 (CYP2E1) and (**D**) IL-1β were detected by qRT-PCR in liver tissues from each group. The data represent mean ± SD ($n = 4$); * $p < 0.05$, ** $p < 0.01$, and *** $p < 0.001$ vs. normal group; # $p < 0.05$ and ### $p < 0.001$ vs. EtOH group.

3.3. SMSP Supplementation Improves Lipid Metabolism in Ethanol-Treated Rats

To examine the underlying molecular mechanism, we examined the effect of SMSP supplementation on hepatic lipid metabolism such as lipogenesis and free fatty acid (FFA) oxidation. AMPK is one of the key activators of the lipogenesis and FFA oxidation catabolism [23]; we tested whether prolonged SMSP supplementation could modulate the AMPK pathways in the liver of alcohol-treated rats. Western blot results showed that SMSP administration significantly increased the phosphorylation of

AMPK reduced by ethanol treatment. However, PPARγ, the well-known lipogenic factor, significantly decreased in the livers of SMSP-fed rats (Figure 5A). The expression of lipogenic nuclear transcription factors, such as sterol regulatory element-binding protein 1c (SREBP1c), PPARγ, and fatty acid synthase (FAS), were down-regulated in the liver of SMSP-treated rats compared with the ethanol-treated rats (Figure 5B). However, SMSP supplementation restored the mRNA levels of adiponectin receptor 1 (AdipoR1) and sirtuin 1 (Sirt1) in the livers of ethanol-treated rats (Figure 5B).

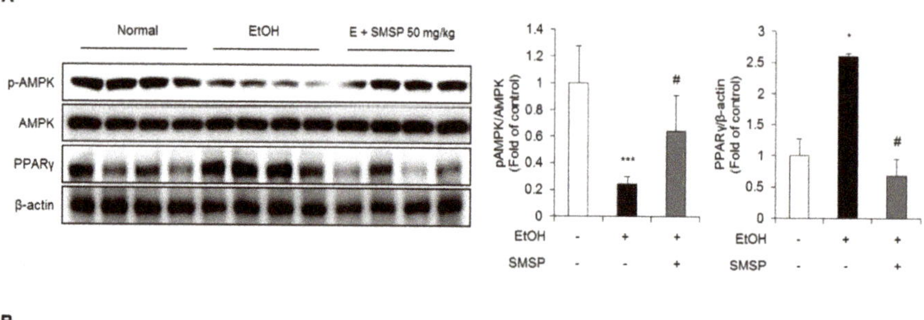

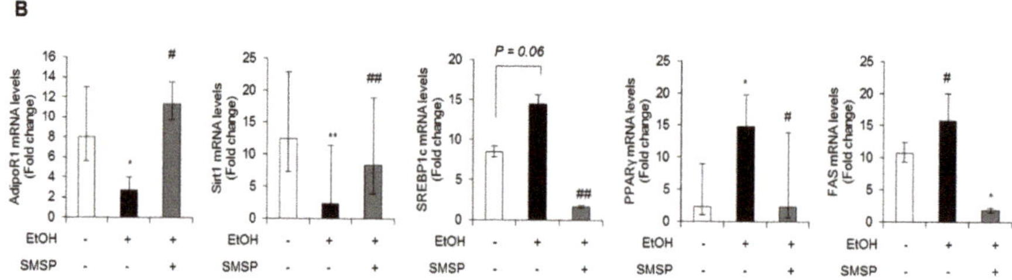

Figure 5. SMSP administration improves lipid metabolism in ethanol-treated rats. (**A**) The expression of phosphorylation of AMP-activated protein kinase (p-AMPK), AMPK and peroxisome proliferator-activated receptor gamma (PPARγ) in the liver tissues were determined by Western blotting and normalized to that of β-actin. (**B**) The mRNA expression of adiponectin receptor 1 (AdipoR1), sirtuin 1 (Sirt1), sterol regulatory element-binding proteins (SREBP1c), PPARγ and fatty acid synthase (FAS) in liver tissues were determined by qRT-PCR and normalized to that of 18s rRNA. The data represent mean ± SD ($n = 4$); * $p < 0.05$, ** $p < 0.01$, and *** $p < 0.001$ vs. normal group; # $p < 0.05$ and ## $p < 0.01$ vs. EtOH group.

The inhibition of ACC1 has been reported to reduce hepatic triglyceride accumulation by decreasing lipogenesis and increasing FFA oxidation [24]. Therefore, we investigated the expression of ACC1 and the phosphorylation. As shown in Figure 6A,B, the phosphorylation of ACC1 was significantly inhibited in ethanol-treated rats while SMSP supplementation recovered the phosphorylation of ACC1. Moreover, the decreased hepatic expressions of ACC1 and peroxisome proliferator-activated receptor-gamma coactivator-1 alpha (PGC-1α) mRNA by ethanol treatment were recovered in SMSP-administered rats, although not statistically significant (Figure 6B,C). These results suggest that SMSP ameliorates EtOH-induced fatty liver through the regulation of lipogenesis and FFA oxidation mediated by AMPK and ACC1.

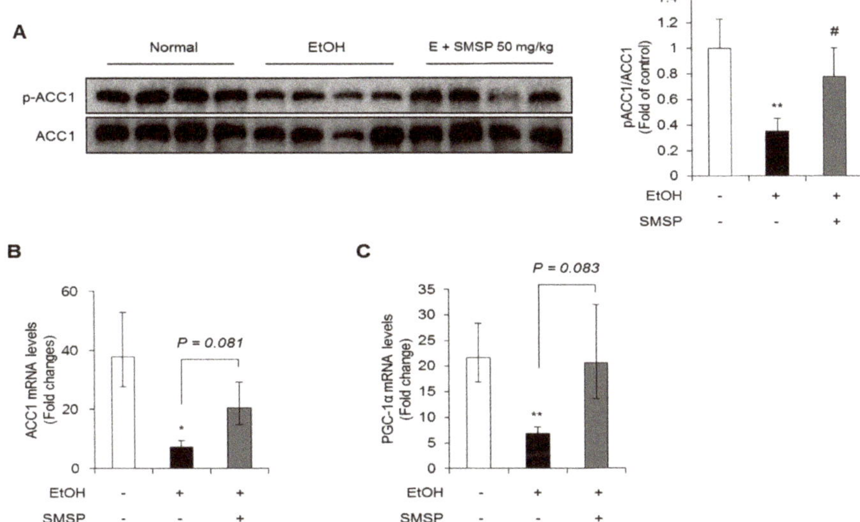

Figure 6. SMSP administration affects the expressions of lipogenesis-related molecules in ethanol-treated rats. (**A**) The phosphorylation of acetyl-CoA carboxylase 1 (ACC1) in the liver tissues was determined by Western blotting and normalized to that of β-actin. The mRNA expression of (**B**) ACC1 and (**C**) peroxisome proliferator-activated receptor gamma coactivator-1 alpha (PGC-1α) in liver tissues was determined by qRT-PCR and normalized to that of 18s rRNA. The data represent mean ± SD ($n = 4$); * $p < 0.05$ and ** $p < 0.01$ vs. normal group; # $p < 0.05$ vs. EtOH group.

4. Discussion

Insects are used for food and animal feed because they are rich in beneficial ingredients, such as proteins, fats, vitamins, minerals, and fiber [25]. In particular, the silkworm *Bombyx mori* has traditionally been used as food in Asia. Our colleagues have compared the nutrient composition of SMSP with those of freeze-dried mature silkworm powder (FMSP) and freeze-dried 3rd day of 5th instar silkworm powder (FDSP) [16]. A proximal analysis revealed that SMSP shows the highest protein contents compared to FMSP and FDSP. In addition, as shown in Table S1, amino acids such as glycine, alanine, and serine, which are the major components of silk protein are very abundant in SMSP. Moreover, the amount of unsaturated fatty acids is more than twice as high than that of saturated fatty acids in SMSP [16]. In addition, the amount of polyphenol, flavonoids, and minerals were examined. Since the amino acids and n-3 fatty acids have been shown to be effective in preventing metabolic diseases and promoting human health [26,27], in line with this notion, we suggest that the administration of SMSP may have beneficial effects on health based on its nutritional composition.

In our previous study, we confirmed the hepatoprotective effects of SMSP 0.1, 1, and 10 g/kg in diethylnitrosamine (DEN)-induced acute liver injury rat model and ethanol-induced liver damage rat model [20,28]. We did not show any side effects such as weight change or decreased dietary intake in rats treated with SMSP. In addition, the previous study with DEN-induced hepatocellular carcinoma rat model have shown that the administration of 1 g/kg of SMSP for 16 weeks does not induce any toxicity compared to the normal group [29]. These results demonstrate that the supplementation of 1–10 g/kg rat body weight of SMSP for 4–16 weeks would be safe. However, 1 g/kg of SMSP in rats corresponds to an intake of approximately 9.73 g/60 kg adult/day, when calculated on the basis of normalization to body surface area [30]. Therefore, this study was conducted to investigate the hepatoprotective effect of the lower dose of SMSP, 50 mg/kg rat body weight, corresponding to an intake of 0.487 kg/60kg in human dose.

Chronic alcohol intake can cause the progress of hepatic steatosis, fibrosis, cirrhosis, and hepatocellular carcinoma (HCC) [17]. Alcohol is a small molecule that can spread easily through the cell membrane [17]. The major enzymes of alcohol metabolism are alcohol dehydrogenase (ADH) and aldehyde dehydrogenase (ALDH) [31]. These enzymes are known as phase 1 xenobiotic metabolize enzymes [31]. ADH rapidly oxidizes alcohol to acetaldehyde, and then ALDH converts acetaldehyde to acetate [32]. This metabolism uses NAD^+ and is based on CYP2E1. The ADH, CYP2E1, and ALDH are mainly produced in hepatocytes [2]. We previously reported that SMSP supplementation alleviated the serum levels of ADH and ALDH in acute alcohol-induced liver injury rat models [33]. In this study, to explore the protective effect of SMSP in rats, we evaluated the anti-hepatosteatotic effect of SMSP using the fatty liver model induced by ethanol (3 g/kg, daily, 4 weeks). Based on our previous study, this dosage was enough to induce hepatic steatosis in SD rats [20]. We found that rats fed the rodent chow diet and oral gavage of SMSP induced no changes, such as in body weight (Figure 1A), but the mass of liver tissue and liver weight/body weight ratio from rats treated SMSP were 8% and 10% respectively lower than those received ethanol only (Figure 1B,C). In addition, SMSP significantly alleviated the levels of liver triglycerides, serum triglycerides, and liver injury markers, such as GGT, ALT, and AST (Figure 2C–E).

Alcoholic liver disease is characterized by the accumulation of large amounts of lipids in the liver with inflammation [11]. In alcoholic liver diseases, chronic ethanol consumption stimulates Kupffer cells to activation by lipopolysaccharides through diverse signal such as Toll-like receptors [11]. This regulation induces the production of various pro-inflammatory cytokines, such as TNF-α, interferon gamma (IFN-γ), and IL-1β, which are critically regulated hepatic inflammation, steatosis, fibrosis, and HCC [34]. It has been reported that both patient and animal model with alcoholic liver disease, the levels of IL-1β are significantly increased in the liver and the serum [35,36]. The present study showed that SMSP treatment can attenuate the serum levels of TNF-α and IL-1β as well as mRNA levels of IL-1β (Figure 4). These results indicate that SMSP can reduce the production of pro-inflammatory cytokines by the chronic alcohol consumption.

In addition, chronic alcohol intake also induces the oxidative stress that induces lipid peroxidation, intracellular membrane damage, and lead to the production of pro-inflammatory cytokines and pro-fibrotic effects. Oxidative stress and ROS are caused through the CYP2E1 signaling [11]. It has been reported that CYP2E1 knock-out mice exhibit ethanol-induced liver disease [37], and we previously demonstrated that SMSP supplementation restored total antioxidant concentration levels and significantly reduced hepatic malondialdehyde levels [20]. In this study, we also found that rats fed SMSP reduced CYP2E1 mRNA levels even the low dosage of SMSP (Figure 4C). These results suggest that low dosage of SMSP can diminish oxidative stress through inhibition of CYP2E1 induced by ethanol.

The accumulation of triglycerides in hepatocytes causes hepatic steatosis, because of the imbalance between lipogenesis and FFA oxidation. Considerable evidence has reported that adiponectin plays a key role in alcoholic fatty liver in several animal models [38–41]. Circulating adiponectin binds to adiponectin receptors in the liver and thereby regulates lipid metabolism through AMPK [42]. In hepatocytes, AMPK is a crucial regulator of intracellular energy sensors that has been reported in the regulation of lipid homeostasis [43]. The inactivation of AMPK negatively regulated FFA oxidation through the inhibition of the transcription factors ACC1 and PGC-1α, and increased activation of SREBP1c, a lipogenic transcription factor [44]. Several studies indicate a relationship between SIRT1 and the AMPK signaling pathway [45]. The upregulation of SIRT1 acts on AMPK upstream, suggesting that the regulation of the SIRT1/AMPK signal may act as a key mechanism for lipid homeostasis in liver. In this study, the protein and mRNA levels of genes involved in SIRT1/AMPK-mediated lipogenesis and FFA oxidation were regulated in rats administered SMSP. These results indicate that SMSP improves lipid metabolism in the livers of rats treated with ethanol.

In summary, the present study suggests that the low dose of SMSP (50 mg/kg) effectively reduced the hepatic steatosis through the activation of SIRT1/AMPK-mediated signaling cascades in ethanol-treated rats. Increased hepatic SIRT1 and AMPK activity appears to be associated with these

Foods **2020**, *9*, 285

beneficial effects of SMSP. Further studies on the effect of SMSP in vitro and in humans should confirm that SMSP could serve as an effective therapeutic agent in treating human alcoholic fatty liver disease.

Supplementary Materials: The following are available online at http://www.mdpi.com/2304-8158/9/3/285/s1, Table S1: The nutrient composition of SMSP.

Author Contributions: Conceptualization, D.-Y.L., K.-S.H., and E.-H.K.; methodology, D.-Y.L., K.-S.H., and E.-H.K.; formal analysis, D.-Y.L., K.-S.H., M.-Y.S., and S.-M.Y.; investigation, D.-Y.L., K.-S.H., M.-Y.S., and S.-M.Y.; resources, S.-D.J. and J.-G.S.; writing—original draft preparation, D.-Y.L.; writing—review and editing, D.-Y.L., and E.-H.K.; project administration, E.-H.K.; funding acquisition, E.-H.K. All authors have read and agreed to the published version of the manuscript.

Funding: This work was supported by Korea Institute of Planning and Evaluation for Technology in Food, Agriculture, Forestry and Fisheries (IPET) through Agri-Bio industry Technology Development Program, funded by Ministry of Agriculture, Food and Rural Affairs (MAFRA) (317004-4).

Conflicts of Interest: The authors declare no conflict of interest.

References

1. Lucey, M.R.; Schaubel, D.E.; Guidinger, M.K.; Tome, S.; Merion, R.M. Effect of alcoholic liver disease and hepatitis C infection on waiting list and posttransplant mortality and transplant survival benefit. *Hepatology* **2009**, *50*, 400–406. [CrossRef] [PubMed]

2. Louvet, A.; Mathurin, P. Alcoholic liver disease: Mechanisms of injury and targeted treatment. *Nat. Rev. Gastroenterol. Hepatol.* **2015**, *12*, 231–242. [CrossRef] [PubMed]

3. Kawaratani, H.; Tsujimoto, T.; Douhara, A.; Takaya, H.; Moriya, K.; Namisaki, T.; Noguchi, R.; Yoshiji, H.; Fujimoto, M.; Fukui, H. The effect of inflammatory cytokines in alcoholic liver disease. *Mediators Inflamm.* **2013**, *2013*, 495156. [CrossRef] [PubMed]

4. Lam, B.; Younossi, Z.M. Treatment options for nonalcoholic fatty liver disease. *Therap. Adv. Gastroenterol.* **2010**, *3*, 121–137. [CrossRef] [PubMed]

5. Nassir, F.; Rector, R.S.; Hammoud, G.M.; Ibdah, J.A. Pathogenesis and Prevention of Hepatic Steatosis. *Gastroenterol. Hepatol. (N. Y.)* **2015**, *11*, 167–175.

6. Hardie, D.G. AMP-activated/SNF1 protein kinases: Conserved guardians of cellular energy. *Nat. Rev.* **2007**, *8*, 774–785. [CrossRef]

7. Zhang, M.; Wang, C.; Wang, C.; Zhao, H.; Zhao, C.; Chen, Y.; Wang, Y.; McClain, C.; Feng, W. Enhanced AMPK phosphorylation contributes to the beneficial effects of Lactobacillus rhamnosus GG supernatant on chronic-alcohol-induced fatty liver disease. *J. Nutr. Biochem.* **2015**, *26*, 337–344. [CrossRef]

8. Park, S.; Gammon, S.; Knippers, J.; Paulsen, S.; Rubink, D.; Winder, W. Phosphorylation-activity relationships of AMPK and acetyl-CoA carboxylase in muscle. *J. Appl. Physiol.* **2002**, *92*, 2475–2482. [CrossRef]

9. Lieber, C.S. Cytochrome P-4502E1: Its physiological and pathological role. *Physiol. Rev.* **1997**, *77*, 517–544. [CrossRef]

10. Lieber, C.S. Alcoholic fatty liver: Its pathogenesis and mechanism of progression to inflammation and fibrosis. *Alcohol* **2004**, *34*, 9–19. [CrossRef]

11. Tilg, H.; Moschen, A.R.; Szabo, G. Interleukin-1 and inflammasomes in alcoholic liver disease/acute alcoholic hepatitis and nonalcoholic fatty liver disease/nonalcoholic steatohepatitis. *Hepatology* **2016**, *64*, 955–965. [CrossRef] [PubMed]

12. You, M.; Matsumoto, M.; Pacold, C.M.; Cho, W.K.; Crabb, D.W. The role of AMP-activated protein kinase in the action of ethanol in the liver. *Gastroenterology* **2004**, *127*, 1798–1808. [CrossRef] [PubMed]

13. Ji, C.; Deng, Q.; Kaplowitz, N. Role of TNF-alpha in ethanol-induced hyperhomocysteinemia and murine alcoholic liver injury. *Hepatology* **2004**, *40*, 442–451. [CrossRef] [PubMed]

14. Navasa, M.; Gordon, D.A.; Hariharan, N.; Jamil, H.; Shigenaga, J.K.; Moser, A.; Fiers, W.; Pollock, A.; Grunfeld, C.; Feingold, K.R. Regulation of microsomal triglyceride transfer protein mRNA expression by endotoxin and cytokines. *J. Lipid Res.* **1998**, *39*, 1220–1230.

15. Kim, D.W.; Hwang, H.S.; Kim, D.S.; Sheen, S.H.; Heo, D.H.; Hwang, G.; Kang, S.H.; Kweon, H.; Jo, Y.Y.; Kang, S.W.; et al. Effect of silk fibroin peptide derived from silkworm Bombyx mori on the anti-inflammatory effect of Tat-SOD in a mice edema model. *BMB Rep.* **2011**, *44*, 787–792. [CrossRef] [PubMed]

16. Ji, S.D.; Kim, N.S.; Kweon, H.Y.; Choi, B.H.; Yoon, S.M.; Kim, K.Y.; Koh, Y.H. Nutrient compositions of Bombyx mori mature silkworm larval powders suggest their possible health improvement effects in humans. *J. Asia-Pacif. Entomol.* **2016**, *19*, 1027–1033. [CrossRef]

17. Tabunoki, H.; Ono, H.; Ode, H.; Ishikawa, K.; Kawana, N.; Banno, Y.; Shimada, T.; Nakamura, Y.; Yamamoto, K.; Satoh, J.; et al. Identification of key uric acid synthesis pathway in a unique mutant silkworm Bombyx mori model of Parkinson's disease. *PLoS ONE* **2013**, *8*, e69130. [CrossRef]

18. Igarashi, K.; Yoshioka, K.; Mizutani, K.; Miyakoshi, M.; Murakami, T.; Akizawa, T. Blood pressure-depressing activity of a peptide derived from silkworm fibroin in spontaneously hypertensive rats. *Biosci. Biotechnol. Biochem.* **2006**, *70*, 517–520. [CrossRef]

19. Ji, S.D.; Kim, N.S.; Lee, J.Y.; Kim, M.J.; Kweon, H.Y.; Sung, G.B.; Kang, P.D.; Kim, K.Y. Development of processing technology for edible mature silkworm. *J. Sericult. Entomol. Sci.* **2015**, *53*, 38–43.

20. Hong, K.S.; Yun, S.M.; Cho, J.M.; Lee, D.Y.; Ji, S.D.; Son, J.G.; Kim, E.H. Silkworm (Bombyx mori) powder supplementation alleviates alcoholic fatty liver disease in rats. *J. Funct. Foods* **2018**, *43*, 29–36. [CrossRef]

21. Bedossa, P.; Poynard, T. An algorithm for the grading of activity in chronic hepatitis C. *Hepatology* **1996**, *24*, 289–293. [CrossRef] [PubMed]

22. Kim, E.H.; Bae, J.S.; Hahm, K.B.; Cha, J.Y. Endogenously synthesized n-3 polyunsaturated fatty acids in fat-1 mice ameliorate high-fat diet-induced non-alcoholic fatty liver disease. *Biochem. Pharmacol.* **2012**, *84*, 1359–1365. [CrossRef] [PubMed]

23. Lee, J.; Narayan, V.P.; Hong, E.Y.; Whang, W.K.; Park, T. Artemisia Iwayomogi Extract Attenuates High-Fat Diet-Induced Hypertriglyceridemia in Mice: Potential Involvement of the Adiponectin-AMPK Pathway and Very Low Density Lipoprotein Assembly in the Liver. *Int. J. Mol. Sci.* **2017**, *18*, 1762. [CrossRef] [PubMed]

24. Kim, C.W.; Addy, C.; Kusunoki, J.; Anderson, N.N.; Deja, S.; Fu, X.; Burgess, S.C.; Li, C.; Ruddy, M.; Chakravarthy, M.; et al. Acetyl CoA Carboxylase Inhibition Reduces Hepatic Steatosis but Elevates Plasma Triglycerides in Mice and Humans: A Bedside to Bench Investigation. *Cell Metab.* **2017**, *26*, 394–406 e396. [CrossRef]

25. Rumpold, B.A.; Schluter, O.K. Nutritional composition and safety aspects of edible insects. *Mol. Nutr. Food Res.* **2013**, *57*, 802–823. [CrossRef]

26. Belitz, H.D.; Grosch, W.; Schieberle, P. Amino acids, peptides, proteins. In *Food Chemistry*; Springer: Berlin/Heidelberg, Germany, 2004; pp. 8–91.

27. Drenjančević, I.; Kralik, G.; Kralik, Z.; Mihalj, M.; Stupin, A.; Novak, S.; Grčević, M. The effect of dietary intake of Omega-3 polyunsaturated fatty acids on cardiovascular health: Revealing potentials of functional food. *Superfood Funct. Food-The Dev. Superfoods Their Roles Med.* **2017**. [CrossRef]

28. Cho, J.M.; Kim, K.Y.; Ji, S.D.; Kim, E.H. Protective effect of boiled and freeze-dried mature silkworm larval powder against diethylnitrosamine-induced hepatotoxicity in mice. *J. Cancer Prev.* **2016**, *21*, 173. [CrossRef]

29. Lee, D.Y.; Yun, S.M.; Song, M.Y.; Ji, S.D.; Son, J.G.; Kim, E.H. Administration of steamed and freeze-dried mature silkworm larval powder prevents hepatic fibrosis and hepatocellular carcinogenesis by blocking TGF-β/STAT3 signaling cascades in rats. *Cells* **2020**, in press. [CrossRef]

30. Reagan-Shaw, S.; Nihal, M.; Ahmad, N. Dose translation from animal to human studies revisited. *FASEB J.* **2008**, *22*, 659–661. [CrossRef]

31. Cequera, A.; Garcia de Leon Mendez, M.C. [Biomarkers for liver fibrosis: Advances, advantages and disadvantages]. *Rev. Gastroenterol. Mex.* **2014**, *79*, 187–199. [CrossRef]

32. Park, J.H.; Kim, Y.; Kim, S.H. Green tea extract (*Camellia sinensis*) fermented by Lactobacillus fermentum attenuates alcohol-induced liver damage. *Biosci. Biotechnol. Biochem.* **2012**, *76*, 2294–2300. [CrossRef] [PubMed]

33. Lee, D.Y.; Hong, K.S.; Yun, S.M.; Song, M.Y.; Ji, S.D.; Son, J.G.; Kim, E.H. Mature silkworm powder reduces blood alcohol concentration and liver injury in ethanol-treated rats. *Int. J. Ind. Entomol.* **2017**, *35*, 123–128.

34. Kubes, P.; Mehal, W.Z. Sterile inflammation in the liver. *Gastroenterology* **2012**, *143*, 1158–1172. [CrossRef] [PubMed]

35. Tilg, H.; Wilmer, A.; Vogel, W.; Herold, M.; Nolchen, B.; Judmaier, G.; Huber, C. Serum levels of cytokines in chronic liver diseases. *Gastroenterology* **1992**, *103*, 264–274. [CrossRef]

36. Petrasek, J.; Bala, S.; Csak, T.; Lippai, D.; Kodys, K.; Menashy, V.; Barrieau, M.; Min, S.-Y.; Kurt-Jones, E.A.; Szabo, G. IL-1 receptor antagonist ameliorates inflammasome-dependent alcoholic steatohepatitis in mice. *J. Clin. Investig.* **2012**, *122*, 3476–3489. [CrossRef]

37. Lu, Y.; Zhuge, J.; Wang, X.; Bai, J.; Cederbaum, A.I. Cytochrome P450 2E1 contributes to ethanol-induced fatty liver in mice. *Hepatology* **2008**, *47*, 1483–1494. [CrossRef]

38. Xu, A.; Wang, Y.; Keshaw, H.; Xu, L.Y.; Lam, K.S.; Cooper, G.J. The fat-derived hormone adiponectin alleviates alcoholic and nonalcoholic fatty liver diseases in mice. *J. Clin. Investig.* **2003**, *112*, 91–100. [CrossRef]

39. You, M.; Considine, R.V.; Leone, T.C.; Kelly, D.P.; Crabb, D.W. Role of adiponectin in the protective action of dietary saturated fat against alcoholic fatty liver in mice. *Hepatology* **2005**, *42*, 568–577. [CrossRef]

40. Chen, X.; Sebastian, B.M.; Nagy, L.E. Chronic ethanol feeding to rats decreases adiponectin secretion by subcutaneous adipocytes. *Am. J. Physiol. Endocrinol. Metab.* **2007**, *292*, E621–E628. [CrossRef]

41. Song, Z.; Zhou, Z.; Deaciuc, I.; Chen, T.; McClain, C.J. Inhibition of adiponectin production by homocysteine: A potential mechanism for alcoholic liver disease. *Hepatology* **2008**, *47*, 867–879. [CrossRef]

42. Yamauchi, T.; Nio, Y.; Maki, T.; Kobayashi, M.; Takazawa, T.; Iwabu, M.; Okada-Iwabu, M.; Kawamoto, S.; Kubota, N.; Kubota, T.; et al. Targeted disruption of AdipoR1 and AdipoR2 causes abrogation of adiponectin binding and metabolic actions. *Nat. Med.* **2007**, *13*, 332–339. [CrossRef] [PubMed]

43. Wang, S.; Li, X.; Guo, H.; Yuan, Z.; Wang, T.; Zhang, L.; Jiang, Z. Emodin alleviates hepatic steatosis by inhibiting sterol regulatory element binding protein 1 activity by way of the calcium/calmodulin-dependent kinase kinase-AMP-activated protein kinase-mechanistic target of rapamycin-p70 ribosomal S6 kinase signaling pathway. *Hepatol. Res.* **2017**, *47*, 683–701. [PubMed]

44. Rogers, C.Q.; Ajmo, J.M.; You, M. Adiponectin and alcoholic fatty liver disease. *IUBMB Life* **2008**, *60*, 790–797. [CrossRef] [PubMed]

45. Hou, X.; Xu, S.; Maitland-Toolan, K.A.; Sato, K.; Jiang, B.; Ido, Y.; Lan, F.; Walsh, K.; Wierzbicki, M.; Verbeuren, T.J. SIRT1 regulates hepatocyte lipid metabolism through activating AMP-activated protein kinase. *J. Biol. Chem.* **2008**, *283*, 20015–20026. [CrossRef]

 *foods*

Article

Could Western Attitudes towards Edible Insects Possibly be Influenced by Idioms Containing Unfavourable References to Insects, Spiders and other Invertebrates?

Victor Benno Meyer-Rochow [1,2,*] **and Aimo Kejonen** [3]

1   Department of Ecology and Genetics, Oulu University, SF-90140 Oulu, Finland
2   Agricultural Science and Technology Research Institute, Andong National University, Andong 36729, Korea
3   Geologian tutkimuskeskus, Itä-Suomen yksikkö, PL 1237, SF-70211 Kuopio, Finland; geologi.kejonen@gmail.com
*   Correspondence: meyrow@gmail.com; Tel.: +82-010-6765-6244

Received: 15 January 2020; Accepted: 8 February 2020; Published: 11 February 2020

Abstract: It is known that idioms, proverbs, and slogans can become integrated into feelings like irritation, contemptuous attitudes, and even anger and disgust. Idioms making reference to insects, spiders, and other invertebrates occur in all languages, but they convey mostly negative content in people of Western cultural orientation. By analyzing a subgroup of insect and spider idioms related to food, eating, and digestion, the authors suggest that mirror neurons are activated in people that are exposed to the largely unfavorable content of such idioms. This could then lead the listener of such idioms to adopt the kind of negative attitude towards insects that is expressed in the idioms and to project it towards edible species.

Keywords: disgust; emotions; entomophagy; sociolinguistics; food choice; mirror neurons

---

1. Introduction

There is no single reason that explains why people would accept or reject insects as food. Vabø and Hansen [1] in connection with food preferences suggested that the latter strongly depended on the perceived healthiness of the food item, the price of the food item and convenience of obtaining it, its sensory appeal, the mood of the consumer as well as familiarity, and ethical concerns. Lensvelt and Steenbekkers [2] added to this list supply and demand, tradition, religious beliefs, etc., while Shouteten et al. [3] and Ghosh et al. [4] focused on sensorial, economic, cultural, and ethnic aspects. Peer pressure and socio-cultural perspectives were investigated by Menozzi et al. [5] and Tan and House [6], respectively, and disgust and neophobia in connection with Westerners' disgust for eating insects formed topics of inquiries for La Barbera et al. [7] and Castro and Chambers [8]. Although researchers could show that the name, type, and role of an insect [9] can affect a species' acceptance or rejection just like the image that an insect's name creates in the consumer does [10], none of these as well as other investigators had previously considered that a rejection of insects as food could also have a linguistic background component.

Idioms and proverbs, for example, occur in all languages and according to Casas et al. [11], constitute categories that "permeate languages at a much deeper level than what is usually taken for granted". The same authors declare that idioms exert an "overriding influence" on society and according to [12] become integrated into popular consciousness to express feelings like irritation, contemptuous attitudes, and even anger and disgust [13]. Unsurprisingly, accepting the validity of this view, idioms and proverbs might then be interpreted as "natural decoders of customs, cultural beliefs, social conventions, and norms" [14] with a profound influence on experience and behaviour [15].

Turning now specifically to idioms that incorporate references to insects and other terrestrial arthropods like spiders, it has repeatedly been shown that amongst people with Western cultural backgrounds these idioms convey predominantly negative attitudes, but that in East Asia more positive attitudes prevail [16–19]. Thus, the question arises as to whether the negative insect idioms stemmed from a genuine, deep-rooted antipathy towards insects as well as terrestrial arthropods generally or whether the negative attitude has been nurtured and strengthened by references to such idioms that are explicitly negative with regard to these invertebrates.

What seemed pertinent in connection with this question is the recent realization of so-called mirror neurons in the human brain, discovered by Rizzolatti and Sinigaglia [20]. Such neurons are implicated in the ability of an individual to understand another individual's emotion and to empathize by activating neuronal circuits, as when the receiving individual experienced the other person's emotion herself or himself directly [21]. Originally seen in connection with visual or odoriferous stimulation, mirror neurons are now also thought to play a role in adopting emotions that are expressed figuratively, i.e., in idioms, as mirror neurons are located near the Broca area of speech development in the human brain and thereby can facilitate the learning process as well as verbal communication [21]. The emotional state of the 'receiver individual' according to Bastiaansen et al. [13] comes to resemble that of the 'sending individual', especially if a facial expression of disgust [22] accompanies the utterance of a particular idiom.

The aim of this paper has been to show that idioms containing negative references to insects and other invertebrates could possibly influence a person's attitude, especially in connection with insects as a novel food item in Western countries. Therefore, this paper must not be seen as a comprehensive and detailed study on insect containing idioms and their effects on people, but is meant to serve as a 'wake-up-call' to conduct further studies into this question.

2. Materials and Methods

The target subgroup of insect, spiders, and other invertebrate-related idioms were those that show a relationship to food, eating, and digestion in the widest sense. Although a variety of such idioms from different languages were investigated by us, the idioms that we have most thoroughly examined belonged predominantly to the Finnish language for three reasons: firstly, Finns have largely been a rural people living in close contact with Nature and their language is therefore particularly rich in idioms that contain references to insects; secondly, the authors' own familiarity with the language provided us with considerable insight into use and meaning of such idioms; thirdly, in agreement with other people of Western cultural backgrounds, Finns until very recently, did not consider insects and other terrestrial arthropods as edible, seeing in fact most of them as pesky and annoying creatures.

Since Finnish idioms and proverbs have been the subject of several books, the ten most relevant books [23–32] were consulted by the authors in order to locate suitable idioms, i.e., those involving insects, spiders, and other invertebrates in connection with food and eating. The idioms were discussed with a small but unspecified number of colleagues, five students of approximately 20 years of age and four elderly relatives, who contributed four additional idioms (referred to as 'other'). The research represents a qualitative study of the relevant literature but does not claim to be exhaustive. As with an earlier study on Korean insect idioms by Meyer-Rochow [19], we are presenting the idioms first in their original form, followed by their translation into English and, where deemed necessary, an explanation as to the idiom's meaning and application in parentheses. Each Finnish idiom's source, i.e., the book it was found in, is indicated at the end of each idiom. How the non-Finnish idioms were found is mentioned in connection with the language they were taken from.

3. Results

Finnish language speakers may find some of the expressions and spellings unusual, because often idioms are connected with local dialects that somewhat differ from the standard Finnish. Moreover, urbanized Finns may not recognize some of the older idioms, especially those used by rural folk.

3.1. Finnish Idioms

1. Kyllä se kärsii, mitä tai järsii [23] = He is in pain, who is eaten by a louse (meaning either that a small thing can cause lasting pain, or that some pains need to be tolerated).
2. Täi lihalla elää, mies muulla ruuvalla [23] = A louse lives by eating flesh/meat, humans eats other foods (humans ought to be more selective in their food choices).
3. Täi aina lihhoo syö [23] = A louse is always eating flesh/meat (see above).
4. Ei sääsken saaresta paljon paisteta [23] = One foot of a mosquito isn't much of a roast (lamenting the lack of food).
5. Kussa vähä kimalaisia, siinaä vähä hunajata [23] = Where there's few bumblebees, there's little honey (with few persons around to be involved in work, life cannot be sweet).
6. Jaakko paarmat kokoo ja keittää niistä hutun [24] = John collects the horseflies and cooks a horsefly-porridge (someone who hasn't got anything better to eat than poor food).
7. Sirkan reisi ja kirpun hanta koyhan miehen rasjas on [24] = The thigh of a cricket and the tail of a flea are in the food box of a poor man (there is not much else available to eat; expressing a severe lack of good food).
8. Tyyristä kärpäsen mesi [25] (kärpänen = fly, but can also be a bee in western Finland like here) = The honey of a fly is expensive (an item, of food perhaps, that is very rare and not easy to obtain by an ordinary person).
9. Minä tome, toucan ruoka, matoni pala maanalainen [25] = I am dust, the food of worms (maggots), the worms' underground bread (a very negative view of one's existence).
10. Yksin tehty työ on kun tervaa, kaksin kun hunajaa [26] = Working alone is tar, working together is honey (cooperation makes things easier).
11. Läksin mina kesäyömä käymään, sano kiljupytty muurahaismaättäässä [26] = I begin to "käydä", here = to ferment, in a summer night, the barrel full of strong home-made beer it says in the anthill (meaning is a little unclear: perhaps to feel relaxed).
12. Ei niin pientä matoo, ettei hiukka rasva [26] = There is no worm so little as not to have a little fat (sarcastic and to mean that even poor food can be acceptable).
13. Niin on koyhä jottei ole täitää [26] = So poor that she/he does not even have lice.
14. Kiitoksia, sano Laurikainen, kun otti madon maasta ja pani suuhunsa [26] = Thank you says Laurikainen as he takes a worm from the earth (grateful for even the poorest food).
15. Sillä välin lapsen nälkä tulee, kun kärpänen pirtin yli lentää [27] = A child is getting hungry as soon as a fly flies over your house (there is so little nourishment at home that even a fly is seen as a desirable food item).
16. Kaunis kakku päältä nähden, mutta jos lie sirkkoja sisällä [27] = The cake may look beautiful, but there could be crickets (grasshoppers) inside (you could be deceived to believe that something is nice by the way it looks).
17. Hyttynen on niin kesy, että heti syö kädestä [other] = The mosquito is so tame, that it eats from your hand at once (sarcastic, meaning that something seemingly harmless, turns out be villainous and exploiting).
18. Parempi mato omenassa kuin puolikas puremassa [other] = Better a whole worm in an apple than half of the worm in the bitten place (jokingly refers to that you would have consumed half a worm, if you see the other half still in the apple).
19. Ei madon syöma makee, eikä hiiren syömä hipee [other] = Worm eating is not sweet and mouse nibbling is not good (having little to eat is not funny).
20. Mehiläinen meidän lintu, herhiläinen hiiden lintu [other] = Bee is our bird, hornet is the Devil's bird (expresses dislike of hornets as evil in comparison with something good like bees).

21. Lisänä rikka rokassa, hämähäkki taikinassa [28] = More means a rock in the soup and a spider in the dough (sarcastic and to mean one should not expect too much and that even something as small as a spider can be an addition).

22. Ole on omituinen kuin olisi perhosia vatsassa [28] = Living is strange like having butterflies in your stomach (an expression of worry, anxiety, or nervousness).

23. Sirkan reisi ja paarman jalka köyhän aitasta löytyy [28] = The thigh of a cricket and the foot of a horsefly are in a poor man's storehouse (there is not much available to eat; no proper food).

24. Kirpum potku ei paljoo tunnu [29] = Cooking a flea does not feel much (a small amount of food does not fill your stomach).

25. Paskakakkokim makosta haukatas sontiaisen suulla [29] = Even shit is good to bite, if you have the mouth of a dung beetle (you are prepared to eat the worst food).

26. Se on niin nuuka et se kirputki halkasee ja pitää itte molemmat puolet [30] = She/he is so stingy that she/he cuts the flea into two pieces and keeps both to herself/himself (someone in such need of nourishment that she/he would not even be willing to share with someone half a flea as food).

27. On niin ahne ett otais kirpultaki kolmppari koippei [30] = So greedy that she/he takes all three pairs of legs from a flea (someone in such need of nourishment that she/he would not even be willing to share one leg of a flea as food with somebody else).

28. Ain se avuks on ikä kärväne kaalfaris [30] = It is always as much help as a fly is meat in the cabbage casserole (the help rendered by someone is worth almost nothing).

29. On niinku kärväne siirapis [30] = To be like a fly in syrup (to be in some sort of serious trouble).

30. Käy niinko mato raatoo [30] = To eat like worms eat the carcass (here: mato = fly larva) (to be non-selective when it comes to food and even accept poor/disgusting food).

31. Meni ku mato rasvonee [30] = She/he went like a worm with its fat (someone did not share or leave anything useful).

32. Syä kun hevonen ja paskantaa kun peronen [30] = Eat like a horse and shit like a butterfly (someone who received a lot of support, but produced a meagre result).

33. Nielöö kun variksenpoeka sittapörrijä [30] = To swallow like a young crow swallows door beetles (when someone eats too much and too hurriedly not caring what she/he's eating).

34. On vatta nin täysi että täin päälla tappas [30] = The stomach is so full, that it could easily kill a louse in it (someone has eaten very well and is satisfied).

35. Syöö kun hevone ja tekköö kun täe [30] = Eat like a horse and work like a louse (a lazy person that works little, but consumes much food).

36. Kun sirkka pääsee suurukselle, niinse laulaa [30] = When a cricket sits at the dinner table it is singing (if there is food around, people are happy).

37. Vennyy kun vesimato piimäsä [30] = To lie or stretch out like water worm in sour milk (a lazy and relaxed person).

38. Niin nuuk, et pannee kirpun poikki ja pittää molemap päätitte [31] = So greedy that she/he cut the flea into two pieces and kept both pieces to herself/himself personally (someone in such need of nourishment that she/he would not be willing to share with someone even half a flea as food).

39. Vattuvaha se on vatun matokii [32] = A raspberry worm is only a part of the raspberry (if one likes something, one needs to accept the problems that come with it).

40. Näläkästä se torakka sillon siälii kun mökinseinästä vellipataan tipahtaa [32] = The cockroach is feeling pity for poor people, when it drops from the wall into the gruel pot (even an insect like the cockroach can be a welcome addition to the food of poor people, because they have nothing much else).

3.2. Non-Finnish Idioms

To show that Finnish is not the only language that possesses idioms in the category of insects and spiders related to food, eating, and digestion, a few examples shall be given of such idioms of other languages.

Lithuanian (L. Panavaite, pers. comm.):
Alkanam šuniui ir grambuolys mėsa = a hungry dog sees a cockchafer as meat.

English [33]:

1.   Flies come to feasts unasked.
2.   A louse in the cabbage is better than no meat at all.
3.   When the bee sucks, it makes honey; when the spider, it's poison.

German [34]:

1.   In der Not frisst der Deubel (= Teufel) Fliegen = in times of need the Devil consumes flies (if there is nothing else to eat, then like the Devil one would even accept flies as food).
2.   Besser eine Laus im Topf als gar kein Fleisch. = Better a louse in the pot than no meat at all.
3.   Da ist der Wurm drin= There's a worm in it (it can be maggot or caterpillar) (something is not right; something is not 'kosher').
4.   Der/sie macht'n Gesicht als haett er'ne Spinne gefressen. = He makes a face as if he'd swallowed a spider (a person's expression when something unpleasant has occurred to him/her).

Korean [19]:

1.   구데기 무서워 장 못담굴가 = Goodeogi mooseoweo jang motdamgeunda (one can make soybean paste sauce even in the presence of maggots).

Japanese [18]:

1.   蓼食うも好き好き Tade kuu mushi mo sukizuki. Literally: There are even bugs that eat knotweed (tastes differ and that there's no accounting for taste).
2.   蟻の甘きに就くがごとし Ari no amakini tsukuga gotoshi. Literally: ants stick to sweet things (people gather together where there is something to be got).

Although assigning idioms to categories designated as 'negative', 'neutral', and 'positive' is a subjective exercise that is likely to show some variation depending on who makes the distinction, our view is that of the 40 Finnish idioms related to insects and eating, feeding and food, 20 contained predominantly negative, 9 neutral, and 7 positive information (idioms with seemingly identical meaning, but slightly different wording were not counted separately). Clearly negative are idioms like "Kirpum potku ei paljoo tunnu" and "On niinku kärväne siirapi" while "Kussa vähä kimalaisia, siinaä vähä hunajata" and "Yksin tehty työ on kun tervaa, kaksin kun hunaja" are seen as neutral and positive, respectively. Although we did not have complete comparative lists of insect idioms in connection with food and eating from other languages, what we can nevertheless state with confidence is, that it is the negative content which is always more dominant in Western than Eastern cultures. One idiom each from Korea and Japan has served as an example of the more positive attitudes towards insects in these countries.

4. Discussion

We need to stress that in Finnish and other languages there are many more idioms referring to insects, spiders, and worms than those we mention in this article. We deliberately selected only idioms that had some connection with food, eating, and digestion. It was, after all, our intention to demonstrate that the use of these figuratively explicit idioms could affect the attitude of people towards accepting insects as food. Therefore: did our study provide evidence for a role of insect idioms in the mindset of people with Western cultural backgrounds?

Most people would agree that if a happy and smiling one-year-old infant sees and hears another, unrelated baby cry, the mood of the former, happy infant will change and it may also start to cry for no apparent reason other than seeing and hearing another baby cry. Likewise, as Stadler [35] explains, when a customer happens to phone a shop owner and talks with an angry voice (irrespective of the call's content (our addition)), it will affect and change the mood of the recipient. Emotions are copied by mirror neurons and learning is not involved. That repeated emotional stimulation can ultimately alter the emotional right hemisphere of the brain and lead to a deeper unconscious reaction in connection with the appropriate stimulation has originally been postulated by Tsunoda [36], who went as far as claiming that the brains of Japanese people differed from those of Westerners in regard to linguistic, musical, and aesthetic perception because of the greater appreciation of insects by the Japanese.

Duda and Brown [37] thoroughly investigated lateral asymmetry of positive and negative emotions and Schapkin [38] confirmed hemispheric asymmetries in connection with emotional words. Holtgraves and Felton [39] studied hemispheric asymmetries in relation to the processing of negative and positive words and Beraha et al. [40], a year later, examined the hemispheric asymmetry for affective stimulus processing. Finally, Gainotti [41] very recently reviewed the evidence for the role of the right hemisphere in emotion processing. It therefore appears highly likely that frequent exposure to idioms expressing a negative attitude towards consuming insects as food can indeed lead to long-lasting effects and a habitual rejection of insects as a food item. Idioms and proverbs, after all, have been credited with an ability to affect the experience and the behaviour of humans [15].

Given that idioms with references to insects and spiders in Western cultures are predominantly negative, but less so in Eastern cultures [16–19,42], there could therefore be one until now overlooked component of the reasons why people with Western cultural orientation hesitate to accept edible insects. Obviously, semantic and scientific knowledge (of the insect mentioned in an idiom) can "affect the outcome of the degree of mirror" [13], irrespective of a person's dislike or even disgust triggered by the person's exposure to the insect idiom. Rational thoughts and learning can override the inherent and unconscious reaction mediated by mirror neurons. A person familiar with cockroaches, knowing the latter do not transmit or carry diseases will in all likelihood be displaying less disgust than someone who knows little more than that the insect in question is a pest, perhaps clumping it together with lice, fleas, and irritating bugs. Thus, educating the public that the vast majority of all insects represents beneficial species would be a step in the right direction. Phasing out negative insect idioms could also be helpful since it has been shown after Citron et al. [43] that negative idioms are rated as more arousing than positive ones, in line with results from single words.

An aspect not to be ignored is 'familiarity' with insects and other arthropods. Urbanization certainly leads to an increasing alienization regarding insects and one consequence of this is that nowadays fewer people know and use idioms that make reference to insects and other arthropods as seen especially in the cases with English and German speakers when compared, for example, with the Japanese [18]. This, however, does not repudiate our suggestion that insect idioms did and still do affect the attitude with which people contemplate insects and spiders, for firstly, historically many more insect and spider idioms were in use than is now the case [44] and secondly, as with taboos [45], once prejudices and antipathies become established, it takes a very long time before they are abandoned. And in the same vein, once a food habit has taken root, it is difficult to eradicate [46].

5. Conclusions

In conclusion, to establish insects as a regular food item (even if it is simply as a component of some other food) in countries that in the distant past had given up the consumption of insects and other arthropods and now enjoy no shortage of other food items, is likely to be an uphill struggle. A number of reasons have been documented and predictions on the intentions of eating insect-based products have recently been formulated [47], but a possible effect of idioms and proverbs in shaping consumers' attitudes towards edible insects and other arthropods has never before been considered as a possible factor—even though the power of words and the role of mirror neurons in connection with emotions has received considerable attention. New idioms appear all the time [48] and perhaps coining some like "forget about the pork when there's a cricket on your fork" or "mealworms and spaghetti is food that makes you happy" can help change attitudes!

Author Contributions: Conceptualization, V.B.M.-R. and A.K.; Methodology V.B.M.-R. And A.K.; Software, not applicable; validation, V.B.M.-R.; Formal analysis, V.B.M.-R.; Investigation, V.B.M.-R. and A.K.; Resources, V.B.M.-R. And A.K.; Data curation, V.B.M.-R.; Writing—original draft preparation, V.B.M.-R.; Writing—review and editing, V.B.M-R.; Visualization, V.B.M.-R.; Supervision, V.B.M.-R.; Project administration, V.B.M.-R. All authors have read and agreed to the published version of the manuscript.

Funding: To complete this study Victor Benno Meyer-Rochow received support from Chuleui Jung of Andong National University's Insect Industry R&D Center via the Basic Science Research Program of the National Research Foundation of Korea (NRF) funded by the Ministry of Education (NRF-2018R1A6A1A03024862).

Acknowledgments: Both authors wish to express their thanks to the many informants who contributed idioms and shared their knowledge of local lore with us. They also appreciated the comments that they had received by the reviewers on the manuscript.

Conflicts of Interest: The authors declare that there is no conflict of interests.

References

1. Vabø, M.; Hansen, H. The relationship between food preferences and food choice: A theoretical discussion. *Int. J. Bus. Soc. Sci.* **2014**, *5*, 145–157.

2. Lensvelt, E.J.S.; Steenbekkers, L.P.A. Exploring consumer acceptance of entomophagy: A survey and experiment in Australia and the Netherlands. *Ecol. Food Nutr.* **2014**, *53*, 543–561. [CrossRef]

3. Shouteten, J.J.; De Steur, H.; De Pelsmaeker, S.; Lagast, S.; Juvinal, J.G.; De Bourdeaudhuij, L.; Verbeke, W.; Gellynck, X. Funcional and sensory profiling of insect-, plant- and meat-based burgers under blind, expected and informed conditions. *Food Qual. Pref.* **2016**, *52*, 27–31. [CrossRef]

4. Ghosh, S.; Jung, C.; Meyer-Rochow, V.B. What governs selection and acceptance of edible insect species? In *Edible Insects in Sustainable Food Systems*; Halloran, A., Flore, R., Vantomme, P., Roos, N., Eds.; Springer Verlag: Berlin, Germany, 2018; pp. 331–351.

5. Menozzi, D.; Sogari, G.; Veneziani, M.; Simoni, E.; Mora, C. Eating novel foods: An application of the theory of planned behavior to predict the consumption of an insect-based product. *Food Qual. Pref.* **2017**, *59*, 27–34. [CrossRef]

6. Tan, H.S.G.; House, J. Consumer acceptance of insects as food: Integrating psychological and socio-cultural perspectives. In *Edible Insects in Sustainable Food Systems*; Halloran, A., Flore, R., Vantomme, P., Roos, N., Eds.; Springer Verlag: Berlin, Germany, 2018; pp. 375–386.

7. La Barbera, F.; Verneau, F.; Amato, M.; Grunert, K. Understanding Westerners' disgust for the eating of insects: The role of food neophobia and implicit associations. *Food Qual. Pref.* **2018**, *64*, 120–125. [CrossRef]

8. Castro, M.; Chambers, E. Consumer avoidance of insect containing foods: Primary emotions, perceptions and sensory characteristics driving consumers' considerations. *Foods* **2019**, *8*, 351. [CrossRef]

9. Schäufele, I.; Barrera Albores, E.; Hamm, U. The role of species for the acceptance of edible insects: Evidence from a consumer survey. *Brit. Food J.* **2019**, *121*, 2190–2204. [CrossRef]

10. Ali, A.E. A semiotic approach to entomophagy. *Perspect. Glob. Dev. Technol.* **2016**, *15*, 391–504. [CrossRef]

11. Casas, R.M.; Hernández Campoy, J.M. A sociolinguistic approach to the study of idioms: Some anthropolinguistic sketches. *Cuad. Filol. Ingl.* **1995**, *4*, 43–61.

12. Cowie, A.P.; Mackin, R.; Mc-Caig, I.R. *Oxford Dictionary of Current Idiomatic English*; Oxford University Press: Oxford, UK, 1983; Volume 2.

13. Bastiaansen, J.A.C.J.; Thioux, M.; Keysers, C. Evidence for mirror systems in emotions. *Philos. Trans. Roy. Soc.* **2009**, *364*, 2391–2404. [CrossRef]

14. Yağiz, O.; Izadpanah, S. Language, culture, idioms, and their relationship with foreign language. *J. Lang. Teach. Res.* **2013**, *4*, 953–957. [CrossRef]

15. Fladerer, M.; Frey, D. Bewusst kommunizieren: Zum Einfluss von Sprichwoertern auf das Erleben und Verhalten. In *Psychologie der Sprichwörter*; Frey, D., Ed.; Springer Verlag: Berlin, Germany, 2017; pp. 263–267.

16. Reye, H. Insekten und Spinnentiere in Sprichwort und Redensart. *Anzeiger für Schädlingskunde Pflanzenschutz und Umweltschutz* **1986**, *59*, 121–122. [CrossRef]

17. Okui, K. Mushi ga kakawaru hyougenkou (A note on the expression to be concerned with "insect"). *Kiyo Kyoyoubu* **1992**, *26*, 137–157. (In Japanese)

18. Meyer-Rochow, V.B.; Okui, K.; Henshall, K.; Brooks, D.L. The mental concept of "worm/bug" in Japanese and English idioms. *J. Finn. Anthropol. Soc.* **2000**, *25*, 29–40.

19. Meyer-Rochow, V.B. Information on therapeutically as well as nutritionally important terrestrial arthropods in North Korea and an examination of the idiomatic roles insects play in North Korean proverbs and sayings. In *North. Korea: Political, Economic and Social Issues*; Harrison, M., Ed.; Nova Sci. Publ.: New York, NY, USA, 2016; pp. 169–185. ISBN 978-1-63485-969-1.

20. Rizzolatti, G.; Sinigaglia, C. Mirror neurons and motor intentionality. *Funct. Neurol.* **2007**, *22*, 205–210.

21. Mara, D. The function of mirror neurons in the learning process. *MATEC Web Conf.* **2017**, *121*, 12012. [CrossRef]

22. Le Goff, G.; Delarue, J. Non-verbal evaluation of acceptance of insect-based products using simple and holistic analysis of facial expressions. *Food Qual. Pref.* **2017**, *56*, 285–293. [CrossRef]

23. Lönnrot, E. *Suomen Kansan Sananlaskuja*; Weilin and Göös: Helsinki, Finland, 1981.

24. Laukkanen, K.; Hakamies, P.; Kuusi, M. *Sananlaskut: 15904 Sananlaskua Kansanrunousarkistosta*; Suomalaisen Kirjallisuuden Seurantoimituksia: Helsinki, Finland, 1978.

25. Kuusi, M. *Vanhan Kansan Sananlaskuviisaus. Suomalaisia Elamanaohjeita, Kansanaforismeja, Lentavia Lauseita ja Kokkapuheita 1544–1826*; WSOY W. Söderström: Porvoo, Finland, 1953.

26. Kuusi, M. *Rapatessa Roiskuu. (Nykys Web of Conferences 121, Suomen Sananparsikirja)*; Suomalaisen Kirjallisuuden Seura: Helsinki, Finland, 1988.

27. Vuorela, T. *Maassa Maan Tavalla: 2000 Suomalaista Kansanviisautta*; WSOY, W. Söderström: Porvoo, Finland, 1979.

28. Kuusi, M. *Suomalaista, Karjalaista Vai Savokarjalaista? Vienan ja Pohjois-Aunuksen Sananlaskut ja Kalevan Runojen Alkuperäkiista*; Suomalaisen Kirjallisuuden Seura: Helsinki, Finland, 1978.

29. Jäeventausta, E. *Suoniemen Sananparsia*; Suomalaisen Kirjallisuuden Seura: Helsinki, Finland, 1974.

30. Kuusi, M. *Suomen Kansan Vertauksia*; Suomalaisen Kirjallisuuden Seura: Helsinki, Finland, 1979.

31. Horila, T. *Someron Sananparsia*; Kansanelämän Kuvauksia 10; Suomalaisen Kirjallisuuden Seura: Helsinki, Finland, 1977.

32. Virtanen, L. *Selevä pyy. Savon Kansan Sananparsia*; Kustannuskiila Oy: Kuopio, Finland, 1990.

33. Govorushko, S. *Human-Insect Interactions*; CRC Taylor and Francis: Oxford, UK, 2017.

34. Wander, K.F.W. *Deutsches Sprachlexikon*; Scientia Verlag: Aalen, Germany, 1963.

35. Stadler, F. Die Macht der Spiegelneuronen. *WIFI Wien.* **2012**, *2012*, 1–16.

36. Tsunoda, T. *Nihonjin no nō—nō no Hataraki to Tōzai no Bunka*; (The Brain of the Japanese and Brain Function in Eastern and Western Cultures); Taishuukan-Shoten: Tokyo, Japan, 1978.

37. Duda, P.D.; Brown, J. Lateral asymmetry of positive and negative emotions. *Cortex* **1984**, *20*, 253–261. [CrossRef]

38. Schapkin, S.A. Hemispheric asymmetry in processing of emotional words: Effects of extraversion and neuroticism. *Vopr. Psichol.* **2000**, *3*, 102–116. (In Russian)

39. Holtgraves, T.; Felton, A. Hemispheric asymmetry in the processing of negative and positive words: A divided field study. *Cogn. Emot.* **2011**, *25*, 691–699. [CrossRef]

40. Beraha, E.; Eggers, J.; Hindi Attar, C.; Gutwinski, S.; Schlagenhauf, F.; Stoy, M.; Sterzer, P.; Kienast, T.; Heinz, A.; Bermpohl, F. Hemispheric asymmetry for affective stimulus processing in healthy subjects- an fMRI study. *PLoS ONE* **2012**, *7*, e46931. [CrossRef]

41. Gainotti, G. Emotions and the right hemisphere: Can new data clarify old models? *Neuroscientist* **2019**, *25*, 258–270. [CrossRef]

42. Brooks, D.; Okui, K. Anthroentomology: A psycholinguistic argument for the importance of insect biodiversity to humans. *Kiyou Bull.* **1998**, *3*, 1–26.

43. Citron, F.M.M.; Cacciari, C.; Kucharski, M.; Beck, L.; Conrad, M.; Jacobs, A.M. When emotions are expressed figuratively: Psycholinguistic and affective norms of 619 idioms for German (PANIG). *Behav. Res.* **2016**, *48*, 91–111. [CrossRef]

44. Piirainen, E. Folk narrative and legends as sources of widespread idioms. *Folklore* **2011**, *48*, 117–143.
45. Meyer-Rochow, V.B. Food taboos: Their origins and purposes. *J. Ethnobiol. Ethnomed.* **2009**, *5*, 18. [CrossRef]
46. Härkönen, M. Use of mushrooms by Finns and Karelians. *Int. J. Circumpolar Health* **1998**, *57*, 40–56.
47. Mancini, S.; Sogari, G.; Menozzi, D.; Nuvoloni, R.; Torracca, B.; Moruzzo, R.; Paci, G. Factors predicting the intention of eating an insect-based product. *Foods* **2019**, *8*, 270. [CrossRef]
48. Fazly, A.; Cook, P.; Stevenson, S. Unsupervised type and token identification of idiomatic expressions. *Comput. Linguist.* **2009**, *35*, 61–103. [CrossRef]

 *foods*

Article

Growth Performance and Nutrient Composition of Mealworms (*Tenebrio Molitor*) Fed on Fresh Plant Materials-Supplemented Diets

Changqi Liu [1], Jasmin Masri [2], Violet Perez [2], Cassandra Maya [1] and Jing Zhao [2,*]

[1] School of Exercise and Nutritional Sciences, San Diego State University, San Diego, CA 92182, USA; changqi.liu@sdsu.edu (C.L.); cmaya6248@sdsu.edu (C.M.)

[2] School of Kinesiology and Nutritional Science, Rongxiang Xu College of Health and Human Services, California State University, Los Angeles, CA 91803, USA; j.masri13@gmail.com (J.M.); vperez59@calstatela.edu (V.P.)

* Correspondence: jzhao30@calstatela.edu; Tel.: +01-323-343-4665

Received: 15 January 2020; Accepted: 3 February 2020; Published: 5 February 2020

Abstract: Mealworms (*Tenebrio molitor*) have a great potential to serve as a sustainable food source for humans due to their favorable nutrient profile and low environmental impact. Feed formulation and optimization are important for mealworm production. The objective of this study was to evaluate the effects of fresh plant materials-supplemented diets on the growth performance and nutritional value of mealworms. Mealworm larvae were grown on wheat bran or wheat bran enriched with carrot, orange, or red cabbage for four weeks. Larval and pupal survival, growth rate, pupating rate, duration of pupal stage, proximate composition, reducing power, metal chelating activity, and radical scavenging activity of the mealworms were analyzed. Dietary supplementation with fresh plant materials did not result in significant changes in mealworm survival, development, proximate composition, or antioxidant activities. However, mealworm larvae fed on carrot-, orange-, and red cabbage-supplemented diets had improved growth rates, and were 40%–46% heavier in week four than those fed on wheat bran only, indicating the supplementation resulted in an increased production efficiency of mealworm larvae. Our findings may help optimize the diet formulation for mealworm mass production.

Keywords: mealworm; feed supplementation; growth performance; nutrient composition; antioxidant activity

---

1. Introduction

Entomophagy, the human consumption of insects, is practiced in many regions including parts of Africa, Asia, and Latin America [1,2]. Although Western acceptance of entomophagy remains low [3], the utilization of edible insects as human food has received increasing attention in recent years, particularly after the Food and Agriculture Organization (FAO) of the United Nations recommended using insects as a sustainable alternative to the traditional livestock [4]. Insects are rich in high quality proteins, polyunsaturated fatty acids, dietary fibers, and a variety of micronutrients [5]. In addition to the nutritional benefits, insects have a high feed conversion rate, low environmental footprints, and are significantly less land-dependent than livestock production [6]. Globally, over 1500 edible insect species are consumed [7], many of which have the potential of becoming part of the Western diet. However, one must exercise caution when selecting nonnative insects for farming, as they may be invasive and detrimental to the environment. For example, palm weevils (*Rhynchophorus* spp.) are insects widely consumed in Africa, South America, Southeast Asia, and New Guinea [7], but are also devastating pests of palm trees in California [8].

Tenebrio molitor L. (Coleoptera: Tenebrionidae), commonly known as mealworm, has been commercially produced in the US for over 70 years [9] and is a favorable candidate for insect rearing due to its high protein content, well-balanced amino acid profile [5], potential health benefits [10–12], efficient feed conversion rate [13], low greenhouse gas emissions [14], low water footprint [15], reduced land usage [9], ability to live on organic by-products [16], and available mass production technology [9]. Although mealworms can be reared exclusively on wheat bran, their diets are often enriched with additional organic matters such as potato, carrot, and cabbage to provide important nutrients [9]. Effects of dietary supplementation on mealworm survival, growth, development, fecundity, feed conversion, and nutrient composition have been reported in previous research works [13,16–21]. However, few studies have focused on antioxidant-rich supplements. Insects have a tracheal respiratory system that directly delivers oxygen to tissues and results in high levels of reactive oxygen species. Therefore, insects are prone to oxidative stress that can negatively impact their growth, development, survival, and fecundity [22]. The objective of this study was to investigate whether dietary supplementation of carrot, orange, and red cabbage as a source of antioxidants can improve growth performance, nutrient composition, and antioxidant activity of mealworm larvae.

2. Materials and Methods

2.1. Materials

Mealworm larvae and wheat bran were purchased from Mulberry Farms (Fallbrook, CA, USA). Shredded carrots, oranges, and red cabbages were purchased from a local grocery store (Los Angeles, CA, USA) and refrigerated (4 °C) until use. Reagent-grade chemicals were purchased from Fisher Scientific (Hampton, NH, USA).

2.2. Feeding

Mealworm larvae were divided into four different dietary groups: (1) 50 g of wheat bran only (control), (2) 50 g of wheat bran supplemented with 20 g of fresh carrot per day, (3) 50 g of wheat bran supplemented with 20 g of fresh orange per day, and (4) 50 g of wheat bran supplemented with 20 g of fresh red cabbage per day. Each group consisted of 60 g of mealworm larvae (approximately 670 mealworms). Mealworms were reared with an ad libitum supply of feed for four weeks. The feed was replaced weekly to rid the environment of insect waste and uneaten feed.

2.3. Growth Performance

Mealworm pupae and dead larvae were removed daily from each group, and the insect numbers were recorded. The separated pupae were observed daily for eclosed beetles. The average weight of the mealworm larvae was recorded each week on 20–30 randomly selected worms. The average weight of the pupae and beetles and the duration of pupal stage were recorded.

2.4. Sample Preparation

Mealworm larvae were sampled each week for analysis. Mealworms were fasted for 24 h and then euthanized by freezing at −40 °C for 10 min. For proximate analyses, mealworms were finely ground using a blender. For antioxidant activity assays, one gram of mealworm larvae was homogenized using a Polytron PT 3000 homogenizer (Kinematica AG, Luzern, Switzerland) in 10 mL of 70% (v/v) ethanol aqueous solution until no large pieces were visible (approximately 20 s). The homogenized samples were centrifuged using a Biofuge Stratos centrifuge (Kendro Laboratory Products, Asheville, NC, USA) at 5000× g for 10 min at 4 °C. The supernatants were collected for the analyses.

2.5. Proximate Analysis

The proximate composition of mealworm larvae was analyzed according to the AOAC Official Methods [23]. Moisture content was measured by drying the samples in a convection oven (Freas,

Marietta, OH, USA) at 125 °C for 90 min. Total crude fat was extracted in petroleum ether using an ST243 Soxtec solvent extraction system (Foss, Hilleroed, Denmark). Nitrogen content was determined using a rapid N exceed nitrogen analyzer (Elementar, Hesse, Germany). A nitrogen-to-protein conversion factor of 4.76 was used for protein quantification [24]. Ash content was measured by incinerating the samples in an Isotemp Programmable muffle furnace (Fisher Scientific, Hampton, NH, USA) at 550 °C for 48 h. Total carbohydrate was calculated by the difference.

2.6. Ferric Reducing Power

The ferric reducing power of the mealworm 70% ethanol extract was analyzed according to Oyaizu [25] with modifications. In microcentrifuge tubes, 15 µL of the sample extract was mixed with 125 µL of 0.2 M sodium phosphate buffer (pH 6.6) and 125 µL of 1% (w/v) potassium ferricyanide solution, then vortexed and incubated for 20 min at room temperature. In a 96-well plate, 125 µL of the mixture was pipetted into each well followed by addition of 125 µL of deionized water and 25 µL of 1% (w/v) ferric chloride solution. Color was developed for 5 min at room temperature and absorbance was read at 700 nm using a Synergy H1 microplate reader (BioTek, Winooski, VT, USA). Ferric reducing power was calculated using a 0.28–2.84 mM ascorbic acid standard curve ($r = 0.99$).

2.7. Ferrous Chelating Activity

Ferrous chelating activity was determined according to Carter [26] with modifications [27]. In a 96-well microplate, 20 µL of the sample extract was mixed with 125 µL of deionized water and 10 µL of 2 mM freshly prepared ferric chloride solution. The plate was set aside for 5 min at room temperature before adding 45 µL of 5 mM Ferrozine into each well. The contents of each well were gently mixed via pipetting and set aside for 10 min to allow for color development. The absorbance was read at 562 nm using a microplate reader and ferrous chelating activity was calculated using a 0.34–1.37 mM ethylenediaminetetraacetic acid (EDTA) standard curve ($r = 0.99$).

2.8. ABTS Radical Scavenging Activity

Seven millimolar 2,2′-azino-di-(3-ethylbenzthiazoline sulfonic acid) (ABTS) was mixed with 2.45 mM potassium persulfate to prepare the $ABTS^{\bullet+}$ stock solution. The reagent was kept in the dark for 12–16 h at room temperature. Before use, the $ABTS^{\bullet+}$ stock solution was diluted in 10 mM sodium phosphate buffer (pH 7.2) to an absorbance of ~0.7 at 734 nm. In a cuvette, 20 µL of the sample extract was mixed with 1.980 mL of the diluted $ABTS^{\bullet+}$ solution and allowed to stand for 10 min. Absorbance was read at 734 nm using a GENESYS 10S spectrophotometer (Thermo Scientific, Waltham, MA, USA) and ABTS radical scavenging activity was calculated using a 0.40–2.40 mM Trolox standard curve ($r = 0.99$) [28].

2.9. Statistical Analysis

All experiments were repeated at least twice with triplicate measurements for each replication. Data were analyzed using R (version 3.6.0) with emmeans and multcomp packages. One-way analysis of variance (ANOVA) and Tukey's honestly significant difference (HSD) test at $\alpha = 0.05$ were used for the overall analysis of variance and mean separation, respectively.

3. Results and Discussion

3.1. Growth Performance

Over four weeks, the larval survival rates of mealworms fed on the control diet (wheat bran only) and diets supplemented with red cabbage, carrot, and orange were 91.7%, 92.5%, 89.3%, and 89.7%, respectively, and pupating rates of the larvae were 15.4%, 12.7%, 15.7%, and 16.2%, respectively (Figure 1). Within each week, no significant difference ($p > 0.05$) in the number of live, dead, and pupated mealworm larvae was detected between diet groups.

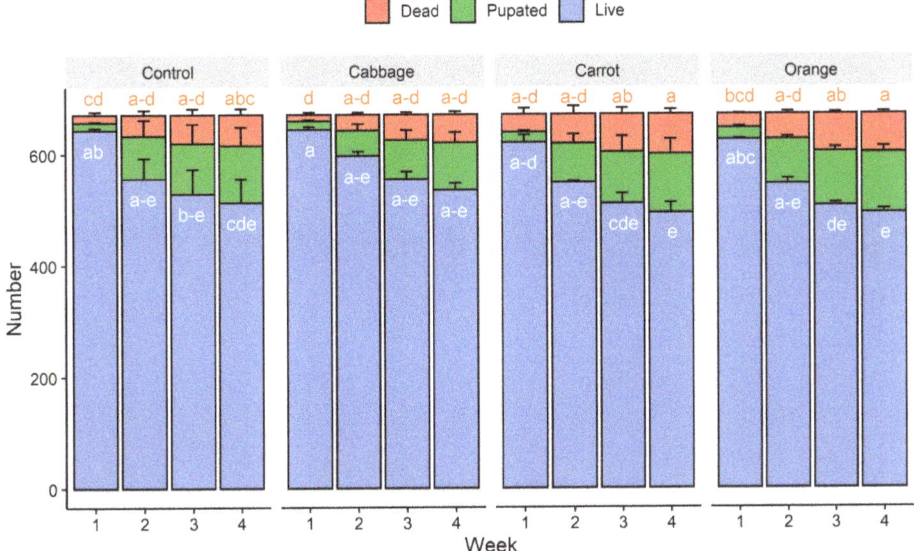

Figure 1. Number of dead, pupated, and live mealworm larvae fed on different diets over four weeks. Numbers of dead or live mealworm larvae sharing no common letters differ significantly ($p < 0.05$). No significant difference was found in the numbers of pupated mealworm larvae.

The survival rates of pupated mealworms fed on the control diet and red cabbage-, carrot-, and orange-enriched diets were 78.8%, 73.9%, 75.1%, and 77.7%, respectively (Figure 2a). The average time required for pupal development and eclosion to beetles was 14 days in all experimental groups (Figure 2b). No significant difference was found in either pupal survival rate or duration of pupal stage between diet groups.

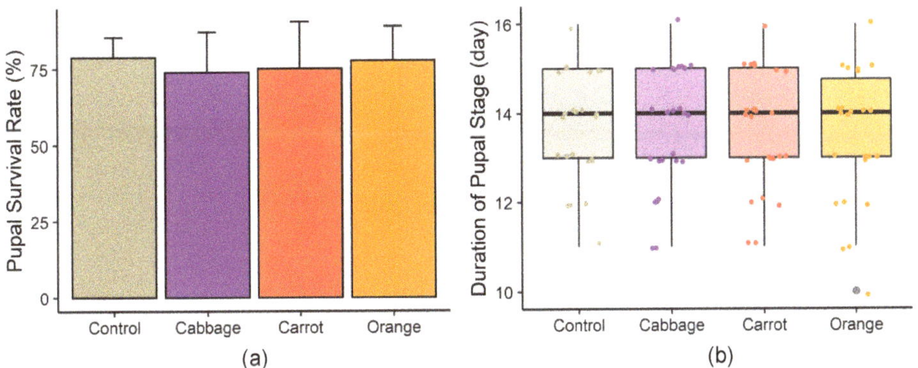

Figure 2. (**a**) Pupal survival rate and (**b**) pupa-to-beetle development time of mealworms fed on different diets.

Average weight of mealworm larvae fed on wheat bran increased 29.3% over four weeks (Figure 3). Dietary supplementation with red cabbage, carrot, and orange improved the growth rates to 37.9%, 49.3%, and 42.5%, respectively. In week four, mealworm larvae fed on fresh plant materials-supplemented diets were 40%–46% heavier ($p < 0.05$) than those fed on wheat bran only. Whether the diet was supplemented with red cabbage, carrot, or orange did not result in a significant difference in the average weight of mealworm larvae. The weight increment as a result of diet enrichment diminished after pupation and eclosion. The only significant differences observed were higher pupae weights in the red cabbage and carrot groups as compared to the control group in week four ($p < 0.05$).

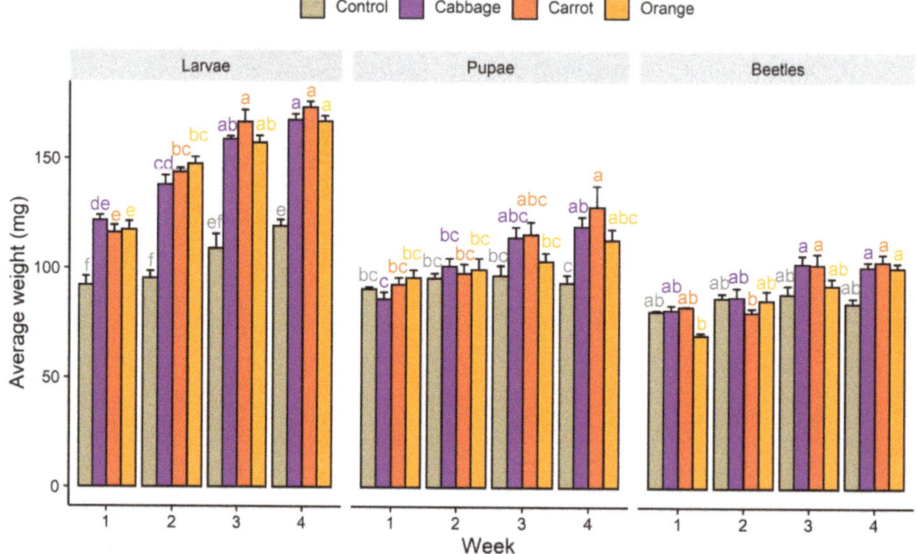

Figure 3. Average weight of mealworm larvae, pupae, and beetles fed on different diets. Means within each metamorphosis stage sharing no common letters differ significantly ($p < 0.05$).

Mealworms can obtain all required nutrients for growth, development, and reproduction from wheat bran [9]. Although feeding experiments have shown improvements in larval survival and development time when additional ingredients were provided, most of the studies have focused on supplementing and balancing macronutrients such as protein, starch, and fat [13,16–20]. The provision of carrot, orange, and red cabbage did not alter the survival or development time of mealworms in our study. This was likely due to the low concentrations of macronutrients (0.93%–1.43% protein, 0.12%–0.24% lipid, and 7.37%–11.75% carbohydrate) in the supplemented fresh plant materials [29]. Our observations were in agreement with a previous study that reported lack of improvement in house cricket (*Acheta domesticus* L.) growth performance after carrot supplementation [30]. However, another study reported that carrot consumption went up in mealworms fed on a diet of poor nutritional quality as compared to those fed on high quality diets, suggesting that carrot was utilized as a source of nutrients to compensate for poor diet quality [16].

The weight of mealworm larvae increased significantly when carrot, orange, and red cabbage were provided. This was likely due to the high moisture content of these plant materials (86.75%–90.39%) [29]. Mealworm larvae are highly drought tolerant and can actively absorb water from air [31]. Yet, they grow faster when water or high moisture foods are provided [9]. This is partly due to the reduced energy requirement for active water vapor absorption [31]. Moreover, the fresh plant materials may provide important micronutrients and bioactive phytochemicals. Orange and red cabbage are rich in ascorbic acid (53.2 mg/100 g and 57 mg/100g, respectively) [29], a vitamin needed for growth, molting, and fertility of many insects [32]. Although ascorbic acid is not an essential nutrient for mealworms [32,33], it may reduce oxidative stress that is known to retard insect growth [22]. Carrot is a good source of carotenoids such as β-carotene (8.29 mg/100 g), α-carotene (3.48 mg/100 g), lutein and zeaxanthin (256 µg/100 g) [29]. These carotenoids are potent antioxidants and can stimulate immune system of invertebrates [34]. However, supplementation with carotenoids may produce mixed results. Lifetime dietary supplementation with 0.1 mg astaxanthin/larva/week has resulted in reduced immunity, growth rate, and survival of mealworm larvae [35]. This was likely due to the interactions between astaxanthin and nitric oxide, a compound involved in both cellular and humoral immunity of insects. Astaxanthin inhibits nitric oxide synthase and scavenges circulating nitric oxide and consequently suppresses the insect immune responses [34]. It is noteworthy that red cabbage is a source of allelochemicals such as glucosinolates and flavonoids (e.g., anthocyanins). Glucosinolates are known to inhibit respiration of several insects [36] including mealworms [37]. Flavonoids can inhibit respiration and growth performance of some insects [36]; however, such detrimental effects have not been reported in mealworms. No deterrent or harmful effects of red cabbage on mealworms were observed in our study.

3.2. Proximate Composition

The larval stage of mealworm is the most favorable for efficient production and safe consumption [38]. Therefore, proximate composition was only determined for mealworm larvae and not for the pupae and beetles. Mealworm larvae fed on the control diet had a proximate composition of 64.1% moisture, 13.8% lipid, 17.6% protein, 1.5% ash, and 3.1% carbohydrate, or on a dry weight basis, 38.3% lipid, 49.1% protein, 4.1% ash, and 8.5% carbohydrate, which was comparable to the values reported in the literature [5]. Due to the presence of non-protein nitrogen in mealworms, a nitrogen-to-protein conversion factor of 4.76 was used for protein quantification [24]. Diet supplementation with fresh plant materials did not result in appreciable changes in proximate composition of mealworms over four weeks (Figure 4). This result was expected as the supplements are not a significant source of major nutrients. Moreover, the nutrient intake of insects is well regulated. When given the opportunity, insects feed selectively to reach their nutrient target. When the diet is restricted as in our experiment, insects employ postingestive regulations to balance the nutrient intake [39]. These mechanisms explain the relatively stable protein content in mealworms fed on diets that differed 2–3 fold in protein content [16]. The excess proteins ingested were eliminated as uric acid and possibly ammonia [16,39].

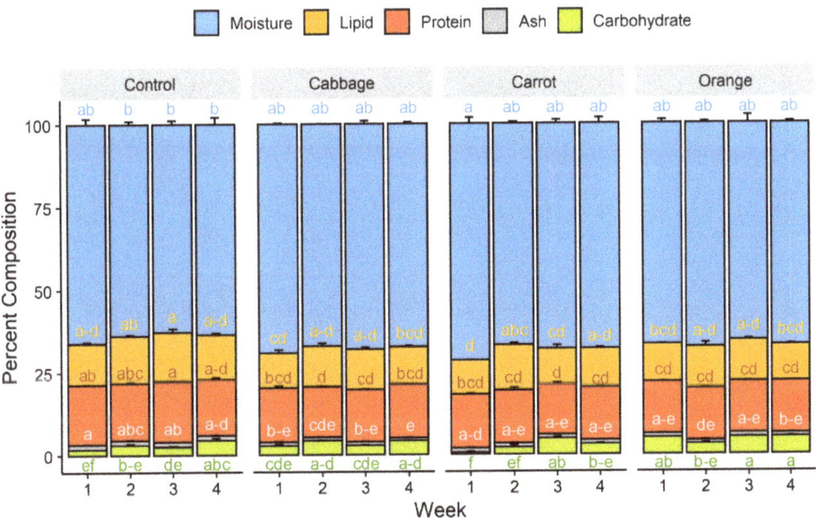

Figure 4. Proximate composition of mealworm larvae fed on different diets over four weeks. Means of each composition sharing no common letters differ significantly (*p* < 0.05).

3.3. Antioxidant Activity

While insects largely rely on endogenous antioxidant enzyme systems to balance the prooxidant-antioxidant homeostasis, dietary supplies of antioxidant can complement this process [22]. The supplemented carrot, orange, and red cabbage are rich in antioxidants such as β-carotene (carrot), ascorbic acid (orange and red cabbage), and anthocyanins (red cabbage) [29,40]. Beta-carotene inhibits lipid peroxidation by quenching singlet oxygens and scavenging peroxyl radicals [41]. Ascorbic acid neutralizes reactive oxygen species and plays pivotal roles in auxiliary antioxidant enzyme systems of insects [22,42]. Anthocyanins scavenge free radicals and inhibit the formation of highly reactive hydroxyl radicals by chelation with ferrous ions [43]. The observed improvement in larvae growth rate in the carrot, orange, and red cabbage groups may be partly attributed to the supplemented antioxidants. Antioxidant activities of the mealworm larvae fed on different diets were tested at the end of each week. Mealworms in all diet groups exhibited little variation in ferric reducing power (7.99–13.10 μmol ascorbic acid equivalent/g dry mass), ferrous chelating activity (34.74–49.96 μmol EDTA equivalent/g dry mass), and ABTS radical scavenging activity (40.40–62.17 μmol Trolox equivalent/g dry mass) over four weeks (Table 1). The ABTS radical scavenging activity was comparable to reported Trolox equivalent antioxidant capacity of mealworm aqueous extract [44]. No clear trend of antioxidant bioaccumulation as a result of dietary supplementation was observed. Although insects are known to accumulate carotenoids [45], such an effect was not noticeable in the mealworm larvae fed on carrot-supplemented diet under the tested conditions.

Table 1. Antioxidant activity of mealworm larvae fed on different diets over four weeks.

Diet	Week	Ferric Reducing Power (μmol Ascorbic Acid Equivalent/g Dry Mass)	Ferrous Chelating Activity (μmol EDTA Equivalent/g Dry Mass)	ABTS Radical Scavenging Activity (μmol Trolox Equivalent/g Dry Mass)
Control	1	13.10 ± 0.71 [a]	36.41 ± 0.06 [d,e]	55.05 ± 1.10 [a,b]
	2	11.86 ± 0.19 [a,b,c]	41.18 ± 0.47 [c]	46.99 ± 0.88 [a,b]
	3	12.97 ± 0.81 [a,b]	36.47 ± 0.22 [d,e]	45.06 ± 2.50 [a,b]
	4	11.66 ± 0.98 [a,b,c]	36.64 ± 0.19 [d,e]	58.80 ± 6.41 [a,b]
Cabbage	1	8.50 ± 1.20 [c]	37.77 ± 0.08 [d]	55.20 ± 4.53 [a,b]
	2	11.94 ± 1.23 [a,b,c]	43.77 ± 0.23 [b]	49.81 ± 1.65 [a,b]
	3	10.52 ± 0.72 [a,b,c]	38.35 ± 0.19 [d]	46.91 ± 3.51 [a,b]
	4	9.82 ± 0.40 [a,b,c]	37.89 ± 0.31 [d]	61.77 ± 7.06 [a]
Carrot	1	10.90 ± 0.58 [a,b,c]	43.74 ± 0.11 [b]	54.22 ± 3.50 [a,b]
	2	8.86 ± 0.85 [b,c]	49.96 ± 1.22 [a]	46.42 ± 3.87 [a,b]
	3	11.98 ± 0.61 [a,b,c]	43.93 ± 0.01 [b]	47.39 ± 1.10 [a,b]
	4	10.55 ± 0.79 [a,b,c]	43.98 ± 0.14 [b]	62.17 ± 1.91 [a]
Orange	1	8.39 ± 0.83 [c]	34.77 ± 0.27 [e]	53.97 ± 5.75 [a,b]
	2	7.99 ± 1.09 [c]	40.42 ± 0.43 [c]	48.65 ± 3.76 [a,b]
	3	9.63 ± 1.01 [a,b,c]	34.74 ± 0.21 [e]	40.40 ± 1.56 [b]
	4	10.80 ± 0.78 [a,b,c]	35.10 ± 0.08 [e]	58.34 ± 7.77 [a,b]

Values are expressed as mean ± standard error. Means in the same column sharing no common letters are significantly different ($p < 0.05$). EDTA: ethylenediaminetetraacetic acid. ABTS: 2,2′-azino-di-(3-ethylbenzthiazoline sulfonic acid).

4. Conclusions

In conclusion, dietary supplementation with fresh carrot, orange, and red cabbage for four weeks improved the growth rate of mealworm larvae without changing their survival rate, development time, or proximate composition. No accumulation of antioxidant activities in mealworm larvae fed on antioxidant-rich fresh plant materials was observed under the tested conditions. Our results suggested that dietary supplementation with fresh plant materials may accelerate the growth of mealworm larvae and shorten the time required to reach a desired weight for harvesting, and thereby improve the efficiency for mass production of mealworms.

Author Contributions: Conceptualization, C.L. and J.Z.; methodology, C.L. and J.Z.; validation, J.M., V.P. and J.Z.; formal analysis, C.L.; investigation, J.M., V.P. and C.M.; resources, J.Z. and C.L.; data curation, J.Z. and C.L.; writing—original draft preparation, C.L. and J.M.; writing—review and editing, J.Z., V.P. and C.M.; visualization, C.L.; supervision, J.Z.; project administration, J.Z.; funding acquisition, C.L. and J.Z. All authors have read and agreed to the published version of the manuscript.

Funding: This research was funded by USDA-NIFA-HSI Program [Grant No. 2017-38422-27108/Project Accession No. 1013661] from the USDA National Institute of Food and Agriculture, by Research, Scholarship and Creative Activities Grant Program from California State University, Los Angeles, and by Education and Research Grant from Southern California Institute of Food Technologists Section. Publication fee was partly supported by Open Access Author Fund from California State University, Los Angeles Library.

Conflicts of Interest: The authors declare no conflict of interest. The funders had no role in the design of the study; in the collection, analyses, or interpretation of data; in the writing of the manuscript, or in the decision to publish the results.

References

1. Liu, C.; Zhao, J. Insects as a Novel Food. In *Encyclopedia of Food Chemistry*; Melton, L., Shahidi, F., Varelis, P., Eds.; Academic Press: Oxford, UK, 2019; pp. 428–436. [CrossRef]

2. Hurd, K.J.; Shertukde, S.; Toia, T.; Trujillo, A.; Pérez, R.L.; Larom, D.L.; Love, J.J.; Liu, C. The cultural importance of edible insects in Oaxaca, Mexico. *Ann. Entomol. Soc. Am.* **2019**, *112*, 552–559. [CrossRef]

3. Woolf, E.; Zhu, Y.; Emory, K.; Zhao, J.; Liu, C. Willingness to consume insect-containing foods: A survey in the United States. *LWT Food Sci. Technol.* **2019**, *102*, 100–105. [CrossRef]

4. van Huis, A.; Van Itterbeeck, J.; Klunder, H.; Mertens, E.; Halloran, A.; Muir, G.; Vantomme, P. *Edible Insects: Future Prospects for Food and Feed Security*; Food and Agriculture Organization of the United Nations: Rome, Italy, 2013.

5. Rumpold, B.A.; Schlüter, O.K. Nutritional composition and safety aspects of edible insects. *Mol. Nutr. Food Res.* **2013**, *57*, 802–823. [CrossRef]

6. van Huis, A. Potential of insects as food and feed in assuring food security. *Annu. Rev. Entomol.* **2013**, *58*, 563–583. [CrossRef]

7. Dossey, A.T.; Morales-Ramos, J.A.; Rojas, M.G. *Insects as Sustainable Food Ingredients: Production, Processing and food Applications*; Elsevier: San Diego, CA, USA, 2016.

8. Milosavljević, I.; El-Shafie, H.A.; Faleiro, J.R.; Hoddle, C.D.; Lewis, M.; Hoddle, M.S. Palmageddon: The wasting of ornamental palms by invasive palm weevils, *Rhynchophorus* spp. *J. Pest Sci.* **2019**, *92*, 143–156. [CrossRef]

9. Cortes Ortiz, J.; Ruiz, A.T.; Morales-Ramos, J.; Thomas, M.; Rojas, M.; Tomberlin, J.; Yi, L.; Han, R.; Giroud, L.; Jullien, R. Insect mass production technologies. In *Insects as Sustainable Food Ingredients*; Elsevier: San Diego, CA, USA, 2016; pp. 153–201.

10. Seo, M.; Goo, T.-W.; Chung, M.; Baek, M.; Hwang, J.-S.; Kim, M.; Yun, E.-Y. *Tenebrio molitor* larvae inhibit adipogenesis through AMPK and MAPKs signaling in 3T3-L1 adipocytes and obesity in high-fat diet-induced obese mice. *Int. J. Mol. Sci.* **2017**, *18*, 518. [CrossRef]

11. Gessner, D.K.; Schwarz, A.; Meyer, S.; Wen, G.; Most, E.; Zorn, H.; Ringseis, R.; Eder, K. Insect meal as alternative protein source exerts pronounced lipid-lowering effects in hyperlipidemic obese Zucker rats. *J. Nutr.* **2019**, *149*, 566–577. [CrossRef]

12. De Carvalho, N.M.; Teixeira, F.; Silva, S.; Madureira, A.R.; Pintado, M.E. Potential prebiotic activity of *Tenebrio molitor* insect flour using an optimized in vitro gut microbiota model. *Food Funct.* **2019**, *10*, 3909–3922. [CrossRef]

13. Oonincx, D.G.; Van Broekhoven, S.; Van Huis, A.; van Loon, J.J. Feed conversion, survival and development, and composition of four insect species on diets composed of food by-products. *PLoS ONE* **2015**, *10*, e0144601. [CrossRef]

14. Oonincx, D.G.; de Boer, I.J. Environmental impact of the production of mealworms as a protein source for humans–a life cycle assessment. *PLoS ONE* **2012**, *7*, e51145. [CrossRef]

15. Miglietta, P.; De Leo, F.; Ruberti, M.; Massari, S. Mealworms for food: A water footprint perspective. *Water* **2015**, *7*, 6190–6203. [CrossRef]

16. Van Broekhoven, S.; Oonincx, D.G.; van Huis, A.; van Loon, J.J. Growth performance and feed conversion efficiency of three edible mealworm species (Coleoptera: Tenebrionidae) on diets composed of organic by-products. *J. Insect Physiol.* **2015**, *73*, 1–10. [CrossRef] [PubMed]

17. Morales-Ramos, J.; Rojas, M.; Shapiro-Ilan, D.; Tedders, W. Developmental plasticity in *Tenebrio molitor* (Coleoptera: Tenebrionidae): Analysis of instar variation in number and development time under different diets. *J. Entomol. Sci.* **2010**, *45*, 75–90. [CrossRef]

18. Morales-Ramos, J.; Rojas, M.; Shapiro-Ilan, D.; Tedders, W. Self-selection of two diet components by *Tenebrio molitor* (Coleoptera: Tenebrionidae) larvae and its impact on fitness. *Environ. Entomol.* **2011**, *40*, 1285–1294. [CrossRef]

19. Morales-Ramos, J.A.; Rojas, M.G.; Shapiro-llan, D.I.; Tedders, W.L. Use of nutrient self-selection as a diet refining tool in *Tenebrio molitor* (Coleoptera: Tenebrionidae). *J. Entomol. Sci.* **2013**, *48*, 206–221. [CrossRef]

20. Davis, G.; Sosulski, F. Nutritional quality of oilseed protein isolates as determined with larvae of the yellow mealworm, *Tenebrio molitor* L. *J. Nutr.* **1974**, *104*, 1172–1177. [CrossRef]

21. Ramos-Elorduy, J.; González, E.A.; Hernández, A.R.; Pino, J.M. Use of *Tenebrio molitor* (Coleoptera: Tenebrionidae) to recycle organic wastes and as feed for broiler chickens. *J. Econ. Entomol.* **2002**, *95*, 214–220. [CrossRef]

22. Felton, G.W.; Summers, C.B. Antioxidant systems in insects. *Arch. Insect Biochem. Physiol.* **1995**, *29*, 187–197. [CrossRef]

23. AOAC. *Official methods of analysis of AOAC International, 20 ed.*; AOAC International: Arlington, TX, USA, 2016.

24. Janssen, R.H.; Vincken, J.-P.; van den Broek, L.A.; Fogliano, V.; Lakemond, C.M. Nitrogen-to-protein conversion factors for three edible insects: *Tenebrio molitor*, *Alphitobius diaperinus*, and *Hermetia illucens*. *J. Agric. Food Chem.* **2017**, *65*, 2275–2278. [CrossRef]

25. Oyaizu, M. Studies on products of browning reaction: Antioxidative activity of products of browning reaction. *Jpn. J. Nutr. Diet.* **1986**, *44*, 307–315. [CrossRef]

26. Carter, P. Spectrophotometric determination of serum iron at the submicrogram level with a new reagent (ferrozine). *Anal. Biochem.* **1971**, *40*, 450–458. [CrossRef]

27. Sansone, K.; Kern, M.; Hong, M.Y.; Liu, C.; Hooshmand, S. Acute effects of dried apple consumption on metabolic and cognitive responses in healthy individuals. *J. Med. Food* **2018**, *21*, 1158–1164. [CrossRef] [PubMed]

28. Zhao, J.; Xiong, Y.L.; McNear, D.H. Changes in structural characteristics of antioxidative soy protein hydrolysates resulting from scavenging of hydroxyl radicals. *J. Food Sci.* **2013**, *78*, C152–C159. [CrossRef] [PubMed]

29. USDA, Agricultural Research Service. FoodData Central. Available online: https://fdc.nal.usda.gov/ (accessed on 9 January 2013).

30. Veenenbos, M.; Oonincx, D. Carrot supplementation does not affect house cricket performance (*Acheta domesticus*). *J. Insects Food Feed* **2017**, *3*, 217–221. [CrossRef]

31. Hansen, L.L.; Ramløv, H.; Westh, P. Metabolic activity and water vapour absorption in the mealworm *Tenebrio molitor* L. (Coleoptera, Tenebrionidae): Real-time measurements by two-channel microcalorimetry. *J. Exp. Biol.* **2004**, *207*, 545–552. [CrossRef]

32. Kramer, K.J.; Seib, P.A. Ascorbic acid and the growth and development of insects. *Adv. Chem.* **1982**, *200*, 275–291.

33. Vanderzant, E.S.; Richardson, C.D. Ascorbic acid in the nutrition of plant-feeding insects. *Science* **1963**, *140*, 989–991. [CrossRef]

34. Vigneron, A.; Jehan, C.; Rigaud, T.; Moret, Y. Immune defenses of a beneficial pest: The mealworm beetle, *Tenebrio molitor*. *Front. Physiol.* **2019**, *10*. [CrossRef]

35. Dhinaut, J.; Balourdet, A.; Teixeira, M.; Chogne, M.; Moret, Y. A dietary carotenoid reduces immunopathology and enhances longevity through an immune depressive effect in an insect model. *Sci. Rep.* **2017**, *7*, 12429. [CrossRef]

36. Després, L.; David, J.-P.; Gallet, C. The evolutionary ecology of insect resistance to plant chemicals. *Trends Ecol. Evol.* **2007**, *22*, 298–307. [CrossRef]

37. Pracros, P.; Couranjou, C.; Moreau, R. Effects on growth and respiration due to the ingestion of the rapeseed meal glucosinolates in young larvae of *Tenebrio molitor*. *Comp. Biochem. Physiol. A Comp. Physiol.* **1992**, *103*, 391–395. [CrossRef]

38. Liu, C.; Zhao, J. *Tenebrio molitor* larvae are a better food option than adults. *J. Insects Food Feed* **2019**, *5*, 241–242. [CrossRef]

39. Behmer, S.T. Insect herbivore nutrient regulation. *Annu. Rev. Entomol.* **2009**, *54*, 165–187. [CrossRef] [PubMed]

40. Wiczkowski, W.; Szawara-Nowak, D.; Topolska, J. Red cabbage anthocyanins: Profile, isolation, identification, and antioxidant activity. *Food Res. Int.* **2013**, *51*, 303–309. [CrossRef]

41. Stahl, W.; Sies, H. Antioxidant activity of carotenoids. *Mol. Asp. Med.* **2003**, *24*, 345–351. [CrossRef]

42. Arrigoni, O.; De Tullio, M.C. Ascorbic acid: Much more than just an antioxidant. *Biochim. Et Biophys. Acta (Bba) Gen. Subj.* **2002**, *1569*, 1–9. [CrossRef]

43. Noda, Y.; Kaneyuki, T.; Mori, A.; Packer, L. Antioxidant activities of pomegranate fruit extract and its anthocyanidins: Delphinidin, cyanidin, and pelargonidin. *J. Agric. Food Chem.* **2002**, *50*, 166–171. [CrossRef]

44. Di Mattia, C.; Battista, N.; Sacchetti, G.; Serafini, M. Antioxidant activities *in vitro* of water and liposoluble extracts obtained by different species of edible insects. *Front. Nutr.* **2019**, *6*, 106. [CrossRef]

45. Duffey, S.S. Sequestration of plant natural products by insects. *Annu. Rev. Entomol.* **1980**, *25*, 447–477. [CrossRef]

 *foods*

Article

Could Defatted Mealworm (*Tenebrio molitor*) and Mealworm Oil Be Used as Food Ingredients?

Yang-Ju Son [1,2], **Soo Young Choi** [3], **In-Kyeong Hwang** [2], **Chu Won Nho** [1] **and Soo Hee Kim** [4,*]

1   Natural Products Research Institute, Korea Institute of Science and Technology, Gangneung Institute of Natural Products, Gangneung, Gangwon-do 25451, Korea; yangjuson@kist.re.kr (Y.-J.S.); cwnho1423@kist.re.kr (C.W.N.)
2   Department of Food and Nutrition and Research Institute of Human Ecology, Seoul National University, Seoul 08826, Korea; ikhwang@snu.ac.kr
3   Sempio, Seoul 04557, Korea; drd0603542@naver.com
4   Department of Culinary Arts, Kyungmin University, Uijeongbu, Gyeonggi-do 11618, Korea
*   Correspondence: kshee@kyungmin.ac.kr; Tel.: +82-31-828-7271; Fax: +82-31-828-7951

Received: 8 December 2019; Accepted: 27 December 2019; Published: 2 January 2020

Abstract: Before edible insects may be used as an alternative food, it is necessary to develop basic product forms and evaluate their characteristics. We made two basic commercial products (defatted powder and oil) from mealworm, a popular edible insect. The defatted mealworm powder possessed a sufficient amount of protein, and it had a savory taste due to plentiful free amino acids. Additionally, it had abundant minor nutrients and bioactive compounds. The physicochemical properties of mealworm oil were very similar to vegetable oil, and mealworm oil was also abundant in bioactive nutrients, especially γ-tocopherol. In addition, the predicted shelf life of mealworm oil was suitable for commercial use. Moreover, mealworm had high antioxidant and anti-inflammation activities, which may arise from functional peptides and glucosamine derivatives such as chitin and chitosan. In short, the defatted mealworm powder and mealworm oil could be successfully used as novel food ingredients.

Keywords: mealworm; edible insect; defatted powder; mealworm oil; characteristics

---

1. Introduction

Although people are apprehensive about consuming insects, edible insects are one of the most promising alternative protein sources. Historically, humans have consumed many types of insects across the globe. The earth has been undergoing environmental change, and one of the major causes of this has been the increase in consumption of meat from vertebrates. However, it is a challenge to replace all animal-based foods with vegetables because of the low content of proteins and limiting amino acids in plants. For this reason, vegetarians may be critically exposed to a paucity of particular vitamins and minerals. In comparison, insects possess sufficient amounts of protein, and their amino acid distribution is better than that of vegetative protein sources [1].

The mealworm (*Tenebrio molitor* L.) is an edible insect that has been popular in the animal feed industry, and mealworm-based feed products have been extensively used from pets to fish farms. Numerous studies have found that mealworm provides moderate nourishment and that there is no noxious effect of consumption [2–5]. Moreover, one of the important advantages of mealworm as a food source is that it is suitable for mass production. The mealworm is tolerant to various environmental conditions and does not require a large area for growth [6]. Additionally, the mealworm has good sensory characteristics. Their texture is crispy, and the mealworm has a savory taste that is very similar to dried shrimp.

A major problem associated with edible insects is an aversion to consumption based on their appearance. Thus, when mealworms are used in the food industry as an ingredient, processing such as pulverization and extraction could be helpful. However, the mealworm is composed of almost 30% fat, and this exceedingly high fat content is troublesome during pulverization processes. The agglomeration of powders, known as caking, is facilitated by high fat content in powders, and this state gives rise to a decrease in powder flowability [7]. For this reason, the separation of protein-rich and oil-rich components could be a good approach for developing food items from mealworms, as is the case for soybean. Hence in this study, defatted mealworm powder, and mealworm oil as a byproduct, were separated and the nutrition value and physicochemical characteristics of each were examined.

The goal of this study was to suggest proper products from mealworm that can be used in the food industry and consumed by public. Considering the psychological hesitation to eat insects, pulverization process was applied for elimination of its appearance. The defatted powder and oil were selected as basic products because the high fat content of mealworms disturbed the powdering process and transport equipment commonly used in the food industry. For defining characteristics of defatted mealworm powders, nutritional values and physicochemical characteristics were determined.

2. Materials and Methods

All chemicals used in this study were purchased from Sigma-Aldrich (St. Louis, MO, USA), and live mealworms were purchased from a farm in Gyeonggi-do, Korea.

2.1. Preparation of Mealworm Powder and Oil Samples

Live mealworms were fasted for three days to empty their guts. Mealworms were washed with tap water three times and blanched in boiling water (1:5, *w/w*) for 3 min. Blanched mealworms were kept on a colander for cooling and removing excess water. After 30 min, the remaining water was removed with a paper towel. An industrial hot-air dryer (LH.FC-PO-150; Lab house, Pocheon, Korea) was used for drying mealworms (12 h, 60 °C). Then, dried mealworms were pulverized with a blender (HR-2860; Philips, Amsterdam, The Netherlands) until resulting powders could pass a 30-mesh sieve (535 μm), and this powder was used as whole-fat mealworm powder (WF-M).

For extracting oils, fivefold n-hexane was added to WF-M, and the mixture was placed on a shaker (SI600R; Lab Companion, Daejeon, Korea) for 6 h at 170 rpm. The extracted liquids were filtered with a No. 1 filter paper (Whatman, Buckinghamshire, UK). After three repetitions of extraction, n-hexane was removed with an evaporator (RE111; Büchi, Flawil, Switzerland) at 34 °C. The remained n-hexane in oils were eliminated with nitrogen gas, and the supernatant was collected as mealworm oil after centrifugation for 10 min at 1350× *g* (Combi-514R; Hanil Science Industrial, Daejeon, Korea). A picture of mealworm oil is shown in Figure S1. The unsaponifiable lipid was made from mealworm oil by applying a saponification process, and its yield was 1.8% ± 0.0%.

After oil extraction, the powder remaining on the filter was collected. The defatted mealworm powder was put in a hood for 24 h at room temperature and further dried using a speed vacuum evaporator (Maxivac alpha; Labogene ApS, Lynge, Denmark) for 24 h at 40 °C. The pressure was adjusted to under 7 mmHg, and its rotation speed was 2000 rpm. Finally, the powder was ground once more with a blender (HR-2860; Philips, The Netherlands) until it could pass through a 42-mesh sieve (355 μm), and it was defatted the mealworm powder (DF-M) sample. The mealworm protein isolate (MPI) was produced from DF-M using the method of Sharma and Singh [8]. First, 20 g of DF-M was mixed with 200 mL of deionized water, and the pH was adjusted to pH 12 with adding 1 N NaOH. After shaking for 1 h at 220 rpm (SI600R; Lab Companion, Korea), the solution was centrifuged for 20 min at 2000× *g* (Combi-514R; Hanil Science Industrial, Korea). The supernatant was collected, and the pH of solution was adjusted for 4.5 using 1 N HCl. After centrifuging for 20 min at 2000× *g* once more, liquid part was eliminated, and residue was collected only. The MPI was prepared by lyophilizing the residue. The yield of MPI was 13.0% ± 0.3% and the crude protein content of MPI was 91.1% ± 1.1%. Pictures of mealworm powder samples are presented in Figure S2.

The 80% methanol extract was obtained from DF-M. 40 g of DF-M was put into a flask, and 80% methanol solution (1:10, *w/v*) was added. At room temperature, the flask was placed in a shaker (SI600R; Lab Companion, Korea) for 12 h at 200 rpm. The solvent was filtered with a No.1 filter paper (Whatman, UK) and remaining powders were extracted twice more. Collected solvents were evaporated with an evaporator (RE111; Büchi, Switzerland), and the extraction yield was 15.5% ± 0.6%.

2.2. Proximate Compositions and Amino Acid Compositions

AOAC methods were used for determining moisture, crude fat, crude protein, and ash contents [9]. The carbohydrate content was calculated based on the sum of other contents. For analyzing the amino acids compositions, 0.5 g of mealworm powder was put in a flask and 20 mL of 6 N HCl was added, heated to 110 °C, and incubated for 24 h. The total volume of the solution was adjusted to 25 mL with distilled water after filtering. For derivatization of amino acids, borate buffer, o-phthaldialdehyde/2-mercaptopropionic acid, and fluorenylmethyloxycarbonyl chloride were mixed just before HPLC injection. The reversed-phase HPLC system (Ultimate 3000; Thermoscientific Dionex, MA, USA) was used for analyzing the content of amino acids, and the analysis conditions are shown in Table S1 [10].

2.3. Protein Solubility

The protein solubility of mealworm powder was determined with little modifications using the methods of Morr et al. [11] Distilled water was added to mealworm powder (1:10, *w/v*), and pH was adjusted to a range of 2–12 with 1 N NaOH and 1 N HCl. After centrifuging for 20 min at 2000× *g* (Combi-514R; Hanil Science Industrial, Korea), the supernatant was collected. The Bradford method was used for measuring the amount of protein in the supernatant, and the protein solubility at each pH was compared with the protein solubility at pH 12, the maximum point of protein solubility.

2.4. Gel Permeation Chromatograph Analysis

Gel permeation chromatography (GPC) was used to observe soluble particle size distributions. Mealworm powder samples were dissolved in 0.02 M $NaNO_3$ (1:10, *w/v*) for gathering hydrophilic particles, and the supernatant was collected by centrifugation (2000× *g*, 20 min). The GPC analysis was conducted with conditions as presented in Table S2. The soluble particle sizes were obtained by comparison with a calibration plot of log (molecular weight) versus retention time, created with polystyrene standards (Figure S3) [12].

2.5. Fatty Acid Composition

A total of 0.25 g of mealworm oil was mixed with 6 mL of 0.5 M methanol sodium hydroxide, and heated using a water bath (80 °C, 10 min) for methylation of the oils. The mixtures were cooled on ice, and 7 mL of 14% boron trifluoride methanol solution was added, then the mixture was reheated (80 °C, 2 min). After cooling on ice, 5 mL of n-hexane was added and the mixture was again heated (80 °C, 1 min). The top layer was collected and examined for fatty acid composition. Gas chromatography (GC) was used for analysis under conditions described in Table S3 [13].

2.6. Physicochemical Properties of Mealworm Oil

The specific gravity of mealworm oil was determined according to ASTM D1298 using a pycnometer. Specific gravity was calculated as the ratio of gravities between that of mealworm oil and distilled water at 15 °C. The viscosity of mealworm oil at 20 °C was measured with a Brookfield DV-IP viscometer (Brookfield, Middleboro, MA, USA) using spindle No. 62 at 60 rpm. Color values were measured with a colorimeter (CM-3500d; Minolta, Tokyo, Japan). A total of 4 g of mealworm oil was put to 35-mm petri dishes, and a light source was set to D65–10°. The Hunter Lab colorimetric system was employed. The peroxide values and acid values were determined by the method of the Ministry of Food and

Drug Safety of Korea [14]. The thiobarbituric acid reactive substances (TBARS) content in oil was measured using the method of Buege and Aust [15]. The induction time of oil was analyzed using rancimat 734 (Metrohm, Herisau, Switzerland). As described in the method of Gómez-Rico et al. [16], the measurement conditions were analogous to AFNOR NF EN ISO 6886. A total of 3 g of mealworm oil was put in a container, and oxidation was induced at 98 °C under 20 L/h air flow.

2.7. Minor Nutrients in Mealworm Oil

The contents of tocopherols in mealworm oil was measured using the method of Gliszczyńska-Świgło and Skiorska [17] with some modifications. First, 1 g of oil was dissolved in 5 mL of n-hexane and filtered with a 0.2 µm syringe filter. Reversed-phase HPLC was used with operation conditions as described in Table S4. For the determination of the content of polyphenol, 10 g of oil was dissolved in 50 mL of n-hexane, and 20 mL of 60% methanol was added. After separating layers, the methanol fractions were collected and evaporated at 40 °C. This condensed polyphenol residue was analyzed using the Folin–Ciocalteu method [18]. The content of squalene and sterols was analyzed in unsaponifiable lipid. The operating conditions for GC-MS are described in Table S5.

2.8. Bioactive Nutrients of Mealworm Powder and Antioxidant Capacity

γ-Aminobutyric acid (GABA) and taurine contents were analyzed using the same methods as amino acid composition analysis. Total glucosamine content was measured by the method of Belcher, Nutten, and Sambrook [19]. The content of polyphenol was determined by the Folin–Ciocalteu method [18], and the flavonoid content was measured by the method of Meda et al. [20] The polyphenol and flavonoid content was calculated with gallic acid equivalent (GAE) and catechin equivalent (CE) using standard curves, respectively. The antioxidant capacity of mealworm powder was confirmed by measuring DPPH radical scavenging activity and ABTS radical scavenging activity [21,22]. The antioxidant capacities of mealworm samples were compared with trolox standard, and were computed using trolox equivalent (TE).

2.9. NO Reduction in Lipopolysaccharide-Induced RAW 264.7 Cell Line

The murine macrophage cell line (RAW 264.7) was purchased from the American Type Culture Collection (ATCC, Rockville, MD, USA). The RAW 264.7 cells were seeded in a 96-well plate with a density of 5×10^4 cells per well. Then, 24 h after plating, the media was removed and cells were washed with PBS. The 80% methanol extract of DF-M and the unsaponifiable lipid of mealworm oil were added to cell cultures in various concentrations at 50 µL per well. An hour later, the 50 µL of media containing lipopolysaccharide (LPS) (2 µg mL^{-1}) was added to each well. To analyze nitrite production, the supernatant of each well was collected after 24 h and mixed with Griess reagent at a 1:1 ratio. After 20 min, samples were analyzed with a microplate reader (SpectraMax 190, Molecular Devices, Sunnyvale, CA, USA) at 550 nm. The concentration of nitric oxide was calculated by comparison with a sodium nitrite standard curve.

2.10. Statistical Analysis

All data are represented as the mean ± standard deviation (SD). Measurements were performed in triplicate. One-way analysis of variance (ANOVA) and Duncan's multiple-range test ($p < 0.05$) were performed using the IBM SPSS statistics program, version 21.0 (IBM Inc., Armonk, NY, USA).

3. Results and Discussion

3.1. Proximate Compositions and Amino Acid Compositions of Mealworm

The mealworm is composed of high fat and high protein with low carbohydrate content (Table S6). The proximate composition of freeze dried WF-M is almost the same as that of Zhao et al. [23] The WF-M had 32.3% ± 1.0% lipid content and this content is much higher than that found in soybean or meat [24,25]. When oils were extracted with n-hexane, DF-M had 70.8% ± 5.8% protein content with 2.0% ± 0.2% lipid content. Therefore, DF-M could be used as a protein rich food source.

The amino acid composition was calculated as a ratio to the crude protein amount (Table 1). The essential amino acid and non-essential amino acid ratio (E/NE) of mealworm was lower than meats such as chicken (0.75) and beef (0.7–0.8) [26,27]. Additionally, the content of branched amino acids (BCAA) was approximately 2% less than that of beef, eggs and chicken and accounts for 20%–22% of all amino acids [26,27].

Table 1. Amino acids composition of whole-fat and defatted mealworm.

	WF-M [†]	DF-M [‡]
Protein (%)	52.2 ± 0.6	70.8 ± 5.8
Essential amino acids (g/100 g protein)		
Histidine	3.1 ± 0.0	2.9 ± 0.2
Lysine	5.1 ± 0.1	5.1 ± 0.4
Methionine	0.5 ± 0.0	1.2 ± 0.1
Phenylalanine	3.9 ± 0.0	3.9 ± 0.2
Threonine	4.8 ± 0.0	4.3 ± 0.2
Isoleucine	4.5 ± 0.0	4.5 ± 0.2
Leucine	7.5 ± 0.0	7.5 ± 0.4
Valine	6.4 ± 0.0	6.4 ± 0.4
Sub total	35.8 ± 0.0	35.6 ± 0.3
Non-essential amino acids (g/100 g protein)		
Alanine	8.1 ± 0.0	8.1 ± 0.6
Aspartic acid	8.4 ± 0.0	8.4 ± 0.4
Arginine	5.7 ± 0.0	5.7 ± 0.3
Cysteine	N.D.	N.D.
Glutamic acid	13.0 ± 0.0	13.2 ± 0.5
Glycine	5.3 ± 0.0	5.3 ± 0.4
proline	5.2 ± 0.0	5.4 ± 0.4
Serine	4.8 ± 0.0	4.8 ± 0.3
Tyrosine	7.3 ± 0.0	7.6 ± 0.5
Sub total	57.7 ± 0.0	58.5 ± 0.4
Total (E + NE) [§]	93.5 ± 0.0	94.1 ± 0.4
E/NE	62.0	60.9
BCAA contents (%) [¶]	19.4 ± 0.0	18.3 ± 0.4

[†] WF-M: Blanched and hot-air dried mealworm, [‡] DF-M: Defatted WF-M with solvent (n-hexane), [§] E: essential amino acids, NE: non-essential amino acids, [¶] BCAA: branched amino acids.

The amino acid scores were calculated with the criteria of FAO/WHO [28] (Table 2). WF-M and DF-M both indicated methionine as a limiting amino acid. The DF-M presented a higher amino acid score than WF-M more than twice, and this may be caused by the difference in the loss-ratio for different types of amino acids. In the study of Jones et al. [29], there were two essential amino acids whose amino acid values were lower than the criteria, and these amino acids were same as those identified in this study (lysine and methionine). Jones et al. [29] said that the amino acid composition of mealworm could vary significantly depending on the mealworm feed, and lysine and methionine are known to be restricted amino acids in grains such as rice and soybean. Therefore, it is deemed necessary to increase the content of methionine in mealworm feed for better amino acid scores.

Table 2. Essential amino acid amounts and amino acid scores of mealworm powders.

	FAO/WHO Ref. (1985)	WF-M [†]	DF-M [‡]
Histidine	20	30.7	28.9
Lysine	55	51.1	50.8
Methionine + Cysteine	35	5.1	11.5
Phenylalanine + Tyrosine	60	111.6	114.5
Threonine	40	47.9	43.3
Isoleucine	40	44.6	44.5
Leucine	70	75.4	74.5
Valine	50	63.9	64.0
Limiting amino acids		Met	Met
Amino Acids Score (AAS)		14.6	32.9

[†] WF-M: Whole-fat mealworm powder, [‡] DF-M: Defatted WF-M.

3.2. Protein Solubility and Soluble Particle Size Distributions of Mealworm Powders

The protein solubility of DF-M and MPI over a range of pH conditions are shown in Figure S4. For DF-M, protein solubility was less than 30% at pH 7 compared to pH 12. It was expected that mealworm protein is mainly composed of hydrophobic rather than hydrophilic protein, which can cause increased turbidity in solution. Therefore, it was deemed that it would be necessary to modify or degrade mealworm protein to enhance its solubility in use. In comparison, the MPI sample had a rapid decrease in solubility only near the isoelectric point, pH 4–6. Therefore, MPI appears to be a more suitable form for use in liquid foods.

Using gel permeation chromatography (GPC), the size distribution of soluble particles of DF-M and MPI were analyzed (Figure 1). The hydrophilic solutions of DF-M and MPI showed peaks at 117–118, 308–311, 958, and 1450 Da. As the average molecular weight of amino acids is approximately 142 Da, peaks that matched single amino acids (117–118 Da) and peptides composed of two (308–311 Da), seven (958 Da), and 10 (1450 Da) amino acids were analyzed. The most conspicuous difference in peak areas between DF-M and MPI was the peak of 308–311 Da. In DF-M, the peak area of the protein fraction with a molecular weight of 308 Da was 42.6%, but when protein was purified to MPI, the peak area of 311 Da was reduced to 18.0%. Therefore, it is assumed that a large number of particles in this range was lost during MPI refining.

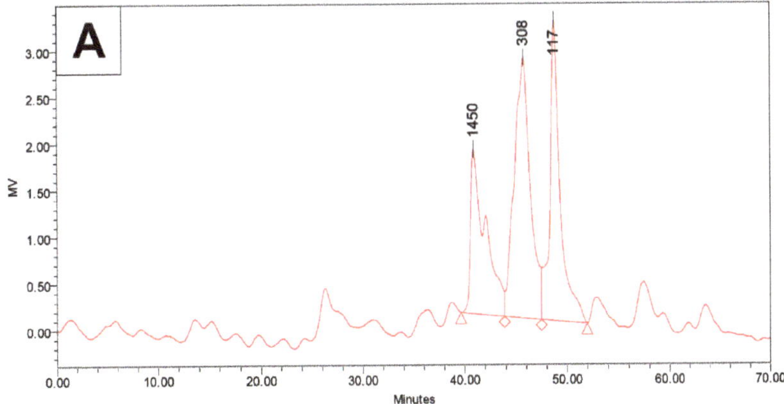

Figure 1. *Cont.*

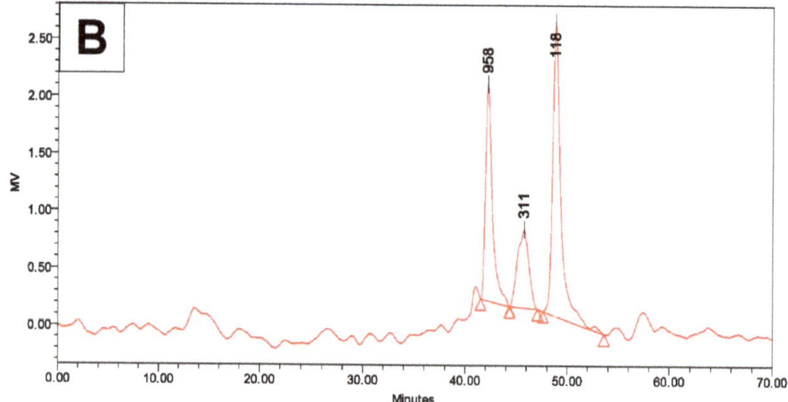

Figure 1. Size distribution of soluble particles of mealworm samples by gel permeation chromatography (GPC) analysis. (**A**) Defatted mealworm powder; (**B**) mealworm protein isolate.

3.3. Fatty Acid Composition of Mealworm Oil

Regarding the composition of fatty acids in mealworm oil (Table 3) oleic acid (44.5% ± 1.1%) was the most abundant fatty acid, and linoleic acid was the (19.5% ± 0.8%) second most abundant. The fatty acid composition results showed were very similar to study of Jeon et al. [13] about oil of roasted or unroasted mealworm oil. Animal lipids contain high levels of saturated fatty acids (SFA) compared to vegetable lipids. Mealworm, however, possessed a much higher amount of unsaturated fatty acid (66.0% ± 0.7%). The content of polyunsaturated fatty acid (PUFA) of mealworm (19.8% ± 0.9%) was also higher than pork (15.7%), lamb (2.3%), and beef (1.6%) [30]. In comparison, mealworm oil had a much higher value (47.0 ± 4.6) for the ratio of n-6 fatty acids and n-3 fatty acids than typical animal lipids. The high ratio of n-6: n-3 is characteristic of vegetative oils, and the ratio for corn oil (31.9) was similar to that of mealworm oil [31]. Therefore, the composition of fatty acids in mealworm oil is believed to be much closer to the characteristics of vegetable oil.

Table 3. Fatty acid composition of mealworm oil.

Fatty Acids	Composition (g/100 g oil)
C4:0	N.D.
C6:0	N.D.
C8:0	N.D.
C10:0	N.D.
C11:0	N.D.
C12:0	0.3 ± 0.0
C13:0	0.1 ± 0.0
C14:0	4.0 ± 0.2
C14:1$_{cis}$	N.D.
C16:0	15.8 ± 0.1
C16:1$_{cis}$	1.8 ± 0.1
C17:0	0.1 ± 0.0
C18:0	2.3 ± 0.2
C18:1$_{cis}$	44.5 ± 1.1
C18:2 (n-6)	19.5 ± 0.8
C18:3 (n-3)	0.4 ± 0.1
C20:0	0.1 ± 0.0
C20:1$_{cis}$	0.1 ± 0.0
C20:2 (n-6)	0.1 ± 0.0

Table 3. *Cont.*

Fatty Acids	Composition (g/100 g oil)
Total	88.6 ± 0.6
SFA [†]	22.6 ± 0.1
MUFA [‡]	46.2 ± 1.6
PUFA [§]	19.8 ± 0.9
USFA [¶]	66.0 ± 0.7
n-6:n-3 ratio	47.0 ± 4.6
P:S ratio	0.9 ± 0.0

Data are expressed as mean ± SD, [†] SFA: Saturated fatty acid, [‡] MUFA: Monounsaturated fatty acid, [§] PUFA: Polyunsaturated fatty acid, [¶] USFA: Unsaturated fatty acid.

3.4. Physicochemical Properties and Minor Nutrients of Mealworm Oil

The physicochemical properties and amounts of minor nutrients of mealworm oil are described in Table 4. Mealworm oil appeared as a light-yellow liquid, such as commonly used edible oils, with an extraction yield of 29.5% ± 1.0%. The peroxide value of mealworm oil was higher than those of commercialized edible oils, which were in the range of less than 2 meq/1000 g oil [31]. A previous study [13] of mealworm oil showed even higher peroxide values (4.98–9.94 meq/1000 g oil) than this study, suggesting that the high peroxide value of extracted oil may be a trait of the mealworm itself.

Table 4. Physicochemical properties and nutrient composition of mealworm oil.

	Mealworm Oil
Extraction Yield (%)	29.5 ± 1.0
Specific gravity (15 °C)	0.8528 ± 0.0032
Viscosity (cP)	324.2 ± 6.3
Color	
L (lightness)	38.9 ± 0.3
a (redness)	−1.9 ± 0.1
b (yellowness)	7.5 ± 0.6
Peroxide value (meq/1000 g oil)	3.5 ± 0.2
Acid value (mg KOH/g oil)	2.6 ± 0.0
TBARS (mg MDA/1000 g oil)	1.8 ± 0.1
The contents of tocopherols (mg/1000 g oil)	
α-Tocopherol	6.3 ± 0.1
β-Tocopherol	8.5 ± 0.6
γ-Tocopherol	123.5 ± 2.7
δ-Tocopherol	6.1 ± 0.3
Total tocopherol	144.3 ± 3.0
Total polyphenol (mg GAE [†]/1000 g oil)	18.0 ± 1.3
Squalene (mg/1000 g oil)	21.1 ± 5.0
Induction time (h)	36.0 ± 0.7
Predicted shelf-life (day) [‡]	305

Data are expressed as mean ± SD, [†] GAE: Gallic acid equivalent, [‡] Shelf-life prediction at 20 °C was calculated by OSI_{20} × (factor of safety), OSI_{20}: induction time at 20 °C (calculated by Q_{10} factor: 2.05), factor of safety: 0.7.

Tocopherol (vitamin E) is one of the most important nutrients in lipids. It is a natural antioxidant that prevents free radical and hydroperoxy radical oxidation of lipids in fat-soluble foods, and it is synthesized in plants and algae [32]. In mealworm oil, the total tocopherol content was 144.3 ± 3.0 mg/1000 g oil, and γ-tocopherol was the main type of tocopherol (123.5 ± 2.7 mg/kg oil), accounting for 85.6% of the total tocopherol. In regard to animal food sources, chicken and pork contained 2.7–4.2 and 9.5–100 mg/1000 g oil total tocopherol, respectively [33,34]. In comparison, the total tocopherol amounts in olive and soybean oils are in the range of 80–1360 mg/1000 g oil [35], so the total tocopherol content of mealworm oil was lower than vegetable sources and higher than animal lipids. Polyphenols are also widely distributed in vegetable foods, and are known to have variable functionalities. The polyphenol content of mealworm oil was measured at 18.0 ± 1.3 mg/1000 g oil. This value was 10%–20% of that in olive oil and grape seed oil (62–172 mg/1000 g oil) [16]. Some vegetable oils contain abundant phytosterols, such as stigmasterol and ergosterol, and these compounds have biological activities when ingested, but neither phytosterol was detected in mealworm oil of this study. Cholesterol is mainly contained in animal foods and is known to be associated with arteriosclerosis, increasing the concentration of lipid globules in the blood. When GC-MS was used for detection of cholesterol, the chromatogram for mealworm oil had extremely small peaks, therefore, we also failed to quantify this. Therefore, mealworm oil was much richer in minor nutrients than animal lipids, and its cholesterol amount was lower than the limit of quantitation and limit of detection (LOQ ≈ 2.6 ppm, LOD ≈ 8.7 ppm).

The induction time of mealworm oil was measured by accelerating oxidation using a rancimat, and the commercial extra-virgin olive oil was used as control in this study. The induction time of mealworm oil was 36.0 ± 0.7 h, and it was lower than that of olive oil (47.6 ± 0.8 h). A higher induction time indicates that they are less oxidized. In the study of Gómez-Rico et al. [16], the induction time of commercial olive oil was in the range of 30–60 h, and this value was similar with that obtained in this study. For the viability of mealworm oil as a commercial product, we predicted the shelf life of oil based on the induction time results. In the study of Farohoosh [36], they calculated Q10 values and the oil stability index (OSI20) that offers a prediction of shelf life of oils at 20 °C based on the induction time. Under these conditions, the shelf life of olive oil in this study was forecasted to 404 d. This value was similar with shelf life of commercial extra virgin olive oil. When using this forecast model, the estimated shelf life of mealworm oil at 20 °C was 305 d, which was approximately 10 months.

3.5. Bioactive Nutrients, Antioxidant Capacity, and Anti-Inflammation Activity of Mealworms

In Table 5 the bioactive nutrients in DF-M are presented. In general, the amount of taurine in animal food sources is higher than that in plants. For example, beef and pork have 40–60 mg/100 g [37]. Although the amount of taurine in mealworms was about half as much as that in beef or pork, the taurine content of DF-M (17.8 ± 0.3 mg/100 g) was higher than that of plant sources. The mealworm is known to contain glucosamine derivatives in its skin, such as chitin and chitosan. Glucosamine is one of the natural amino sugars of the hexosamine family and it is a basic monomer of chitin (N-acetyl-D-glucosamine) and chitosan. Chitin and chitosan relieve degenerative arthritis by inhibiting inflammatory reactions in the joint region [38]. The total glucosamine content of DF-M was measured as 7.0% $\pm$ 0.7%.

Table 5. The γ-Aminobutyric acid (GABA), taurine, glucosamine, polyphenol, and flavonoid contents and radical scavenging activities of defatted mealworm.

	DF-M [†]
GABA (mg/100 g dry basis)	3.5 ± 0.1
Taurine (mg/100 g dry basis)	17.8 ± 0.3
Glucosamine (g/100 g dry basis)	7.0 ± 0.7
Polyphenol (mg GAE [‡]/100 g dry basis)	8.2 ± 0.2
Flavonoid (mg CE [§]/100 g dry basis)	2.0 ± 0.2
DPPH (mg TE [¶]/g dry basis)	21.5 ± 0.5
ABTS (mg TE [¶]/g dry basis)	12.3 ± 0.5

Data are expressed as mean $\pm$ SD, [†] DF-M: Defatted mealworm powder, [‡] GAE: Gallic acid equivalent, [§] CE: Catechin equivalent, [¶] TE: Catechin equivalent.

The content of polyphenols and flavonoids in DF-M were 8.2 ± 0.2 mg GAE/100 g and 2.0 ± 0.2 mg CE/100 g, respectively, and these amounts were lower than those of vegetables such as onions, tomatoes, and cabbage, and fruits such as cherries and plums (13–100 mg GAE/100 g) [39]. However, the DPPH and ABTS radical scavenging powers of mealworm extract were not much less than vegetable foods. Thus, it was thought that there are additional substances involved in antioxidant properties besides polyphenols and flavonoids. Many studies have shown that some peptides in protein foods have excellent antioxidant properties [40]. Therefore, it is expected that some specific peptides in mealworm also affect in its antioxidant capacity.

In addition, as shown in Figure 2, the NO production of LPS-induced RAW 264.7 cells were lowered by treatment with 80% MeOH extract of DF-M (25–500 µg/mL) and unsaponifiable lipid (0.05–5 µg/mL). As suggested for antioxidant activities, functional peptides and glucosamine derivatives such as chitin and chitosan may strongly affect the anti-inflammation activity of mealworm extract [38,41]. Additionally, the unsaponifiable lipid of mealworm oil decreased NO production significantly, demonstrating that the lipid portion of mealworm could alleviate inflammation as well.

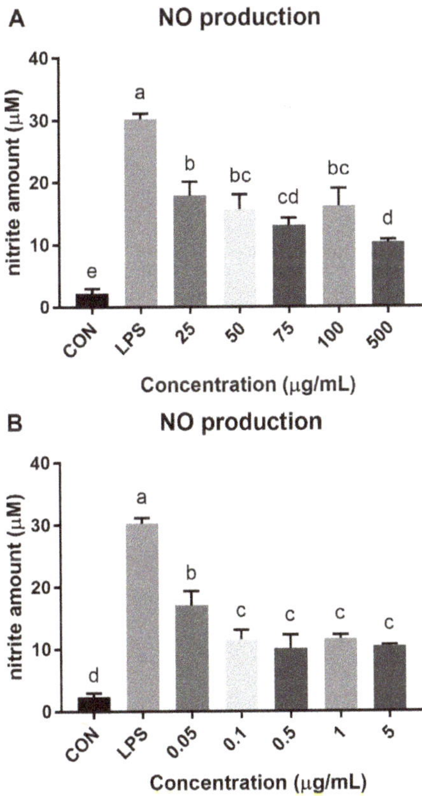

Figure 2. Reduce of NO production on lipopolysaccharide (LPS)-induced (1 µg/mL) RAW 264.7 cell line. (**A**) The 80% methanol extract of defatted mealworm powder; (**B**) unsaponifiable lipid of mealworm oil. Different superscripts (a–e) represent significant differences at $p < 0.05$.

4. Conclusions

In this study, two basic types of products (defatted powder and oil) were prepared from whole fat mealworm, and both products were characterized for use as food ingredients. DF-M was light-brownish powder rich in protein (70.8% ± 5.8%). Methionine was a limiting amino acid of DF-M, and the proportion of lysine in DF-M was also lower than established criteria. Therefore, it would be necessary to increase methionine in mealworm feed in order to improve amino acid scores. Meanwhile, free amino acids were very abundant in DF-M, which could strongly affect the taste of mealworm products. The protein solubility in water was not great with DF-M, so some processing would be needed for preventing turbidity and precipitation. The refining of DF-M to MPI was able to enhance the protein solubility. For mealworm oil, oleic acid was the most abundant fatty acid, and the fatty acid composition of mealworm oil was very similar to vegetable oils. Mealworm oil contained plentiful tocopherols and other minor nutrients, but only trace levels of cholesterol. The predicted shelf life was almost 10 months. Therefore, mealworm oil has appropriate characteristics for general-purpose use. Moreover, the antioxidant capacity and anti-inflammation activity of DF-M extract were very high, likely as a result of abundant glucosamine derivatives and functional peptides. In short, defatted mealworm powder has adequate nutritional value and is rich in protein, minor nutrients, and bioactive compounds. Mealworm oil, a byproduct of DF-M, also had good nutritional value and characteristics for versatile use as a food ingredient. In addition, the DF-M and mealworm oil had antioxidant and anti-inflammation activities. Thus, mealworm could replace meat from vertebrates, but there was one limitation with respect to a specific amino acid, i.e., methionine.

Supplementary Materials: The following are available online at http://www.mdpi.com/2304-8158/9/1/40/s1, Figure S1: Pictures of mealworm oil extracted with n-hexane, Figure S2: Pictures and color values of whole- or defatted-mealworm powders and mealworm protein isolate (MPI), Figure S3: GPC calibration plot, Figure S4: Protein solubility of mealworm powder, Table S1: HPLC operating conditions for analyzing amino acid contents, Table S2: Operating conditions for GPC analysis, Table S3: GC analysis conditions for examining fatty acids composition, Table S4: Operating conditions of HPLC for analyzing tocopherol contents, Table S5: Analysis conditions of squalene and sterols by GC-MS analysis, Table S6: Proximate compositions of mealworm powders.

Author Contributions: The study was designed by Y.-J.S., I.-K.H. and S.H.K. Y.-J.S. and S.Y.C. conducted experiments and analyses of this study. Y.-J.S., S.Y.C. and C.W.N. organized figures and tables with data. The manuscript was written by Y.-J.S., I.-K.H., C.W.N. and S.H.K. All authors have read and agreed to the published version of the manuscript.

Funding: This work was supported by "Development of recipes and manufacturing technology for promoting the consumption of edible insects (Project No. 315060-3)" by the Ministry of Agriculture, Food and Rural Affairs, Republic of Korea.

Conflicts of Interest: The authors declare no conflict of interest.

References

1. Oo. incx, D.G.A.B.; van Itterbeeck, J.; Heetkamp, M.J.W.; van den Brand, H.; van Loon, J.J.A. An exploration on greenhouse gas and ammonia production by insect species suitable for animal or human consumption. *PLoS ONE* **2010**, *5*, e14445. [CrossRef] [PubMed]

2. Li, L.Y.; Zhao, Z.R.; Liu, H. Feasibility of feeding yellow mealworm (*Tenebrio molitor* L.) in bioregenerative life support systems as a source of animal protein for humans. *Acta Astronaut.* **2013**, *92*, 103–109. [CrossRef]

3. Siemianowska, E.; Kosewska, A.; Aljewicaz, M.; Skibniewska, K.A.; Polak-Juszczak, L. Larvae of mealworm (*Tenebrio molitor* L.) as European novel food. *Agric. Sci.* **2013**, *4*, 287–291.

4. Han, S.R.; Lee, B.S.; Jung, K.J.; Yu, H.J.; Yun, E.Y.; Hwang, J.S.; Moon, K.S. Safety assessment of freeze-dried powdered *Tenebrio molitor* larvae (yellow mealworm) as novel food source: Evaluation of 90-day toxicity in Sprague-Dawley rats. *Regul. Toxicol. Pharmacol.* **2016**, *77*, 206–212. [CrossRef]

5. Ravazanaadii, N.; Kim, S.H.; Choi, W.H.; Hong, S.J.; Kim, N.J. Nutritional value of mealworm, *Tenebrio molitor* as food source. *Int. J. Indus. Entomol.* **2012**, *25*, 93–98. [CrossRef]

6. Makkar, P.S.H.; Tran, G.; Heuzé, V.; Ankers, P. State-of-the-art on use of insects as animal feed. *Anim. Feed Sci. Technol.* **2014**, *197*, 1–33. [CrossRef]

7. Foster, D.K.; Bronlund, E.J.; Paterson, T.A.H.J. The contribution of milk fat towards the caking of dairy powders. *Int. Dairy J.* **2005**, *15*, 85–91. [CrossRef]

8. Sharma, L.; Singh, C. Sesame protein based edible films: Development and characterization. *Food Hydrocoll.* **2016**, *61*, 139–147. [CrossRef]

9. AOAC. Methods 932.06, 925.09, and 923.03. In *Official Method of Analysis of AOAC International*, 16th ed.; Association of Official Analytical Communities: Arlington, VA, USA, 1995.

10. Son, Y.J.; Lee, J.C.; Hwang, I.K.; Nho, C.W.; Kim, S.H. Physicochemical properties of mealworm (*Tenebrio molitor*) powders manufactured by different industrial processes. *LWT-Food Sci. Technol.* **2019**, *116*, 108514. [CrossRef]

11. Morr, C.V.; German, B.; Kinsella, J.E.; Regenstein, J.M.; van Buren, J.P.; Kilara, A.; Lewis, B.A.; Mangino, M.E. A collaborative study to develop a standardized food protein solubility procedure. *J. Food. Sci.* **1985**, *50*, 1715–1718. [CrossRef]

12. Grubisic, Z.; Rempp, P.; Benoit, H. A universal calibration for gel permeation chromatography. *Polym. Lett.* **1967**, *5*, 753–759. [CrossRef]

13. Jeon, Y.H.; Son, Y.J.; Kim, S.H.; Yun, E.Y.; Kang, H.J.; Hwang, I.K. Physicochemical properties and oxidative stabilities of mealworm (*Tenebrio molitor*) oils under different roasting conditions. *Food Sci. Biotechnol.* **2016**, *25*, 105–110. [CrossRef] [PubMed]

14. Ministry of Food and Drug Safety. *Korean Food Standards Codex II*; Ministry of Food and Drug Safety: Seoul, Korea, 2015; p. 48.

15. Buege, J.A.; Aust, S.D. Microsomal lipid peroxidation. *Methods Enzymol.* **1978**, *52*, 302–310. [PubMed]

16. Gómez-Rico, A.; Salvador, M.D.; Moriana, A.; Pérez, D.; Olmedilla, N.; Ribas, F.; Fregapane, G. Influence of different irrigation strategies in a traditional Cornicabra cv. olive orchard on virgin olive oil composition and quality. *Food Chem.* **2007**, *100*, 568–578. [CrossRef]

17. Gliszczyńska-Świgło, A.; Skiorska, E. Simple reversed-phase liquid chromatography method for determination of tocopherols in edible plant oils. *J. Chromatogr.* **1992**, *1048*, 1184–1188. [CrossRef]

18. Singleton, V.L.; Rossi, J.A. Colorimetry of total phenolics with phosphomolybdic-phosphotungstic acid reagents. *Am. J. Enol. Viticult.* **1965**, *16*, 144–158.

19. Belcher, R.; Nutten, A.J.; Sambrook, C.M. The determination of glucosamine. *Analyst* **1954**, *79*, 201–208. [CrossRef]

20. Meda, A.; Lamien, C.E.; Romito, M.; Millogo, F.; Nacoulma, O.G. Determination of the total phenolic, flavonoid and proline content in Burkina Fasan honey, as well as their radical scavenging acitivity. *Food Chem.* **2005**, *91*, 571–577. [CrossRef]

21. Brand-Williams, W.; Cuvelier, M.E.; Berset, C. Use of a free radical method to evaluate antioxidant activity. *LWT-Food Sci. Technol.* **1995**, *28*, 25–30. [CrossRef]

22. Re, R.; Pellegrini, N.; Proteggente, A.; Pannala, A.; Yang, M.; Rice-Evans, C. Antioxidant activity applying an improved ABTS radical cation decolorization assay. *Free Radic. Biol. Med.* **1999**, *26*, 1231–1237. [CrossRef]

23. Zhao, X.; Vázquez-Gutiérrez, J.L.; Johansson, D.P.; Landberg, R.; Langton, M. Yellow mealworm protein for food purposes-Extraction and functional properties. *PLoS ONE* **2016**, *11*, e0147791. [CrossRef] [PubMed]

24. Friedrich, J.P.; List, G.R. Characterization of soybean oil extracted by supercritical carbon dioxide and hexane. *J. Agric. Food Chem.* **1982**, *30*, 192–193. [CrossRef]

25. Brewer, M.S. Reducing the fat content in ground beef without sacrificing quality: A review. *Meat. Sci.* **2012**, *91*, 385–395. [CrossRef] [PubMed]

26. Lee, K.C.; Lee, S.K.; Kim, H.K. Chemical compositions of the four lines of Korean native chickens. *Korean J. Poult. Sci.* **2016**, *43*, 119–128. [CrossRef]

27. Franco, D.; González, L.; Bispo, E.; Rodríguez, P.; Garabal, J.I.; Moreno, T. Study of hydrolyzed protein composition, free amino acid, and taurine content in different muscles of galician blonde beef. *J. Muscle Foods* **2010**, *21*, 769–784. [CrossRef]

28. FAO/WHO/UNU. *Energy and Protein Requirements*; World Health Organization Technical Report Series 724; FAO/WHO/UNU: Geneva, Switzerland, 1985.

29. Jones, L.D.; Cooper, R.W.; Harding, R.S. Composition of mealworm *Tenebrio molitor* larvae. *J. Zoo Anim. Med.* **1972**, *3*, 34–41. [CrossRef]

30. Enser, M.; Hallett, K.; Hewitt, B.; Fursey, G.A.J.; Wood, J.D. Fatty acid content and composition of English beef, lamb and pork at retail. *Meat. Sci.* **1996**, *42*, 443–456. [CrossRef]

31. Kamal-Eldin, A. Effect of fatty acids and tocopherols on the oxidative stability of vegetable oils. *Eur. J. Lipid. Sci. Technol.* **2006**, *58*, 1051–1061. [CrossRef]

32. Bramley, P.M.; Elmadfa, I.; Kafatos, A.; Kelly, F.J.; Manios, Y.; Roxborough, H.E.; Schuch, W.; Sheehy, P.J.A.; Wagner, K.H. Review: Vitamin E. *J. Sci. Food Agric.* **2000**, *80*, 913–938. [CrossRef]

33. Maraschiello, C.; Sárraga, C.; Regueiro, J.A.G. Glutathione peroxidase activity, TBARS, and α-tocopherol in meat from chickens fed different diets. *J. Agric. Food Chem.* **1999**, *47*, 867–872. [CrossRef]
34. Surai, P.F.; Sparks, N.H.C. Tissue-specific fatty acid and α-tocopherol profiles in male chickens depending on dietary tuna oil and vitamin E provision. *Poult. Sci.* **2000**, *70*, 1132–1142. [CrossRef] [PubMed]
35. Okogeri, O.; Tasioula-Margari, M. Changes occurring in phenolic compounds and α-tocopherol of virgin olive oil during storage. *J. Agric. Food Chem.* **2002**, *50*, 1077–1080. [CrossRef] [PubMed]
36. Farhoosh, R. The effect of operational parameters of the rancimat methods on the determination of the oxidative stability measures and shelf-life prediction of soybean oil. *J. Am. Oil Chem. Soc.* **2007**, *84*, 205–209. [CrossRef]
37. Lourenço, R.; Camilo, M.E. Taurine: A conditionally essential amino acid in humans? An overview in health and disease. *Nutr. Hosp.* **2002**, *17*, 262–270.
38. Largo, R.; Alvarez-Soria, M.A.; Díez-Ortego, I.; Calvo, E.; Sánchez-Pernaute, O.; Egido, J.; Herrero-Beaumont, G. Glucosamine inhibits IL-1β-induced NFκB activation in human osteoarthritic chondrocytes. *Osteoarthr. Cartil.* **2003**, *11*, 290–298. [CrossRef]
39. Bahorun, T.; Luximon-Ramma, A.; Crozier, A.; Aruoma, O. Total phenol, flavonoid, proanthocyanidin and vitamin C levels and antioxidant activities of Mauritian vegetables. *J. Sci. Food Agric.* **2004**, *84*, 1553–1561. [CrossRef]
40. Je, J.Y.; Park, P.J.; Kim, S.K. Antioxidant activity of a peptide isolated from Alaska pollack (*Theragra chalcogramma*) frame protein hydrolysate. *Food Res. Int.* **2005**, *38*, 45–50. [CrossRef]
41. Zielińska, E.; Baraniak, B.; Karaś, M. Identification of antioxidant and anti-inflammatory peptides obtained by simulated gastrointestinal digestion of three edible insects species (*Gryllodes sigillatus, Tenebrio molitor, Schistocerca gragaria*). *Int. J. Food Sci. Technol.* **2018**, *53*, 2542–2551. [CrossRef]

 *foods*

Article

Diversity and Use of Edible Grasshoppers, Locusts, Crickets, and Katydids (Orthoptera) in Madagascar

Joost Van Itterbeeck [1,2,*], Irina N. Rakotomalala Andrianavalona [3], Faneva I. Rajemison [3], Johanna F. Rakotondrasoa [4], Valisoa R. Ralantoarinaivo [4], Sylvain Hugel [5] and Brian L. Fisher [6]

[1] Center for Asian Area Studies, Rikkyo University, 3-34-1 Nishi-Ikebukuro, Toshima, Tokyo 171-8501, Japan
[2] Department of Plant Medicals, Andong National University, 1375 Kyungdong-Street, Andong, Gyeongbuk 36729, Korea
[3] Madagascar Biodiversity Center, Parc Botanique et Zoologique de Tsimbazaza, BP 6257, Antananarivo 101, Madagascar; irinaandrianavalona@gmail.com (I.N.R.A.); iharantsoa.faneva@gmail.com (F.I.R.)
[4] Department of Entomology, University of Antananarivo, BP 906, Antananarivo 101, Madagascar; faniryjohanna@gmail.com (J.F.R.); valisoaral96@gmail.com (V.R.R.)
[5] Institute of Cellular and Integrative Neuroscience, National Center for Scientific Research (CNRS), University of Strasbourg, 5 rue Blaise Pascal, 67084 Strasbourg, France; hugels@inci-cnrs.unistra.fr
[6] California Academy of Sciences, San Francisco, CA 94118, USA; bfisher@calacademy.org
* Correspondence: joostvanitterbeeck@hotmail.com

Received: 2 November 2019; Accepted: 25 November 2019; Published: 10 December 2019

Abstract: Madagascar has a long history of using Orthoptera as food and feed. Our understanding of the biological diversity of this resource, its contemporary use, and its future potentials in Madagascar is extremely limited. The present study contributes basic knowledge of the biological diversity and local uses of edible Orthoptera in Malagasy food cultures. Data was collected with key informants in 47 localities covering most of the ecoregions of Madagascar and corresponding to 12 of the 19 ethnic groups. Orthoptera are consumed throughout Madagascar. We report 37 edible Orthoptera species, of which 28 are new species records of edible Orthoptera in Madagascar and 24 are new species records of edible Orthoptera in the world. Most species are endemic and occur in farming zones. Children are the primary collectors and consumers of edible Orthoptera. The insects are eaten both as snacks and main meals. Edible Orthoptera are primarily collected casually and marketing is rare, with the notable exceptions of the large cricket *Brachytrupes membranaceus colosseus* and during locust outbreaks (e.g., *Locusta migratoria*). The use of Orthoptera as feed seems rare. Further investigations of cultural and personal preferences are required to assess the future potential roles of Orthoptera in Malagasy food habits.

Keywords: insect; entomophagy; biodiversity; food; bioresource; culture

---

1. Introduction

Madagascar has a long history of using Orthoptera as food and feed. However, our understanding of the biological diversity of this resource, its contemporary use, and its future potentials in Madagascar is extremely limited.

As early as 1617, missionaries and other visitors to Madagascar reported the regular consumption of Orthoptera, especially locusts, by the Malagasy people, and received invitations from the Malagasy people to join them in eating the insects [1]. Some visitors consumed the insects with as good an appetite as the Malagasy [2]. The Malagasy claimed to revenge themselves on these Orthoptera that devastated rice fields and plantations by eating the insects [2,3]. During outbreaks, locusts provided the necessary nutrition for people and livestock (e.g., pigs). Additionally, between harvests, when food supplies were low, Orthoptera and other insects were consumed regularly [1,3]. Locusts were sundried for later use. Malagasy royals consumed the locusts as well. Their most appreciated recipe involved soaking sundried locusts for half an hour in salted water and then frying them in fat. Queen Ranavalona II, who reigned Madagascar in the nineteenth century, had female servants travel to the countryside specifically to collect locusts for her [1,3]. Camboué [3], as well as others [1], considered Orthoptera a potential good source of nutrition for the Malagasy people, and/or their livestock, citing the consumption of Orthoptera in Leviticus in the Bible.

The recent surge of interest in the use of nutritious insects as food and feed may benefit Madagascar. Supporting and advancing the traditional inclusion of insects in Malagasy food cultures, as well as the raising of livestock with insects, may aid the country's nutritional and economic development, potentially in ecologically sustainable ways [4–6]. Edible Orthoptera are among the candidate insects for innovative action. Madagascar's food production relies heavily on rice cultivation, supplemented with vegetable and fruit crops. Improved agricultural productivity will help to reduce high poverty and food insecurity rates [7]. Rice field ecosystems and other agricultural fields are habitats that globally support high diversities of edible Orthoptera and other edible insects [4,8,9]. One potential strategy is to advance the mechanical control of edible Orthopteran pest insects, which may simultaneously support crop protection and provide a 'second crop' in the form of the edible pest insects for food and/or feed [10–12]. Another potential strategy is to develop farming technologies to produce local edible Orthoptera for food and/or feed [13–15].

The present study contributes to the exploration of edible Orthoptera as candidate insects for edible insect-based nutritional, economic, and ecologically sustainable food production programs. Very few studies have addressed the biological diversity and local uses of edible Orthoptera in contemporary Malagasy food cultures, and these have probably underestimated the diversity of edible Orthopterans [16,17]. Such basic knowledge is of major importance to understand the future potential of edible Orthoptera in Madagascar [18]. Hence, this study aims to (i) identify the edible Orthopteran diversity and explore its geographical distributions and (ii) explore the basic uses of Orthoptera as food and feed in different ethnic groups and across different ecoregions of Madagascar.

2. Materials and Methods

2.1. Localities and Key Informants

Field work was conducted from February to July 2018, in December 2018, and from February to April 2019. We surveyed for edible Orthopteran diversity in 47 localities covering most of the ecoregions of Madagascar and corresponding to 12 of the 19 ethnic groups (Figure 1). All localities were rural villages with rice cultivation as the main subsistence activity. Data was collected with key informants, i.e., rural villagers, who are highly knowledgeable about which insects are edible, where they can be found, how to collect them, and how they are used as food and feed. The key informants were identified via intermediaries, i.e., market vendors, hotel staff, taxi drivers, and agricultural extension offices in rural towns. We explained our research objectives to the intermediaries who then suggested potential key informants and localities and provided contact details and directions. We then met with the potential key informants, thoroughly discussed our research objectives with them, and made a basic assessment of their knowledge by discussing edible insect collection and consumption. Whenever we had doubts about the quality of a potential key informant, we identified another potential key informant (this was rarely necessary). Each locality was then surveyed with the key informants for 1–2 days. All informants gave their consent to participate in the research. Key informants were 28 adults (23 males and 5 females, between about 20 and 60 years old); children sometimes assisted by collecting specimens. Key informants received monetary rewards equivalent to daily wages of manual labor. Research permissions were provided by the Ministry of Environment and Sustainable Development, local authorities (e.g., mayoral offices), and village leaders. A higher number of localities were surveyed in east Madagascar (assessed from February to July 2018) than elsewhere, for two reasons. Firstly, more researchers were available for field work during February–July 2018, which allowed them to split into two teams daily, each accompanied by a key informant, and conduct field work in different sections of a single village and in neighboring villages. Secondly, safety issues and scarcity of transportation limited data collection in central and west Madagascar.

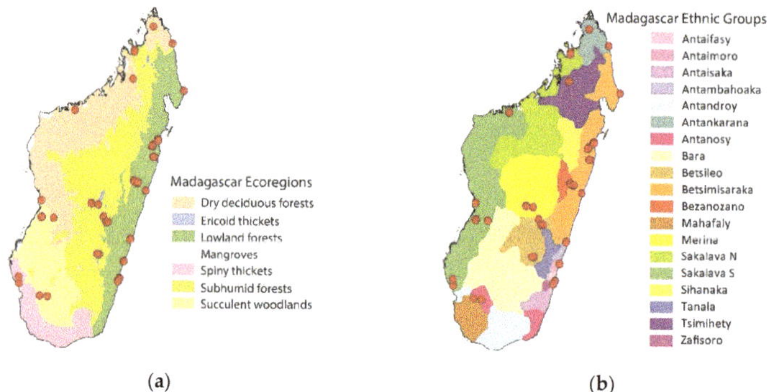

Figure 1. (**a**) Map of Madagascar ecoregions; (**b**) map of Madagascar ethnic groups. Red dots mark the surveyed localities. Particularly in east Madagascar, red dots overlap due to map scale.

2.2. Insect Specimens

Edible Orthoptera specimens were sampled during field trips with key informants in natural areas, agricultural areas, and home gardens. The insects were photographed both in their natural habitat and close up (profile, head, and dorsal). Specimens were sampled by hand, with tweezers, and with sweep nets, and stored in 95% alcohol in vials and Whirl-Pak® bags (Nasco, Fort Atkinson, WI, USA). Specimens are permanently stored in the entomology collections of the California Academy of Sciences, San Francisco, USA. Specimens were identified by Sylvain Hugel, co-author of the present study, with data from Dirsh and Descamps [19]. Geographical distributions and threats to crops were determine by referring to Braud et al. [20]. Calculations were performed with Microsoft Excel.

2.3. Edible Orthoptera Use

Key informants were interviewed on edible Orthoptera use during the edible insect specimen collection field trips. We recorded: (i) vernacular names, (ii) collection methods, (iii) gender and age of collectors and consumers, (iv) consumption mode (i.e., snack or main meal), (v) food preservation methods, (vi) frequencies and quantities of collection and consumption, (vii) marketing, (viii) animal feed uses, (ix) medicinal uses, and (x) taboos. We emphasize that we report a qualitative overview only. Preferences in the use of edible Orthoptera vary among ethnic groups and individuals within ethnic groups. Due to the use of a limited number of informants, it is not in the scope of the present study to make inferences of cultural and personal preferences.

3. Results

3.1. Edible Orthopteran Diversity

We recorded 37 species of edible Orthoptera, of which 31 were able to be identified to species level and six to either genus or subfamily level (Table 1). This edible Orthopteran diversity corresponds to 3% of the recorded Orthopteran diversity in Madagascar (ca. 700 species, likely underestimated). These 37 species are comprised of 26 species of grasshoppers (Caelifera) and 11 species of crickets and katydids (Ensifera; Figure 2). Most of the edible Caelifera belong to the Acrididae (23 out of 26 species). The edible Ensifera includes Tettigonioidea (bush crickets/katydids, six species) and Grylloidea (crickets, five species).

Table 1. Overview of the 37 edible Orthoptera species in Madagascar recorded in the present study.

Genus and Species	Threat Level to Crops	Antaifasy	Antaimoro	Antankarana	Antanosy	Betsileo	Betsimisaraka	Bezanozano	Merina	Sakalava N	Sakalava S	Tsimihety	Zafisoro	Total	Vernacular Names
CAELIFERA (grasshoppers, locusts)															
Locusta migratoria	+++	X	X	X	X	X			X		X		X	8	Valala, Kijeja, Mendry
Nomadacris septemfasciata	+++			X					X					2	Valala, Sipanga
Gastrimargus africanus	++		X			X	X		X			X	X	6	Valala
Paracinema tricolor	++	X	X			X	X	X	X	X	X		X	9	Valala
Acorypha decisa	+					X			X					2	Valala
Acrida madecassa	+				X	X	X		X		X		X	6	Valala, Kesoly
Catantopsis malagassus	+					X	X		X					3	Valala
Catantopsis sacalava	+								X	X	X			3	Valala
Cyrtacanthacris tatarica	+	X		X	X	X			X		X	X		7	N/A
Eyprepocnemis smaragdipes	+	X	X			X	X		X		X		X	7	Valala
Finotina radama	+								X					1	Valala
Oedaleus virgula	+	X		X		X	X		X		X			6	Valala
Oxya hyla	+	X	X		X	X	X	X	X		X		X	9	Valala
Rhadinacris schistocercoides	+								X		X			2	Valala, Sipanga
Rubellia nigrosignata	+						X							1	N/A
Acrida sp.						X								1	Valala
Acrida subtilis						X								1	Valala
Aiolopus thalassinus rodericensis									X		X			2	Valala
Atractomorpha acutipennis			X			X	X	X	X				X	6	Valala, Sompatra
Calephorus ornatus		X				X			X					3	Valala
Duronia chloronota											X			1	Kelimavo
Gelastorhinus edax						X			X					2	Valala
Gymnobothrus madacassus							X							1	Valala
Gymnobothrus variabilis		X	X			X								3	Valala
Heteracris nigricornis									X					1	Valala
Lemuracris longicornis							X							1	N/A
Trilophidia cinnabarina						X			X					2	Valala
ENSIFERA (crickets, katydids)															
Brachytrupes sp.										X		X		2	Sahobaka
Brachytrupes membranaceus colosseus											X			1	Sahobaka
Colossopus sp.						X			X					2	Sakova
Conocephalus cf. *affinis*			X				X							2	Valala

Table 1. *Cont.*

Genus and Species	Threat Level to Crops	Ethnic Groups												Total	Vernacular Names
		Antaifasy	Antaimoro	Antankarana	Antanosy	Betsileo	Betsimisaraka	Bezanozano	Merina	Sakalava N	Sakalava S	Tsimihety	Zafisoro		
Phaneropterinae sp.							X							1	Kisovasova
Fryerius sp.							X							1	N/A
Gryllus sp.							X							1	N/A
Modicogryllus sp.						X								1	N/A
Phaneroptera sparsa						X								1	Kisovasova
Pteronemobius malagachirus							X							1	N/A
Ruspolia differens							X		X				X	3	Valala, Sakoririka
Ruspolia sp.			X			X	X				X			4	Valala, Sakoririka
Total 37		8	9	4	4	20	18	3	22	2	13	2	8		

Bold = new species records of edible Orthoptera in Madagascar; bold and underlined = new species records of edible Orthoptera in the world (see references in text). Threat level to crops: +++ = high, ++ = moderate, + = low, and blank = no threat. *Acrida* sp. and *Brachytrupes* sp. have been excluded from the total counts of edible Orthoptera species since we have not counted them as different species from *Acrida* subtilis and *A. madecassa*, and from *Brachytrupes membranaceus colosseus*, respectively (see text).

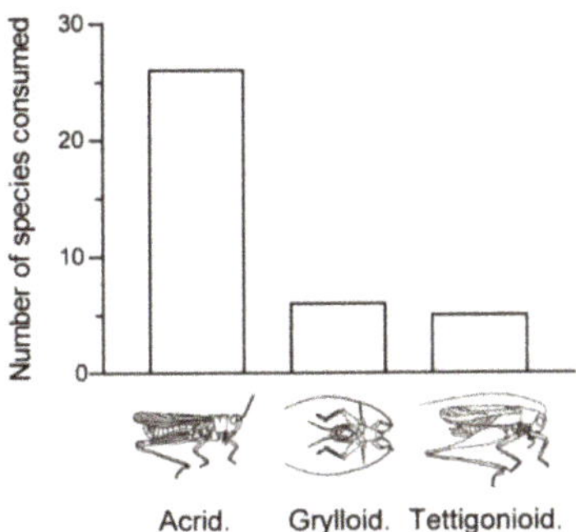

Figure 2. Number of edible Orthoptera species as a function of systematics. Legend: Acrid., Acrididae; Grylloid., Grylloidea; Tettigonioid., Tettigonioidea.

The five most recorded species were found to be *Oxya hyla* (25 localities in three ecoregions: lowland forests, subhumid forests, and succulent woodlands; nine ethnic groups), *Paracinema tricolor* (22 localities in three ecoregions: lowland forests, subhumid forests, and succulent woodlands; nine ethnic groups), *Locusta migratoria* (17 localities in five ecoregions: lowland forests, subhumid forests, succulent woodlands, dry deciduous forests, and spiny thickets; eight ethnic groups), *Acrida madecassa* (13 localities in five ecoregions: lowland forests, subhumid forests, succulent woodlands, dry deciduous forests, and spiny thickets; six ethnic groups), and *Cyrtacanthacris tatarica* (12 localities in five ecoregions: lowland forests, subhumid forests, succulent woodlands, dry deciduous forests, and spiny thickets; seven ethnic groups). Of the 37 recorded species, 14 species were sampled only once (i.e., one locality per species; Figure 3). An annotated list of edible Orthoptera is provided in Section 3.3.

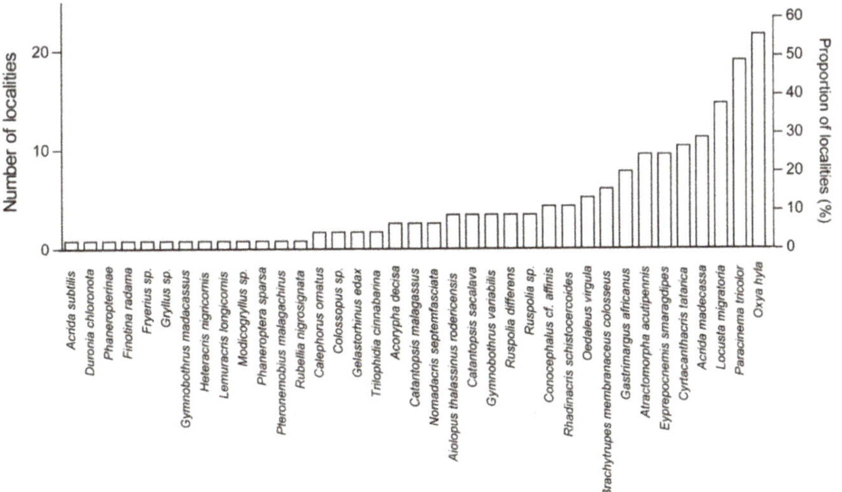

Figure 3. Number of localities per edible Orthoptera species.

The total count of 37 species is conservative. In cases of unidentified species, all specimens belonging to a single genus or subfamily were counted as a single species. We thus have not counted *Acrida* sp. and *Brachytrupes* sp. as a different species from *Acrida subtilis* and *A. madecassa*, and from *Brachytrupes membranaceus colosseus*, respectively, since only juveniles were collected and could not be identified to species level. *Ruspolia* sp., however, was considered a separate species since some adults were collected and the specimens were definitely not *Ruspolia differens*. Both juveniles and adults are edible. Among the 279 species-sorted batches examined for the present work, 87% contained at least some juveniles. Juveniles are smaller than adults but cannot escape by flying.

Most of the recorded species are endemic to the Malagasy area: 18 species (56%) are endemic to Madagascar and five species (16%) occur also in nearby islands, whereas nine species (28%) are widely distributed across Africa ($n = 32$ species; Figure 4).

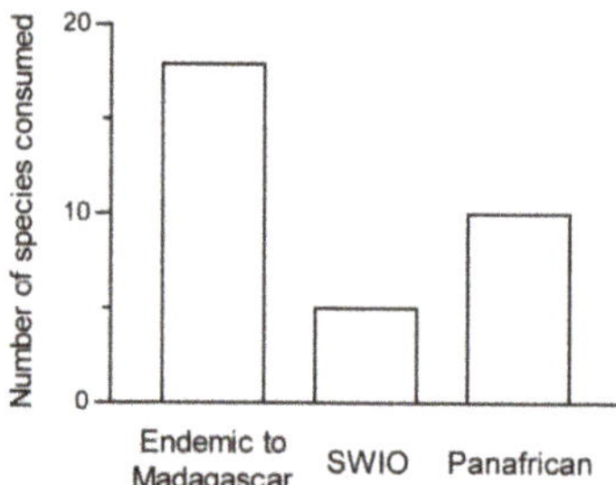

Figure 4. Number of edible Orthoptera species as a function of their distribution in Madagascar, South Western Indian Ocean (SWIO), and African areas.

Somewhat surprisingly, two Caelifera species, *Rubellia nigrosignata* and *Atractomorpha acutipennis*, belong to the Pyrgomorphidae, a family with many species considered to be unpleasant-tasting or -smelling, or even toxic.

Most edible Orthoptera were found to occur in habitats with herbaceous plant cover, including farming zones (83%); few were found to occur in areas mixed with taller plant cover (17%); and none were found to occur in forest habitat ($n = 36$; Figure 5). The majority of Orthoptera species recorded as edible were no threat to crops (22 species, 59% of recorded edible Orthoptera in the present study; Figure 6). Fifteen species were found to be a threat to crops, of which two species posed a high threat (5%), two species a moderate threat (5%), and 11 species a low threat (30%; $n = 37$ species). Our record indicates that threat level is associated with use as food: all species representing a high threat (100%) and a moderate threat (100%) to crops were reported as edible, whereas 73% of species posing a low threat were reported as edible and 3% of species posing no threat were reported as edible ($n = 37$ species). Rice field ecosystems, other agricultural fields, and fallow land near agricultural fields are the main sources of the edible Orthoptera bioresource used in Madagascar.

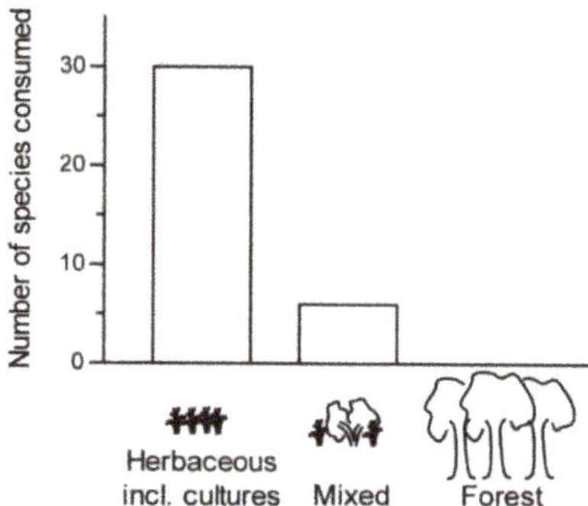

Figure 5. Number of edible Orthoptera species as a function of their habitat.

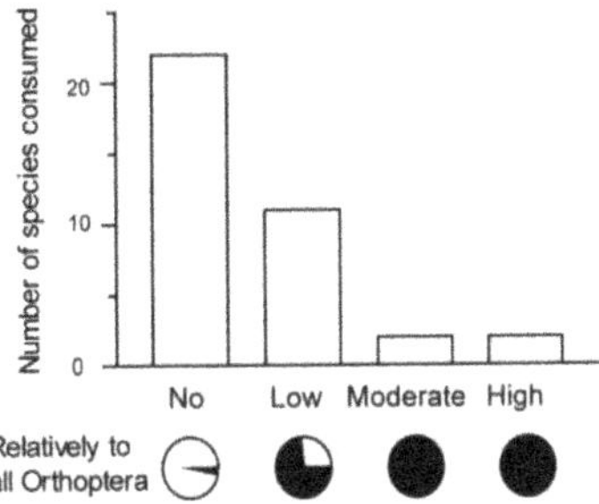

Figure 6. Number of edible Orthoptera species as a function of their threat to crops.

3.2. Edible Orthoptera Uses

The highest edible Orthopteran diversity was recorded among the Merina in the subhumid forests ecoregion (22 species, 59% of recorded edible Orthoptera in the present study), the Betsileo in the subhumid forests ecoregion (20 species, 54%), and the Betsimsaraka in the lowland forest ecoregion (18 species, 49%). The lowest edible Orthopteran diversity was recorded among the Sakalava N (North) and Tsimihety, both in the dry deciduous forests ecoregion (two species each, 5%), the Bezanozano in the lowland forests ecoregion (three species, 8%), the Antankarana in the dry deciduous forests ecoregion (four species, 11%), and the Antanosy in the succulent woodlands ecoregion (four species, 11%). The most widespread used edible Orthoptera were found to be *P. tricolor* and *O. hyla* (nine ethnic groups each), *L. migratoria* (eight ethnic groups), and *C. tatarica* and *Eyprepocnemis smaragdipes* (seven ethnic groups each). The most common vernacular name observed for edible Orthoptera was *valala*, with some other names also mentioned (Table 1).

Edible Orthoptera are present throughout the year with roughly two peak abundance periods from October–January and March–June. Edible Orthoptera are primarily collected casually in Madagascar. All ethnicities collect the insects only by hand. Children are the primary collectors and consumers of Orthoptera in all ethnicities, although adults may also collect and eat them. No gender differences seem to be present in the collection and consumption of Orthoptera. Collecting edible Orthoptera is a common game for children playing in the rice fields or walking home from school. The children then prepare and eat the Orthoptera themselves or have their mothers prepare the insects for them (mothers may then snack on the Orthoptera with their children). Adults mostly collect Orthoptera during their peak availability but may also collect them outside of the peak availability periods, notably while working in the rice fields. When adults collect edible Orthoptera, they may prefer to give them to their children for playing and for eating. Orthoptera are eaten as a snack and as part of a main meal. The wings, legs, head, and intestines may be removed before the insects are prepared by grilling, frying, and frying after being cooked in water. The insects may be seasoned with salt. Some recipes to prepare Orthoptera in Madagascar are provided in the Supplementary File. We do not have clear information about collection and consumption frequencies and quantities. Orthoptera may be used as animal feed (e.g., poultry) but this seems rare. Preservation of Orthoptera is very rare. Only locusts in south Madagascar are dried when outbreaks occur.

Marketing is rare due to low numbers of Orthoptera that are collected casually and due to time constraints to increase Orthoptera yields. Some exceptions occur. Locusts are marketed during outbreaks in south and central Madagascar, and adults may then collect locusts even daily and opportunistically to eat and market them. The large-sized *B. membranaceus colosseus* may be intentionally collected for marketing purposes in west Madagascar. This cricket may also be specifically sought after when meat is expensive and when no other rice accompaniments are available. The cricket is sold for 100–200 Malagasy Ariary (0.027–0.055 US$) per individual body and for 500–1500 Malagasy Ariary (0.14–0.41 US$) per *kapoaka*, a tin can of about 400 mL commonly used as a unit of measurement for selling produce. The level of edible Orthoptera marketing seems to be lower than in some other countries with Orthoptera consumption, such as Togo [21], Laos [22], and Japan [23,24].

Malagasy Muslims, who live primarily in north-west Madagascar, do not eat any insects for religious reasons. The eating of *sakoririka*, a vernacular name for Orthoptera that at least includes the genus *Ruspolia*, is taboo among the Antaisaka ethnic group: the insect is only consumed as a medicine by infants who refrain from speech, or are incapable of speech, to help them to speak. This taboo may not be shared by all ethnicities. The taboo was mentioned by members of the Antaisaka ethnic group of southeast Madagascar who had migrated to the regions commonly associated with the Sakalava ethnic groups in west Madagascar. No other medicinal use of Orthoptera was mentioned in any surveyed localities. Mole crickets are not considered edible due to a widespread cultural belief. They are called "zazavery", which means "lost child". The Malagasy people regularly encounter mole crickets in wet soils, e.g., along riverbanks while cultivating vegetables or cutting wood. However, these mole crickets quickly disappear as they dig themselves into the wet soil to hide. The Malagasy liken this act of disappearing to the behavior of children who may suddenly wander off to play without notifying their parents. The large Pyrgomorphidae grasshopper *Phymateus saxosus* is renowned as inedible in Madagascar, as are some other Pyrgomorphidae. It has aposematic coloration and produces an unpleasant smell when disturbed. This grasshopper is named "dog's grasshopper" (*valalan'alika* or *valalan'amboa*) and Malagasy people say that "even dogs don't eat them".

3.3. Annotated List of Edible Orthoptera

3.3.1. Caelifera, Acridinae

Acrida madecassa (Brancsik, 1892) (Figure 7) and *Acrida subtilis* (Burr, 1902)

These elongated grasshoppers measure 35–75 mm and are endemic to Madagascar. They are widespread in most of Madagascar but are absent from the northern tip of the island. They live in tall grassy areas, including within or near crops. These species are active by daylight and pose no threat to crops. They are consumed in the southern half of Madagascar from the coasts to the central plateau (Figure 3). Despite their strong sexual dimorphism (males measuring 35–45 mm and females 60–75 mm), both males and females are consumed.

(a)

(b)

Figure 7. (**a**) *Acrida madecassa* (picture courtesy of Faneva I. Rajemison); (**b**) distribution map—red dots mark the localities where *A. madecassa* specimens were sampled in the present study and the grey background indicates the ecoregions where *A. madecassa* occurs.

Aiolopus thalassinus rodericensis (Butler, 1876)

This grasshopper measures 15–25 mm. It is present in virtually every non-forested area of Madagascar except the east coast and surrounding islands. *Aiolopus thalassinus rodericensis* live in bare soil and grassy areas. This species is active by daylight and poses no threat to crops. It is consumed in the central part of Madagascar (from the west coast to the central plateau).

Duronia chloronota (Stål, 1876)

This grasshopper measures 20–45 mm. It is present in non-forested areas of Madagascar, except along the east coast. It lives in grassy areas. This species is active by daylight and poses no threat to crops. The consumption of this species was recorded for one single locality in west Madagascar.

Calephorus ornatus (Walker, 1870)

This small grasshopper measures 10–25 mm. It is widespread in wet grassy areas of Madagascar, particularly near rivers and lakes and in wetlands, including rice fields. This species is active by daylight and poses no threat to crops. It is consumed in the southeast of Madagascar.

Gymnobothrus madacassus (Bruner, 1910) and Gymnobothrus variabilis (Bruner, 1910)

Gymnobothrus species measure 12–25 mm. The genus *Gymnobothrus* is found widely across Madagascar and the Comoros. They live in bare soil and grassy areas. These species are active by daylight and pose no threat to crops. These species are consumed in central and east Madagascar.

3.3.2. Caelifera, Calliptaminae

Acorypha decisa (Walker, 1870)

This grasshopper measures 20–35 mm. It is endemic to Madagascar and widespread across most of the island save the northern tip. It lives in grassy areas. The species is active by daylight and poses no threat to crops. This species is consumed in central Madagascar between Antananarivo and Fianarantsoa.

3.3.3. Caelifera, Catantopinae

Catantopsis malagassus (Karny, 1907) and *Catantopsis sacalava* (Brancsik, 1892)

Catantopsis species measure 15–30 mm. The genus *Catantopsis* is endemic to Madagascar and the Comoros, where it is widely present. These grasshoppers live in grassy areas as well as shrublands. These species are active by daylight and pose no threat to crops. Localities where *Catantopsis* are consumed are scattered across Madagascar.

3.3.4. Caelifera, Cyrtacanthacridinae

Cyrtacanthacris tatarica (Linnaeus, 1758)

This large locust measures 40–65 mm. It is widespread in Madagascar, the Comoros, and continental Africa. It lives in areas with tall grass as well as shrublands. This species is active by daylight. It represents a low threat to crops. Localities where *C. tatarica* is consumed are scattered across Madagascar, in both lowlands and highlands (Figure 8).

Rhadinacris schistocercoides (Brancsik, 1892)

This large locust measures 30–50 mm. It is endemic to Madagascar. It is widespread across the island, except along the east coast. It lives in areas with tall grass as well as shrublands. This species is active by daylight and poses no threat to crops. This species is consumed in central Madagascar, from the east to the central plateau.

Figure 8. (a) *Cyrtacanthacris tatarica* (picture courtesy of Sylvain Hugel); (b) distribution map—red dots mark the localities where *C. tatarica* specimens were sampled in the present study, and the grey background indicates the ecoregions where *C. tatarica* occurs.

Finotina radama (Brancsik, 1892)

This large locust measures 38–60 mm. It is endemic to Madagascar. It is widespread across the island except along the east coast. It lives in areas with tall grass as well as shrublands. This species is active by daylight and poses no threat to crops. The consumption of this species was recorded for one locality in central Madagascar.

Nomadacris septemfasciata (Serville, 1838)

This large plague locust measures 60–75 mm. It is widespread in continental Africa, Madagascar, and islands around Madagascar. It lives in areas with tall grass as well as shrublands. This species is active by daylight and poses a high threat to crops. Localities where *N. septemfasciata* is consumed are scattered across Madagascar.

3.3.5. Caelifera, Eyprepocnemidinae

Heteracris nigricornis (Saussure, 1899)

This grasshopper measures 28–50 mm. It is endemic to Madagascar. It is widespread across the island except in the north and southwest. It lives in areas with tall grass as well as shrublands. This species is active by daylight and poses no threat to crops. The consumption of this species was recorded for one locality in central Madagascar.

Eyprepocnemis smaragdipes (Bruner, 1910)

This grasshopper measures 20–40 mm. It is widespread in continental Africa, Madagascar, and islands around Madagascar. It lives in areas with tall grass. This species is active by daylight and poses a low threat to crops. This species is consumed in central Madagascar, from the east to the west coast (Figure 9).

3.3.6. Caelifera, Gomphocerinae

Gelastorhinus edax (Saussure, 1888)

This grasshopper measures 22–45 mm. It is endemic and widespread in Madagascar. It lives in grassy areas. This species is active by daylight and poses no threat to crops. This species is consumed in central Madagascar, between Antananarivo and Fianarantsoa.

(a)

(b)

Figure 9. (**a**) *Eyprepocnemis smaragdipes* (picture courtesy of Sylvain Hugel); (**b**) distribution map—red dots mark the localities where the current study sampled *E. smaragdipes* specimens, and the grey background indicates the ecoregions where *E. smaragdipes* occurs.

3.3.7. Caelifera, Oedipodinae

Paracinema tricolor (Thunberg, 1815)

This grasshopper measures 15–40 mm. It is widespread in continental Africa and Madagascar. It is among the most common grasshoppers found in wet, grassy areas of Madagascar, particularly near rivers and lakes and in wetlands, including rice fields. This species is active by daylight and poses a moderate threat to crops. This species is among the most widely consumed in Madagascar (Figure 10).

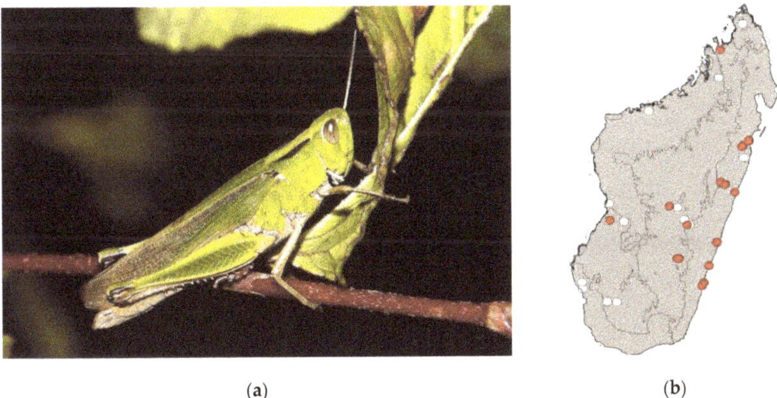

(a)

(b)

Figure 10. (**a**) *Paracinema tricolor* (picture courtesy of Sylvain Hugel); (**b**) distribution map—red dots mark the localities where the current study sampled *P. tricolor* specimens, and the grey background indicates the ecoregions where *P. tricolor* occurs.

Gastrimargus africanus (Saussure, 1888)

This relatively large locust measures 22–40 mm. It is widespread in continental Africa, Madagascar, and the southwestern Indian Ocean islands. In Madagascar it is widespread in grassy areas, including savannas maintained by fires. This species is active by daylight and poses a moderate threat to crops. Localities where *G. africanus* is consumed are mostly in the central and eastern parts of Madagascar (Figure 11).

(a)

(b)

Figure 11. (a) *Gastrimargus africanus* (picture courtesy of Sylvain Hugel); (b) distribution map—red dots mark the localities where the current study sampled *G. africanus* specimens, and the grey background indicates the ecoregions where *G. africanus* occurs.

Lemuracris longicornis (Dirsh, 1966)

This grasshopper measures > 30 mm. It is endemic to the east of Madagascar. It lives in areas with tall grass as well as shrublands. This species is active by daylight and poses no threat to crops. The consumption of this species was recorded for one locality in east Madagascar.

Locusta migratoria (Linnaeus, 1758)

This large plague locust measures 35–55 mm. It is widespread in continental Africa, Madagascar, and the southwestern Indian Ocean islands. It occurs in grassy areas of Madagascar and is frequently found in and around crops. This species is active by daylight and poses a high threat to crops. *Locusta migratoria* is mostly consumed in the southern half of Madagascar (Figure 12).

(a)

(b)

Figure 12. (a) *Locusta migratoria* (picture courtesy of Sylvain Hugel); (b) distribution map—red dots mark the localities where the current study sampled *L. migratoria* specimens, and the grey background indicates the ecoregions where *L. migratoria* occurs.

Oedaleus virgula (Snellen van Vollenhoven, 1869)

This grasshopper measures 13–32 mm. It is widespread in continental Africa and Madagascar. It occurs in bare soil and grassy areas of Madagascar. It is frequently found in and around crops. This species is active by daylight and poses no threat to crops. *Oedaleus virgula* is mostly consumed in the eastern half of Madagascar, from the central plateau to the east coast.

Trilophidia cinnabarina (Brancsik, 1892)

This small grasshopper measures 14–25 mm. It is endemic and widespread in Madagascar. It occurs in bare soil and grassy areas of Madagascar and is frequently found in and around crops. This species is active by daylight and poses no threat to crops. Consumption of *T. cinnabarina* was recorded for two localities in central Madagascar.

3.3.8. Caelifera, Oxyinae

Oxya hyla (Serville, 1831)

This grasshopper measures 20–31 mm. It is widespread in continental Africa and Madagascar. It is among the most common grasshoppers in wet, grassy areas of Madagascar, particularly near rivers and lakes and in wetlands, including rice fields. This species is active by daylight. It represents a low-to-moderate threat to crops. This species is consumed in the central plateau and along the east coast. This species is among the most widely consumed in Madagascar (Figure 13).

(a)

(b)

Figure 13. (a) *Oxya hyla* (picture courtesy of Sylvain Hugel); (b) distribution map—red dots mark the localities where the current study sampled *O. hyla* specimens, and the grey background indicates the ecoregions where *O. hyla* occurs.

3.3.9. Caelifera, Pyrgomorphinae

Members of Pyrgomorphidae feed on toxic plants such as Apocynaceae and sequester secondary metabolites which have significant effects on the physiology of vertebrates via cardenolide steroids and pyrrolizidine alkaloids [25–29]. These Pyrgomorphidae often display aposematic coloration and some possess specialized glands or devices allowing the release of these metabolites upon disturbance.

Atractomorpha acutipennis (Guérin-Méneville, 1844)

This grasshopper measures 15-40 mm. It is widespread in continental Africa, Madagascar, and the Comoros. This grasshopper lives in areas with tall grass as well as shrublands. The species is active by daylight and poses no threat to crops. This species is very widely consumed in the central plateau and along the east coast (Figure 14). *Atractomorpha acutipennis* does not have an aposematic coloration, emits an odor less strong than that of most Pyrgomorphidae, and does not feed primarily on toxic plants.

(a)

(b)

Figure 14. (**a**) *Atractomorpha acutipennis* (picture courtesy of Sylvain Hugel); (**b**) distribution map—red dots mark the localities where the current study sampled *A. acutipennis* specimens, and the grey background indicates the ecoregions where *A. acutipennis* occurs.

Rubellia nigrosignata (Stål, 1875)

This grasshopper measures 15–30 mm. It is endemic to Madagascar. This grasshopper lives in areas with tall grass as well as shrublands. The species is active by daylight and poses a low threat to crops. Consumption of *R. nigrosignata* was recorded for one locality in east Madagascar. The insect produces an unpleasant smell when manipulated. *Rubellia nigrosignata* feeds on a variety of plants, though not primarily toxic ones.

3.3.10. Ensifera, Grylloidea

Gryllus sp.

The consumption of a juvenile of one large ground cricket (20 mm) was recorded for one locality on the east coast of Madagascar. These crickets are nocturnal and represent no threat to crops. Juveniles of *Gryllus* sp. Cannot be identified at the species level.

Modicogryllus sp.

The consumption of a medium-sized female ground cricket (10 mm) was recorded for one locality on the east coast of Madagascar. These crickets are nocturnal and represent no threat to crops. Females of *Modicogryllus* sp. Cannot be identified at the species level.

Brachytrupes membranaceus colosseus (Saussure, 1899)

These large ground crickets are widespread in sandy areas of the west of Madagascar. Several endemic species of *Brachytrupes* sp. Occur in Madagascar and are in need of taxonomic review. These large crickets may represent a low threat to crops. *Brachytrupes* sp. Dig burrows at the entrance of which males produce a loud call at dusk. These crickets are consumed in various localities in the west of Madagascar (Figure 15).

(a)

(b)

Figure 15. (a) *Brachytrupes membranaceus colosseus* (picture courtesy of Faneva I. Rajemison); (b) distribution map—red dots mark the localities where the current study sampled *B. membranaceus colosseus* specimens, and the grey background indicates the ecoregions where *B. membranaceus colosseus* occurs.

Fryerius sp.

The consumption of a large female tree cricket (>20 mm) was recorded for one locality. These crickets are nocturnal and represent no threat to crops. Females of *Fryerius* sp. cannot be identified at the species level.

Pteronemobius malagachus (Saussure, 1877)

The consumption of a small ground cricket (>10 mm) was recorded for one locality on the east coast of Madagascar. *Pteronemobius malagachus* is frequently found in relatively wet grassy areas of Madagascar and its surrounding islands, including in and around rice fields. This cricket is both diurnal and nocturnal and represents no threat to crops.

3.3.11. *Ensifera, Tettigoniidae*

Conocephalus affinis (Redtenbacher, 1891)

This katydid measures 20–30 mm. It occurs in areas with tall grass all over Madagascar, including cultivated lands. This katydid is both diurnal and nocturnal and poses no threat to crops. *Conocephalus affinis* is consumed in various localities in the east of Madagascar.

Ruspolia differens and other *Ruspolia* sp.

Ruspolia sp. measure 30–40 mm. They occur in areas with tall grass all over Madagascar, including cultivated lands. These katydids are mostly nocturnal and may represent a low threat to crops (particularly *R. differens*). *Ruspolia* sp. are consumed in various localities from the central plateau to the east of Madagascar.

Colossopus sp.

Colossopus sp. are large nocturnal katydids living in areas with shrubs and trees. Adults can reach up to 60 mm. Some species occur in and around cultivated areas. Consumption of *Colossopus* sp. has been recorded in various localities from the central plateau. The collected samples were juveniles and not identifiable at the species level.

Phaneroptera sparsa (Stål, 1857) and other Phaneropterinae

Phaneroptera sparsa is a small katydid. It is widespread in tropical Africa and across Madagascar and the Comoros. It is mostly nocturnal and lives in areas with tall grass and/or shrubs, including cultivated lands. Consumption of *Phaneroptera sparsa* and another non-adult Phaneropterinae were recorded for two localities in the central plateau.

Odontolakis sp.

We have not recorded the consumption of *Odontolakis* species in the present work, but *Odontolakis sexpunctata* bush crickets have previously been recorded as edible by Randrianandrasana and Berenbaum [17].

4. Discussion

In this work, we contribute 28 new species records of edible Orthoptera in Madagascar. Of our 31 edible Orthoptera identified to species level, only three species have been reported as edible in the literature on Madagascar: *L. migratoria*, *Nomadacris septemfasciata*, and *B. membranaceus* [1,16,17]. Of all reported edible Orthoptera in Madagascar, only *Odontolakis sexpunctata*, reported by Randrianandrasana and Berenbaum [17], is not recorded in the present study. Additionally, of those 28 new species records, 24 are new species records of edible Orthoptera in the world and have not been reported as edible in the world list of edible insects by Jongema [30].

There are 14 species that we only sampled in one locality per species (Figure 2). This may be due to a generally low abundance of the species in an area and/or due to a low abundance of the species at the time of sampling (e.g., sampling may have occurred outside the peak of seasonal abundance).

The order Orthoptera is divided into two sub-orders: Caelifera (including grasshoppers) and Ensifera (including katydids and crickets). Most edible Orthoptera of Madagascar belong to Caelifera Acridoidea (grasshoppers). These represent 70% of the edible species, with the other 30% corresponding to Ensifera. This may be linked to the predominant timing of the activity of each group. All edible Caelifera recorded in the present study are active by daylight and are therefore easier to spot, while most Ensifera are concealed at that time. Although we focused on sampling insects during daylight hours, a methodological bias is unlikely. Our key informants indicate that edible Orthoptera are mostly collected while playing in the fields and while walking home from school in the case of children, and when outbreaks occur and when working in the fields in the case of adults. These activities occur during daylight hours. In addition, the large cricket *B. membranaceus colosseus* is collected during the day by locating entrance holes and digging it out of the ground. During the last hours of the day, entrance holes are sometimes spotted by listening for calling males. Edible insect collection after sundown does not seem to occur.

Rice field and grassland ecosystems are abundant in Madagascar and provide contemporary Malagasy rural people with the bioresource of edible Orthoptera, as they have throughout Malagasy history. Little may have changed about Orthoptera consumption habits in comparison to a century ago [1–3]. Edible Orthoptera have a place in Malagasy cultures throughout the entire country. What remains to be explored is the significance of edible Orthoptera in contemporary Malagasy food cultures and whether edible Orthoptera can continue to play a role in Malagasy food cultures in the future. Key aspects that require investigation include: (i) the determinants of casual collecting as the main contemporary strategy to collect edible Orthoptera (e.g., ecological influences, cultural influences, and personal influences); (ii) contemporary dietary contributions of edible Orthoptera in terms of consumption frequencies, consumption quantities, and nutrients, in comparison to other foods; (iii) contemporary cultural preferences and personal preferences with regard to Orthopteran species as food and with regard to the place of edible Orthoptera within dietary habits (e.g., snack or main meal); (iv) desirability for a more enterprising edible Orthoptera collection strategy for human food and/or animal feed (e.g., increased time allocation, improved harvesting techniques, and adoption of farming techniques); and (v) desirability for a prominent role of edible Orthoptera in local food cultures (e.g., increased consumption frequency and increased consumption quantity). These key aspects need to be investigated in all 19 different ethnicities and the ecoregions they inhabit. We used a small number of key informants per locality, ethnicity, and ecoregion, and therefore, generalizations for ethnicities cannot and should not be made based on our findings. We emphasize the qualitative character of the information provided.

The present study contributes to future research and development programs investigating the use of insects as human food and animal feed by increasing our knowledge of edible Orthoptera diversity, the geographical distribution of edible Orthoptera species, and the basic uses of edible Orthoptera species, by different ethnic groups and in different ecoregions in Madagascar.

Supplementary Materials: The following are available online at http://www.mdpi.com/2304-8158/8/12/666/s1. The Supplementary File provides recipes used in Madagascar to prepare Orthoptera.

Author Contributions: Conceptualization, J.V.I. and B.L.F.; methodology, J.V.I.; formal analysis, S.H., F.I.R., and J.V.I.; investigation, J.V.I., I.N.R.A., J.F.R., V.R.R., and F.I.R.; data curation, I.N.R.A. and F.I.R.; writing—original draft preparation, J.V.I. and S.H.; writing—review and editing, all authors; visualization, S.H.; funding acquisition, B.L.F.

Funding: This research was supported in part by awards from the Critical Ecosystem Partnership Fund (CEPF), and National Academy of Sciences (NAS) and United States Agency for International Development (USAID) PEER program AID-OAA-A-11-00012.

Acknowledgments: Administrative assistance and assistance in data collection was provided by Balsama Rajemison, Jean-Jacques Rafanomezantsoa, Claver E. Randrianandrasana, and Chrislain Ranaivo of the Madagascar Biodiversity Center, Parc Botanique et Zoologique de Tsimbazaza, Antananarivo 101, Madagascar. Jane Hooper, George Mason University, United States of America, is thanked for providing literature on the historical use of Orthoptera as food and feed in Madagascar. All necessary permits were obtained for the collection of insects in Madagascar. Permits to research, collect, and export arthropods were obtained from the Ministry of Environment and Sustainable Development as part of an ongoing collaboration between the California Academy of Sciences and the Ministry of Environment and Sustainable Development, Madagascar National Parks, and Parc Botanique et Zoologique de Tsimbazaza.

Conflicts of Interest: The authors declare no conflict of interest.

References

1. Decary, R. L'entomophagie chez les indigènes de Madagascar. *Bull. Soc. Entomol. Fr.* **1937**, *42*, 168–171.
2. Dellon, G. *A Voyage to the East-Indies: Giving an Account of the Isles of Madagascar, and Mascareigne, of Suratte, the Coast of Malabar, of Goa, Gameron, Ormus, and the Coast of Brasil, with the Religion, Manners and Customs of the Inhabitants, etc. As Also a Treatise*; HardPress Publishing: Wahroonga, NSW, Australia, 2019; pp. 25–27.
3. Camboué, R.P. Les sauterelles à Madagascar sur le riz malgache. *Bull. Mens. Soc. Natl. Acclim. Fr.* **1886**, *33*, 168–172.
4. Van Huis, A.; Van Itterbeek, J.; Klunder, H.; Mertens, E.; Halloran, A.; Muir, G.; Vantomme, P. *Edible Insects: Future Prospects for Food and Feed Security*; Food and Agriculture Organization: Rome, Italy, 2013; p. 187.
5. Van Huis, A. Potential of insects as food and feed in assuring food security. *Annu. Rev. Entomol.* **2013**, *58*, 563–583. [CrossRef] [PubMed]
6. Meyer-Rochow, V.B. Can insects help to ease the problem of world food shortage? *Search* **1975**, *6*, 261–262.
7. Minten, B.; Barrett, C.B. Agricultural technology, productivity, and poverty in Madagascar. *World Dev.* **2008**, *36*, 797–822. [CrossRef]
8. Choulamany, X. Traditional use and availability of aquatic biodiversity in rice-based ecosystems. III. Xieng Khouang and Houa Phanh provinces, Lao PDR. In *Aquatic Biodiversity in Rice-Based Ecosystems. Studies and Reports from Cambodia, China, Lao People's Democratic Republic and Vietnam*; Halwart, M., Bartley, D., Funge-Smith, S., Eds.; FAO: Rome, Italy, 2005.
9. Nonaka, K. Resource use in wetland and paddy field in Vientiane Plain, Lao PDR. *Tropics* **2008**, *17*, 325–334. [CrossRef]
10. Cerritos, R.; Cano-Santana, Z. Harvesting grasshoppers *Sphenarium purpurascens* in Mexico for human consumption: A comparison with insecticidal control for managing pest outbreaks. *Crop Prot.* **2008**, *27*, 473–480. [CrossRef]
11. Cerritos Flores, R.; Ponce-Reyes, R.; Rojas-García, F. Exploiting a pest insect species *Sphenarium purpurascens* for human consumption: Ecological, social, and economic repercussions. *J. Insects Food Feed* **2014**, *1*, 75–84. [CrossRef]
12. Payne, C.R.; Van Itterbeeck, J. Ecosystem services from edible insects in agricultural systems: A review. *Insects* **2017**, *8*, 24. [CrossRef] [PubMed]

13. Hanboonsong, Y.; Durst, P. *Edible Insects in Lao PDR: Building on Tradition to Enhance Food Security*; FAO: Bangkok, Thailand, 2014; p. 55.

14. ICIPE. Available online: http://www.icipe.org/research/plant-health/insects-food-and-feed/projects/entonutri-development-and-implementation-insect (accessed on 26 September 2019).

15. IPSIO. Available online: https://www.ipsio.org/valala-farms.html (accessed on 26 September 2019).

16. Van Huis, A. Insects as food in Sub-Saharan Africa. *Insect Sci. Appl.* **2003**, *23*, 163–185. [CrossRef]

17. Randrianandrasana, M.; Berenbaum, M.R. Edible non-crustacean arthropods in rural communities of Madagascar. *J. Ethnobiol.* **2015**, *2*, 354–383. [CrossRef]

18. Yen, A.L.; Van Itterbeeck, J. No taxonomists? No progress. *J. Insects Food Feed* **2016**, *2*, 223–224. [CrossRef]

19. Dirsh, V.M.; Descamps, M. Insectes Orthoptères Acridoidea. *Faune de Madagascar* **1968**, *26*, 1–312.

20. Braud, Y.; Franc, A.; Gay, P.E. *Les Acridiens des Formations Herbeuses de Madagascar*; FAO: Rome, Italy, 2014; p. 134.

21. Badanaro, F.; Komina, A.; Courdjo, L. Edible *Cirina forda* (Westwood, 1849) (Lepidoptera: Saturniidae) caterpillar among Moba people of the Savannah Region in North Togo: From collector to consumer. *Asian J. Appl. Sci. Eng.* **2014**, *3*, 275–286. [CrossRef]

22. Boulidam, S. Edible insects in a Lao market economy. In *Forest Insects as Food: Humans Bite Back, Proceedings of the a Workshop on Asia-Pacific Resources and Their Potential for Development, Chiang Mai, Thailand, 19–21 February 2008*; Durst, P.B., Johnson, D.V., Leslie, R.N., Shono, K., Eds.; FAO: Bangkok, Thailand, 2010; pp. 131–140.

23. Nonaka, K. Cultural and commercial roles of edible wasps in Japan. In *Edible Forest Insects: Humans Bite Back, Proceedings of the a Workshop on Asia-Pacific Resources and Their Potential for Development, Chiang Mai, Thailand, 19–21 February 2008*; Durst, P.B., Johnson, D.V., Leslie, R.N., Shono, K., Eds.; FAO: Bangkok, Thailand, 2010; pp. 123–130.

24. Van Itterbeeck, J. Traditional edible insect products and the shops that sell them in central Japan. *Najima Accessible Asia* **2019**, *9*, 24–26.

25. Pugalenthi, P.; Livingstone, D. Cardenolides (heart poisons) in the painted grasshopper *Poecilocerus pictus* F. (Orthoptera: Pyrgomorphidae) feeding on the milkweed *Calotropis gigantea* L. (Asclepiadaceae). *J. N. Y. Entomol. Soc.* **1995**, *103*, 191–196.

26. Seibt, U.; Kasang, G.; Wickler, W. Suggested pharmacophagy of the African bushhopper *Phymateus leprosus* (Fabricius) (Pyrgomorphidae, Orthoptera). *Z. Naturforsch. C J. Biosci.* **2000**, *55*, 442–448. [CrossRef] [PubMed]

27. Al-Robai, A.A. Different ouabain sensitivities of Naþ/Kþ-ATPase from *Poekilocerus bufonius* tissues and a possible physiological cost. *Comp. Biochem. Physiol.* **1993**, *B 106*, 805–812.

28. Mathen, C.; Peter, S.M.; Hardikar, B.P. Comparative evaluation of the cytotoxic and apoptotic potential of *Poecilocerus pictus* and *Calotropis gigantea*. *J. Environ. Pathol. Toxicol. Oncol.* **2011**, *30*, 83–92. [CrossRef] [PubMed]

29. Sreenivasulu, Y.; Riyaz Basha, M.; Rajarami Reddy, G.J. Inhibitory action of the defensive discharge of the grasshopper, *Poecilocerus pictus*, on certain enzymes in the lizard, *Calotes nemoricola*. *Enzyme Inhib.* **1996**, *11*, 135–140. [CrossRef] [PubMed]

30. Jongema, Y. List of Edible Insects of the World. 2015. Available online: http://www.wur.nl/en/ExpertiseServices/Chair-groups/Plant-Sciences/Laboratory-of-Entomology/Edible-insects/Worldwide-specieslist.htm (accessed on 1 September 2016).

Article

Microbiological Profile and Bioactive Properties of Insect Powders Used in Food and Feed Formulations

Concetta Maria Messina [1], Raimondo Gaglio [2], Maria Morghese [1], Marco Tolone [2], Rosaria Arena [1], Giancarlo Moschetti [2], Andrea Santulli [1], Nicola Francesca [2,*] and Luca Settanni [2]

[1] Laboratory of Marine biochemistry and ecotoxicology, Department of Earth and Marine Science DISTEM, University of Palermo, Via G. Barlotta 4, 91100 Trapani, Italy

[2] Department of Agricultural, Food and Forestry Science, University of Palermo, Viale delle Scienze 4, 90128 Palermo, Italy

*   Correspondence: nicola.francesca@unipa.it; Tel.: +39-091-2389-6043

Received: 23 July 2019; Accepted: 6 September 2019; Published: 9 September 2019

Abstract: Microbiological, nutritional and bioactive properties of edible powders obtained from *Acheta domesticus* (house cricket) and *Tenebrio molitor* (mealworm) were investigated. Except for the enterobacteria, viable bacteria were at a higher concentration in mealworm flour. The diversity evaluation carried out using MiSeq Illumina that mainly identified *Citrobacter* and *Enterobacteriaceae* in mealworm powder and members of the *Porphyromonadaceae* family in house cricket powder. Enterococci were identified and characterized for their safety characteristics in terms of the absence of antibiotic resistance and virulence. Both powders represent a good source of proteins and lipids. The fatty acid profile of mealworm powder was characterized by the predominance of the monounsaturated fatty acids and house cricket powder by saturated fatty acids. The enzymatic hydrolysis produced the best results in terms of percentage of degree of hydrolysis with the enzyme Alcalase, and these data were confirmed by SDS-PAGE electrophoresis. Furthermore, the results showed that the protein hydrolysate of these powders produces a significant antioxidant power.

Keywords: Alcalase; insect powders; *Acheta domesticus*; *Tenebrio molitor*; *Enterococcus*; antioxidant activity

---

1. Introduction

Insect consumption occurs almost worldwide, and this practice would represent a potential solution to food shortages and famine [1,2]. The nutritional relevance of insects is mostly represented by their high digestible protein content [3]. Compared to conventional livestock, their breeding systems are characterized by fewer environmental issues, including lower water consumption [4], greenhouse gases and ammonia generation [5].

Insect consumption in Western countries is still limited [6,7]. First of all, because of the unpleasant perception that the majority of consumers have towards such foods, which are not considered as conventional [8]. The consumption of edible insects is also hampered by regulations regarding hygiene and safety issues. Furthermore, religious concerns should also be considered in future. In areas of Asia, Africa and South America, where insects are eaten daily, they are commonly collected from natural environments [9]. Thus, the microbiological load implications of these foods might be relevant. Durst et al. [10] reported some cases of botulism and other foodborne illnesses due to the consumption of insects stored in Africa. The major risks derive from the ingestion of the gastrointestinal tract of the insects [11]. However, several commercial insect farms are keeping the growth of edible insects under controlled hygiene conditions.

In Western countries, insects are often consumed as flour added to some traditional food ingredients in several formulations. Insect powders were mixed with maize flour to produce tortillas [12] and emulsion sausages [13]. Commonly used species is the mealworm *Tenebrio molitor* (*T. molitor*) [14], but house cricket *Acheta domesticus* (*A. domesticus*) also showed a high potential for the enrichment of food products, including fermented ones [11]. *T. molitor* and *A. domesticus* and their powders have been studied for microbiological aspects showing a consistent presence of *Enterobacteriaceae* family [11,14–17], techno-functionality, chemical and nutritional composition showing high quantity of crude proteins, micronutrients, B-group vitamins and low crude fat [14,18,19].

Currently, there is a growing interest in the applications of food proteins and peptides in the form of functional foods or nutraceuticals as alternatives ingredients to conventional treatments. Enzymatic modification of proteins is useful to improve functionality [20]. Some peptides obtained from dietary proteins using enzymatic hydrolysis have been demonstrated to be antioxidant, antimicrobial, antidiabetic, antihypertensive, antithrombotic and immunomodulating [21,22].

Edible insects are viable sources of bioactive peptides owing to their high protein content and sustainable production [20,23]. A multidisciplinary approach consisting of chemical/nutritional, biochemical and microbiological investigations has been applied to characterize mealworm and house cricket powders.

2. Materials and Methods

2.1. Raw Materials and Microbiological Analyses

The powders analyzed were prepared from *T. molitor* and *A. domesticus* insects (two samples for each species) and were provided by Kreca Ento-Food (Harderwijk, Gheldria, The Netherlands). As reported in labels, both insect powders are important sources of proteins, since they contain all the nine essential amino acids. Furthermore, both powders have a high digestibility, possess a low carbohydrate profile and are free of preservatives, antibiotics, hormones and pesticides. The powders were kept under refrigeration in the dark as suggested by the supplier. To evaluate the changes of the chemical/nutritional, biochemical and microbiological characteristics of mealworm and cricket powder during storage, the powders were analyzed after 12 months of refrigerated storage (4 °C).

The insect powders were subjected to decimal serial dilution in Ringer's solution (Sigma-Aldrich, Milan, Italy). The first dilution was obtained using a Stomacher (BagMixer® 400, Interscience, Saint Nom la Bretèche, Yvelines, France) at the highest speed (260 rpm) for 2 min. Cell suspensions were measured using plate count for the enumeration of the following microbial groups: total mesophilic microorganisms (TMM) on plate count agar (PCA) incubated aerobically at 30 °C for 72 h; mesophilic lactic acid bacteria (LAB) rods and cocci on de Man-Rogosa-Sharpe (MRS) and M17 agar, incubated anaerobically at 30 °C for 48 h; enterococci on kanamycin esculin azide (KAA) agar incubated aerobically at 37 °C for 24 h; members of the *Enterobacteriaceae* family on violet red bile glucose agar (VRGBA) incubated aerobically at 37 °C for 24 h; coagulase-positive staphylococci (CPS) on Baird-Parker (BP) agar supplemented with rabbit plasma fibrinogen (RPF), incubated aerobically at 37 °C for 48 h; pseudomonads on *Pseudomonas* agar base (PAB) supplemented with cephaloridine sodium fusidate cetrimide (CFC), incubated aerobically at 25 °C for 48 h; yeasts and moulds on malt agar (MA) with chloramphenicol (0.1 g/L), incubated aerobically for 48 h and 7 days, respectively, at 28 °C; spore-forming aerobic bacteria were investigated after heating of cell suspensions at 85 °C for 15 min and then spread plated on nutrient agar (NA) before aerobic incubation at 32 °C for 48 h. Anaerobiosis occurred in hermetically sealed jars with the AnaeroGen AN25 system (Oxoid, Milan, Italy). All media were purchased from Oxoid. Microbiological counts were carried out in triplicates.

2.2. V3-V4 Amplification and Illumina Data Analysis

To maximize the effective length of the MiSeq's 300PE sequencing reads, the region encompassing the V3 and V4 hypervariable regions of the 16S rRNA gene (approximately 469 bp) was targeted for sequencing. Genomic DNA was extracted from insect powder samples using a QIAamp DNA Mini Kit (Qiagen, Hilden, Düsseldorf, Germany) and diluted to 5 ng/μl in 10 mM Tris pH 8.5 as indicated using the Illumina protocol 16S Metagenomic Sequencing Library Preparation, 15044223 Rev.

B. Briefly, to amplify and sequence the V3-V4 hypervariable region of the 16S rRNA gene, primers were designed that had overhang adapter sequences that must be appended to the primer pair sequences for compatibility using the Illumina index (San Diego, CA, USA) and sequencing adapters. The libraries were sequenced using the MiSeq Reagent Kit v3, 600 Cycles Sequencing kit (MS-102-3003) on the MiSeq System (Illumina).

Sequences obtained from Illumina Sequencing were processed using the QIIME2 software package version 2018.4 [24]. The reads were assigned to each sample according to the unique index; pairs of reads from the original DNA fragments were first merged using an import tool implemented in QIIME2. Quality check and trimming were done to trim sequences where the Phred quality score was < 20 using the DADA2 a R packages [25] wrapped in QIIME2. The Phred quality score is a measure of the quality of the identification of the nucleobases generated by automated DNA sequencing. Moreover, to remove chimeras from the Illumina sequenced FASTQ files the "consensus" method implemented in DADA2 was used. For taxa comparisons, we used the QIIME2 q2-feature-classifier plugin and the Naïve Bayes classifier that was trained on the Greengenes 13.8 database with a 99% Operational Taxonomic Units (OTUs) full-length sequences. QIIME2 taxa barplot command and ggplot2 were used for visualization of the taxonomic composition of the samples. Alpha diversity analysis was done with the q2-diversity plugin in QIIME2. In particular, Chao1 [26] metric that is a nonparametric abundance-based estimator of species richness and observed OTU were used to study diversity within each sample. Finally, to compare the relative abundance of microbial communities between the two samples, a Kruskal-Wallis test was done [27].

2.3. Phenotypic and Genotypic Characterization of LAB

Some colonies of presumptive LAB (Gram-positive, determined using the Gregersen KOH method [28], and catalase negative, determined by addition of fresh colonies from the agar media to 5%, *w/v*, H_2O_2) from the highest plated dilutions of the microbial cells on MRS, M17 and KAA agar were collected for all different morphologies recognized considering color, shape, edge, and surface (smooth or jagged). Gram-positive and catalase-negative cultures were purified through successive sub-culturing in the same media used for plate counts. All cultures were characterized for their cell morphology determined using an optical microscope at 100 × (Zeiss, Oberkochen, Stuttgart, Germany), growth at 15 and 45 °C, metabolism type, testing the ability to produce CO_2 from glucose and growth in the presence of a mixture of pentose carbohydrates (xylose, arabinose and ribose; 8 g/L each) in place of glucose [29]. The coccus-shaped isolates were finally tested for their growth at pH 9.2 and in the presence of 6.5% (*w/v*) NaCl.

Genomic DNA from the PCR assay was prepared using the InstaGene Matrix kit (Bio-Rad, Hercules, CA, USA) as described by the manufacturer. Cells were harvested from insect flour isolated cultures grown overnight in MRS or M17 broths at 30 °C, and genomic DNAs were extracted using the Instagene Matrix kit (Bio-Rad), as described by the manufacturer. Crude cell extracts were used as templates for the polymerase chain reaction (PCR).

Strain differentiation was done using random amplification of polymorphic DNA (RAPD)-PCR analysis using the single primers M13, AB111, and AB106 as previously described by Gaglio et al. [30] using a T1 Thermocycler (Biometra, Göttingen, Germany) to generate amplicons. The software package Gelcompare II Version 6.5 (Applied Maths, Sint-Martens-Latem, East Flanders, Belgium) was used to analyze the LAB profiles.

Gene sequencing of 16S rRNA was using as reported by Weisburg et al. (1991) with the primers rD1 (5'-AAGGAGGTGATCCAGCC-3') and fD1 (5'-AGAGTTTGATCCTGGCTCAG-3') for LAB identification at the species level. DNA fragments of about 1600 bp were purified and sequenced at Eurofins Genomics (Milan, Italy). The sequences obtained were compared with those available in the EzTaxon-e database (http://eztaxon-e.ezbiocloud.net/) with the sequences of the type strains only and the GenBank/EMBL/DDBJ (http://www.ncbi.nlm.nih.gov). The unequivocal identification of the *Enterococcus faecium* was further verified using the sodA gene-based multiplex PCR described by Jackson et al. [31] using the primers FM1 (5′-GAAAAAACAATAGAAGAATTAT-3′) and FM2 (5′-TGCTTTTTTGAATTCTTCTTTA-3′).

PCR mixture (22.5 µL total volume) included 20 µL of master mix and 2.5 µL of whole-cell template. The PCR program applied for all primers comprised 30 cycles of denaturation for 4 min at 95 °C, annealing for 1 min at 55 °C, and elongation for 1 min at 72 °C. Amplification was followed by a final extension at 72 °C for 7 min. The amplifications were performed using a T1 Thermocycler.

(Biometra) and the amplicons were separated by electrophoresis on a 2% (*w/v*) agarose gel (Gibco BRL, Cergy Pontoise, Val-d′Oise, France), stained with SYBR® Safe DNA gel stain (Molecular Probes, Eugene, OR, USA), and subsequently visualized by UV transillumination.

2.4. Safety Aspects of Dominant Insect Powder LAB

The antimicrobial susceptibility of enterococci was evaluated through the standard disk diffusion method of Kirbye-Bauer according to the Clinical and Laboratory Standards Institute guidelines [32] on Mueller Hinton Agar (Oxoid) incubated at 37 °C for 18 h. The following antimicrobials were tested: penicillin—10 units, ampicillin—10 µg, vancomycin—30 µg, erythromycin—15 µg, tetracycline—30 µg, ciprofloxacin—5 µg, levofloxacin—5 µg, chloramphenicol—30 µg, quinupristin-dalfopristin—15 µg, linezolid—30 µg, high-level gentamicin—120 µg and high-level streptomycin—300 µg. All antimicrobial compounds are commonly used for the treatment of human and animal infections. *Enterococcus faecalis* ATCC 29212 was used as the quality control strain for performing antimicrobial testing. All antimicrobial compounds were purchased from Oxoid.

The phenotypic assay of gelatinase production by *Enterococcus* strains was done on a plate containing gelatin agar as described by [33]. The gelatinase production was classified as positive when a clear zone of hydrolysis was detected around the colonies. The production of haemolytic activity was determined by streaking the bacterial cultures onto Columbia blood agar supplemented with 5% (*v/v*) horse blood (Becton Dickinson, Franklin Lakes, NJ, USA). Plates were incubated at 37 °C for 24–48 h with anaerobic conditions, after which the plates were examined for haemolysis. The hemolytic reactions were classified as total or β-hemolysis (clear zone of hydrolysis around the colonies), partial or α-hemolysis (green halo around the colonies) and absent or γ-hemolysis.

2.5. Proximate Composition

The proximate composition was measured as follows: moisture and ash content using the AOAC method [34], total nitrogen using the Kjeldahl method [35]; crude protein (P) and chitin (Q) content were determined applying the following equation used by Díaz-Rojas et al. [36]:

$$P = ((Nt \times Cq + K - 100) \times Cp)/(Cq - Cp) \tag{1}$$

$$Q = ((Nt \times Cp + K - 100) \times Cq)/(Cp - Cq) \tag{2}$$

where Nt was the total nitrogen content. K was the sum of total lipid, moisture and ash. Cp and Cq were conversion coefficients that relate the mass fraction of nitrogen with protein and chitin. The protein content of different insect species in the literature is mainly based on nitrogen content using the nitrogen to protein conversion factor (Cp) of 6.25 [37,38] while the value of Cq is 14.5 [36].

The total fatty acid (FA) methyl esters were determined from the total lipid [37] of insect powders according to Lepage and Roy [38] and analyzed using the conditions described by Messina et al. [39] employing a Perkin Elmer (Waltham, MA, USA) autosystem XL instrument equipped using a silica capillary column (30 m × 0.32 mm, d_f 0.25 µm, Omegawax 320, Supelco, Bellefonte, PA, USA). Individual FAME were measured by comparison of known standards (mix of PUFA 1, PUFA 2 and PUFA 3 mixed oil, Supelco).

Caloric content was measured as total energy content (kcal/100 g) using an isoperibolic oxygen bomb calorimeter (model 6200, Parr Instrument Co., Moline, IL, USA).

2.6. Enzymatic hydrolysis

The samples were subjected to enzymatic hydrolysis in distilled water (1:1 *w/v*), using three different proteases (peptidases) (Protamex, Flavourzyme and Alcalase, Sigma-Aldrich). The hydrolysis reaction was performed according to Messina et al. [40] at 60 °C keeping the pH at 8.0 with the addition of NaOH 5M. These conditions are optimal for enzymatic activity [40]. The degree of hydrolysis (DH%) of each enzyme was determined directly every 15 min for 195 min, and applying the equation used by Dumay et al. [41]. The enzymatic activity was stopped at 90 °C for 5 min and the samples were centrifuged at 7142 g force for 15 min at 4 °C. The supernatants were lyophilized for further determinations and stored at 4 °C [42].

2.7. Sodium Dodecyl Sulphate–Polyacrylamide Gel Electrophoresis (SDS-PAGE)

The hydrolysates were separated using SDS-PAGE (SDS-PAGE, Bio-Rad). The concentration of total proteins in all samples (powder and hydrolysates) done using the Lowry et al. [43] method, using BSA as the standard assuming it was 100% pure. Aliquots of 100 μg of protein, diluted with Laemmli buffer (1970) (Sigma-Aldrich) and denatured for 5 min at 90 °C, were loaded on a gradient polyacrylamide minigel (4–15%) (Bio-Rad) and subjected to electrophoresis at 20 mA for about 2 h. A mix of standards proteins, having relative molecular mass varying between 250 and 14 kDa (Bio-Rad) was run simultaneously into the gel. After the electrophoretic run, the gel was stained with a reagent which uses the reference protocol of staining with Coomassie Blue (GelCode Blue Stain Reagent, Pierce, Rockford, IL, USA). The image of the gel was acquired and elaborated using the software Image Lab 4.1 (Bio-Rad).

2.8. DPPH Radical Scavenging Activity

The total antioxidant power of the hydrolysates obtained at the end of the enzymatic processes for both insect powders was measured using the DPPH assay [44]. DPPH (1.1-diphenyl-2-picryhydrazyl, Sigma-Aldrich) is a stable free radical widely used in the detection of scavenging activity of hydrolysates for screening antioxidant compounds. Different aliquots of the sample were taken and the volume was made to 1.0 mL with ethanol. The reaction was started by the addition of 1.0 mL of 200 μM DPPH solution in 96% ethanol. The reaction mixture was kept at ambient temperature (25 °C) for 30 min and the absorbance was measured at 517 nm. Gallic acid (Sigma-Aldrich) was used as a positive control. The scavenging activity was determined using the following equation by Manuguerra et al. [45].

$$\text{Scavenging activity (\%)} = [1 - (\text{Absorbance sample}/\text{Absorbance control})] \times 100 \qquad (3)$$

2.9. Statistical Analyses

Microbiological data were subjected to one-way analysis of variance (ANOVA). Pair comparison of treatment means was done using Duncan procedure at $p < 0.05$. Differences between *T. molitor* and *A. domesticus* powders were evaluated using the generalized linear model (GLM) procedure. The linear dose-dependent relationship between the scavenging properties of the DPPH radical and the various concentrations tested has been tested through the linear regression test. The statistical analysis was done with Statistical Analysis System 9.2 software (SAS Institute, Cary, NC, USA).

3. Results and Discussion

3.1. Microbial Loads of Insect Flours

Insect powders have been recently investigated for their microbiological/safety aspects by different research groups, but so far, very little is known about their characteristics at the expiry date. The microbiological investigation of *T. molitor* and *A. domesticus* powders in this work was carried out after 12 months of storage. Mealworm powder hosted 7 microbial populations, while house cricket powder was characterized by a higher microbial diversity, since the *Pseudomonas* group was also detected, forming a total of 8 microbial populations (Figure 1).

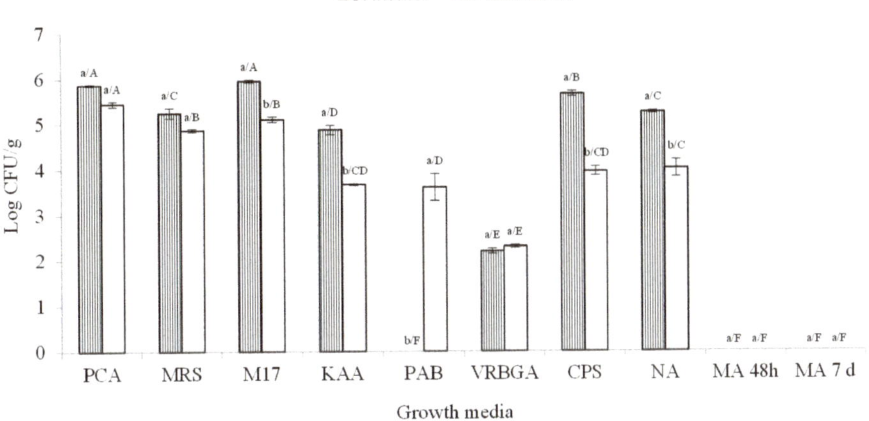

Figure 1. Microbial loads (Log CFU/g) of insect powders. Abbreviation: PCA, plate count agar for detection of total mesophilic microorganism; MRS, de Man-Rogosa-Sharpe agar for detection of mesophilic rod LAB; M17, medium 17 agar for detection of mesophilic coccus LAB; KAA, kanamycin esculin azide agar for detection of enterococci; PAB, *Pseudomonas* agar base for detection of pseudomonads; VRBGA, violet red bile glucose agar for detection of *Enterobacteriaceae*; CPS, coagulase-positive staphylococci; NA, nutrient agar for detection of spore-forming aerobic bacteria; MA, malt agar for detection of yeasts and moulds incubated for 48 h and 7 days, respectively. Results indicate mean values and standard deviation of three plate counts. Different lowercase letters indicate significant differences on microbial concentrations performed according to Duncan test between insect powders for $p < 0.05$ while different uppercase letters indicate significant differences on microbial concentrations between different growth media for $p < 0.05$.

Statistically significant differences were observed for the levels of LAB cocci, enterococci, pseudomonads, CPS and members of the Bacillaceae family between the powders of *T. molitor* and *A. domesticus*. Yeasts and moulds were undetectable for both matrices, while only mealworm powder showed pseudomonads below the detection level. The levels of TMM in both powders were a little lower than 106 CFU/g. The highest levels in mealworm powder were reached by LAB cocci (5.95 log CFU/g) followed by CPS, members of Bacillaceae family and LAB rods, which were all at cell densities above 105 CFU/g, while house cricket powder showed only LAB cocci at these levels. In general, *A. domesticus* powder showed lower levels of all microbial groups than *T. molitor* powder except the members of *Enterobacteriaceae* family.

The microbiology of insect powders soon after production has been studied by other authors considering different microbial groups. Bußler et al. [14] evaluated the total viable counts (on PCA) of *T. molitor* powder and observed levels of 7.72 log CFU/g of dry matter. Klunder et al. [11] analyzed the microbial loads of *T. molitor* powder after the insect were subjected to boiling and crushing reporting levels of 4.8 log CFU/g. The same authors also analyzed house crickets after boiling and stir-frying showing levels of 2.7 log CFU/g on PCA. In the same study, the members of *Enterobacteriaceae* family were 2.6 log CFU/g for mealworm and below the detection level in house cricket powder, while bacterial endospores were detected only in *A. domesticus* powder and counted at 1.5 log CFU/g. LAB were also the object of investigation by Klunder et al. [11], but their levels (ranging from 7.9 to 8.9 log CFU/g) were evaluated only after fermentation of the mixture mealworm powder/water at 30 °C. Information about the microbiology of *T. molitor* and *A. domesticus* are also available for the entire insects after freeze-drying [15]. Furthermore, Osimani et al. [46] also investigated on the hygiene of these insects reared under controlled conditions.

Only the study of Klunder et al. [11] considered the microbiological changes occurring during the refrigeration (4 °C) and ambient temperature (25 °C) storage of *A. domesticus* powder, but the monitoring period lasted 16 days. Thus, due to the different preparation and storage duration/conditions and the samples analyzed, a real comparison of data with those available in the literature is difficult. In general, the levels of TMM and enterobacteria registered for *A. domesticus* and *T. molitor* powders after 12 months of storage were comparable to those reported for the insects raised in the open field [11].

3.2. Culture-Independent Microbiological Analysis

After processing of the demultiplexed FASTQ files using DADA2 package, we obtained 69,818 and 133,305 reads for house cricket and mealworm powder, respectively. Only taxonomic groups with, at least, two representative sequences per taxonomic unit were retained and the relative abundances were reported in Figure 2.

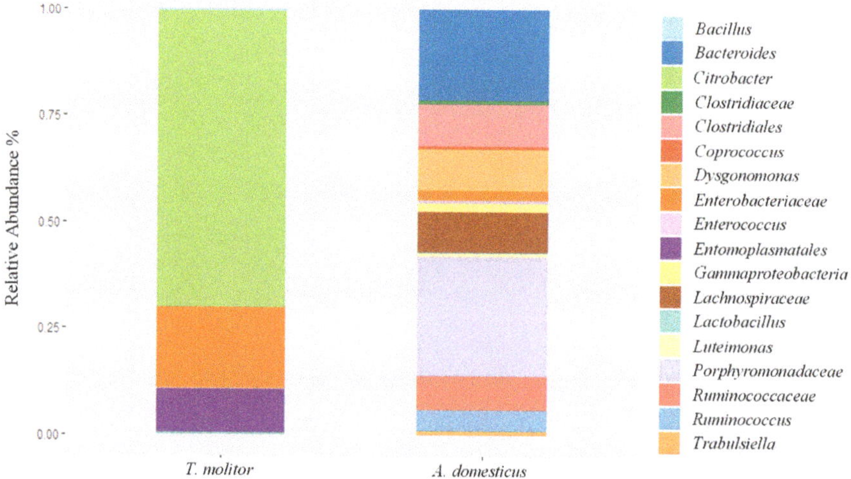

Figure 2. Relative abundances (%) of bacterial groups identified by MySeq Illumina in insect powders. Only taxonomic groups with at least two representative sequences per taxonomic unit were retained.

The diversity evaluation done using MiSeq Illumina identified members of the *Citrobacter* genus as the major components of the mealworm powder, which are commonly associated with the mid-gut of Lepidoptera insects [47], followed by *Enterobacteriaceae*. Members of *Porphyromonadaceae* family constituted the major bacterial group of house cricket powder. Many species of the family *Porphyromonadaceae* are part of the indigenous microbiota of the human and animal gastrointestinal tract and oral cavity [48]. To retrieve information at the species level, the most representative sequences of the two samples were manually blasted against the NCBI database. All *Enterobacteriaceae* OTUs were identified as belonging to *Salmonella enterica/Pseudocitrobacter faecalis/Cronobacter sakazakii*, while *Porphyromonadaceae* OTUs belonged to uncultured bacteria. No significant differences were found between observed and predicted (Chao1estimator) OTUs. Therefore, it is possible to capture the majority of OTUs present in each sample. Statistical analysis using the Kruskal-Wallis test revealed significant differences among *A. domesticus* and *T. molitor* ($p < 0.05$). This analysis showed a highest biodiversity in terms of bacterial species for *A. domesticus* powder. Regarding the lactic acid bacteria group, Illumina identified *Enterococcus* and *Lactobacillus* genera, generally found in entire mealworms [46].

Even though NGS analysis, as performed in this study, was based on DNA and this approach does not provide any indication on the viability of the detected species some safety issues arose from the composition of the microbiotas of the two insect powders analysed. In particular, the major groups belonged to enteric bacteria that are commonly found in several raw materials, such as *Enterobacteriaceae* family members in raw milk [49], meat [50] and vegetables [51]. Thus, considering insect powders as raw materials to be added as ingredient in food matrix formulations rather than as foods themselves, no particular food safety alerts concerning the major foodborne pathogens were shown by this study, even though the presence of *Salmonella* spp. deserves deepen investigations. However, several enterococci were isolated in viable form. Due to the antibiotic resistance gene transfer that could occur in some *Enterococcus* strains [52], a more comprehensive investigation on the isolates of this study should be done once insects flours will be used in food and feed formulations.

3.3. Characterization of LAB

The colonies grown on the media (MRS, M17 and KAA) specifically used for mealworm and house cricket powder LAB enumeration were collected and characterized. Barely 40 cultures were still considered presumptive LAB after Gram determination and catalase test. The microscopic investigation showed a coccus shape for all bacteria even though MRS allows the growth of rod LAB. The genetic typing showed nine different RAPD profiles, which corresponded to nine distinct strains (Figure 3). The sequencing of rRNA genes allotted all strains into *Enterococcus* genus.

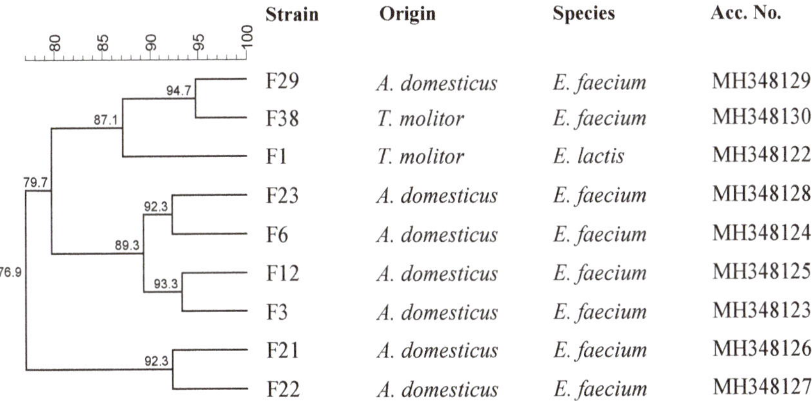

Strain	Origin	Species	Acc. No.
F29	*A. domesticus*	*E. faecium*	MH348129
F38	*T. molitor*	*E. faecium*	MH348130
F1	*T. molitor*	*E. lactis*	MH348122
F23	*A. domesticus*	*E. faecium*	MH348128
F6	*A. domesticus*	*E. faecium*	MH348124
F12	*A. domesticus*	*E. faecium*	MH348125
F3	*A. domesticus*	*E. faecium*	MH348123
F21	*A. domesticus*	*E. faecium*	MH348126
F22	*A. domesticus*	*E. faecium*	MH348127

Figure 3. Dendrogram obtained with combined random amplification of polymorphic DNA (RAPD)-PCR patterns of *Enterococcus* strains.

In particular, *Enterococcus faecium* were identified from *A. domesticus* powder, while *E. faecium* and *Enterococcus lactis* were observed in mealworm powder as confirmed by species-specific multiplex PCR.

Although some information is available on the presence of LAB in edible insects [46], this is the first study aimed at increasinging the characterization of insect powder LAB. Mealworm larvae and their frass analyzed for the presence of LAB by culture-independent tools were found to host *Lactobacillus*, *Pediococcus*, and *Leuconostocaceae* when the investigation was done through Illumina, while *Lactococcus* spp., *Enterococcus* spp. and *Lactobacillus* spp. when denaturing gradient gel electrophoresis was used [17], but no information on their viability was reported. As anticipated above, Klunder et al. [11] investigated LAB during the fermentation of mealworm powder, starting from a batch prepared mixing the insect powder with water at a ratio of 40:60, which was subjected to 5 fermentation cycles using 10% inoculums from the previous cycle. LAB increased in time, but the acidifying species were not identified.

It was seen that the dominant LAB of *T. molitor* and *A. domesticus* powders were all members of *Enterococcus* genus. Enterococci are bacteria of intestinal origin often associated with food matrices, but they rarely represent starter cultures for the fermentation processes [53]. The presence of *Enterococcus* in food products is a direct consequence of faecal contaminations [53]. In the case of insects, they are transferred from their intestinal tracts. In general, the enterococci from insect powder did not show dangerous risk factors since all of them were not virulent and were sensitive to the 12 antibiotics commonly used for the treatment of human and animal infections.

3.4. Proximate Composition, Energy and Fatty Acid Profile of Insect Powders

The proximate composition of insect powders is summarized in Table 1.

Table 1. Proximate composition (%), chitin (%), energy content (kJ/100g) and fatty acid composition (%) of the insect powders. Values are mean (three replications) ± standard deviation.

Proximate Composition	*Tenebrio molitor* (*T. molitor*)	*Acheta domesticus* (*A. domesticus*)
Moisture	3.12 ± 0.39	4.72 ± 0.15
Ash	3.32 ± 0.04	4.49 ± 0.04
Lipid	26.17 ± 0.21	21.66 ± 0.13
Protein	52.95 ± 0.33	63.62 ± 0.5
Chitin	14.42 ± 0.33	5.50 ± 0.5
Energy	1868.59 ± 2.26	1882.88 ± 3.65
Fatty acids		
Myristic acid (14:0)	2.35 ± 0.07	1.66 ± 0.07
Palmitic acid (16:0)	17.96 ± 0.14	25.56 ± 0.44
Palmitoleic acid (16:1 *n*-7)	1.67 ± 0.01	0.81 ± 0.02
9,12-Hexadecadienoic acid (16:2 *n*-4)	0.02 ± 0.01	0.08 ± 0.00
6,9,12-Hexadecatrienoic acid (16:3 *n*-4)	0.15 ± 0.00	0.11 ± 0.00
Stearic acid (18:0)	3.34 ± 0.05	12.47 ± 0.10
Oleic acid (18:1 *n*-9)	45.75 ± 0.38	22.59 ± 0.04
Vaccenic acid (18:1 *n*-7)	0.50 ± 0.66	0.93 ± 0.02
Linoleic acid (18:2 *n*-6)	25.73 ± 0.18	32.35 ± 0.42
γ-linolenic acid (18:3 *n*-6)	n.d.	n.d.
8,11,14-Octadecatrienoic acid (18:3 *n*-4)	0.01 ± 0.02	n.d.
α-Linolenic acid (18:3 *n*-3)	2.30 ± 0.03	1.75 ± 0.01
Stearidonic acid (18:4 *n*-3)	n.d.	0.04 ± 0.04
Eicosenoic acid (20:1 *n*-9)	0.13 ± 0.04	0.21 ± 0.00
Arachidonic acid (20:4 *n*-6)	n.d.	0.14 ± 0.01
Eicosatetraenoic acid (20:4, *n*-3)	n.d.	n.d.
Eicosapentaenoic acid (20:5, *n*-3)	0.01 ± 0.01	0.48 ± 0.04
Cetoleic acid (22:1 *n*-11)	0.01 ± 0.02	0.11 ± 0.00
Erucic acid (22:1 *n*-9)	0.01 ± 0.02	0.01 ± 0.00
Adrenic acid (22:4 *n*-6)	0.03 ± 0.05	n.d.
Osbond acid (22:5 *n*-6)	0.00 ± 0.01	n.d.
Docosapentaenoic acid (22:5 *n*-3)	n.d.	0.10 ± 0.01
Docosahexaenoic acid (22:6 *n*-3)	0.02 ± 0.02	0.63 ± 0.02
Nervonic acid (24:1 *n*-9)	2.35 ± 0.07	0.00 ± 0.00
Saturated	23.65 ± 0.12	39.68 ± 0.41
Monounsaturated	48.06 ± 0.20	24.66 ± 0.00
Polyunsaturated	28.28 ± 0.31	35.66 ± 0.41
Total *n*-3	2.33 ± 0.07	2.99 ± 0.02
Total *n*-6	25.77 ± 0.24	32.49 ± 0.43

n.d. = not detected.

The lipid value is 26.17% (Table 1) in *T. molitor* and 21.66% in *A. domesticus*. As expected, both insect species showed a high proportion of protein, 52.95% for *T. molitor* and 63.62% for *A. domesticus*. Insects contain chitin, a primary component of the exoskeleton of arthropods [54]; our analyses showed a chitin percentage of 14.42 for *T. molitor* and 5.50 for *A. domesticus*. Regarding moisture and ash, the values were between 3 and 5% for *T. molitor* and *A. domesticus*, respectively. The energy content

(Table 1) varied slightly between the species, ranging from 1868.59 ± 2.26 kJ/100g for *T. molitor* to 1882.88 ± 3.65) kJ/100 g for *A. domesticus*. These results are comparable to the energy supplied by beef (1735 kJ/100 g) or fish (1662 kJ/100 g) [55]. The results of the proximate composition were comparable to those reported in literature [18,54–56] and put in evidence the high potential of insects as alternative sources of new and renewable animal proteins and fat [54,57].

Table 1 shows the fatty acid composition of edible insect's powder. The amount of saturated fatty acids (SFA) ranged from 23.65% for *T. molitor* to 39.68% for *A. domesticus*. The two main components of the SFA were palmitic acid (C16:0) and stearic acid (C18:0). The highest amount of these fatty acids was observed in *A. domesticus*. Values detected in *T. molitor* powder were in agreement to those reported from Zielińska et al. [55] that showed 18% of palmitic acid and 3.8% of stearic acid in the larvae of *T. molitor*. The content of monounsaturated fatty acids (MUFA) varied from 24.66% in *A. domesticus* to 48.06% in *T. molitor*. The major MUFA of edible insect's powder is oleic acid (C18:1 *n*-9). The highest content of oleic acid was observed in *T. molitor*. The fraction of polyunsaturated acids (PUFA) ranged from 28.28% in *T. molitor* to 35.66% (Table 1) in *A. domesticus*. Yang et al. [58] obtained similar values for crickets (33.8%). In particular, in this study, in *T. molitor* powder, the omega-6 (*n*-6) PUFA were higher than omega-3 (*n*-3) (2.3%). For the n-6 class, the most abundant FA was linoleic acid. Regarding *A. domesticus* powder, n-6 PUFA were higher than omega-3 (*n*-3). Linoleic acid was the most abundant fatty acid of the n-6 class similar to values observed by Yang et al. [58] in ground crickets (32.2%) and by Osimani et al. [18] in the same species. Furthermore, various data regarding the body composition of insects showed the variable composition of FA between species, origin and developmental stages [56].

3.5. Enzymatic Hydrolysis

The enzyme selection is the most important factor in protein hydrolysis affecting the yield and physico-chemical properties of the final product. From the comparison of the three types of commercial proteases used, Alcalase from *Bacillus licheniformis* had provided the best results in hydrolyzing the total proteins of the *T. molitor* powder (Figure 4a) in terms of DH %, with increasing trend, followed by Protamex and Flavourzyme, which instead showed rather stable values over time ($p < 0.05$). The maximum DH% (between 20.1% of the enzyme Protamex and 25.8% of the enzyme Alcalase) were reached after 180 min of reaction.

The enzymatic reaction of hydrolysis to the powder of *A. domesticus* (Figure 4b) was better in terms of DH%, also in this case, with the enzyme Alcalase ($p < 0.05$). The plateau phase was reached at 195 min, where the maximum value of 24.6% (DH) (Figure 4d) was observed, which remains unchanged in subsequent measurements. Several authors had reported that, compared to other proteolytic enzymes, Alcalase allows superior protein recovery and provide hydrolysates with good functional properties [59–63]. Generally, alkaline proteases, including Alcalase, exhibit greater proteolytic activity than acid or neutral proteases such as Flavourzyme [62]. Yang et al. [64] reported that, among the proteases used, Alcalase had higher DH during the hydrolysis period, which suggested that Alcalase is more efficient than the other enzymes for preparing protein hydrolysates from edible insects. Tang et al. [23] obtained better results in terms of DH%, in *T. molitor* larvae, with a combination of Alcalase and Flavourzyme followed by slightly lower values obtained only with the enzyme Alcalase. Mizani et al. [60] reported that Alcalase, when used in combination with sodium sulphite and triton x-100, increased the yield of protein hydrolysates from *Penaeus semisulcatus* shrimp waste (heads) from 45.1% to 62–65%. This is related to a reduction of disulphide bonds and increased the solubility of proteins, as has been previously demonstrated in the case of soy products [65]. Recently, Alcalase has been used extensively in the hydrolysis of plant and animal proteins [20,21,66].

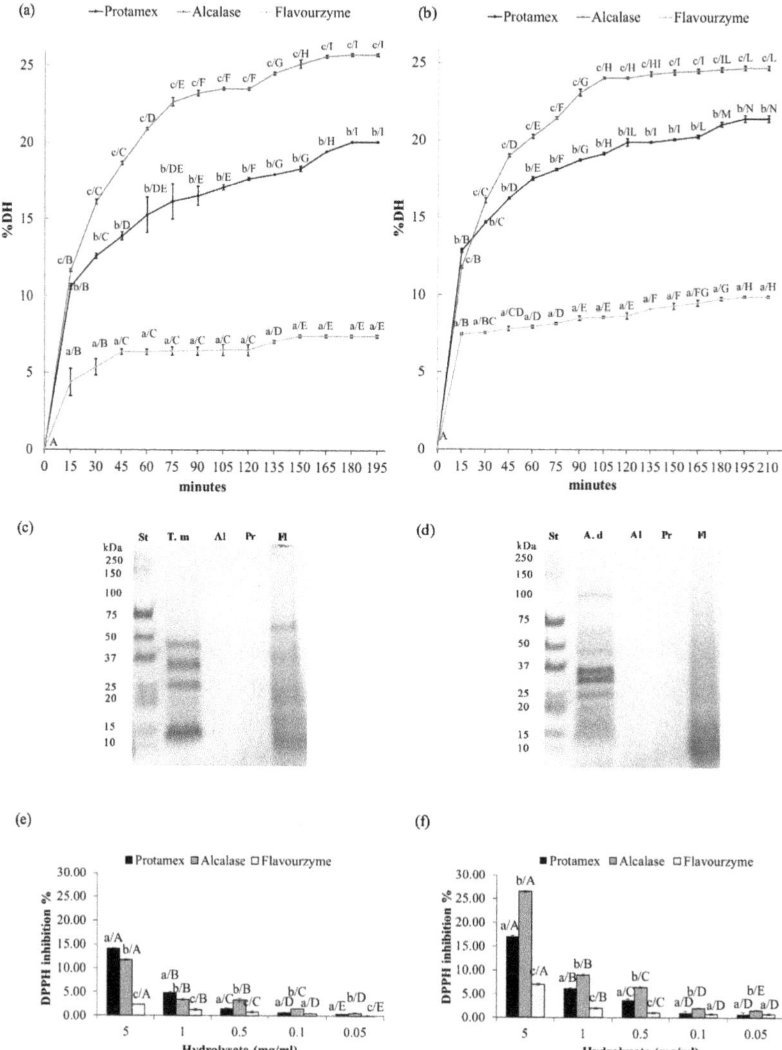

Figure 4. Physico-chemical properties: (**a**) Degree of hydrolysis of *T. molitor* powder; (**b**) Degree of hydrolysis of *A. domesticus* powder, different lowercase letters indicate significant differences between different enzymes (a, b, c...: $p < 0.05$), different uppercase letters indicate significant differences between different time points for each enzyme (A; B; C ... $p < 0.05$); (**c**) SDS-PAGE of the total proteins (TM) and proteins hydrolisates, obtained with the enzymes alcalase AL, protamex Pr and flavourzyme Fl, from *T. molitor* powder at the end of the enzymatic process; (**d**) SDS-PAGE of the total proteins (TM) and proteins hydrolisates, obtained with the enzymes alcalase AL, protamex Pr and flavourzyme Fl, from *A. domesticus* powder at the end of the enzymatic process; (**e**) DPPH radical scavenging activity of protein hydrolysates of *T. molitor* powder; (**f**) DPPH radical scavenging activity of protein hydrolysates of *A. domesticus* powder. Different lowercase letters indicate significant differences between different enzymes at the same concentration (a, b, c...: $p < 0.05$), different uppercase letters indicate significant differences between different concentrations tested for each enzyme (A; B; C ... $p < 0.05$); values are mean (three replications) ± standard deviations.

3.6. SDS PAGE

The SDS-PAGE electrophoresis (Figure 4c,d) showed that the electrophoretic pattern of the protein hydrolysates with respect to the total proteins extracted from the whole insect meals (T. m. and A. d. in Figure 4c,d), had a progressive loss of the bands at higher molecular mass, attesting to the efficiency of the reaction. According to the low DH obtained, the electrophoretic profile of Flavourzyme hydrolysates showed protein bands with a higher relative molecular mass than the other two enzymes, comparable to the profile of the total proteins of the whole samples (T. m. and A. d. in Figure 4c,d). In the study done by Kristinsson and Rasco [66], peptides with a lower molecular weight were observed with a higher DH. The enzyme Alcalase gave peptides at 65 kDa, while with Flavourzyme most of the peptides present a molecular mass of 70 kDa. At a higher degree of hydrolysis the bands with greater mass begin to disappear; this confirmed the fact that the molecular weight of the peptides formed by hydrolysis is associated with the degree of hydrolysis.

3.7. DPPH Radical Scavenging Activity

The DPPH assay showed that the protein hydrolysate of *T. molitor* powder (Figure 4e) had antioxidant properties, attested by inhibition of the DPPH radical up to 14.0% (Figure 4e). In particular, the hydrolysate obtained from the enzyme Protamex showed a higher antioxidant power in respect to the hydrolysate obtained with Alcalase and Flavourzyme only at the highest concentrations tested (5 and 1 mg/mL) (Figure 4e) ($p < 0.05$).

The DPPH assay on *A. domesticus* powder hydrolysate (Figure 4f) showed that the best antioxidant properties, resulting in an inhibition of DPPH radical up to 26.5%, were observed in the hydrolysates obtained using the enzyme Alcalase ($p < 0.05$). In Figure 4f it is evident a linear and dose-dependent relationship between the scavenging properties of the DPPH radical and the various concentrations tested, the linearity has been tested through the linear regression test obtaining the following values of R^2: *A. domesticus* Alcalase 0.982 ($p > 0.001$); *A. domesticus* Protamex 0.976 ($p > 0.001$); *A. domesticus* Flavourzyme 0.994 ($p > 0.001$). These results were comparable to previous reports [23,64].

Similar results have already been reported in the literature, confirming that the degree of hydrolysis can significantly influence the antioxidant activity of the resulting hydrolysates, probably due to the presence of a high amount of low molecular weight active peptides [67–69]. Ahn et al., [70] stated that the molecular weight of peptides was related to their functional properties, with greater efficacy in bioactive peptides at a molecular weight of about 1.0–3.0 kDa. Taheri et al., [71] showed that protein hydrolysate fractions between 1.0 and 10 kDa had higher antioxidant power than higher molecular weight fractions. Therefore, since high DH means that more peptide bonds were cleaved, the protein would release lower molecular weight peptides to the hydrolysates, endowing the hydrolysates with high antioxidant activity, indicating that a certain degree of DH is necessary to the physico-chemical activity of hydrolysates. Our future analyses will be directed towards the specific amino-acid analysis of the hydrolysates obtained.

4. Conclusions

Insect powders have been subjected to a deep microbiological characterization in view of their application in wheat powder fermentations to obtain fortified products. The fortification of traditional food products represents a successful strategy to provide the necessary nutrients without substantial modification of the alimentary habits [72]. Generally, insects were found to be highly nutritious and to represent good sources of protein, fat, minerals, vitamins, and energy [54]. They are traditionally used as a food source in different countries, but nowadays, they are becoming globally increasingly attractive as a protein and fat source for humans and many types of pet and farm animals [73]. Insects are useful not only for their nutritional composition [74] but also for the transfer of other indispensable nutrients and micronutrients to the recipients [75]. Insect protein is readily available with protein quality values similar to, or slightly higher than, fish meat or soybean powder [74]. Moreover, proteins analyzed showed to be a suitable source of biologically active peptides to generate multifunctional hydrolysates that could be incorporated into functional foods or used as nutraceuticals or as natural alternatives to synthetic antioxidants. Further study is also needed in the characterization of edible insect peptides, optimization of functional properties, sensory evaluation, and establishing applications of these hydrolysates in food formulations [20,21].

Author Contributions: Conceptualization, C.M.M., N.F. and L.S.; methodology, R.G., M.M. and R.A.; software, R.G. and M.T.; formal analysis, R.G., M.M. and R.A.; investigation, R.G., M.M. and R.A.; data curation, R.G., M.M., M.T. and N.F.; writing—original draft preparation, C.M.M. and L.S.; writing—review and editing, C.M.M., R.G., M.M., M.T. and L.S.; supervision, G.M. and A.S.

Funding: This research received no external funding.

Acknowledgments: The authors are grateful to Emilio Calligaris and Riccardo Zamponi for their precious collaboration during samples collection.

Conflicts of Interest: All authors certify that there is no conflict of interests in this study.

References

1. Belluco, S.; Losasso, C.; Ricci, A.; Maggioletti, M.; Alonzi, C.; Paoletti, M.G. Edible insects: A food security solution or a food safety concern? *Anim. Front.* **2015**, *5*, 25–30.
2. van Huis, A. Potential of Insects as Food and Feed in Assuring Food Security. *Annu. Rev. Entomol.* **2013**, *58*, 563–583. [CrossRef] [PubMed]
3. Ramos-Elorduy, J.; Moreno, J.M.P.; Prado, E.E.; Perez, M.A.; Otero, J.L.; de Guevara, O.L. Nutritional Value of Edible Insects from the State of Oaxaca, Mexico. *J. Food Compos. Anal.* **1997**, *10*, 142–157. [CrossRef]
4. Chapagain, A.K.; Hoekstra, A.Y. *Virtual Water Flows between Nations in Relation to Trade in Livestock and Livestock Products Value of Water*; UNESCO-IHE: DA Delft, The Netherlands, 2003.
5. Oonincx, D.G.A.B.; van Itterbeeck, J.; Heetkamp, M.J.W.; van den Brand, H.; van Loon, J.J.A.; van Huis, A. An Exploration on Greenhouse Gas and Ammonia Production by Insect Species Suitable for Animal or Human Consumption. *PLoS ONE* **2010**, *5*, e14445. [CrossRef] [PubMed]
6. Woolf, E.; Zhu, Y.; Emory, K.; Zhao, J.; Liu, C. Willingness to consume insect-containing foods: A survey in the United States. *LWT-Food Sci. Technol.* **2019**, *102*, 100–105. [CrossRef]
7. Orsi, L.; Voege, L.L.; Stranieri, S. Eating edible insects as sustainable food? Exploring the determinants of consumer acceptance in Germany. *Food Res. Int.* **2019**, *125*, 108573. [CrossRef]
8. Tan, H.S.G.; Fischer, A.R.H.; Tinchan, P.; Stieger, M.; Steenbekkers, L.P.A.; van Trijp, H.C.M. Insects as food: Exploring cultural exposure and individual experience as determinants of acceptance. *Food Qual. Prefer.* **2015**, *42*, 78–89. [CrossRef]
9. Nonaka, K. Feasting on insects. *Entomol. Res.* **2009**, *39*, 304–312. [CrossRef]
10. Durst, P.B.; Johnson, D.V.; Leslie, R.N.; Shono, K. *Forest Insects as Food: Humans Bite back Proceedings of a Workshop on Asia-Pacific Resources and Their Potential for Development*; Food and Agriculture of the United Nations: Rome, Italy, 2010; ISBN 9789251064887.
11. Klunder, H.C.; Wolkers-Rooijackers, J.; Korpela, J.M.; Nout, M.J.R. Microbiological aspects of processing and storage of edible insects. *Food Control* **2012**, *26*, 628–631. [CrossRef]
12. Aguilar-Miranda, E.D.; López, M.G.; Escamilla-Santana, C.; Barba de la Rosa, A.P. Characteristics of Maize Flour Tortilla Supplemented with Ground *Tenebrio molitor* Larvae. *J. Agric. Food Chem.* **2001**, *50*, 192–195. [CrossRef]
13. Kim, H.-W.; Setyabrata, D.; Lee, Y.J.; Jones, O.G.; Kim, Y.H.B. Pre-treated mealworm larvae and silkworm pupae as a novel protein ingredient in emulsion sausages. *Innov. Food Sci. Emerg. Technol.* **2016**, *38*, 116–123. [CrossRef]
14. Bußler, S.; Rumpold, B.A.; Jander, E.; Rawel, H.M.; Schlüter, O.K. Recovery and techno-functionality of flours and proteins from two edible insect species: Meal worm (*Tenebrio molitor*) and black soldier fly (*Hermetia illucens*) larvae. *Heliyon* **2016**, *2*, e00218. [CrossRef] [PubMed]
15. Megido, R.C.; Desmedt, S.; Blecker, C.; Béra, F.; Haubruge, É.; Alabi, T.; Francis, F. Microbiological load of edible insects found in Belgium. *Insects* **2017**, *8*, 1–8.

16. Garofalo, C.; Osimani, A.; Milanović, V.; Taccari, M.; Cardinali, F.; Aquilanti, L.; Riolo, P.; Ruschioni, S.; Isidoro, N.; Clementi, F. The microbiota of marketed processed edible insects as revealed by high-throughput sequencing. *Food Microbiol.* **2017**, *62*, 15–22. [CrossRef] [PubMed]

17. Osimani, A.; Milanović, V.; Cardinali, F.; Garofalo, C.; Clementi, F.; Pasquini, M.; Riolo, P.; Ruschioni, S.; Isidoro, N.; Loreto, N.; et al. The bacterial biota of laboratory-reared edible mealworms (*Tenebrio molitor* L.): From feed to frass. *Int. J. Food Microbiol.* **2018**, *272*, 49–60. [CrossRef] [PubMed]

18. Osimani, A.; Garofalo, C.; Milanovic´, V.; Milanovic´, M.; Taccari, M.; Cardinali, F.; Aquilanti, L.; Pasquini, M.; Mozzon, M.; Raffaelli, N.; et al. Insight into the proximate composition and microbial diversity of edible insects marketed in the European Union. *Eur. Food Res. Technol.* **2017**, *243*, 1157–1171. [CrossRef]

19. Osimani, A.; Milanović, V.; Cardinali, F.; Roncolini, A.; Garofalo, C.; Clementi, F.; Pasquini, M.; Mozzon, M.; Foligni, R.; Raffaelli, N.; et al. Bread enriched with cricket powder (*Acheta domesticus*): A technological, microbiological and nutritional evaluation. *Innov. Food Sci. Emerg. Technol.* **2018**, *48*, 150–163. [CrossRef]

20. Hall, F.G.; Jones, O.G.; O'Haire, M.E.; Liceaga, A.M. Functional properties of tropical banded cricket (*Gryllodes sigillatus*) protein hydrolysates. *Food Chem.* **2017**, *224*, 414–422. [CrossRef]

21. Hall, F.; Johnson, P.E.; Liceaga, A. Effect of enzymatic hydrolysis on bioactive properties and allergenicity of cricket (*Gryllodes sigillatus*) protein. *Food Chem.* **2018**, *262*, 39–47. [CrossRef]

22. Li-Chan, E.C. Bioactive peptides and protein hydrolysates: Research trends and challenges for application as nutraceuticals and functional food ingredients. *Curr. Opin. Food Sci.* **2015**, *1*, 28–37. [CrossRef]

23. Tang, Y.; Debnath, T.; Choi, E.-J.; Kim, Y.W.; Ryu, J.P.; Jang, S.; Chung, S.U.; Choi, Y.-J.; Kim, E.-K. Changes in the amino acid profiles and free radical scavenging activities of *Tenebrio molitor* larvae following enzymatic hydrolysis. *PLoS ONE* **2018**, *13*, e0196218. [CrossRef]

24. Caporaso, J.G.; Kuczynski, J.; Stombaugh, J.; Bittinger, K.; Bushman, F.D.; Costello, E.K.; Fierer, N.; Peña, A.G.; Goodrich, J.K.; Gordon, J.I.; et al. QIIME allows analysis of high-throughput community sequencing data. *Nat. Methods* **2010**, *7*, 335–336. [CrossRef] [PubMed]

25. Callahan, B.J.; McMurdie, P.J.; Rosen, M.J.; Han, A.W.; Johnson, A.J.A.; Holmes, S.P. DADA2: High-resolution sample inference from Illumina amplicon data. *Nat. Methods* **2016**, *13*, 581–583. [CrossRef] [PubMed]

26. Chao, A.; Bunge, J. Estimating the Number of Species in a Stochastic Abundance Model. *Biometrics* **2002**, *58*, 531–539. [CrossRef] [PubMed]

27. Zhang, X.; Shen, D.; Fang, Z.; Jie, Z.; Qiu, X.; Zhang, C.; Chen, Y.; Ji, L. Human Gut Microbiota Changes Reveal the Progression of Glucose Intolerance. *PLoS ONE* **2013**, *8*, e71108. [CrossRef] [PubMed]

28. Gregersen, T. Rapid method for distinction of gram-negative from gram-positive bacteria. *Eur. J. Appl. Microbiol. Biotechnol.* **1978**, *5*, 123–127. [CrossRef]

29. Alfonzo, A.; Ventimiglia, G.; Corona, O.; Di Gerlando, R.; Gaglio, R.; Francesca, N.; Moschetti, G.; Settanni, L. Diversity and technological potential of lactic acid bacteria of wheat flours. *Food Microbiol.* **2013**, *36*, 343–354. [CrossRef] [PubMed]

30. Gaglio, R.; Francesca, N.; Di Gerlando, R.; Mahony, J.; De Martino, S.; Stucchi, C.; Moschetti, G.; Settanni, L. Enteric bacteria of food ice and their survival in alcoholic beverages and soft drinks. *Food Microbiol.* **2017**, *67*, 17–22. [CrossRef]

31. Jackson, C.R.; Fedorka-Cray, P.J.; Barrett, J.B. Use of a genus- and species-specific multiplex PCR for identification of enterococci. *J. Clin. Microbiol.* **2004**, *42*, 3558–3565. [CrossRef]

32. Clinical Laboratory Standards Institute. *Performance Standards for Antimicrobial Susceptibility Testing*; Twenty-seventh Informational Supplement (M100); CLSI: Wayne, PA, USA, 2017.

33. Lopes, M.D.F.S.; Simões, A.P.; Tenreiro, R.; Marques, J.J.F.; Crespo, M.T.B. Activity and expression of a virulence factor, gelatinase, in dairy enterococci. *Int. J. Food Microbiol.* **2006**, *112*, 208–214. [CrossRef]

34. AOAC Official methods of analysis. *J. Pharm. Sci.* **1990**, *1*, 1230.

35. AOAC Association of Official Analytical Chemists Official Method, 981.10 Crude protein in meat block digestion method. *J. AOAC Int.* **1992**, *65*, 1339.

36. Díaz-Rojas, E.I.; Argüelles-Monal, W.M.; Higuera-Ciapara, I.; Hernández, J.; Lizardi-Mendoza, J.; Goycoolea, F.M. Determination of chitin and protein contents during the isolation of chitin from shrimp waste. *Macromol. Biosci.* **2006**, *6*, 340–347. [CrossRef] [PubMed]

37. Folch, J.; Lees, M.; Stanley, G.H.S. A simple method for the isolation and purification of total lipids from animal tissues. *J. Biol. Chem.* **1957**, *226*, 497–509. [PubMed]

38. Lepage, G.; Roy, C.C. Improved recovery of fatty acid through direct transesterification without prior extraction or purification. *J. Lipid Res.* **1984**, *25*, 1391–1396. [PubMed]

39. Messina, C.M.; Renda, G.; La Barbera, L.; Santulli, A. By-products of farmed European sea bass (*Dicentrarchus labrax* L.) as a potential source of n-3 PUFA. *Biologia* **2013**, *68*, 288–293. [CrossRef]

40. Messina, C.M.; Renda, G.; Randazzo, M.; Laudicella, V.A.; Gharbi, S.; Pizzo, F.; Morghese, M.; Santulli, A. Extraction of bioactive compounds from shrimp waste. *Bull. Inst. Natl. Sci. Tech. Mer de Salammbô* **2015**, *42*, 27–29.

41. Dumay, J.; Donnay-Moreno, C.; Barnathan, G.; Jaouen, P.; Bergé, J.P. Improvement of lipid and phospholipid recoveries from sardine (*Sardina pilchardus*) viscera using industrial proteases. *Process Biochem.* **2006**, *41*, 2327–2332. [CrossRef]

42. de Holanda, H.D.; Netto, F.M. Recovery of Components from Shrimp (*Xiphopenaeus kroyeri*) Processing Waste by Enzymatic Hydrolysis. *J. Food Sci.* **2006**, *71*, C298–C303. [CrossRef]

43. Lowry, O.H.; Rosebrough, N.J.; Farr, A.L.; Randall, R.J. Protein measurement with the Folin phenol reagent. *J. Biol. Chem.* **1951**, *193*, 265–267.

44. Sharma, O.P.; Bhat, T.K. DPPH antioxidant assay revisited. *Food Chem.* **2009**, *113*, 1202–1205. [CrossRef]

45. Manuguerra, S.; Caccamo, L.; Mancuso, M.; Arena, R.; Rappazzo, A.C.; Genovese, L.; Santulli, A.; Messina, C.M.; Maricchiolo, G. The antioxidant power of horseradish, *Armoracia rusticana*, underlies antimicrobial and antiradical effects, exerted in vitro. *Nat. Prod. Res.* **2018**, 1–4. [CrossRef] [PubMed]

46. Osimani, A.; Milanović, V.; Cardinali, F.; Garofalo, C.; Clementi, F.; Ruschioni, S.; Riolo, P.; Isidoro, N.; Loreto, N.; Galarini, R.; et al. Distribution of transferable antibiotic resistance genes in laboratory-reared edible mealworms (*Tenebrio molitor* L.). *Front. Microbiol.* **2018**, *9*, 1–11. [CrossRef] [PubMed]

47. Motcha Anthony Reetha, B.; Mohan, M. Diversity of commensal bacteria from mid-gut of pink stem borer (*Sesamia inferens* [Walker])-Lepidoptera insect populations of India. *J. Asia. Pac. Entomol.* **2018**, *21*, 937–943. [CrossRef]

48. Sakamoto, M. The Family Porphyromonadaceae. In *The Prokaryotes*; Springer: Berlin/Heidelberg, Germany, 2014; pp. 811–824.

49. Gaglio, R.; Cruciata, M.; Di Gerlando, R.; Scatassa, M.L.; Cardamone, C.; Mancuso, I.; Sardina, M.T.; Moschetti, G.; Portolano, B.; Settanni, L. Microbial activation of wooden vats used for traditional cheese production and evolution of neoformed biofilms. *Appl. Environ. Microbiol.* **2016**, *82*, 585–595. [CrossRef] [PubMed]

50. Gaglio, R.; Francesca, N.; Maniaci, G.; Corona, O.; Alfonzo, A.; Giosuè, C.; Di Noto, A.; Cardamone, C.; Sardina, M.T.; Portolano, B.; et al. Valorization of indigenous dairy cattle breed through salami production. *Meat Sci.* **2016**, *114*, 58–68. [CrossRef] [PubMed]

51. Alfonzo, A.; Gaglio, R.; Miceli, A.; Francesca, N.; Di Gerlando, R.; Moschetti, G.; Settanni, L. Shelf life evaluation of fresh-cut red chicory subjected to different minimal processes. *Food Microbiol.* **2018**, *73*, 298–304. [CrossRef] [PubMed]

52. Gaglio, R.; Couto, N.; Marques, C.; de Fatima Silva Lopes, M.; Moschetti, G.; Pomba, C.; Settanni, L. Evaluation of antimicrobial resistance and virulence of enterococci from equipment surfaces, raw materials, and traditional cheeses. *Int. J. Food Microbiol.* **2016**, *236*, 107–114. [CrossRef]

53. Foulquié Moreno, M.R.; Sarantinopoulos, P.; Tsakalidou, E.; De Vuyst, L. The role and application of enterococci in food and health. *Int. J. Food Microbiol.* **2006**, *106*, 1–24. [CrossRef]

54. Rumpold, B.A.; Schlüter, O.K. Nutritional composition and safety aspects of edible insects. *Mol. Nutr. Food Res.* **2013**, *57*, 802–823. [CrossRef]

55. Zielińska, E.; Baraniak, B.; Karaś, M.; Rybczyńska, K.; Jakubczyk, A. Selected species of edible insects as a source of nutrient composition. *Food Res. Int.* **2015**, *77*, 460–466. [CrossRef]

56. Finke, M.D. Complete nutrient composition of commercially raised invertebrates used as food for insectivores. *Zoo Biol.* **2002**, *21*, 269–285. [CrossRef]

57. Sánchez-Muros, M.J.; Barroso, F.G.; Manzano-Agugliaro, F. Insect meal as renewable source of food for animal feeding: A review. *J. Clean. Prod.* **2014**, *65*, 16–27. [CrossRef]

58. Yang, L.-F.; Siriamornpun, S.; Li, D. Polyunsaturated fatty acid content of edible insects in Thailand. *J. Food Lipids* **2006**, *13*, 277–285. [CrossRef]

59. Baek, H.H.; Cadwallader, K.R. Enzymatic Hydrolysis of Crayfish Processing By-products. *J. Food Sci.* **1995**, *60*, 929–935. [CrossRef]

60. Mizani, M.; Aminlari, M.; Khodabandeh, M. An Effective Method for Producing a Nutritive Protein Extract Powder from Shrimp-head Waste. *Food Sci. Technol. Int.* **2005**, *11*, 49–54. [CrossRef]
61. Quaglia, G.B.; Orban, E. Influence of the degree of hydrolysis on the solubility of the protein hydrolysates from sardine (*Sardina pilchardus*). *J. Sci. Food Agric.* **1987**, *38*, 271–276. [CrossRef]
62. Rebeca, B.D.; Peña-Vera, M.T.; Díaz-Castañeda, M. Production of Fish Protein Hydrolysates with Bacterial Proteases; Yield and Nutritional Value. *J. Food Sci.* **1991**, *56*, 309–314. [CrossRef]
63. Shahidi, F.; Han, X.-Q.; Synowiecki, J. Production and characteristics of protein hydrolysates from capelin (*Mallotus villosus*). *Food Chem.* **1995**, *53*, 285–293. [CrossRef]
64. Yang, R.; Zhao, X.; Kuang, Z.; Ye, M.; Luo, G.; Xiao, G.; Liao, S.; Li, L.; Xiong, Z. Optimization of antioxidant peptide production in the hydrolysis of silkworm (*Bombyx mori* L.) pupa protein using response surface methodology. *J. Food Agric. Environ.* **2013**, *11*, 952–956.
65. Abtahi, S.; Aminlari, M. Effect of Sodium Sulfite, Sodium Bisulfite, Cysteine, and pH on Protein Solubility and Sodium Dodecyl Sulfate–Polyacrylamide Gel Electrophoresis of Soybean Milk Base. *J. Agric. Food Chem.* **1997**, *45*, 4768–4772. [CrossRef]
66. Kristinsson, H.G.; Rasco, B.A. Fish Protein Hydrolysates: Production, Biochemical, and Functional Properties. *Crit. Rev. Food Sci. Nutr.* **2000**, *40*, 43–81. [CrossRef] [PubMed]
67. Picot, L.; Ravallec, R.; Fouchereau-Péron, M.; Vandanjon, L.; Jaouen, P.; Chaplain-Derouiniot, M.; Guérard, F.; Chabeaud, A.; LeGal, Y.; Alvarez, O.M.; et al. Impact of ultrafiltration and nanofiltration of an industrial fish protein hydrolysate on its bioactive properties. *J. Sci. Food Agric.* **2010**, *90*, 1819–1826. [CrossRef] [PubMed]
68. Raghavan, S.; Kristinsson, H.G. Antioxidative Efficacy of Alkali-Treated Tilapia Protein Hydrolysates: A Comparative Study of Five Enzymes. *J. Agric. Food Chem.* **2008**, *56*, 1434–1441. [CrossRef] [PubMed]
69. Thiansilakul, Y.; Benjakul, S.; Shahidi, F. Compositions, functional properties and antioxidative activity of protein hydrolysates prepared from round scad (*Decapterus maruadsi*). *Food Chem.* **2007**, *103*, 1385–1394. [CrossRef]
70. Ahn, C.-B.; Kim, J.-G.; Je, J.-Y. Purification and antioxidant properties of octapeptide from salmon byproduct protein hydrolysate by gastrointestinal digestion. *Food Chem.* **2014**, *147*, 78–83. [CrossRef] [PubMed]
71. Taheri, A.; Sabeena Farvin, K.H.; Jacobsen, C.; Baron, C.P. Antioxidant activities and functional properties of protein and peptide fractions isolated from salted herring brine. *Food Chem.* **2014**, *142*, 318–326. [CrossRef]
72. Allen, L.; De Benoist, B.; Dary, O.; Hurrell, R. *Guidelines on Food Fortification with Micronutrients*; WHO: Geneva, Switzerland, 2006; ISBN 9241594012.
73. Starčević, K.; Gavrilović, A.; Gottstein, Ž.; Mašek, T. Influence of substitution of sunflower oil by different oils on the growth, survival rate and fatty acid composition of Jamaican field cricket (Gryllus assimilis). *Anim. Feed Sci. Technol.* **2017**, *228*, 66–71. [CrossRef]
74. Wang, D.; Zhai, S.W.; Zhang, C.X.; Zhang, Q.; Chen, H. Nutrition value of the Chinese grasshopper Acrida cinerea (Thunberg) for broilers. *Anim. Feed Sci. Technol.* **2007**, *135*, 66–74. [CrossRef]
75. Mlcek, J.; Borkovcova, M.; Rop, O.; Bednarova, M. Biologically active substances of edible insects and their use in agriculture, veterinary and human medicine—A review. *J. Cent. Eur. Agric.* **2014**, *15*, 225–237. [CrossRef]

Article

Consumer Avoidance of Insect Containing Foods: Primary Emotions, Perceptions and Sensory Characteristics Driving Consumers Considerations

Mauricio Castro and Edgar Chambers IV *

Center for Sensory Analysis and Consumer Behavior, Kansas State University, 1310 Research Park Dr., Ice Hall, Manhattan, KS 66502, USA
* Correspondence: cciv@ksu.edu

Received: 1 July 2019; Accepted: 12 August 2019; Published: 17 August 2019

Abstract: Why do many human beings find bugs repulsive? Disgust, a psychological factor, is believed to be the main reason why consumers would not consider eating foods containing insect ingredients. This study aimed to understand specific consumers' behaviors toward insect based products. A global survey was launched in 13 different countries. The participants ($n = 630$ from each country) completed the survey that included demographic questions and questions about why they would or would not eat insect-based products. The results show, particularly for some of the Asian countries, that it is necessary to start exposing and familiarizing the populations about insects in order to diminish the disgust factor associated with insects. It is strongly recommended that an insect-based product should not contain visible insect pieces, which trigger negative associations. The exceptions were consumers in countries such as Mexico and Thailand, evaluated in this study, which did not show significant negative beliefs associated with including insects in their diets. Additional research to promote insect-based product consumption with popular product types might be the first strategy to break the disgust barriers and build acquaintance about insect-based products. The need to educate consumers that not all insects are unhygienic is crucial to eliminating the potentially erroneous concepts from consumer mindsets.

Keywords: insect; food; avoid; disgust; sensory; attitude; psychology; willingness to eat

---

1. Introduction

Why do consumers find insects disgusting? By research and definition, disgust is an emotional response of rejection or revulsion to something potentially contagious or something considered offensive, distasteful, or unpleasant [1,2]. Some authors report that it is not the taste that makes food disgusting, but rather the nature and origin of the food that triggers the disgust emotion [3]. Different negative perceptions toward insects, such as being disease transmitters, filthy, unhealthy and unhygienic [4] and the lack of accurate consumer information about insects have built a foundation of disgust. Despite the many excellent reasons to introduce insects to our diets, the current social paradigm is likely to undergo rather drastic alterations before consumers decide to get a side of crickets with their meal or eat other foods containing insect powders as an ingredient [5]. "It's disgusting" was the main reason, other than not knowing what it is, that consumers said that insect powder was not "natural" [6].

Food perceptions can change, especially if nutritional factors are involved, as reported by Bech-Larsen et al. [7]. In this age of environmental concerns, people are viewing products in new ways. There are other external variables, such as religion and allergic reactions that also can contribute to an increase in rejection to eating insects [5], but disgust has been perceived as the primary motivator. Unfortunately, there is little data showing the actual reasons. Baker et al. [8] suggested that it is necessary to understand more about all the barriers and the food neophobia challenges to fully understand how to reduce the consumer's negative perceptions and attitudes.

In recent years, different studies have covered the consumer acceptance of insect based products [9], entomophagy [10] and the willingness to eat food produced using insects as an ingredient [11] in one or two countries or regions. However, only one study was found that looked more globally at the issue of insect-based food consumption [5] and that study only examined the issue in terms of willingness to eat such foods and the impact on brand image if companies chose to use such an ingredient. The research helping to understand the actual barriers to insect-based food is minimal. However, Lorenz et al. [12] indicated that the simple fact of contemplating the idea of eating insects provokes an immediate disgust response to the general public. What reasons do consumers from various parts of the world and from different cultures, backgrounds and languages give to eating or not eating insect-based food products? There may be a compendium of thoughts, such as those associated with the consumption of wine (feeling smart and sophisticated), another product that must be learned and is not immediately accepted by most people [13]. If there is only a feeling of fear or disgust preventing people from eating insect-based products, a different challenge is presented than if more extensive concerns must be alleviated.

In consequence, this study aims to understand more thoroughly the psychological and sensory reasons for not eating insects. It does not seek to understand physical or social factors such as allergies or religious restrictions. The study was conducted in 13 countries to provide a somewhat global perspective.

2. Materials and Methods

This research was conducted in conjunction with a previously published project [5] where a detailed description of the survey methodology can be found.

2.1. Participant Profile

The respondents (n = 630 per country) were recruited from existing databases by Qualtrics, one of the world's largest on-line survey companies. The company's and its contractor databases include more than 30 million people. Approximately 100 participants of each gender (male, female) and age (18–34 years old; 35–54 years old; 55+ years old) combination completed the questionnaire per country (630 participants per country). A total of 7560 consumers completed the questionnaire.

The participants were from 13 countries (United States (USA), Mexico, Peru, Brazil, United Kingdom (UK), Spain, Russia, India, China, Thailand, Japan, South Africa, and Australia). The differences in the cultures, languages, traditions and religions make this a broad-based multi cross-cultural international survey. However, two additional countries, Egypt and Ghana, were eliminated from the study because not enough participants in each category completed the survey (for all categories in Ghana and for the 55+ age group in Egypt) after multiple attempts. This shows that there were not enough willing consumers with access to an appropriate mobile device to complete the survey in those countries, a limitation of this type of research.

2.2. Survey

The global willingness to eat insect products research study was divided in phases [5]. This portion of the survey focused only on the phase that covered the psychological and sensory reasons for not eating insect-based products. The participants indicated their agreement/disagreement on each of the reasons (Table 1) using a 7-point Likert-type scale with 7 as strongly agree to 1, strongly disagree.

Table 1. The reasons for not eating foods containing insect powder as an ingredient.

Reasons for not Eating Insect-Based Products Scale (7-Point Likert Type)
1. The idea is disgusting
2. I do not think it would taste good
3. Insects are not safe to eat
4. The texture would be bad
5. Just the thought makes me sick
6. Insects are dirty/filthy
7. Color would not be good
8. I do not want insect pieces in my foods

Scale: 1 = Strongly disagree, 2 = Disagree, 3 = Somewhat disagree, 4 = Neither agree nor disagree, 5 = Somewhat agree, 6 = agree, 7 = Strongly agree.

The survey was developed based on interviews and focus groups with participants from various countries and many prior studies that identified various issues associated with eating new food products. The final statements were developed based on the aggregate information provided in those focus groups, which was grouped into key topic areas. This portion of the survey focused exclusively on psychological or attitudinal reasons why a person might not eat such products. The English version of the survey was tested for face-validity using four professionals with an understanding of sensory and consumer behavior. The survey was tested for the correctness, use, and timing by seven students of various backgrounds and was pre-tested again using 50 consumers whose data were checked and analyzed to ensure the questions were understandable and did not lead to answers that were inappropriate or unreasonable.

For statistical modeling (regression analysis), a question on willingness to eat a food product that included an insect-based ingredient was used from Castro and Chambers [5]. The question read: "If a major worldwide company; e.g., Nestle, Coca-Cola, KFC, Starbucks, etc., introduces a new product similar to one you currently buy that contains insect powder, how willing would you be to try this product?". The number of consumers who were "willing to try", "not willing to try", or "not sure" is given in Table 2.

Table 2. The number (*n*) of consumers in each country who were not, were, and not sure about eating an insect-based product from a well-known manufacturer (total *n* = 630/country; sorted by not willing).

Country	Not Willing, *n*	Willing, *n*	Not Sure, *n*
Russia	399	203	28
Japan	390	131	109
India	364	205	61
Spain	356	205	69
USA	324	218	88
South Africa	311	248	71
England	307	227	96
Australia	294	216	120
Brazil	278	282	70
China	216	278	136
Peru	203	363	64
Thailand	186	353	91
Mexico	131	450	49

The survey was translated into nine languages (English, Spanish, Portuguese, Russian, Hindi, Mandarin Chinese, Thai, Japanese, and Afrikaans). The inspection of the translations was either by back translation or multiple translation, both with discussion afterwards by the translators to resolve any problems. The single translated versions were offered in some countries (e.g., Russia, UK), but multiple translations appropriate for the country were offered in others (e.g., South Africa, India).

2.3. Data Analysis

For each country, the data initially were simply categorized and described using percentages for each potential answer. The next step was to combine the three disagree scores into a category of "disagree" and the same procedure was implemented to the three agree choices obtaining the "agree" category. Score "4", neither agree nor disagree remained a separate category.

The statistical analyses using multiple regression with a stepwise elimination were performed using MiniTab-18 (Minitab Inc. State College, PA, USA) and SAS 9.4 (SAS Institute Inc., Cary, NC, USA) to estimate the impact of the reasons for not eating products containing insect powder over the dependent variable (willingness to eat food containing insect powder). For every country, the following regression equation was executed:

Y (*Willing to eat products made with insect powder*) = β_0 + β_1 Idea Disgusting + β_2 Taste Not Good + β_3 Insects Not Safe to Eat + β_4 Bad Texture + β_5 Thought makes me Sick + β_6 Insects are Dirty/Filthy to + β_7 Color Not Good + β_8 No Insect Pieces in my Food + ε.

For all the countries, the significant coefficients were described in a bar graph for a better interpretation. Additionally, *r*-squared and *p*-value summaries were noted to understand the percentage of the variation by the models.

This study was approved by the Committee on Research with Human Subjects at Kansas State University.

3. Results and Discussion

3.1. Psychological/Sensory Reasons for Not Eating Insects Consumers Who Are Unwilling to Eat Insect Powder—Based Products

Figure 1 shows the percentages of the consumers who were unwilling to try insect powder-based products in each country who selected each psychological or sensory reason for not eating those products. The consumers in all thirteen countries agreed that the most important reason is related to appearance, consumers do not want to see insect pieces in their food, followed by the "Idea is disgusting" or "The thought makes me sick". The two least important reasons were "The color would not be good" and "Insects are not safe to eat", although those reasons still averaged approximately 50% of consumers. More than 70% of the participants from each country agreed that appearance is extremely important. It is not appealing to consumers to see insect pieces in their food or snacks. Similar information on the visual appearance was noted by Meyer-Rochow and Hakko [14] who commented that using powders or pastes where the insect was not obvious was a better way to introduce the use of insects in food.

The primarily English-speaking countries (i.e., USA, Australia and United Kingdom) and South Africa generally were the top countries whose consumers strongly agreed that the reasons for not eating insects were "I do not want to see insect pieces in my food" and "Just the thought makes me sick"/"Idea is disgusting". Although the number of people who would not eat an insect-based product was high in India to begin with (>65%), with many saying they would not eat such foods based on religious constraints [5], Indian consumers also selected all eight psychological and sensory reasons for not eating insect-based products at a high percentages (65–90%). This probably is because the concept of eating a meat-containing food is anathema to most Hindus and any reason would be a reason not to eat such foods. Mexican and Thai consumers were the least likely to reject an insect-based product [5], but those consumers who said they would not try such a product did not agree on reasons for not eating insect-based foods. The percentage of consumers in Mexico and Thailand who chose particular reasons for not eating insect-based foods were among the lowest of all countries for all reasons except the color. Fu and others [15] showed that cultural values modulate consumer beliefs. Therefore, this study hypothesized that when a product is acceptable to the culture (i.e., more people are willing to eat than not eat that food), it is likely that individual beliefs related to food become the key decision criteria. In the case of Mexico and Thailand, it is likely that the reasons for not eating a food are more individualized and, thus, more likely to be spread among many factors than when overarching cultural or societal norms are present.

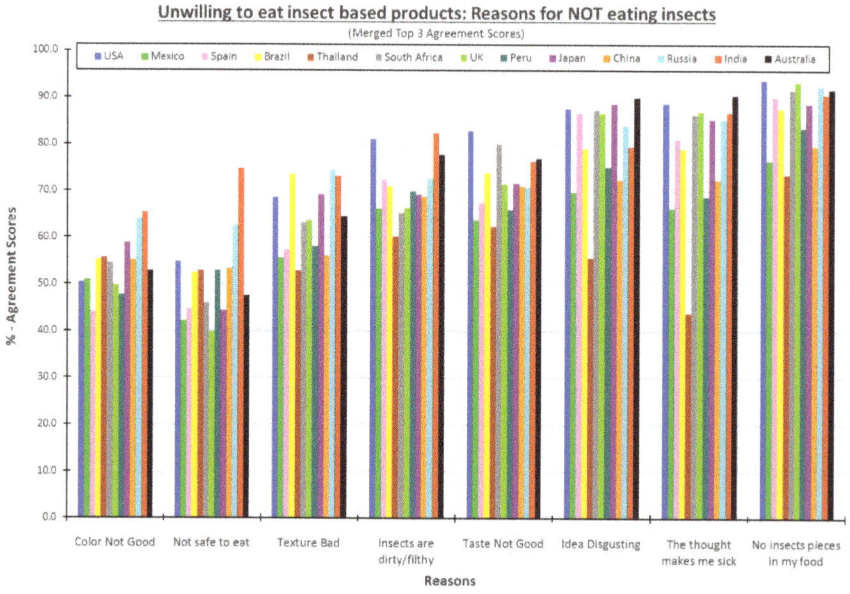

Figure 1. Graph of the reasons for not eating insect based products—Consumers unwilling to try.

3.2. Consumers Who Are Willing to Eat Insect Based Products

Figure 2 shows the percentages of the consumers who were willing to try insect products in each country who selected each reason for not eating an insect powder-based food product. Not surprisingly, the results showed that for most countries except India, the consumers who were willing to try insect-based foods generally did not have reasons for not choosing such foods. In most cases, less than 40% of consumers in those countries chose a specific reason for not choosing an insect-based food product. There were two major exceptions. More than 40% of consumers willing to try insect-based product in most countries indicated they still would not choose a food if there were insect pieces in it. Second, even those Indian consumers who were willing to try insect-based products found many reasons not to eat such products. Almost every reason, except disgust, was chosen by 50% or more of willing Indian consumers as a reason not to eat insect-based products. It is not surprising that even those who say they are willing to try a food, find reasons for not eating such foods. From other research [16–19], it is known that liking and pleasure, habits, convenience, hunger, health, and a myriad of other reasons motivate people to eat the foods they choose.

When considering the reasons for willing consumers, the same patterns overall were observed in reasons as the unwilling consumers. The main reason for not eating such products being the appearance of pieces. Mexico and Thailand rated the lowest agreement scores across all the reasons. Over 40% of Chinese and Japanese consumers agreed to the following reasons: Color not good, the thought makes me sick, taste not good, the idea is disgusting, insects are dirty/filthy, and no insect pieces in my food. The results were expected to show low agreement scores because these consumers were willing to eat food obtained from insects. However, it is interesting that the pattern of responses is similar to those from unwilling consumers. This information suggests that the decision to eat or not eat a particular food made from insects has the same barriers regardless of the individuals' willingness to try such products. This is good news for those who wish to promote the use of insects as a food ingredient because some types of information provided to consumers have been found to increase likelihood to eat insect-based products [20,21].

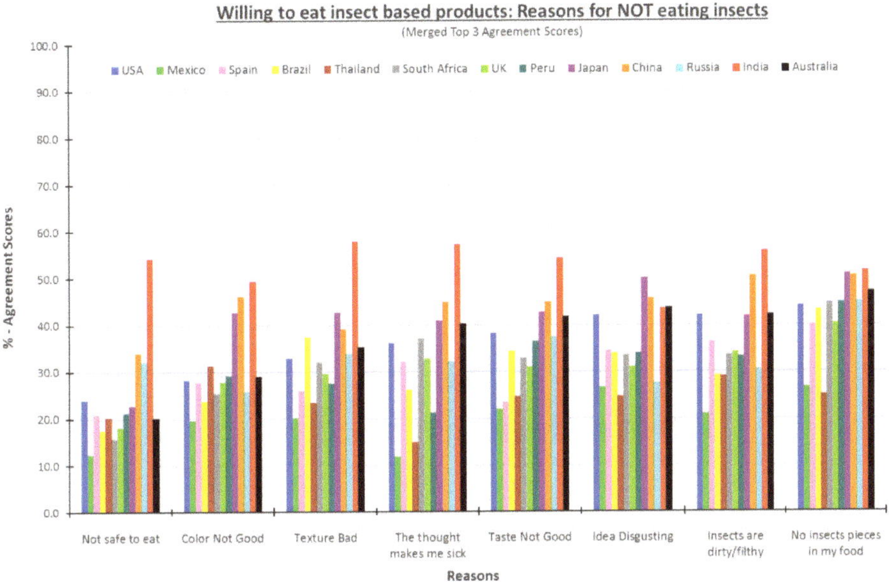

Figure 2. A graph of the reasons for not eating insect based products—consumers willing to try.

3.3. Regression Analysis—Reasons

The participants' responses from all countries showed similarity in which responses were less important to them. Before conducting the multiple linear regression analysis, a covariance was detected between two of the reasons. The idea is disgusting and the thought makes me sick were highly correlated and therefore the "thought makes me sick" was dropped from the models because the term "disgust" was commonly used to describe this emotional construct related to insects [12]. Figure 3 shows the regression coefficients for the seven reasons for not considering eating foods with insect-based components as an ingredient.

"The texture would be bad", was removed from further analysis during the multilinear regression analysis (stepwise procedure) because this variable was not significant and, ultimately, was eliminated from the equation in all but one country (USA). The remaining six reasons were compared and analyzed with the appearance and disgust being the two independent variables that were present in most of the countries' regression equations. Thus, consumers emphasized a sensory factor—the visual appearance of "insect pieces", an emotional factor—disgust, and a psychological belief/trust factor—"insects are not safe to eat", as primary motivations for not eating insect-based products. Recent research has found similar constructs to be barriers for the consumption of insect-based foods (new FOODS). The rest of the variables, insects are dirty/filthy, taste not good and color, not good were small and generally irrelevant reasons that either were co-dependent on other reasons or did not affect the willingness to eat insect-based food products once other considerations were noted.

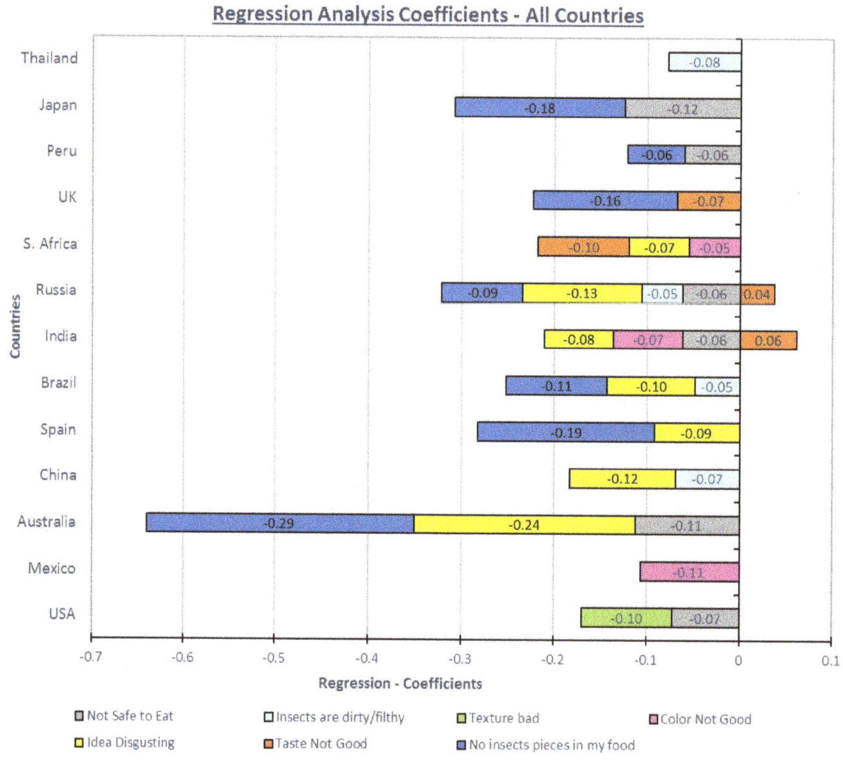

Figure 3. Regression analysis coefficients—the stepwise method.

3.4. Specific Reasons for Not Eating Insect-Based Foods

3.4.1. No Insect Pieces in My Food—All Consumers Considered

When consumers were asked for the reasons that they would not consider eating foods containing insect powder as an ingredient, over 60% of the participants in China, Peru, Australia, UK, Brazil, Russia, South Africa, Spain, India, Japan, and USA strongly agree, agree or somewhat agree that the appearance was extremely important and did not want insect pieces in the food. In Mexico and Thailand, the percentages were also considered high. Over 40% of the respondents agreed with the statement, "no insect pieces in my food" (Figure 4).

In most of the countries except for South Africa ($\rho = 0.38$) and Thailand ($\rho = 0.35$), the reason "I do not want insect pieces in my food" was highly correlated to "The idea is disgusting" with correlation scores over 0.50. When consumers can see insects or pieces in their food, the food becomes more disgusting. This is a visual sensory cue that should be easy to correct (not seeing insect pieces in my food) by grinding the insects into a powder (flour). Grinding would avoid any visual parts of the insects in the food to be prepared. During the regression analysis, the texture reason was eliminated which confirms that consumers were focusing more into the appearance/visual aspect of the product.

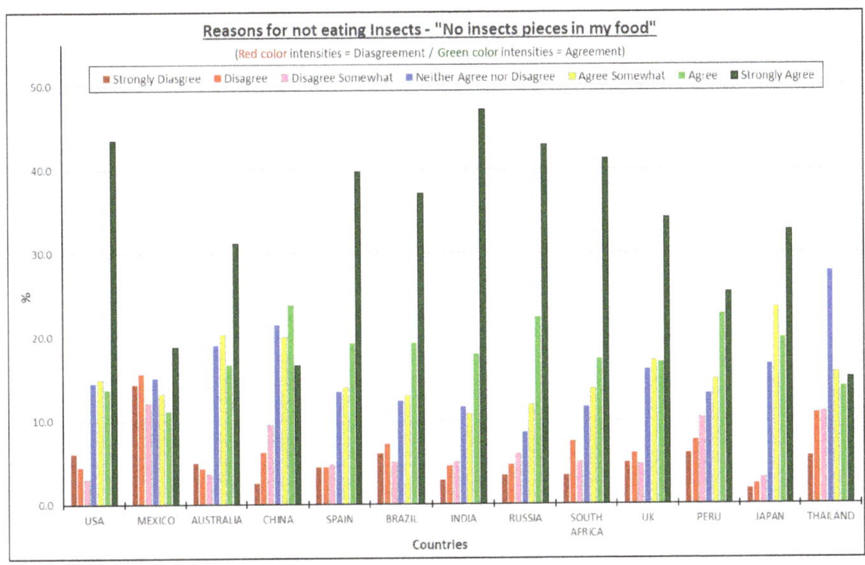

Figure 4. The reasons for not eating insect products—"Do not want insect pieces in my food".

3.4.2. The Idea Is Disgusting—All Consumers Considered

More than 50% of the participants in each of the countries, except for Mexico and Thailand, shared the idea that using insects as an ingredient in food is disgusting. Japanese consumers scored this reason the highest out of the six reasons and the highest of all the thirteen countries, that is, 77% of the respondents agree that the idea of eating insect-based products is disgusting. The USA (68%) and Spain (67%) completed the top three countries for this concept (Figure 5).

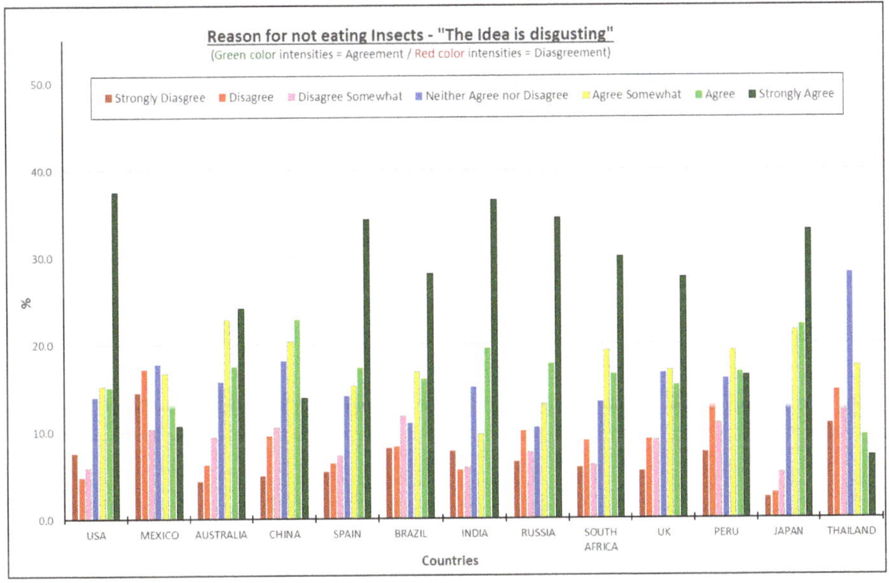

Figure 5. The reasons for not eating insect products—"The idea is disgusting".

The high percentages for disgust align with the regression coefficients showing that the disgust factor is the second most significant reason for not eating insects after the reason "No insect pieces in my food". Therefore, for such products to be successful, it is essential to begin breaking this emotional barrier by developing insect-based products that are familiar to consumers to show that insects can simply be another ingredient as opposed to a contaminant.

There are many foods whose ingredients alone are not appealing to consumers either because they may not be natural, organic, GMO free, etc., [6,22] but may not cause the same level of emotional response in an actual food product. The more exposure to these kinds of typical products made with insect ingredients and education about the benefits of insects as food, the more probability there is to decrease the disgust factor.

3.4.3. Insects Are Not Safe—All Consumers Considered

For all the countries, except for Mexico and India, the statement "Insects are not safe to eat" was largely considered to be a neutral statement (Figure 6), although it appeared in some regression coefficients as a negative factor (Figure 3). Most consumers in those countries scored it neither, agree nor disagree. However, 65% of consumers in India agreed to this concept, while in Mexico barely 20% disagreed with that statement. The uncertainty about the safety of insects is a topic that needs further research both from the standpoint of how it impacts the potential use of insect-based foods and the human health perspective. Castro et al. [5] recently reported that people believe insects carry diseases and some people believe themselves to be allergic to insects. Those are powerful reasons to question the safety of insect-based foods for those consumers. Studies have identified that consumers might experience similar allergic reactions to seafood when insects are consumed [23].

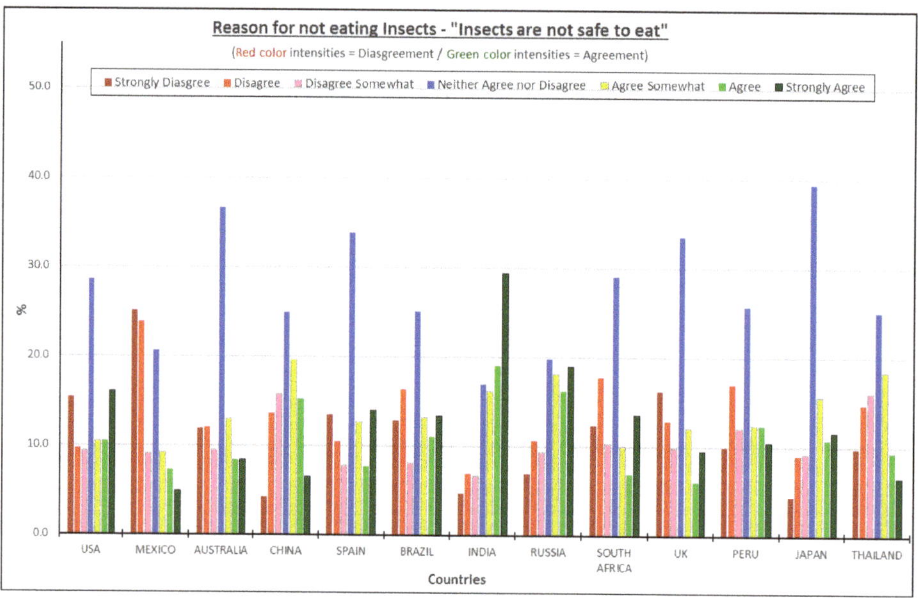

Figure 6. The reasons for not eating insect products—"Insects are not safe to eat".

Furthermore, research in conjunction with clinical studies is necessary from the human standpoint to prove which specific diseases insects might transmit to humans and what chemical components and parts of the insects could provoke allergic reactions [24]. In addition, conclusive zoonotic diseases need extensive research to diminish the concept that "Insects are not safe".

3.4.4. Insects Are Dirty/Filthy—All Consumers Considered

The percentages of consumers who agreed that insects are dirty-filthy trends lower than the first three concepts discussed (Figure 7). India, USA and Japan, are the top three countries associating insects with filth or dirt.

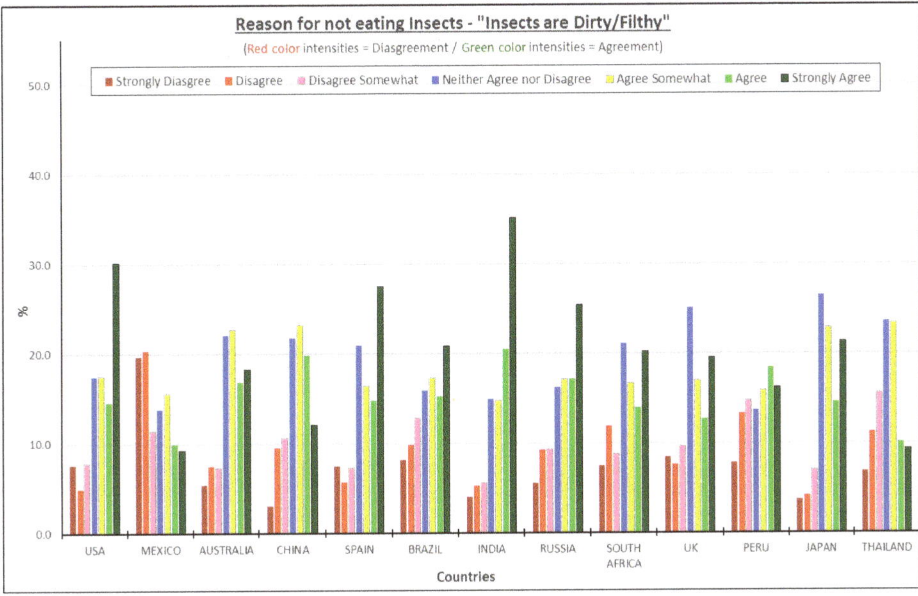

Figure 7. The reasons for not eating insect products—"Insects are dirty/filthy".

For those three countries, the percentages of consumers, ranged from approximately 60–70%. Less than 35% of Mexican consumers agreed with that statement, the lowest of all the countries. Peru and Thailand round out the three countries with the lowest agreement on this statement. This reason might be associated with previous perceptions or misconceptions about intoxicating bacteria, viruses, and parasites or the fact that some insects are connected to waste or decay material [25].

3.4.5. Taste Not Good—All Consumers Considered

The consumers' responses to the agreement "I do not think it would taste good" (Figure 8) resemble results from the reason "Insects are dirty/filthy" (Figure 7). India, Japan, USA, and Russia are the countries with the highest percentage of people agreeing with this reason, all slightly higher than 60%. Mexico and Thailand showed the lowest percentage of people agreeing with that statement. Mexico was the only country where the disagreement response was higher than the agreement rate. It is important to highlight that "Taste not good" is the highest reason related to ingested sensory properties although the visual perception of insects or insect pieces may imply ingested effects. One caution is that a number of consumers choose "neither agree nor disagree", suggesting that consumers are not sure of how insect-based products would taste and are not confident that an insect based product would taste good. This may suggest that insects are not a barrier from an ingested sensory standpoint, per se, but that the visual perception of insects promotes associations with disgust or textures that would not be desirable. This is a great opportunity to conduct sensory discrimination tests to evaluate if consumers can differentiate between a regular product and an insect product and if they can to conduct descriptive sensory studies to determine the actual sensory differences in those products.

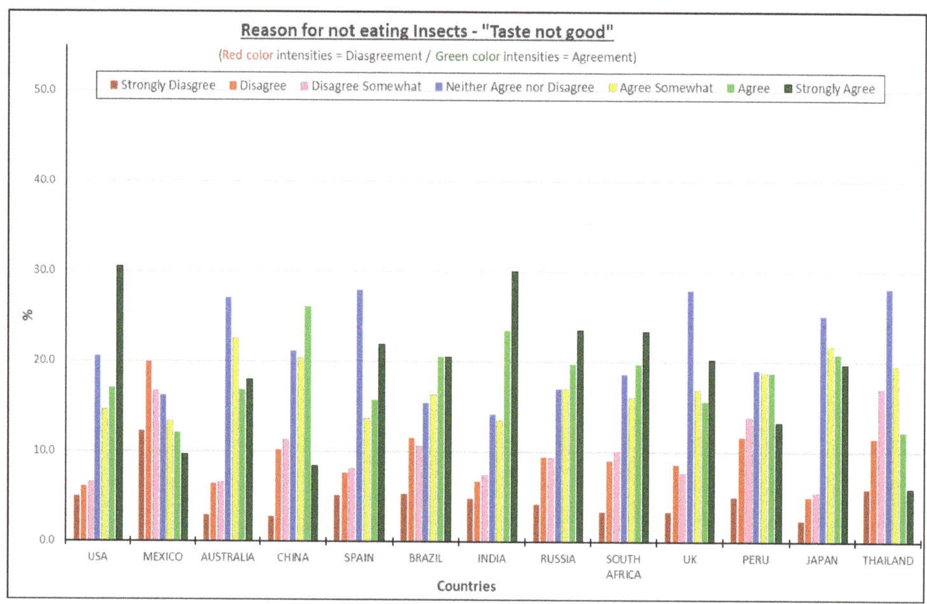

Figure 8. Reasons for not eating insect products—"I do not think it would taste good".

Several studies focused on the sensory aspect of overall liking to determine the consumer's behavior towards insect products [26–28]. The conclusions from those studies showed that the participants' overall liking was influenced by the appearance and taste. In addition, it was noted in some studies that the insect parts needed to be invisible for acceptance, which is similar to the findings related to appearance considerations in this study. In a small study with 26 students who were blindfolded and held their noses, slightly more than half could identify the processed insect samples from cheese, dried fish or bread [14]. The authors concluded that this indicated potential acceptance if consumers ate products containing insect flours or pastes where the consumers could not see the insects. In one study [26] of burgers, men rated the insect burger between the beef and lentil burger, with a preference for the mealworm and beef burger. It should be noted in all the prior studies, only those consumers who indicated they were willing and interested in eating insect-based products were used, which may skew results more positively.

3.4.6. Color Would Not Be Good—All Consumers Considered

A further sensory consideration is color. Over 50% of the participants in India, Russia and Japan agreed that the color of an insect based product would not be good (Figure 9). Interestingly, the Latin American countries (Mexico, Peru, and Brazil) were the only countries where more or almost more consumers disagreed with that statement than agreed. It must be noted that this statement generated uncertainty (high percentages of neither agree nor disagree) similar to the statement "Insects are not safe to eat", suggesting that consumers simply were not sure what the impact of insect ingredients would be on color. Of course, if insect ingredients do result in a color issue, that color might be altered or fixed by adding natural and familiar colors to imitate the original color of a specific product category.

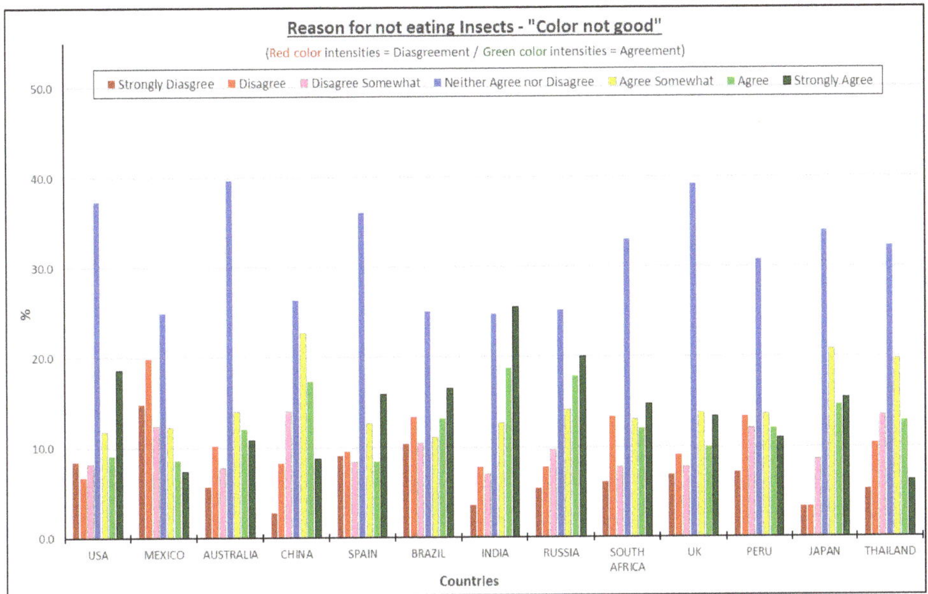

Figure 9. The reasons for not eating insect products—"Color would not be good".

4. Conclusions

This study provides a better understanding of the reasons that consumers would not consider eating foods containing insect ingredients. The appearance is a critical issue and a high priority for consumers. There is no doubt that the fragments or pieces of the insect cannot be present in the final product. The emotional and psychological issues represented by the statements "The idea is disgusting"/"Just the thought makes me sick" and the potential misconception that all "insects are not safe to eat" are as crucial as the visual factor.

Of lesser impact in this study, but potentially related to the key factors are such aspects as the misbelief that all insects are dirty/filthy, which may cause consumers to avoid insect-based products. The two sensory characteristics that concerned some participants were the potential impact of the insect ingredients on the taste and texture, but those should be able to be overcome with the adequate selection and formulation of the food product. Certainly, those two aspects are barriers for any new or revamped products that research and development groups need to carefully consider before the creation of a new product.

This research suggests that the use of insect-based powders/flours to avoid the appearance and textural issues and education that overcomes the disgust and safety concerns of consumers are key to the introduction of insect-based food products in many countries. In some countries, such as Mexico and Thailand, the sensory issues may be of more concern than the disgust issues because insect-based products are already known, although not necessarily widely eaten.

Author Contributions: Both authors contributed to all parts of the project including conceptualization, methodology, formal analysis, investigation, resources, data curation, and writing and editing the paper.

Funding: This material is based upon work that is supported by the National Institute of Food and Agriculture, U.S. Department of Agriculture, Hatch under accession number 1016242.

Conflicts of Interest: The authors declare no conflicts of interest.

References

1. Curtis, V. Disgust. In *Encyclopedia of Human Behavior*, 2nd ed.; Ramachandran, V.S., Ed.; Elsevier: Amsterdam, The Netherlands, 2012; pp. 702–709.
2. Rozin, P.; Fallon, A. A perspective on disgust. *Psychol. Rev.* **1978**, *94*, 23–41. [CrossRef]
3. Rozin, P.; Wright, J. Psychology of disgust. In *International Encyclopedia of the Social & Behavioral Sciences*, 2nd ed.; Wright, J., Ed.; Elsevier: Amsterdam, The Netherlands, 2015; Volume 6, pp. 546–549.
4. Van Huis, A.; van Itterbeeck, J.; Klunder, H.; Mertens, E.; Halloran, A.; Muir, G.; Vantomme, P. *Edible Insects: Future Prospects for Food and Feed Security*; FAO: Rome, Italy, 2013; ISBN 978-92-5-107595-1.
5. Castro, M.; Chambers, E., IV. Willingness to eat an insect based product and impact on brand equity: A global perspective. *J. Sens. Stud.* **2019**, *34*, e12486. [CrossRef]
6. Chambers, E.V.; Tran, T.; Chambers, E., IV. Natural: A $75 billion word with no definition—Why not? *J. Sens. Stud.* **2019**, *34*, e12501. [CrossRef]
7. Bech-Larsen, T.; Grunert, K.G. The perceived healthiness of functional foods. A conjoint study of Danish, Finnish and American consumers' perception of functional foods. *Appetite* **2003**, *40*, 9–14. [CrossRef]
8. Baker, M.; Shin, J.T.; Wook, Y. An exploration and investigation of edible insect consumption: The impacts of image and description on risk perceptions and purchase intent. *Psychol. Market.* **2016**, *33*, 94–112. [CrossRef]
9. House, J. Consumer acceptance of insect-based foods in The Netherlands: Academic and commercial implications. *Appetite* **2016**, *107*, 47–58. [CrossRef] [PubMed]
10. Roberts, W. Eating insects: Waiter, there's no fly in my soup. *Alter. J.* **2008**, *34*, 8–10.
11. Gmuer, A.; Guth, J.N.; Hartmann, C.; Siegrist, M. Effects of the degree of processing of insect ingredients in snacks on expected emotional experiences and willingness to eat. *Food Qual. Pref.* **2016**, *54*, 117–127. [CrossRef]
12. Lorenz, A.; Libarkinb, J.; Ording, G. Disgust in response to some arthropods aligns with disgust provoked by pathogens. *Global Ecol. Conserv.* **2014**, *2*, 248–254. [CrossRef]
13. Thompson, K.; Barret, E. The millennial generation and wine purchasing behaviors in casual dining restaurants. *J. Foodserv. Bus. Res.* **2016**, *19*, 525–535. [CrossRef]
14. Meyer-Rochow, V.B.; Hakko, H. Can edible grasshoppers and silkworm pupae be tasted by humans when prevented to see and smell these insects? *J. Asia Pac. Entomol.* **2018**, *21*, 616–619. [CrossRef]
15. Fu, P.P.; Kennedy, J.; Tata, J.; Yukl, G.; Bond, M.H.; Peng, T.K.; Srinivas, E.S. The impact of societal cultural values and individual social beliefs on the perceived effectiveness of managerial influence strategies: A meso approach. *J. Intern. Bus. Stud.* **2004**, *35*, 284–305. [CrossRef]
16. Phan, U.X.T.; Chambers, E., IV. Application of an eating motivation survey to study eating occasions. *J. Sens. Stud.* **2016**, *31*, 114–123. [CrossRef]
17. Phan, U.X.T.; Chambers, E., IV. Motivations for choosing various food groups based on individual foods. *Appetite* **2016**, *105*, 204–211. [CrossRef] [PubMed]
18. Chambers, D.; Phan, U.X.T.; Chanadang, S.; Maughan, C.; Sanchez, K.; Di Donfrancesco, B.; Gomez, D.; Higa, F.; Li, H.; Chambers, E. Motivations for food consumption during specific eating occasions in Turkey. *Foods* **2016**, *5*, 39. [CrossRef] [PubMed]
19. Phan, U.X.T.; Chambers, E., IV. Motivations for meal and snack times: Three approaches reveal similar constructs. *Food Qual. Pref.* **2018**, *68*, 267–275. [CrossRef]
20. Mancini, S.; Sogari, G.; Menozzi, D.; Nuvoloni, R.; Torracca, B.; Moruzzo, R.; Paci, G. Factors predicting the intention of eating an insect-based product. *Foods* **2019**, *8*, 270. [CrossRef] [PubMed]
21. Mancini, S.; Moruzzo, R.; Riccioli, F.; Paci, G. European consumers' readiness to adopt insects as food. A review. *Food Res. Int.* **2019**, *122*, 661–678. [CrossRef] [PubMed]
22. Chambers, E.V.; Chambers, E., IV; Castro, M. What is natural? Consumer responses to selected ingredients. *Foods* **2018**, *7*, e65. [CrossRef]
23. Witteman, A.M.; Akkerdaas, J.H.; van Leeuwen, J.; van der Zee, J.S.; Aalberse, R.C. Identification of a cross-reactive allergen (Presumably Tropomyosin) in Shrimp, Mite and Insects. *Int. Arc. Allergy Immun.* **1994**, *105*, 56–61. [CrossRef]
24. Dossey, A.; Morales-Ramos, J.; Rojas, M. *Insects as Sustainable Food Ingredients. Production, Processing and Food Applications*, 1st ed.; Academic Press: Cambridge, MA, USA, 2016; ISBN 978-0-12-802856-8.

25. Marshall, D.; Dickson, J.; Nguyen, N. Ensuring food safety in insect based foods: Mitigating microbiological and other foodborne hazards. In *Insects as Sustainable Food Ingredients*; Dossey, A., Morales-Ramos, J., Rojas, M., Eds.; Academic Press: Cambridge, MA, USA, 2016; pp. 223–253.
26. Caparros Megido, R.; Gierts, C.; Blecker, C.; Brostaux, Y.; Haubruge, É.; Alabi, T.; Francis, F. Consumer acceptance of insect-based alternative meat products in Western countries. *Food Qual Pref.* **2016**, *52*, 237–243. [CrossRef]
27. Hartmann, C.; Siegrist, M. Insects as food: Perception and acceptance. *Ernaehrungs Umschau* **2016**, *64*, 44–50.
28. Tan, H.S.G.; Verbaan, Y.T.; Stieger, M. How will better products improve the sensory-liking and willingness to buy insect-based foods? *Food Res. Int.* **2017**, *92*, 95–105. [CrossRef] [PubMed]

Article

Mealworms as Food Ingredient—Sensory Investigation of a Model System

Karin Wendin [1,2,*], Viktoria Olsson [1] and Maud Langton [3]

[1]   Department of Food and Meal Science, Kristianstad University, SE-291 88 Kristianstad, Sweden
[2]   Department of Food Science, University of Copenhagen, DK-1958 Frederiksberg C, Denmark
[3]   Department of Molecular Sciences, SLU—Swedish University of Agricultural Sciences,
      SE-750 07 Uppsala, Sweden
*   Correspondence: karin.wendin@hkr.se

Received: 25 June 2019; Accepted: 2 August 2019; Published: 6 August 2019

Abstract: The use of insects as food is a sustainable alternative to meat and as a protein source is fully comparable to meat, fish and soybeans. The next step is to make insects available for use in the more widespread production of food and meals. Sensory attributes are of great importance in being able to increase the understanding of insects as an ingredient in cooking and production. In this pilot study, mealworms were used as the main ingredient in a model system, where the aim was to evaluate the impact on sensory properties of changing particle size, oil/water ratio and salt content of the insects using a factorial design. Twelve different samples were produced according to the factorial design. Further, the effect of adding an antioxidant agent was evaluated. Sensory analysis and instrumental analyses were performed on the samples. Particle size significantly influenced the sensory attributes appearance, odor, taste and texture, but not flavor, whereas salt content affected taste and flavor. The viscosity was affected by the particle size and instrumentally measured color was affected by particle size and oil content. The addition of the antioxidant agent decreased the changes in color, rancidity and separation.

Keywords: mealworm; *Tenebrio molitor*; insects; sensory; model system

---

1. Introduction

The world's population is increasing and is expected to reach around 11 billion in 2050 [1], thus increasing pressure on the earth's resources. FAO (Food and Agriculture Organization of the United Nations) estimates that food production must increase by 70 percent by 2050 to ensure an adequate food supply for the increasing population [1]. However, food production currently has a huge impact on the environment [2] and eating habits need to change, in particular by replacing meat with other protein sources [3,4]. It is necessary to find and use new production sources of protein and other important nutrients. Further, consumers need to be informed about what alternative protein sources that may be available [5]. Common alternatives include the use of plant protein found for example in peas and legumes, another alternative could be to use insect protein. According to the EAT Lancet commission (EAT is the science-based global platform for food system transformation), there is an urgent need for change regarding food and the report concludes that food will be a defining issue of the 21st century [4].

Using insects as food is a sustainable alternative because their production exerts less ecological pressure than that of conventional livestock such as cattle, pigs and poultry as it requires less feed, soil and water [6–8]. However, the environmental impact varies greatly between different insect species [9]. Nutritionally, insects are usually fully comparable to other protein sources, such as fish, soybeans or meat e.g., beef, although there is a large variation between species. Insects have been indicated as one of the alternative food sources that could cover the future need for protein and other nutrients [10,11].

Insects contain protein, fat, vitamins and minerals. The nutritional content varies between species, stage of growth and breeding factors [12–14]. The protein content is generally high, 20–70% of dry matter, with a high proportion of essential amino acids and good digestibility [15,16]. Insect protein has been shown to be of higher quality than soy protein due to the amino acid composition [17]. The fat content of different insect species varies greatly but most species contain a high percentage of polyunsaturated fatty acids [13,18]. The content of minerals and vitamins varies widely between species, however it should be noted that vitamin B12 can be found in many species, e.g., mealworms [14,19].

It is beneficial, both from sustainability and nutritional perspectives, to include insects as a food ingredient [3]. However, a very important criterion for accepting insects as food is that they taste good. The three so-called gateway insects—mealworms, crickets and locusts—are all described as tasty [20]; the flavor of mealworms may be described as nutty with a taste of umami and cereals, cricket resembles popcorn with a hint of chicken and umami, while locust has the flavor of shrimp, nuts and vegetables [21,22]. Insects may be eaten raw, dried or cooked, and also whole, in pieces or milled. They can be cooked in a variety of ways [23], although there is a lack of knowledge regarding how to use insects as an ingredient in everyday cooking.

Despite all the positive factors, the vast majority of people in Western cultures are reluctant to eat insects and acceptance is generally low. The underlying causes may be attributed to aversion, which can be the result of various sensory signals, such as the ugliness or bad odor of insects, and this is often associated with feelings of anxiety [24]. A negative attitude towards invertebrates, both generally but especially as food, is deeply rooted in Western culture [25]. However, in countries where eating insects is the norm they are seen as a delicacy and a valuable protein source. Knowledge about which species are edible and how they should be prepared is considered important [10].

Internationally, there are more cultures that include insects in their diet than those who do not. It is common to consume insects as food in approximately 120 countries around the world. Further, it is estimated that more than 2000 insect species are edible [26]. Several studies indicate that it is easier for people in western cultures to accept insects as food when the insect ingredient is neutral and less visible or minced than in dishes or products including whole and visible insects [27,28].

The Dutch Council of Affairs describes an increased interest in the consumption of insects and proposes large-scale production in Europe [29]. Dobermann et al. (2017) [10] argue that the economic value of insect production may be higher than that of conventional meat production.

Through a growing awareness of the importance of sustainable food production, the interest in insects as food is increasing. Despite the widespread aversion towards insects as food, interest and acceptance is increasing, not the least since more people consider them as nutritious, sustainable and tasty [20]. The time has come to take the next step by making insects available not only as delicious restaurant food, but also in the more widespread production of food and meals based on insects. The sensory attributes are of great importance in being able to increase the understanding of insects as a main ingredient in cooking and food production.

2. Aim

Mealworms were used as the main ingredient in a model system where the aim was to evaluate the impact on sensory properties by changing particle size, oil/water ratio and salt content through the use of factorial design. The effect of an antioxidant agent was also analyzed in samples of differing particle size.

3. Material and Methods

3.1. Material

Water boiled fresh mealworms (*Tenebrio molitor*, small scale rearer in Sweden) cut/ground into different particle sizes (small: Max 1 mm, medium: Max 3 mm, large: Max 5 mm). The mealworms larvae were fed mainly on oat flakes and carrots, which may affect the flavor. The cut mealworms, sunflower oil (Farm, SR&F, Sweden), water and salt (NaCl, Nordfalks, Sweden) were blended according to a factorial design. The samples were prepared in duplicates. The resulting products were evaluated by descriptive sensory analysis in addition to instrumental measurements of viscosity and color. Nutritional contents were calculated in the software Dietist Net (Kost och Näringsdata, Sweden) by adding values from Finke (2002) [30] to the database. The samples and the ingredients are presented in Tables 1 and 2. The calculated nutritional values for samples 1–12 are found in Table 3.

Table 1. Sample preparation with mealworm (50 g) particle size, oil/water ratio of 50 g mixture and NaCl contents.

Sample	Sample Identifier	Particle Size	Oil/Water Ratio (g/g)	NaCl
1	SOS	Small	High/Low	High
2	SOs	Small	High/Low	Low
3	SoS	Small	Low/High	High
4	Sos	Small	Low/High	Low
5	MOS	Medium	High/Low	High
6	MOs	Medium	High/Low	Low
7	MoS	Medium	Low/High	High
8	Mos	Medium	Low/High	Low
9	LOS	Large	High/Low	High
10	LOs	Large	High/Low	Low
11	LoS	Large	Low/High	High
12	Los	Large	Low/High	Low

Samples Identifier: 1st letter refers to particle size (S, M, L), 2nd letter refers to oil/water ratio with o = small amount of oil and O = high amount of oil, 3rd letter refers to NaCl content with S = high amount and s = low amount. Particle size: Small = max 1 mm, medium = max 3 mm, large = max 5 mm. Oil/water ratio: High = 37.5 g, low = 12.5 g. High NaCl: High = 2.0 g, low = 0.5 g.

Table 2. Samples with added antioxidant agent. Sample with added rosemary where r = 0.1% and R = 0.3% added rosemary.

Sample Identifier	Sample with Added Rosemary	Rosemary (%)
SoS	SoS	0
SoS	SoSr	0.1
SoS	SoSR	0.3
LOs	LOs	0
LOs	LOsr	0.1
LOs	LOsR	0.3

Table 3. Calculated nutritional content per 100 g sample (calculations on mealworms larvae was based on Finke, 2002 [30] and performed in Software Dietist Net, Kost och Näringsdata, Sweden).

Sample	1	2	3	4	5	6	7	8	9	10	11	12
Macro Nutrients												
Protein (g)	9.2	9.3	9.2	9.3	9.2	9.3	9.2	9.3	9.2	9.3	9.2	9.3
Fat (g)	43.4	44.0	18.9	19.1	43.4	44.0	18.9	19.1	43.4	44.0	18.9	19.1
Fiber (g)	3.8	3.8	3.8	3.8	3.8	3.8	3.8	3.8	3.8	3.8	3.8	3.8
Energy (kJ)	1795	1820	888	900	1795	1820	888	900	1795	1820	888	900
Energy (kcal)	429	435	212	215	429	435	212	215	429	435	212	215
Minerals												
Calcium (mg)	11.5	9.6	12.4	10.6	11.5	9.6	12.4	10.6	11.5	9.6	12.4	10.6
Iron (mg)	1.0	1.0	1.0	1.0	1.0	1.0	1.0	1.0	1.0	1.0	1.0	1.0
Magnesium (mg)	40.0	40.1	40.2	40.3	40.0	40.1	40.2	40.3	40.0	40.1	40.2	40.3
Sodium (mg)	754	212	724	205	754	212	724	205	754	212	724	205
Zinc (mg)	2.6	2.6	2.6	2.6	2.6	2.6	2.6	2.6	2.6	2.6	2.6	2.6

Table 3. *Cont.*

Sample	1	2	3	4	5	6	7	8	9	10	11	12
Fatty acids (%)												
Saturated	5.3	5.4	2.8	2.8	5.3	5.4	2.8	2.8	5.3	5.4	2.8	2.8
Monounsaturated	13.5	13.6	6.4	6.5	13.5	13.6	6.4	6.5	13.5	13.6	6.4	6.5
Polyunsaturated	22.5	22.8	8.7	8.8	22.5	22.8	8.7	8.8	22.5	22.8	8.7	8.8
Vitamins												
Vitamin A IU	162	164	162	164	162	164	162	164	162	164	162	164
Vitamin D IU	0.3	0.3	0.3	0.3	0.3	0.3	0.3	0.3	0.3	0.3	0.3	0.3
Vitamin E IU	22.4	22.7	7.6	7.7	22.4	22.7	7.6	7.7	22.4	22.7	7.6	7.7
Vitamin C (mg)	2.7	2.7	2.7	2.7	2.7	2.7	2.7	2.7	2.7	2.7	2.7	2.7
Thiamine (mg)	0.1	0.1	0.1	0.1	0.1	0.1	0.1	0.1	0.1	0.1	0.1	0.1
Riboflavin (mg)	0.4	0.4	0.4	0.4	0.4	0.4	0.4	0.4	0.4	0.4	0.4	0.4
Folic acid (μg)	0.1	0.1	0.1	0.1	0.1	0.1	0.1	0.1	0.1	0.1	0.1	0.1
Vitamin B12 (μg)	0.2	0.2	0.2	0.2	0.2	0.2	0.2	0.2	0.2	0.2	0.2	0.2

The room tempered (19–21 °C) ingredients for samples 1–12 were blended directly before analysis and the samples were then assessed at an ambient temperature of approximately 20 °C (19–21 °C) during a period of 10–40 min after blending.

In addition, Rosemary (Duralox Oxidation Management Blend NM-45, HT, NS, product code: 62-103-03) was added in different amounts to samples 3 and 10. The added amounts were 0%, 0.1% or 0.3% (calculated on weight), Table 2. Five mL portions of each sample were packaged in 15 vacuum-sealed plastic bags and stored in a fridge (+5 °C) for up to 15 days. One bag to be sensorially assessed each day.

3.2. Methods

Sensory analysis and instrumental analyses were performed on the samples.

3.2.1. Sensory Analysis

The sensory panel was selected and trained according to the ISO (International Standard Organization) standard 8586-2:2008 [31]. The panel consisted of eight assessors who were first trained in how to perform testing and rate intensity on a numerical scale from 0 to 100, where 0 = no intensity and 100 = highest intensity possible. References and a few selected odor test sample extracts were then presented and assessed in order to reach consensus, i.e., consensus of how to assess the samples and where it was on the intensity scale. Each assessor signed up for participation after being informed about the products and the terms of participation: Voluntary participation, freedom to leave the test without giving a reason and the right to decline to answer specific questions.

Using a slightly modified version of the Flavor Profile Method® [32], the samples in Table 1 were analyzed. The sensory panel was instructed to identify and describe the sensory attributes appearance, odor, taste, flavor and texture. Consensus regarding each attribute should be reached and defined, Table 4. Each attribute was then assessed for each sample by using the intensity scale. Each assessor needed to agree upon the placement of each sample along the intensity axis. The sensory panel members took a break between each testing to refresh their senses. They were instructed to use water and neutral wafers to clean the palate and neutralize the senses. The assessors were informed not to swallow the samples but to spit them out after assessment.

Table 4. Sensory attributes and their definitions.

Attribute	Abbreviation	Definition
Appearance		
Coarseness	Acoarse	Level of coarseness of mealworm
Yellowness	Ayellow	Intensity of yellow color
Brown-greyish	Abrowngrey	Intensity of brow-grey color
Odor		
Oat	Ooat	Odor of oats
Nutty	Onutty	Odor of nuts, no specific type
Taste/Flavor		
Salt	Tsalt	Saltiness
Oat	Foat	Flavor of oats
Nuts	Fnutty	Flavor of nuts, no specific type
Texture (in mouth)		
Oily	Toil	Fatty, oily sensation
Crispy	Tcrisp	Crisp/crunchy on the first bite
Gritty	Tgritty	Grittiness from peel/shell

To get an indication of the impact of the added antioxidant, the samples with the added antioxidant agent, rosemary (Table 2), were assessed by two food experts for rancidity, color and texture measured on a five-point scale. The assessments were made each day for 15 days.

3.2.2. Instrumental Measurements

Instrumental measurements were performed on all twelve samples, see Table 1. Color measurements were performed using a spectrophotometer (Konica Minolta CM-700d) by measuring L*-, a*- and b*-values, where L* corresponds to the lightness, a* to the red-green scale (red is positive and green negative) and b* to the yellow-blue scale (yellow is positive and blue is negative) [33]. Rheology is the study of the deformation and flow of materials and can describe the properties of materials ranging from liquids to elastic solids. One common method is to measure viscosity. In this study, the viscosities of the samples were measured using a rotational viscometer (Myr VR 3000) equipped with a S1 spider.

3.2.3. Presentation and Analysis of Data

Data was analyzed by calculating mean values and standard deviations. Results were correlated by Pearson correlation (Excel, Microsoft Office 2016). Data were also analyzed using the factorial design parameters, linear regression was performed and included all main effects. In order to find trends for interaction effects univariate of analysis of variance was performed with the factorial design parameters two by two. Sensory attributes and instrumental parameters were dependent factors and the factorial design variables were independent (IBM SPSS, version 23). Principal component analysis (PCA; Panel Check V 1.4.2, Nofima, Norway) was performed to give an overview of results.

4. Results

Results from the sensory analysis are shown in Figure 1, where the results are divided into the three groups of mealworm particle size: Small, medium and large. An increased particle size increased the yellowness and the perceived coarseness as well as the viscosity and crispness. An increased particle size also resulted in a decreased odor. Increased salt content did, as expected, increase saltiness. It also increased the nutty flavor. Different ratios of oil/water did not seem to impact the sensory properties.

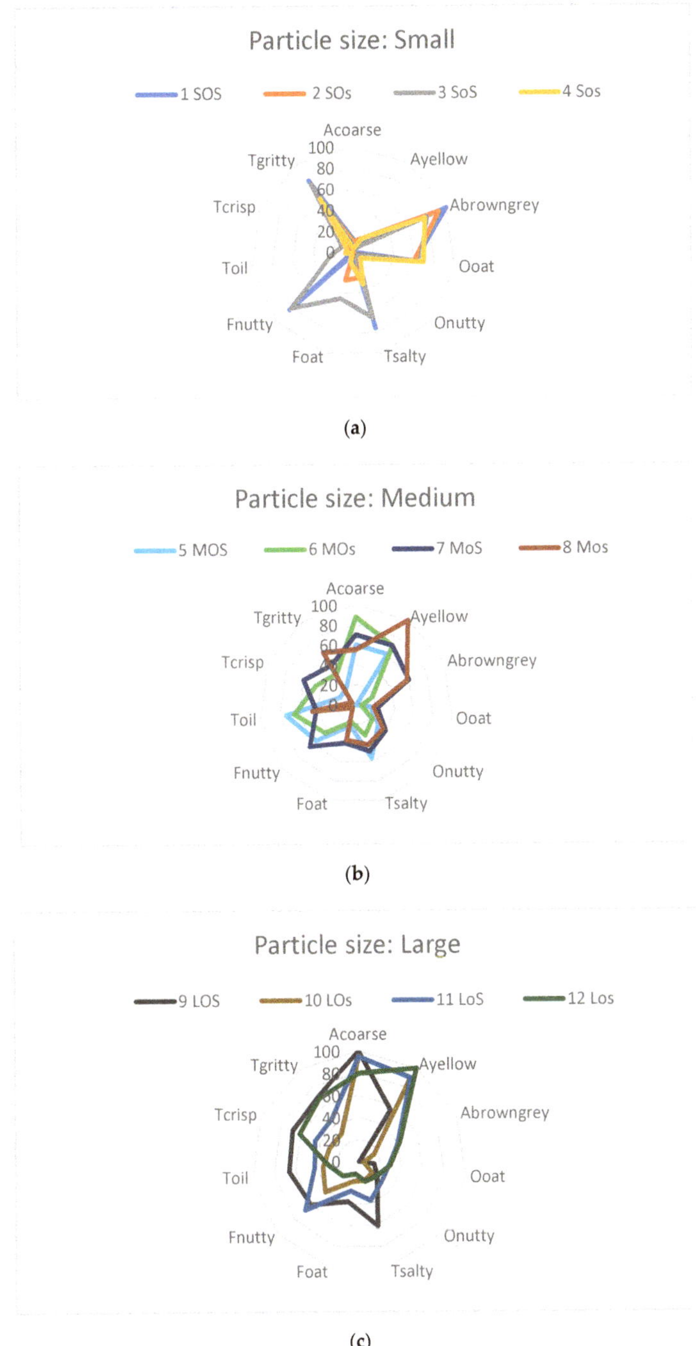

Figure 1. Sensory results displayed in spider plots showing each category of particle size; (**a**) Small particle size, (**b**) Medium particle size, (**c**). Large particle size. Code numbers and identifiers refer to samples in Table 1.

The results from the instrumental analyses are shown in Table 5 where it can be seen that the measurements could be related to the design factors by results varying in accordance with the design factors. Results from the regression analysis showed that the particle size of the mealworms had a significant impact on the sensory attributes of most of the samples. The results showed that salt content had an impact on the salty taste and nutty flavor, and oil content had an impact on color (Table 6). No significant interaction effects were obtained; however, the interaction of the particle size and oil content was close to being significant, $p = 0.068$, which can also be seen as a trend in Table 5, i.e., viscosity in samples 1–6.

Table 5. Mean values (± SD) of L*, a*, b* color space and viscosity of the sample prepared, $n = 2$.

Sample	L* (m + sd)	a*(m + sd)	b*(m + sd)	Viscosity (mpas; m + sd)
1 SOS	52 + 1.1	4.7 + 0.16	16 + 0.11	1200 + 52
2 SOs	54 + 0.2	4.5 + 0.10	16 + 0.53	1590 + 62
3 SoS	62 + 0.3	3.4 + 0.18	14 + 0.47	400 + 7
4 Sos	60 + 0.2	3.6 + 0.07	14 + 0.14	360 + 14
5 MOS	41 + 1.0	7.1 + 0.28	29 + 1.19	5050 + 1600
6 MOs	37 + 0.2	6.3 + 0.11	24 + 0.04	9700 + 420
7 MoS	45 + 0.2	7.0 + 0.91	25 + 3.80	10,000 + 0
8 Mos	44 + 1.6	5.3 + 0.13	18 + 0.22	10,000 + 0
9 LOS	44 + 0.2	8.7 + 0.95	35 + 4.33	8780 + 630
10 LOs	44 + 0.7	6.0 + 0.09	27 + 0.98	10,000 + 0
11 LoS	46 + 0.2	5.4 + 0.29	22 + 1.63	10,000 + 0
12 Los	46 + 0.1	5.9 + 0.09	25 + 0.15	10,000 + 0

Table 6. Regression analysis showing the significant ($p \leq 0.05$) the main effects of the parameters in the factorial design.

	Particle Size	Oil Content	Salt Content
L*	0.017	ns	ns
a*	0.008	ns	ns
b*	0.001	0.037	ns
viscosity	0.001	ns	ns
Acoarse	0.000	ns	ns
Ayellow	0.001	ns	ns
Abrowngrey	0.004	ns	ns
Ooat	0.004	ns	ns
Onutty	ns	ns	ns
Tsalty	0.05	ns	0.001
Foat	ns	ns	ns
Fnutty	ns	ns	0.001
Toil	0.05	ns	ns
Tcrisp	0.008	ns	ns
Tgritty	ns	ns	ns

The PCA plot in Figure 2 shows the influence of particle size. It also shows that "particle size" is situated far away from origo along PC1, with this first PC accounting for 69% of the total variation in data. It can, for example, be seen that particle size is positively correlated to viscosity and "Acoarse" and negatively correlated to "Ooat"; the results are supported by the correlations shown in Table 7. Salt content had a positively significant impact on perceived saltiness ("Tsalty") and Fnutty (Tables 6 and 7). This can be seen in the PCA plot (Figure 2) where all odd sample numbers (with higher salt content) are located in the upper half together with the Tsalt and salt content, and almost all even sample numbers in the lower half. The ratio of oil/water had no significant impact on perceived sensory parameters (Table 6).

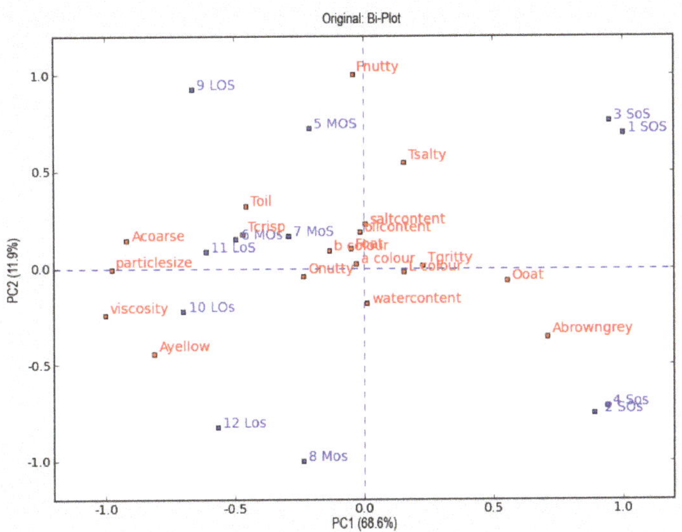

Figure 2. Principal component analysis (PCA) plot giving an overview of all results, visualizing the impact of the design parameters in the factorial design.

Table 7. Pearson correlation coefficients ≥0.7 are given in the table.

	Particle Size	Salt Content	L*	a*	b*	Viscosity	Acoarse	A Yellow	ABrown Grey	Ooat	Onutty	Tsalty
a*			0.79									
b*	0.79		0.75	0.96								
viscosity	0.87		0.83									
Acoarse	0.94		0.84	0.79	0.84	0.93						
Ayellow	0.85		0.76			0.94	0.84					
Abrown Grey	−0.77		0.76	0.79	0.89							
Ooat	−0.79		0.94	0.80	0.81	−0.89	−0.91	−0.83	0.86			
Onutty						0.78		0.78		−0.73		
Tsalty		0.82										
Fnutty		0.86										0.71
Toil			0.79	0.79	0.78		0.74		−0.89	−0.84	0.73	
Tcrisp	0.77			0.73	0.76	0.73	0.80					

The results obtained from the sensory testing and the instrumental analyses were correlated by Pearson calculations of correlation (Table 7) and these confirm the impact of the particle size on viscosity, color, texture and odor parameters. Further, a high salt content was highly correlated with nutty flavor.

Samples with added rosemary were assessed for rancidity, visual color change and separation. The samples changed slightly and it could be seen that the addition of rosemary had a positive impact on shelf life by slower changes. Samples with larger particles changed color, had a higher rancidity and separated earlier than those with small particle size.

5. Discussion

This study makes an important contribution to knowledge about how to formulate the use of insects as a food ingredient. Sensorially appealing products are necessary for their acceptance [20]. From cookbooks we learn that insects may be cooked in a variety of small scale, gastronomic ways [23], but knowledge about how to use insects as an ingredient in industrial food and meal product development is lacking. Ways of rational inclusion of insects in industrial scale product development, to our knowledge, have not systematically been studied. This study evaluated the impact on sensory properties of mealworms in a blend of oil/water and salt where, particle size of the mealworms, oil/water ratio and salt were blended according to a factorial design. The effect of an antioxidant agent was also analyzed in samples differing in particle size. All samples can be considered as high in protein and thereby relevant as a sustainable protein source. The mineral content is high and, according to International Plattform of Insects for Food and Feed -IPIFF (2018) [34], some mineral deficiencies, e.g., iron, may be tackled through the use of insects as food. Contrary to many other sustainable protein sources, e.g., soy beans, insects contain vitamin B12, which may be of importance in future products [14,19].

The flavor of mealworm has been described as nutty with a taste of umami and cereals [21,22]. The result from this study revealed that the nutty flavor might increase with increased salt content, probably due to the polarity of sodium chloride. The results also showed that the particle size of the mealworms had a great impact on appearance and texture, such that an increased particle size increased the yellowness and perceived coarseness. Further, both viscosity and crispness increased. Increased particle size also resulted in decreased odor. We assumed that larger surface area would give larger possibilities for flavor compounds to leak out of the mealworm particle. Increased salt content did not only increase the nutty flavor, but as expected, also increased the saltiness. Different ratios of oil/water did not seem to influence the sensory properties to the same extent as particle size and the addition of salt. Interestingly, the perceived yellowness was not positively correlated to the instrumentally measured b* value. This indicates that humans seem to perceive something different regarding the term "yellow" than that measured by the instrument. All instrumentally measured color values, L*, a* and b* are situated close to origo in Figure 2, indicating little influence on these measures. It can also be observed that higher oil content had a tendency to increase saltiness, as salt is dissolved in water, thus higher oil content results in lower water content e.g., less water for the salt to dissolve in. All samples were prepared using cooked mealworms, thus the protein could already be denatured and less affected by the salt addition, which may not be the case when using unheated mealworms or extracted mealworm protein.

This model system cannot be regarded as a complete (commercial) food product, since the stability is not high, both in terms of phase separation and in rancidity. With reference to the antioxidative effects of carnosic acid and carnosol [35,36], the addition of rosemary had a significant impact on shelf life in terms of decreased rancidity and color changes. In this study, only two experts evaluated the stability over time and the results are to be regarded as an indication. However, the two assessors might have become too expert in the detection of a rancid flavor to then be able to detect this without bias. After five days there was separation, which has to be considered when converting it to a commercial product. This is a common issue in many food products, however by using hydrocolloids such as starch and solutions this could then be a complete food product [37].

In order to make it possible to incorporate insects in food, it is important to better understand what might evoke disgust in relation to insect consumption [38]. Information and education are keys to give objective insight into the connection between food behavior, sustainability and nutrition. Increased knowledge through framed messages might appeal to moral issues such as a person's value orientation, moral obligation and environmental concerns regarding food choices [38] and overrule factors such as neophobia and disgust that may vary in individuals over the course of a lifetime [39]. Acceptance of edible insects may thus be enhanced by the provision of the right messages. However, sensory and visual aspects of food are important criteria for consumers when deciding on the overall acceptability of a dish [40] and have to be taken into account when developing messages about using insects as food. Our study shows that nutty and oat were the two attributes that described the flavor of the products, which may be regarded as important and appealing attributes in the future development of food products based on mealworms.

6. Conclusions

Particle size and salt content significantly influenced the sensory attributes of mealworm samples in a model system. Particle size is therefore an important factor to consider in the development of insects as food products. The addition of an antioxidant agent decreased changes in color, rancidity and separation. This needs to be further considered in future product development.

Author Contributions: Conceptualization, K.W., M.L. and V.O.; Methodology, K.W. and V.O.; Validation, M.L.; Formal Analysis, K.W.; Investigation, K.W.; Resources, K.W.; Data Curation, K.W. and M.L.; Writing-Original Draft Preparation, K.W.; Writing-Review & Editing, K.W., M.L. and V.O.; Visualization, K.W.; Project Administration, K.W.; Funding Acquisition, K.W.

Funding: This research was funded by KK-stiftelsen (The Knowledge-foundation) grant number20170141*.

Acknowledgments: Thanks to the project group members for discussion, analyses and inspiration: Peter Andersson, Johan Berg, Karina Birch, Fanny Cedergaardh, Fredrik Davidsson, Sarah Forsberg, Johanna Gerberich, Åsa Josell, Ingemar Jönsson, Susanne Rask and Sofia Stüffe.

Conflicts of Interest: The authors declare no conflict of interest. The funders had no role in the design of the study; in the collection, analyses, or interpretation of data; in the writing of the manuscript, and in the decision to publish the results.

References

1. FAO. *Report: State of World Fisheries and Aquaculture. Rome: Food and Agriculture Organization of the United Nations*; FAO: Rome, Italy, 2014; p. 243.
2. Tilman, D.; Clark, M. Global diets link environmental sustainability and human health. *Nature* **2014**, *515*, 518–522. [CrossRef] [PubMed]
3. Yen, A. Edible insects: Traditional knowledge or western phobia? *Entomol. Res.* **2009**, *39*, 289–298. [CrossRef]
4. Willett, W.; Rockström, J.; Loken, B.; Springmann, M.; Lang, T.; Vermeulen, S.; Jonell, M. Food in the Anthropocene: The EAT–Lancet Commission on healthy diets from sustainable food systems. *Lancet* **2019**, *393*, 447–492. [CrossRef]
5. Schösler, H.; De Boer, J.; Boersema, J.J. Can we cut out the meat of the dish? Constructing consumer-oriented pathways towards meat substitution. *Appetite* **2012**, *58*, 39–47. [CrossRef] [PubMed]
6. Oonincx, D.G.A.B.; de Boer, I.J.M. Environmental Impact of the Production of Mealworms as a Protein Source for Humans—A Life Cycle Assessment. *PLoS ONE* **2012**, *7*, e51145. [CrossRef]
7. Miglietta, P.; De Leo, F.; Ruberti, M.; Massari, S. Mealworms for food: A water footprint perspective. *Water* **2015**, *7*, 6190–6203. [CrossRef]
8. Abbasi, T.; Abbasi, S.A. Reducing the global environmental impact of livestock production: The minilivestock option. *J. Clean. Prod.* **2016**, *112*, 1754–1766. [CrossRef]
9. Lundy, M.E.; Parrella, M.P. Crickets Are Not a Free Lunch: Protein Capture from Scalable Organic Side-Streams via High-Density Populations of Acheta domesticus. *PLoS ONE* **2015**, *10*, e0118785. [CrossRef]
10. Dobermann, D.; Swift, J.; Field, L.M. Opportunities and hurdles of edible insects for food and feed. *Nutr. Bull.* **2017**, *42*, 293–308. [CrossRef]
11. Rumpold, B.A.; Schlüter, O.K. Nutritional composition and safety aspects of edible insects. *Mol. Nutr. Food Res.* **2013**, *57*, 802–823. [CrossRef]
12. Finke, M.D.; Oonincx, D. Insects as Food for Insectivores. In *Mass Production of Beneficial Organisms*; Elsevier: Amsterdam, The Netherlands, 2014; p. 583.
13. Nowak, V.; Persijn, D.; Rittenschober, D.; Charrondiere, U.R. Review of food composition data for edible insects. *Food Chem.* **2016**, *193*, 39–46. [CrossRef] [PubMed]
14. Zielińska, E.; Baraniak, B.; Karaś, M.; Rybczyńska, K.; Jakubczyk, A. Selected species of edible insects as a source of nutrient composition. *Food Res. Int.* **2015**, *77*, 460–466. [CrossRef]
15. Bukkens, S.G.F. The nutritional value of edible insects. *Ecol. Food Nutr.* **1997**, *36*, 287–319. [CrossRef]
16. Ramos-Elorduy, J.; Moreno, J.; Prado, E. Nutritional value of edible insects from the state of Oaxaca, Mexico. *J. Food Compos. Anal.* **1997**, *10*, 142–157. [CrossRef]
17. Belluco, S.; Losasso, C.; Maggioletti, M.; Alonzi, C.C.; Paoletti, M.G.; Ricci, A. Edible insects in a food safety and nutritional perspective: A critical review. *Compr. Rev. Food Sci. Food Saf.* **2013**, *12*, 296–313. [CrossRef]

18. Zhao, X.; Vázquez-Gutiérrez, J.L.; Johansson, D.P.; Landberg, R.; Langton, M. Yellow Mealworm Protein for Food Purposes-Extraction and Functional Properties. *PLoS ONE* **2016**, *11*, e0147791. [CrossRef] [PubMed]

19. Christensen, D.L.; Orech, F.O.; Mungai, M.N.; Larsen, T.; Friis, H.; Aagaard-Hansen, J. Entomophagy among the Luo of Kenya: A potential mineral source? *Int. J. Food Sci. Nutr.* **2006**, *57*, 198–203. [CrossRef] [PubMed]

20. Elhassan, M.; Wendin, K.; Olsson, V.; Langton, M. Review paper: The appeal of insects as human food-with emphasis on mealworm texture, taste, and flavor. *Foods* **2019**, *8*, 95. [CrossRef] [PubMed]

21. Evans, J.; Flore, R.; Frøst, M.B. *On Eating Insects Essays, Stories and Recipes*; Phaidon: London, UK, 2017.

22. Albrektsson, O. It Really Bugs Me En deskriptiv Sensorisk Analys Av Sju Ätbara Insekter. Bachelor's Thesis, Örebro University, Örebro, Sweden, 2017.

23. Huis, A.; van Gurp, H.; Dick, M. *The Insect Cookbook-Food for a Sustainable Planet*; Colombia University Press: New York, NY, USA, 2014.

24. Terrizzi, J.A.; Shook, N.J.; McDaniel, M.A. The behavioral immune system and social conservatism: A meta-analysis. *Evol. Hum. Behav.* **2013**, *34*, 99–108. [CrossRef]

25. Looy, H.; Dunkel, F.V.; Wood, J.R. How then shall we eat? Insect-eating attitudes and sustainable foodways. *Agric. Hum. Values* **2014**, *31*, 131. [CrossRef]

26. Jongema, Y. List of Edible Insects of the World. Available online: http://www.wageningenur.nl/en/Expertise-Services/Chair-groups/Plant-Sciences/Laboratory-of-Entomology/Edible-insects/Worldwide-species-list.htm (accessed on 4 April 2012).

27. Wendin, K.; Norman, C.; Forsberg, S.; Langton, M.; Davidsson, F.; Josell, Å.; Prim, M.; Berg, J. Eat'em or not? Insects as a Culinary Delicacy. In *Proceedings of the 10th International Conference on Culinary Arts and Sciences, 6–7 July 2017, Aalborg University, Copenhagen Denmark*; Mikkelsen, B., Ofei, K.T., Olsen Tvedebrink, T.D., Quinto Romano, A., Sudzina, F., Eds.; Publisher Aalborg University: Copenhagen, Denmark, 2017; pp. 100–106.

28. Tan, H.; Verbaan, Y.; Stieger, M. How will better products improve the sensory-liking and willingness to buy insect-based foods? *Food Res. Int.* **2017**, *92*, 95–105. [CrossRef] [PubMed]

29. Council of Animal Affairs. *Report: The Emerging Insect Industry*; RDA Advisory Report: The Hague, The Netherlands, 2018.

30. Finke, M.D. Complete Nutrient Composition of Commercially Raised Invertebrates Used as Food for Insectivores. *Zoo Biol.* **2002**, *21*, 269–285. [CrossRef]

31. ISO Standard 8586-2:2008. *Sensory Analysis-General Guidance for the Selection, Training and Monitoring of Assessors-Part 2: Expert Sensory Assessors*; International Organization for Standardization: Genève, Switzerland, 2008.

32. Lawless, H.; Heymann, H. *Sensory Evaluation of Food–Principles and Practices*, 2nd ed.; Springer: New York, NY, USA, 2010.

33. CIE. *Technical Report: Colorimetry*; CIE: Washington DC, USA, 2004.

34. Internationa Platform of Insects for Food and Feed. Edible Insects and Human Nutrition. Available online: http://www.ipiff.org/ (accessed on 7 June 2018).

35. Erkan, N.; Ayranci, G.; Ayranci, E. Antioxidant activities of rosemary (*Rosmarinus officinalis* L.) extract, blackseed (*Nigella sativa* L.) essential oil, carnosic acid, rosmarinic acid and sesamol. *Food Chem.* **2008**, *110*, 76–82. [CrossRef] [PubMed]

36. Aruoma, O.I.; Halliwell, B.; Aeschbach, R.; Löligers, J. Antioxidant and pro-oxidant properties of active rosemary constituents: Carnosol and carnosic acid. *Xenobiotica* **1992**, *22*, 257–268. [CrossRef] [PubMed]

37. Gao, Z.; Fang, Y.; Cao, Y.; Liao, H.; Nishinari, K.; Phillips, G.O. Hydrocolloid-food component interactions. *Food Hydrocoll.* **2017**, *68*, 149–156. [CrossRef]

38. Leijdekkers, S. Value Orientation and the Effect of Framing: The Acceptance of Edible Insects. Master's Thesis, Wageningen University, Wageningen, The Netherlands, 2016.

39. Pliner, P.; Salvy, S. *Food Neophobia in Humans in The Psychology of Food Choice*; Shepherd, R., Raats, M., Eds.; Cabi: Guildford, UK, 2006.

40. Halloran, A.; Flore, R.; Mercier, C. Notes from the 'Insects in a gastronomic context' workshop in Bangkok, Thailand. *J. Insects Food Feed.* **2015**, *1*, 241–243. [CrossRef]

 *foods*

Article

Factors Predicting the Intention of Eating an Insect-Based Product

Simone Mancini [1,*]**, Giovanni Sogari** [2]**, Davide Menozzi** [2]**, Roberta Nuvoloni** [1,3]**,
Beatrice Torracca** [1]**, Roberta Moruzzo** [1] **and Gisella Paci** [1,3]

[1] Department of Veterinary Sciences, University of Pisa, Viale delle Piagge 2, 56124 Pisa, Italy
[2] Department of Food and Drug, University of Parma, Parco Area delle Scienze 27/A, 43124 Parma, Italy
[3] Interdepartmental Research Center Nutrafood "Nutraceuticals and Food for Health", University of Pisa,
Via del Borghetto 80, 56124 Pisa, Italy
* Correspondence: simone.mancini@for.unipi.it; Tel.: +39-050-2216-803

Received: 26 June 2019; Accepted: 18 July 2019; Published: 19 July 2019

Abstract: This study provides a framework of the factors predicting the intention of eating an insect-based product. As part of the study, a seminar was carried out to explore how the provision of information about ecological, health, and gastronomic aspects of entomophagy would modify consumer beliefs regarding insects as food. Before and after the informative seminar, two questionnaires about sociodemographic attributes and beliefs about the consumption of insects as food were given. Participants were then asked to carry out a sensory evaluation of two identical bread samples, but one was claimed to be supplemented with insect powder. Results showed that perceived behavioral control is the main predictor of the intention, followed by neophobia and personal insect food rejection. The disgust factor significantly decreased after the participants attended the informative seminar. Sensory scores highlighted that participants gave "insect-labelled" samples higher scores for flavor, texture, and overall liking, nevertheless, participants indicated that they were less likely to use the "insect-labelled" bread in the future. Our findings provide a better understanding of insect food rejection behavior and help to predict the willingness to try insect-based products based on some important individual traits and information.

Keywords: entomophagy; novel food; neophobia; disgust; edible insects

---

1. Introduction

In Europe, the habit of eating insects, or entomophagy, has not become widespread, with the exception of some specific and local cases [1,2]. Traditionally, most people in Western countries consider insects to be a threat and a health risk rather than a food source [3], despite the fact that insect protein and fat contents, mineral contents, and amino acids profiles confirm that they are a good source of nutrients [4].

Most of the studies carried out in Europe about consumers' acceptance of insects as food illustrate a low consumer willingness to try eating insects [5], with the exception of the Netherlands, Belgium, and few other Northern countries, where some insect-based products have been commercially available in the past few years [6–8]. Particularly, in urban and Western societies, insects are rarely eaten or their consumption is perceived to be culturally inappropriate [2]. As reported by Deroy et al. [9], one of the greatest barriers to adopting entomophagy is cognitive disgust linked to the perception that insects are related to undesirable thoughts such as dirt, death, disease, and contamination [10]. Only a few works have investigated how potential consumers from Southern European countries might respond to entomophagy [11].

The theory of planned behavior (TPB) [12], has been proven to be a proper theoretical framework for understanding sustainable and ethical consumer behaviors concerning food [13,14]. According to

TPB, the perceived behavioral control (PBC), along with subjective norms (SN) and attitude, impact a person's intention. Perceived behavioral control is the measure of perceived control over the behavior (i.e., how easy or difficult performing the behavior will be). Perceived behavioral control can also have a direct impact on behavior because performance of a behavior not only depends on motivation, but also the individual control of the behavior. If an individual has limited control over an activity, the activity might not be implemented, even in the presence of strong motivational factors.

Menozzi et al. [15], examined whether the TPB could be employed to understand young adults' behavior when faced with the prospect of eating products containing insect flour, and suggested that attitude and, to a lesser extent, PBC play a significant role in affecting the intention of performing the behavior, while the SN is not a significant factor in forming the behavioral intention. This result is common in several other studies [16], and the SN is generally a weak predictor of intentions [17].

From another point of view, when a new food product is introduced into a culture, it generally induces feelings of fear and refusal that is called food neophobia [18], which refers to the unwillingness to try and the tendency to avoid novel food [19].

Food neophobia is expected to reduce the likelihood of readiness to incorporate insects into the diet [20–22]. Pliner and Hobden [19] developed an instrument to measure food neophobia which consisted of ten statements, five positively worded (neophilic) and five negatively worded (neophobic), and was called a food neophobia scale (FNS).

Recently, many studies have indicated that food neophobia, as an individual trait, is one of the most important predictors for understanding consumer willingness to try insects [22–25]. The relationship between entomophagy and food neophobia may depend on the knowledge of their origin and habitat or may be based on negative consequences following their consumption [26].

Food exposure increases product familiarity and influences the willingness to accept new food, and thus reduces neophobic reactions [21,27–29]. At the same time, information is a factor that plays a role in consumers' acceptance [30,31] and can be used to promote entomophagy [22]. Information about insects, such as production technology, safety risk, and scientific-like information [32,33], tailored to consumers can modify their perception of items, both novel and familiar.

Despite the huge interest in factors specifically related to the intention of eating insects, there is a lack of research that investigates the link between the mechanism of rejection of insects as food (both at a personal and social level), the level of food neophobia for unfamiliar food, and the perceived behavior control of introducing insects to the diet.

On the basis of these assumptions, the first aim of this study was to evaluate the willingness of consumers to adopt insects as food and to investigate the main factors (e.g., sociodemographic variables, food neophobia, and other behavioral constructs) that affect the intention to eat insects. The second aim was to investigate whether and how information can influence the willingness to taste insects and their acceptability. This was achieved by comparing consumers' willingness to adopt insects before and after a seminar on insects, and investigating how an informative seminar on the benefits of including insects in human diets influenced the consumer acceptance. In addition, an analysis about the perception of the sensory properties of two identical breads ("insect-labelled" vs. "no insect-labelled") was performed.

2. Material and Methods

2.1. Theoretical Model

The model includes two concepts, food neophobia and insect food rejection, with the introduction of some elements of the TPB.

In accordance with previous studies, given the premises mentioned in the introduction, we tested the following hypotheses (Figure 1).

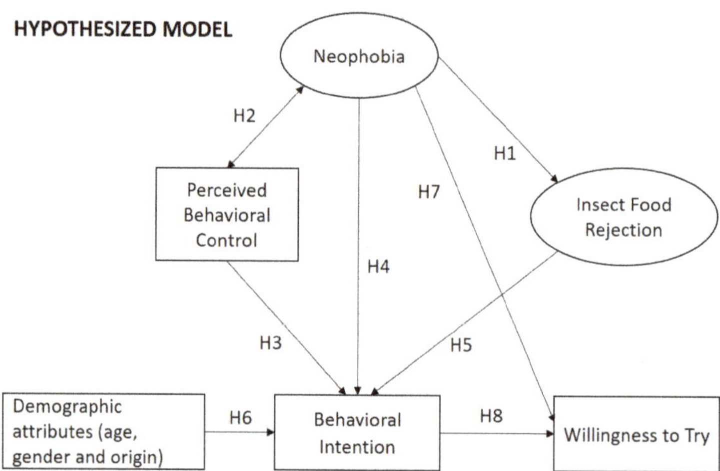

Figure 1. The conceptual model corresponding to the current research. Our study tested the following hypotheses (H): H1, food neophobia will positively affect rejection factors of insect as food (e.g., disgust) [10]; H2, food neophobia and the perceived behavioral control (PBC) will be correlated; H3, PBC will have a significant positive effect on the intention of eating an insect-based food in the coming months [15]; H4, a higher level of food neophobia will have a strong negative effect on the behavioral intention of eating an insect-based food in the coming months [21]; H5, a higher level of insect food rejection will have a strong negative effect on the intention of eating an insect-based food in the coming months; H6, demographic attributes will have a moderate effect on the intention of eating an insect-based food in the coming months [34]; H7, food neophobia will negatively affect the willingness to eat an insect-based product; H8, intention of eating an insect-based food in the next months will be a strong predictor of the willingness to eat an insect-based product [15].

2.2. Participants and Data Collection

The survey was conducted at the University of Pisa (Italy) in March 2018. Participants were students attending a seminar entitled "Insects as Food and Feed: Future Prospects". The three-hour seminar was structured as oral presentations from different speakers with visual support. The aim of the seminar was to describe benefits, disadvantages, and future prospects of insects as food and feed. In particular, aspects were delivered related to species and breeding technologies, food safety and regulatory concerns, and consumers' approach.

A total of 165 students with a bachelor's degree and a master's degree participated in this study. They did not receive monetary compensation for their participation. Prior to the seminar session, participants were informed about the questionnaires and an informed consent form was signed.

Our protocol was based on previous studies where an informative session about entomophagy was held among university students followed by an insect tasting session [34,35]. Two questionnaires were proposed chronologically to the students: one before (questionnaire 1, Q1) and another after (questionnaire 2, Q2) the seminar. Furthermore, at the end of Q2, the participants were invited to take part in a tasting session immediately after the seminar ended.

2.3. Questionnaire Design and Measures

The structure of the questionnaire for both Q1 and Q2 was based on previous study designs [15,34]. The first part of the questionnaire (Q1) included three questions concerning the demographic profile of the respondents (i.e., age, gender, and Italian region of origin) and 20 individual statements, divided into two different sections. Section A included 10 items about a person's tendency to reject or avoid eating unfamiliar food (neophobia). Section B, which followed A, included 10 items focused on beliefs towards the consumption of insects (i.e., disgust, distaste, fear, and social acceptance). For all the statements, responses were given on a seven-point agreement scale, ranging from "1 = strongly disagree" to "7 = strongly agree".

In addition, participants were asked whether they had ever eaten insects, and if yes, they were asked to indicate on what occasion and their liking of the experience (i.e., "If you have already tried eating insects in the past, how much did you like it?"). Next, they were asked to evaluate the importance of possible barriers to the consumption of insects (e.g., negative taste and texture, low level of food safety, being vegetarian).

The second questionnaire (Q2) was composed of three sections. The first section was composed of three items that measured behavioral beliefs toward eating products containing insect powder. The aim was to evaluate whether this practice would have: (1) "positive effects on health", (2) "positive effects on the environment", and (3) "a familiar taste compared to known products". The second part of the questionnaire was structured in the same way as section B of Q1 in order to compare the participants' beliefs before and after the informative seminar. The last part included the intention of eating insect-based food in the coming months (behavioral intention) and the perceived behavioral control, i.e., "I believe that eating products containing insect powder in the coming months is possible." Finally, participants were asked about their willingness to eat insects (i.e., whether they wanted take part in the tasting session). If participants decided to not attend the tasting section, they were asked why (e.g., short of time, afraid to being allergic, afraid of health risks, vegetarian or other).

2.4. Tasting Session

Although the students who agreed to take part in the tasting session had been told that edible insects would be included, no further detail was given about what food they were going to taste at the end of the session, as suggested by Spence [36]. Participants were asked to score five sensory attributes of each sample (appearance, odor, flavor, texture, overall liking) using a nine-point balanced hedonic scale (1, extremely negative; 5, neither negative nor positive; 9, extremely positive) and to express the probability of consuming the product in the future using a similar nine-point scale (1, extremely improbable; 9, extremely probable). The two bread samples were identical, except one was claimed to be supplemented with insect powder, i.e., "insect-labelled" bread, although it did not contain any insect ingredients. Bread was chosen as the food product because consumers were familiar with it, an attribute which has been suggested to reduce food neophobia among Western consumers [23,27,35]. The tasting session was organized in a separate room, and once they had tasted the bread samples, participants were asked to not share information or impressions with others that had not yet tried them. Participants were also informed about the safety of the bread preparations and the allergenic potential of insects, even though the "insect-labelled" bread did not contain insect.

The two types of samples came from the same pre-sliced wholemeal bread (Lidl Stiftung & Co. KG, Neckarsulm, Germany), and they were the same size (5.7 cm × 5.7 cm × 1.1 cm, 9 g), measuring one quarter of a whole slice.

Samples were presented monadically in white plastic dishes, and the samples of "insect-labelled" bread were clearly indicated to the participants. Participants were instructed to drink water between sample tasting to neutralize the taste. Ordering of samples was randomized, so that approximately half of participants tasted the "control" bread first while the other half tasted the "insect-labelled" bread first (32 and 34 participants, respectively).

2.5. Statistical Analyses

A comparison of mean scores, i.e., independent samples *t*-test, was used to assess associations between behavioral beliefs about eating products containing insect powder (interval variables, seven-point scale) and willingness to try an insect-based food (dummy variable: yes or no).

Additionally, correlational analyses (Pearson correlation) between behavioral beliefs and behavioral intention to eat a product containing insect powder in the coming months was conducted.

Table 3. Confirmatory factor analysis of the food neophobia (FN) construct.

	Items	Cronbach's Alpha	Factor Loadings
	Food Neophobia	0.862	
FN1	I am constantly sampling new and different foods (R)		0.613
FN2	I like foods from different cultures (R)		0.909
FN3	Ethnic food looks too weird to eat		0.719
FN4	At dinner parties, I will try new foods (R)		0.568
FN5	I am afraid to eat things I have never had before		0.579
FN6	I like to try new ethnic restaurants (R)		0.874

Model fit: χ^2 (5) = 7.002; CFI = 0.996; TLI = 0.987; RMSEA (CI 90%) = 0.049 (0.000, 0.081). R-reverse coded.

On the other hand, the EFA results suggest that a multidimensional structure of the food rejection construct may be more appropriate. The analysis yielded two factors explaining a total of 70.4% of the variance for the entire set of variables (Table 4).

Table 4. Confirmatory factor analysis of the insect food rejection (FR) construct.

	Items	Cronbach's Alpha	Factor Loadings
	Social Insect Food Rejection	0.81	
FR1	I believe that insect-based food implies a poor hygiene		0.75
FR2	I believe that eating insects is not part of our diet		0.872
FR3	Eating insects is not socially acceptable		0.684
	Personal Insect Food Rejection	0.758	
FR4	The idea of eating insects provokes my disgust		0.59
FR5	I fear that insect-based foods have negative texture properties		0.816
FR6	I fear that insect-based foods have negative taste properties		0.776

Model fit: χ^2 (8) = 7.207; CFI = 1.000; TLI = 1.004; RMSEA (CI 90%) = 0.000 (0.000, 0.086).

The first factor, explaining 52.8% of the variance, includes three items related to danger and social inappropriateness of insects as food. This factor was labelled as "social insect food rejection" due to the high loadings by the following items: "I believe that insect-based food implies a poor hygiene", "I believe that eating insects is not part of our diet", and "eating insects is not socially acceptable". The items communalities, representing the proportion of the variance in that variable that can be accounted for by the extracted factors, are all above the 0.5 threshold, indicating that the extracted factors account for a big proportion of the items' variance. The second factor was labelled "personal insect food rejection" due to the high loadings by the following items: "the idea of eating insects provokes my disgust", "I fear that insect-based foods have negative texture properties", and "I fear that insect-based foods have negative taste properties". This factor explained 17.7% of the variance. Overall, the Cronbach's alpha of the two factors above the 0.7 value, the factor loadings, and the communalities all above the 0.50 threshold, indicate a good representation and internal consistency of this two-factor model. The CFA was performed to test for this two-factor structure, showing a good model fit to the data (model fit: χ^2 (8) = 7.207; CFI = 1.000; TLI = 1.004; RMSEA (CI 90%) = 0.000 (0.000, 0.086)).

The prediction of eating insect-based food was examined using structural equation modelling (SEM) techniques. The results are reported in the tested model (Figure 2).

TESTED MODEL

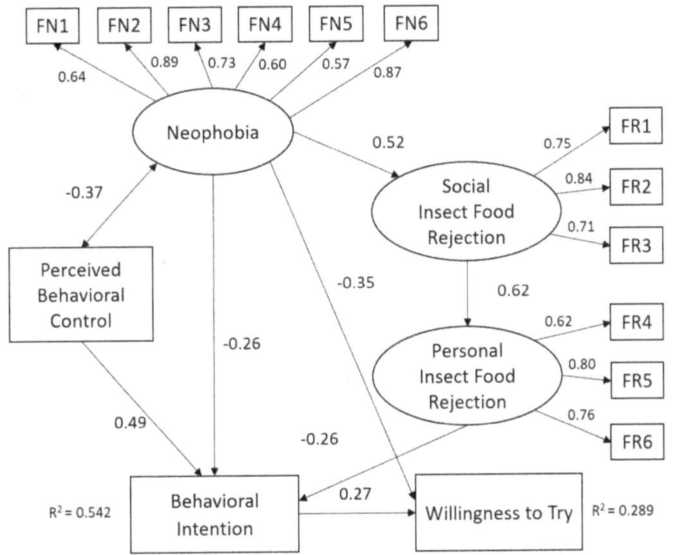

FN: Food Neophobia items; FR: Food Rejection items.
Fit measures: χ^2 (82) = 150.066, CFI = 0.938, TLI = 0.921, RMSEA (CI 90%) = 0.071 (0.053; 0.089); SRMR = 0.080
All paths have a p-value < 0.001

Figure 2. The standardized regression coefficients, explained variance (R^2), and fit measures.

The model fits the data satisfactorily as documented by the fit indices (χ^2 (82) = 150.066; CFI = 0.938; TLI = 0.921; RMSEA (CI 90%) = 0.071 (0.053, 0.089); SRMR = 0.080) and explains a substantial amount of variance of the behavioral intention (R^2 = 0.54) and a relative smaller variance of the willingness to eat insect-based food (R^2 = 0.29).

Confirming H2, a statistically significant negative correlation between neophobia and PBC was detected (p < 0.001), suggesting that respondents with lower phobia of new food believe that they can better accomplish the behavior of eating insect-based food and vice versa. PBC is the main predictor of the intention (p < 0.001), followed by neophobia (p < 0.001) and personal insect food rejection (p < 0.001). Therefore, these results confirmed H3, H4, and H5. The willingness to eat insect-based food is negatively affected by food neophobia (p < 0.001), and positively influenced by the intention to eat products containing insect powder in the coming months (p < 0.001). These results confirmed, respectively, H7 and H8.

Moreover, the results indicate that the social insect food rejection construct mediates the effect of neophobia with personal insect food rejection. In other words, this indicates that a higher neophobia positively affected the perceived social rejection of insects as food which, in turn, influenced the personal rejection. Therefore, the hypothesis suggesting that food neophobia positively influenced factors of the rejection of insects as food (H1) is also confirmed.

Finally, sociodemographic characteristics, such as age, gender, and region of origin, were neither significantly correlated with the intention nor with the willingness to eat insects, and therefore have not been included in the final model. This suggests that H6 was not supported by the data.

Since food disgust sensitivity and neophobia traits can be strongly influenced by familiarity and previous experience, a question about past exposure of insect eating was asked. However, considering the low number of people that had already tasted insects before the study, this question was not included in the model.

3.4. Tasting Session

Out of the total 165 participants, 66 (40%) took part in the tasting session; 11 were males and 55 were females, corresponding to 39.29% and 39.86% of total males and females, respectively. As such, no statistical difference was found between the genders in their willingness to participate in the tasting session. The scores of the tasting session are reported in Table 5.

Table 5. Results of tasting section reported as mean (standard deviation).

Items	"control" Sample	"insect labelled" Sample	p
Appearance	7.2 (1.32)	7.29 (1.19)	0.407
Odor	6.64 (1.74)	6.7 (1.53)	0.626
Flavor	7.06 (1.12)	7.26 (1.14)	0.044
Texture	6.77 (1.38)	7.12 (1.42)	<0.001
Overall liking	7.18 (1.14)	7.39 (1.09)	0.045
Probability of consuming	7.52 (1.51)	7.21 (1.67)	0.107

The results of the Wilcoxon rank sum test on the sensory scores indicated that participants gave "insect-labelled" samples higher scores for flavor, texture, and overall liking ($p = 0.044$, $p < 0.001$, $p = 0.045$, respectively). This data did not reflect on the score given to the probability of future consuming, since 28.79% of participants indicated that they was less likely to use the "insect-labelled" bread in the future, while only 18.18% stated the opposite (although no statistical difference was found at the Wilcoxon rank sum test). Figure 3 shows the delta scores obtained for each characteristic. Only 13.64% of the participants gave the two bread samples identical scores for every evaluated parameter. In two cases, the delta score for probability of future use was ≤-6 and represented the biggest score difference registered.

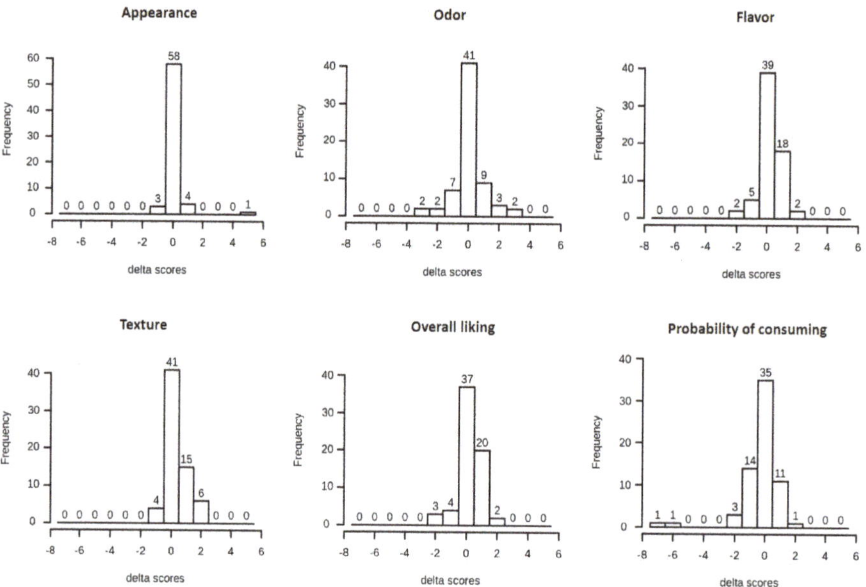

Figure 3. Delta scores between "insect labelled" sample and "control" bread.

4. Discussion

Reactions towards Insects

Our research findings suggest that a variety of factors need be taken into consideration when assessing the willingness to eat insect-based food, from which several implications can be derived both theoretically and practically.

As reported by Hartmann and Siegrist [10], disgust sensitivity prevents consumers from eating harmful substances (e.g., contamination of toxic ingredients), however, it can also become a strong barrier to trying new food and/or unfamiliar food products like insects. The results have shown that the relationship between neophobia and personal factors of insect food rejection is mediated by the social factors of insect food rejection construct, i.e., the perceived danger and social inappropriateness of insects as food. This mediation effect suggests a more detailed relationship between food neophobia and disgust than the one depicted by Hartmann and Siegrist [10]. As expected, neophobic individuals have a higher social rejection (i.e., insects are not part of my diet, not acceptable as food) which is strongly correlated with a greater personal insect rejection (i.e., negative sensory expectations). However, the mediation effect suggests how the personal factors of insect food rejection can be more easily influenced, for instance, by an increasing social appeal of insects as food driven by positive information about entomophagy and the tasting experience. This enhanced the social acceptance of insects as food which may improve the personal factors, and therefore decrease the disgust and distaste. As reported by Barsics et al. [35], long-term exposure to information and positive effects (mainly entomophagy familiarity and experience) could reduce disgust and distance and make better assimilation of the new further information. In addition, other studies reported how information about the sustainability and environmental perspectives of edible insects, could positively influences consumers' willingness to consume insect-based products [22,30,40–42].

According to the literature, since disgust influences people's food preferences, higher food disgust and distaste sensitivity (personal insect food rejection) are associated with a lower behavioral intention. This has been confirmed in our study, since we have found a significant negative effect of personal insect food rejection on behavioral intentions. Nevertheless, our results also suggest how the insect food rejection construct helps to explain the relationship between neophobia and intention. This is, however, a partial mediation since the direct effect of neophobia on behavioral intention is still significant. The PBC has been found to be the main predictor of the intention, as also suggested by the theory [12].

The willingness to try an insect-based product is positively affected by behavioral intention and negatively influenced by food neophobia. This result is consistent with previous studies [6,10,22,43] indicating that the trait of food neophobia is an important predictor of consumers' willingness to try insect-based food products. In addition, the positive and significant correlation between behavioral intention and beliefs suggests that consumers believe that this practice will have positive effects on their health and on the environment, as also reported by Sogari [44] and Menozzi et al. [15].

The informative seminar positively influenced all the opinions about entomophagy, improving the consumers' attitude towards eating insects while lowering insect food rejection. The greatest change after the seminar was found for the disgust factor, which strongly decreased after participants were informed about the technological, social, and cultural context of entomophagy. The communication about the gastronomic aspects of insects influenced only the texture properties and not the taste. This might be explained by the fact that it might be easier to convince people that the texture is similar to known food products (not disgusting) rather than the organoleptic attributes. In addition, the fear regarding poor hygiene decreased following the explanation about the long-time habits of entomophagy around the world and the food safety production practices which will be used before these novel products can be commercialized in the Western countries. Barsics et al. [35] reported that familiarity with entomophagy, intrinsically related with the culture of the consumers, plays a major role in acceptance and a positive attitude toward edible insects. Information could enhance familiarity with entomophagy, nevertheless, it seems more effective if it is continuous with a long-term impact.

The role of communication in decreasing the rejection of new food can be explained by the two hypotheses of Rozin and Fallon [26], who explained this tendency might be related to the origin of the food and/or the anticipated negative post-ingestion consequences. Positive information on eating insects can address these two barriers. Given the strong positive correlations between the positive health effects of eating insects and the behavioral intention, our findings suggest that consumer-targeted communication of health aspects would increase the intention to eat.

Findings of this work have several implications on sensory aspects of product development and the role of information on influencing consumer acceptance.

Our data indicate that the provision of information plays a role in decreasing the rejection of insects as food, both at a personal and social level. To our knowledge, this implication has been previously investigated by only a few other authors [20,35].

We are aware that the likelihood of trying insects will strongly depend on the appropriateness of the preparation method [45–47]. If edible insects are to become more widely consumed in the Western countries, positive thoughts and sensory-liking properties need to be associated with familiar products [15,27]. As reported by Schouteten et al. [48], consumers have to associate (positive) emotions with the products (based on the satisfaction of sensory attributes) to replace the negative expected emotions prior to consumption [49].

Although our study used an "insect-labelled" product, the greater sensory appreciation as compared with the similar "no insect" product confirms that integrating insects into the Western diet could be done using a transitional phase [20,50]. Moreover, familiar products could help to increase the likelihood of having the first bite, but to be regularly consumed, they must be appropriate and taste good [47,49].

5. Conclusions

Understanding which factors could affect consumers' perception of edible insects plays a key role in the future prospect of entomophagy and, in general, on novel food protein production and consumption. In summary, the participants' willingness to eat insect-based products in the coming months appears to be mainly influenced by neophobia and behavioral intention. Communication becomes crucial to achieve this aim and informative tasting sessions are a good approach to influence people to try insect-based foods for the first time. The current study suggests that an educational approach paired with a tasting session could be a good strategy to increase objective knowledge while reducing the insect food rejection mechanism and enhancing the positive evaluation of sensory properties.

Author Contributions: Conceptualization, S.M., G.S., and D.M.; data curation, G.S., D.M., and B.T.; investigation, S.M., D.M., R.N., B.T., and R.M.; methodology, S.M., G.S., B.T., and R.M.; supervision, G.P.; writing—original draft, S.M., G.S., B.T., and R.M.; writing—review and editing, S.M., G.S., D.M., R.N., R.M., and G.P.

Funding: This research received no external funding.

Acknowledgments: We want to thank the seminar organizers and the lecturers who made the oral presentations. Special thanks also go to the students, without whom this article could not be done.

Conflicts of Interest: The authors declare no conflict of interest.

References

1. Evans, J.; Alemu, M.H.; Flore, R.; Frøst, M.B.; Halloran, A.; Jensen, A.B.; Maciel-Vergara, G.; Meyer-Rochow, V.B.; Münke-Svendsen, C.; Olsen, S.B.; et al.. 'Entomophagy': An evolving terminology in need of review. *J. Insects Food Feed* **2015**, *1*, 293–305. [CrossRef]
2. van Huis, A.; Van Itterbeeck, J.; Klunder, H.; Mertens, E.; Halloran, A.; Muir, G.; Vantomme, P. *Edible Insects. Future Prospects for Food and Feed Security*; FAO: Rome, Italy, 2013.
3. Looy, H.; Dunkel, F.V.; Wood, J.R. How then shall we eat? Insect-eating attitudes and sustainable foodways. *Agric. Hum. Values* **2014**, *31*, 131–141. [CrossRef]
4. Zielińska, E.; Karaś, M.; Baraniak, B. Comparison of functional properties of edible insects and protein preparations thereof. *LWT* **2018**, *91*, 168–174. [CrossRef]

5. Hartmann, C.; Siegrist, M. Consumer perception and behaviour regarding sustainable protein consumption: A systematic review. *Trends Food Sci. Technol.* **2017**, *61*, 11–25. [CrossRef]

6. House, J. Consumer acceptance of insect-based foods in the Netherlands: Academic and commercial implications. *Appetite* **2016**, *107*, 47–58. [CrossRef] [PubMed]

7. Van Thielen, L.; Vermuyten, S.; Storms, B.; Rumpold, B.; Van Campenhout, L. Consumer acceptance of foods containing edible insects in Belgium two years after their introduction to the market. *J. Insects Food Feed* **2018**, *1*, 1–10. [CrossRef]

8. Caparros Megido, R.; Haubruge, É.; Francis, F. Insects, The Next European Foodie Craze? In *Edible Insects in Sustainable Food Systems*; Halloran, A., Flore, R., Vantomme, P., Roos, N., Eds.; Springer: Cham, Switzerland, 2018; pp. 353–361.

9. Deroy, O.; Reade, B.; Spence, C. The insectivore's dilemma, and how to take the West out of it. *Food Qual. Prefer.* **2015**, *44*, 44–55. [CrossRef]

10. Hartmann, C.; Siegrist, M. Development and validation of the Food Disgust Scale. *Food Qual. Prefer.* **2018**, *63*, 38–50. [CrossRef]

11. Mancini, S.; Moruzzo, R.; Riccioli, F.; Paci, G. European consumers' readiness to adopt insects as food. A review. *Food Res. Int.* **2019**, *122*, 661–678. [CrossRef]

12. Ajzen, I. The theory of planned behavior. *Organ. Behav. Hum. Decis. Process.* **1991**, *50*, 179–211. [CrossRef]

13. Chen, M.-F. Extending the theory of planned behavior model to explain people's energy savings and carbon reduction behavioral intentions to mitigate climate change in Taiwan–moral obligation matters. *J. Clean. Prod.* **2016**, *112*, 1746–1753. [CrossRef]

14. Ricci, E.C.; Banterle, A.; Stranieri, S. Trust to Go Green: An Exploration of Consumer Intentions for Eco-friendly Convenience Food. *Ecol. Econ.* **2018**, *148*, 54–65. [CrossRef]

15. Menozzi, D.; Sogari, G.; Veneziani, M.; Simoni, E.; Mora, C. Eating novel foods: An application of the Theory of Planned Behaviour to predict the consumption of an insect-based product. *Food Qual. Prefer.* **2017**, *59*, 27–34. [CrossRef]

16. Pambo, K.O.; Okello, J.J.; Mbeche, R.M.; Kinyuru, J.N.; Alemu, M.H. The role of product information on consumer sensory evaluation, expectations, experiences and emotions of cricket-flour-containing buns. *Food Res. Int.* **2018**, *106*, 532–541. [CrossRef]

17. Armitage, C.J.; Conner, M. Efficacy of the Theory of Planned Behaviour: A meta-analytic review. *Br. J. Soc. Psychol.* **2001**, *40*, 471–499. [CrossRef]

18. Dovey, T.M.; Staples, P.A.; Gibson, E.L.; Halford, J.C.G. Food neophobia and 'picky/fussy' eating in children: A review. *Appetite* **2008**, *50*, 181–193. [CrossRef]

19. Pliner, P.; Hobden, K. Development of a scale to measure the trait of food neophobia in humans. *Appetite* **1992**, *19*, 105–120. [CrossRef]

20. Caparros Megido, R.; Gierts, C.; Blecker, C.; Brostaux, Y.; Haubruge, É.; Alabi, T.; Francis, F. Consumer acceptance of insect-based alternative meat products in Western countries. *Food Qual. Prefer.* **2016**, *52*, 237–243. [CrossRef]

21. Sogari, G.; Menozzi, D.; Mora, C. The food neophobia scale and young adults' intention to eat insect products. *Int. J. Consum. Stud.* **2019**, *43*, 68–76. [CrossRef]

22. Verbeke, W. Profiling consumers who are ready to adopt insects as a meat substitute in a Western society. *Food Qual. Prefer.* **2015**, *39*, 147–155. [CrossRef]

23. Hartmann, C.; Shi, J.; Giusto, A.; Siegrist, M. The psychology of eating insects: A cross-cultural comparison between Germany and China. *Food Qual. Prefer.* **2015**, *44*, 148–156. [CrossRef]

24. Hartmann, C.; Siegrist, M. Becoming an insectivore: Results of an experiment. *Food Qual. Prefer.* **2016**, *51*, 118–122. [CrossRef]

25. Verneau, F.; La Barbera, F.; Kolle, S.; Amato, M.; Del Giudice, T.; Grunert, K. The effect of communication and implicit associations on consuming insects: An experiment in Denmark and Italy. *Appetite* **2016**, *106*, 30–36. [CrossRef]

26. Rozin, P.; Fallon, A. The psychological categorization of foods and non-foods: A preliminary taxonomy of food rejections. *Appetite* **1980**, *1*, 193–201. [CrossRef]

27. Cavallo, C.; Materia, V.C. Insects or not insects? Dilemmas or attraction for young generations: A case in Italy. *Int. J. Food Syst. Dyn.* **2018**, *9*, 226–239.

28. Laureati, M.; Bertoli, S.; Bergamaschi, V.; Leone, A.; Lewandowski, L.; Giussani, B.; Battezzati, A.; Pagliarini, E. Food neophobia and liking for fruits and vegetables are not related to Italian children's overweight. *Food Qual. Prefer.* **2015**, *40*, 125–131. [CrossRef]

29. Woolf, E.; Zhu, Y.; Emory, K.; Zhao, J.; Liu, C. Willingness to consume insect-containing foods: A survey in the United States. *LWT* **2019**, *102*, 100–105. [CrossRef]

30. Laureati, M.; Proserpio, C.; Jucker, C.; Savoldelli, S. New sustainable protein sources: consumers' willingness to adopt insects as feed and food. *Ital. J. Food Sci.* **2016**, *28*, 652–668.

31. Laureati, M.; Jabes, D.; Russo, V.; Pagliarini, E. Sustainability and organic production: How information influences consumer's expectation and preference for yogurt. *Food Qual. Prefer.* **2013**, *30*, 1–8. [CrossRef]

32. Lobb, A.E.; Mazzocchi, M.; Traill, W.B. Modelling risk perception and trust in food safety information within the theory of planned behaviour. *Food Qual. Prefer.* **2007**, *18*, 384–395. [CrossRef]

33. Waldman, K.B.; Kerr, J.M. Does safety information influence consumers' preferences for controversial food products? *Food Qual. Prefer.* **2018**, *64*, 56–65. [CrossRef]

34. Sogari, G.; Menozzi, D.; Mora, C. Exploring young foodies' knowledge and attitude regarding entomophagy: A qualitative study in Italy. *Int. J. Gastron. Food Sci.* **2017**, *7*, 16–19. [CrossRef]

35. Barsics, F.; Caparros Megido, R.; Brostaux, Y.; Barsics, C.; Blecker, C.; Haubruge, E.; Francis, F. Could new information influence attitudes to foods supplemented with edible insects? *Br. Food J.* **2017**, *119*, 2027–2039. [CrossRef]

36. Spence, C. On the psychological impact of food colour. *Flavour* **2015**, *4*, 21. [CrossRef]

37. Byrne, B.M. *Structural Equation Modeling with AMOS: Basic Concepts, Applications and Programming*; Routledge/Taylor & Francis Group: New York, NY, USA, 2010.

38. Hu, L.; Bentler, P.M. Cutoff criteria for fit indexes in covariance structure analysis: Conventional criteria versus new alternatives. *Struct. Equ. Model. A Multidiscip. J.* **1999**, *6*, 1–55. [CrossRef]

39. Kline, R.B. *Principles and Practice of Structural Equation Modeling*; Little, T.D., Ed.; The Guilford Press: New York, NY, USA, 2011.

40. Lombardi, A.; Vecchio, R.; Borrello, M.; Caracciolo, F.; Cembalo, L. Willingness to pay for insect-based food: The role of information and carrier. *Food Qual. Prefer.* **2019**, *72*, 177–187. [CrossRef]

41. Lensvelt, E.J.S.; Steenbekkers, L.P.A. Exploring Consumer Acceptance of Entomophagy: A Survey and Experiment in Australia and the Netherlands. *Ecol. Food Nutr.* **2014**, *53*, 543–561. [CrossRef]

42. Berger, S.; Bärtsch, C.; Schmidt, C.; Christandl, F.; Wyss, A.M. When Utilitarian Claims Backfire: Advertising Content and the Uptake of Insects as Food. *Front. Nutr.* **2018**, *5*, 88. [CrossRef]

43. Shelomi, M. Why we still don't eat insects: Assessing entomophagy promotion through a diffusion of innovations framework. *Trends Food Sci. Technol.* **2015**, *45*, 311–318. [CrossRef]

44. Sogari, G. Entomophagy and Italian consumers: An exploratory analysis. *Prog. Nutr.* **2015**, *17*, 311–316.

45. Balzan, S.; Fasolato, L.; Maniero, S.; Novelli, E. Edible insects and young adults in a north-east Italian city an exploratory study. *Br. Food J.* **2016**, *118*, 318–326. [CrossRef]

46. Tan, H.S.G.; Fischer, A.R.H.; Tinchan, P.; Stieger, M.; Steenbekkers, L.P.A.; van Trijp, H.C.M. Insects as food: Exploring cultural exposure and individual experience as determinants of acceptance. *Food Qual. Prefer.* **2015**, *42*, 78–89. [CrossRef]

47. Tan, H.S.G.; Verbaan, Y.T.; Stieger, M. How will better products improve the sensory-liking and willingness to buy insect-based foods? *Food Res. Int.* **2017**, *92*, 95–105. [CrossRef]

48. Schouteten, J.J.; De Steur, H.; De Pelsmaeker, S.; Lagast, S.; Juvinal, J.G.; De Bourdeaudhuij, I.; Verbeke, W.; Gellynck, X. Emotional and sensory profiling of insect-, plant- and meat-based burgers under blind, expected and informed conditions. *Food Qual. Prefer.* **2016**, *52*, 27–31. [CrossRef]

49. Sogari, G.; Menozzi, D.; Mora, C. Sensory-liking expectations and perceptions of processed and unprocessed insect products. *Int. J. Food Syst. Dyn.* **2018**, *9*, 314–320.

50. Tao, J.; Davidov-Pardo, G.; Burns-Whitmore, B.; Cullen, E.M.; Li, Y.O. Effects of edible insect ingredients on the physicochemical and sensory properties of extruded rice products. *J. Insects Food Feed* **2017**, *3*, 263–278. [CrossRef]

 *foods*

Article

Product Quality during the Storage of Foods with Insects as an Ingredient: Impact of Particle Size, Antioxidant, Oil Content and Salt Content

Karin Wendin [1,2,*], Lennart Mårtensson [1], Henric Djerf [1] and Maud Langton [3]

[1] Faculty of Natural Sciences, Kristianstad University, SE-291 88 Kristianstad, Sweden; lennart.martensson@hkr.se (L.M.); henric.djerf@hkr.se (H.D.)

[2] Department of Food Science, University of Copenhagen, DK-1958 Frederiksberg C, Denmark

[3] Department of Molecular Sciences, SLU, Swedish University of Agricultural Sciences, SE-750 07 Uppsala, Sweden; maud.langton@slu.se

* Correspondence: karin.wendin@hkr.se

Received: 8 May 2020; Accepted: 13 June 2020; Published: 16 June 2020

Abstract: To increase the acceptability of insects as food in Western culture, it is essential to develop attractive, high-quality food products. Higher acceptability of insect-based food has been shown if the insects are "invisible". Mincing or chopping the insect material could be a first processing step to reduce the visibility of the insects. In this work, we processed yellow mealworms by using traditional food techniques: chopping, mixing and heat treatment in a retort. The results show that all factors in the experimental design (particle size, oil content, salt content and antioxidant) influenced the products to a larger extent than the storage time. The results, measured by sensory analysis, TBAR values (Thiobarbituric acid reactive substances), colourimetry and viscosity, show clearly that the food products packaged in TRC (Tetra recart cartons) 200 packages and processed in a retort stayed stable during a storage time of 6 months at room temperature.

Keywords: yellow mealworm; *Tenebrio molitor*; insects; sensory; processed; model system; shelf life

---

1. Introduction

Globally, insects are popular as human food in many countries [1,2]. Edible insects, such as yellow mealworms, may be a sustainable source of protein and fat since they are efficient feed converters with a high nutritional value [1,2]. Yellow mealworms have been shown to contain many of the essential amino acids, such as histidine, isoleucine, leucine, lysine, methionine + cysteine, phenylalanine + tyrosine, threonine and valine [3]. There are numerous studies on insects as human food, however very few focus on their taste and flavour or show their potential to become tasty foods [4].

The development of attractive, high-quality food products is essential if the consumption of insect-based food products is to increase in Western culture. Tan et al. [5] pointed out that insect-based food products are unlikely to be accepted if they do not meet the consumers' expectations. The use of "insect-flour" containing non-visible insects had a significantly higher acceptance than the use of whole and visible insects in studies in both Sweden [6] and Denmark [7]. Mincing or chopping the insect material could be a first processing step to reduce the visibility of the insects. Another way of reducing visibility is to use insects as an ingredient; for example, Azzollini et al. [8] added ground yellow mealworm larvae blended with wheat flour to produce extruded cereal snacks. Extraction of the insect proteins may also be an alternative to the use of whole insects [9].

There is a need to understand the mechanisms affecting the sensory properties and oxidation stability of insects when used as food. The primary attributes of the sensory quality of a food product depend on the product type and may be due to appearance (colour, size, shape), odour, flavour, taste and texture [10]. In a previous study, it was shown that the salt content and particle size had a high impact on the sensory attributes of mealworm samples in a model food system [11]. Salt level and particle size are therefore important factors to evaluate in the development of food products based on insects. Salt usually affects protein aggregation, solubility and rheological properties [12]. Soy protein isolate was found to form a gel upon heating, which is common for many proteins. Chen et al. [12] also found that adding salt increased the storage modulus of soy protein isolate gels. However, in a previous study where extracted mealworm protein gels were analysed, the addition of 1.0% NaCl was found to affect the storage modulus G′, which decreased. This was explained by the change in solubility [3]. In contrast to many other proteins, the storage modulus G′ decreased during heating in extracted mealworm protein [3]. Thus, mealworm protein could exhibit different behaviours compared to many other proteins. The addition of an antioxidant agent, extract of rosemary, seemed to affect the colour, rancidity and separation; however, further studies are needed [11]. For this purpose, sensory characteristics can by evaluated by a trained sensory panel and results can then be related to instrumental analysis of viscosity and colour. Rancidity is often measured using measurements of Thiobarbituric acid reactive substances (TBAR), i.e., the degradation products of lipids, which can be detected through the use of a reagent consisting of thiobarbituric acid [13,14] Another common way to measure TBAR values is through the use of a commercially available Lipid Peroxidation (Malondialdehyde (MDA)) Assay Kit [15].

Food products need to be processed and packaged in order to maintain sensory characteristics and safety, and to increase the shelf life of the food. Dobermann [16] suggests that clear processing and storage methodologies should be established for insects as food. Heat treatment is often used to prevent microbial growth and to produce safe products. A retort is commonly used for the preservation of many food products [17]. The packaging is also of importance and must include barriers to oxygen and light to prevent the development of rancidity during storage [18].

Product browning occurs in minced mealworm, possibly due to enzymes, according to Tonneijck-Srpová et al. [19]. They showed that inactivation of the enzymes responsible for browning could be achieved by blanching and/or high-pressure processing at 400 and 500 MPa. Nevertheless, this also resulted in a decrease in the texturizing properties of the minced mealworms, such as lower values for fracture stress and Young's modulus, particularly when blanching was used.

In this study, a food model system was produced according to an experimental design [20]. In order to create model food products similar to real food, oil, water, salt and antioxidant were added to finely grated or coarsely chopped mealworms. These additions can affect the mealworm products during storage due to stability problems that may occur. The model products can be compared to food products such as paté or purée, or Swedish meatballs. Patés typically have a fat content around 25% but may vary between 10% and 30% [21] and Swedish meatballs also vary between 10% and 30%, with commercial products containing 17% [22]. We used sunflower oil in the model products. Both yellow mealworms and sunflower oil contain unsaturated fatty acids, which make the products susceptible to oxidation and a rancid off-flavour can therefore occur during storage. Rosemary extract was used as an antioxidant and combined with packaging with highly protective barriers, which was used to prevent oxidation and increase the shelf life of the model products based on insects.

2. Aim

Food model products based on yellow mealworms were put into packages and processed in a retort to evaluate the impact of particle size, oil/water ratio, salt content and amount of antioxidant on the sensory properties, viscosity, rancidity and colour during storage, compared to freshly produced products.

3. Materials and Methods

3.1. Material

The fresh Yellow Mealworms (*Tenebrio molitor*) were reared on a small scale in Sweden. The breeding of the mealworms took place in plastic boxes, about 300 mm × 300 mm × 100 mm in size and kept at an ambient temperature (approximately 22 °C). Humidity was approximately 50%. The time from egg to mealworm was 12 weeks. The mealworms were fed on oat bran and carrot. Sunflower oil (Farm, Swedish Fine Rice and Food AB, Huddinge, Sweden), salt (NaCl, Nordfalks, Sweden) and an antioxidant based on rosemary (Duralox Oxidation Management Blend NM-45, HT, NS, product code: 62-103-03) were also used. The products were packaged in 200 mL Tetra Recart cartons (TRC 200). The nutrient composition was analysed by ALS Scandinavia AB, Danderyd, Sweden (Table 1).

Table 1. Chemical analysis of raw material, *Tenebrio molitor*, analysed by ALS Scandinavia AB.

Parameter	Mean Value (±) Standard Deviation
TS at 105 °C (%) [1]	36.6 ± 2.0
As (mg/kg) [2]	<0.006
Cd (mg/kg) [2]	0.0329 ± 0.0062
Pb (mg/kg) [2]	<0.01
Hg (mg/kg) [2]	<0.006
Fe (mg/kg) [2]	9.61 ± 2.09
Zn (mg/kg) [2]	33.8 ± 7.1
Energy (kJ/100 g) [3]	679 ± 48
Energy (kcal/100 g) [3]	162 ± 11
Fat (g/100 g) [3]	10.0 ± 0.50
Saturated fat (g/100 g) [3]	2.58 ± 0.77
Monounsaturated fatty acids (g/100 g) [3]	3.79 ± 1.14
Polyunsaturated fatty acids (g/100 g) [3]	3.17 ± 0.95
Carbohydrate (g/100 g) [3]	1.98 ± 0.14
Fibre (g/100 g) [3]	0.676 ± 0.135
Protein (g/100 g) [3]	15.8 ± 0.793
Salt (g/100 g) [3]	0.0892 ± 0.02
Water (g/100 g) [3]	70.5 ± 0.70
Myristic acid (C14:0), (g/100 g fat) [4]	2.95 ± 0.88
Palmitic acid (C16:0), (g/100 g fat) [4]	18.8 ± 5.64
Steric acid (18:0), (g/100 g fat) [4]	3.32 ± 1.00
Oleic acid (C18:1n9c), (g/100 g fat) [4]	36.2 ± 10.9
Linoleic acid (C18:2n6c), (g/100 g fat) [4]	30.1 ± 9.02
Linolenic acid (C18:3n3), (g/100 g fat) [4]	1.61 ± 0.48
Omega 3 fatty acids, total fat (g/100 g fat) [4]	1.61 ± 0.48
Omega 6 fatty acids, total fat (g/100 g fat) [4]	30.1 ± 9.02
Omega 3 fatty acids, total (g/100 g) [4]	0.16 ± 0.05
Omega 6 fatty acids, total (g/100 g) [4]	3.01 ± 0.90

[1] Dry matter according to SS028113 (Water content by gravimetric method at 105 °C. Ash content after 550 °C). [2] Analysis with inductively coupled plasma sector field mass spectrometry (ICP-SFMS) according to standards: SS EN ISO 17294-1, 2 (mod) and EPA- method 200.8 (mod). [3] Fat content was analysed by using nuclear magnetic resonance (NMR). [4] The composition of fatty acids was analysed by using gas chromatography–flame ionization detection (GC-FID).

The composition of the mealworm is such that it could provide a useful contribution to the intakes of zinc and iron, especially zinc, since the zinc content of mealworm is high at 34 mg/kg. Thus, the intake of this type of food could be a useful addition to the daily intake. The recommended daily intakes of zinc and iron are both around 10 mg, depending on age and gender [23,24]. The product also contains omega 3 and omega 6 fatty acids, with high levels of omega 6, and therefore could contribute to the intake of this essential fatty acid. The level of cadmium is low (Table 1) and therefore only a high consumption of the product would exceed the toxic level.

3.2. Experimental Design

Four different samples were produced according to a factorial design (Figure 1). The recipes for each of the samples were the following:

1. 4 kg meal worm, 1 kg oil, 3 kg water, 80 g salt, 0.36 g rosemary extract, resulting in an oil content around 12.5%.

2. 4 kg meal worm, 3 kg oil, 1 kg water, 80 g salt, 11.2 g rosemary extract, resulting in an oil content around 37.5%.

3. 4 kg mealworm, 1 kg water, 3 kg oil, 20 g salt, 0.36 g rosemary extract, resulting in an oil content around 37.5%.

4. 4 kg mealworm, 3 kg water, 1 kg oil, 20 g salt, 11.2 g rosemary extract, resulting in an oil content around 12.5%.

Sample No	Meal Worm	o/w ratio (%)	Salt (%)	Rosemary (%)
1	Finely grated (≤1 mm)	25/75	1	0.0045
2	Finely grated (≤1 mm)	75/25	1	0.14
3	Coarsely chopped (≤5 mm)	75/25	0.25	0.0045
4	Coarsely chopped (≤5 mm)	25/75	0.25	0.14

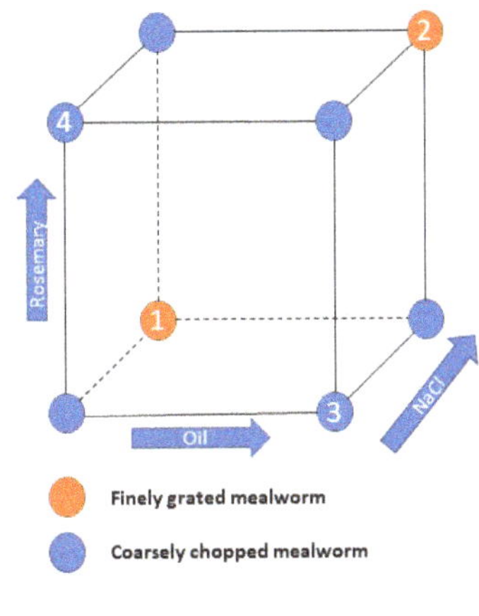

Figure 1. Illustration of the experimental design. To the **left** an overview of the design and to the **right** an illustration of the design oil content ~12.5% or ~37.5%, NaCl content ~0.25% or ~1%, and rosemary content ~0.0045% or ~0.14%.

3.3. Production

A total of 16 kg of mealworms were provided, allowing a maximum of 45 packages of each recipe to be prepared. The manufacturing process for the products will now be described.

Frozen mealworms were blanched for 2–3 min and then chopped by hand or grated using a kitchen blender before the packages were filled. The ingredients for the liquid were mixed and then added to the chopped or grated mealworms. Each package was filled to the top manually before sealing.

A standard sterilisation process was selected for the heat treatment of the products at 122 °C while monitoring the F-value of the product. The four samples were processed in a JBT Retort: AR092-T JBT Mobile Test Unit (Belgium). Production, packaging and processing took place at Tetra Pak, Lund Sweden.

The four different products were run 2 by 2 in the retort, thus allowing all products to be processed on the same day. Samples with a similar oil content were put into the same retort since differences had been found when monitoring product-specific F-values during a pre-trial. Recipes 1 and 4 were therefore run together, as were recipes 2 and 3. Data from the heating process are presented in Table 2. Each package was labelled with a date of withdrawal for shelf life analysis, which was 6 months after production day. All sample batches were tested using microbiological analysis: 5 packages incubated at 30 °C for 7 days and 4 packages incubated at 55 °C for 7 days. The food products were evaluated on Plate Count Agar after package incubation. Results for all samples were <1 Colony Forming Unit (CFU)/10 microlitre. After microbiological approval, the samples were released for use in the project.

Table 2. Retort process.

Process Data	Sample 1	Sample 4	Sample 2	Sample 3
Come-Up Time (min)	21		21	
Holding Time (min)	48		55	
Cooling Time (min)	49		52	
Total Process Time (min)	118		128	
F-value before cooling *	20	49	30	30
F-value Total *	31	56	40	40
Initial Temperature (°C)	23		23	
End Temperature (°C)	30		30	

* The F-value or the thermal death time is used for comparing heat sterilization procedures. It represents the total time–temperature combination received by the food, thus the time required to achieve a specified reduction in microbial numbers at a given temperature [25].

3.4. Storage

Packaged samples were stored at an ambient temperature of approximately 21 °C. The samples were retrieved 7 times, first directly after preparation and then each month for 6 months, and then put into a freezer (Vestfrost solutions VTS 098, Esbjerg, Denmark) and kept at −80 °C until analysis. All samples were thawed before analysis and analysed at the same time. The samples were labelled as shown in Table 3.

Table 3. Sample code for all four samples during 6 months of storage.

Storage	Sample 1	Sample 2	Sample 3	Sample 4
Fresh	181106-1	181106-2	181106-3	181106-4
After 1 month	181206-1	181206-2	181206-3	181206-4
After 2 months	190107-1	190107-2	190107-3	190107-4
After 3 months	190206-1	190206-2	190206-3	190206-4
After 4 months	190306-1	190306-2	190306-3	190306-4
After 5 months	190405-1	190405-2	190405-3	190405-4
After 6 months	190506-1	190506-2	190506-3	190506-4

3.5. Sensory Analysis

The sensory panel was selected and trained according to ISO (International Standard Organisation) standard 8586-2:2008 [26]. The panel consisted of eight assessors who were trained to perform a quantitative descriptive analysis. During training, reference samples were presented, and their sensory attributes were defined (Table 4). The panel was further trained in how to rate attribute intensities on a numerical scale from 0 to 100. The assessments were then performed in triplicate, in a randomised order. The panel was instructed to use water and neutral wafers to cleanse the palate and neutralise the senses. The assessors were informed not to swallow the samples but to spit them out after assessment. The software Eye Question, Netherlands, was used for sensory data collection. Each assessor signed up for participation after being informed about the products and the terms of participation: voluntary participation, freedom to leave the test without giving a reason and the right to decline to answer specific questions.

Table 4. Sensory attributes and their definitions.

ATTRIBUTE	ABBREVIATION	DEFINITION
Appearance		
Roughness	Aroughness	Level of roughness of mealworm
Yellow	Ayellow	Intensity of yellow colour
Brownish-grey	Abrownishgrey	Intensity of brown-grey colour
Glossy	Aglossy	Intensity of glossiness
Odour		
Oat	Ooat	Odour of oats
Chickpea	Ochickpea	Odour of boiled chickpea
Roasted-Sesame-Seed	Oroastedsesameseed	Odour of roasted sesame seed
Popcorn	Opopcorn	Odour of fresh popcorn
Taste/Flavour		
Sweetness	Tsweetness	Basic taste sweet
Sourness	Tsourness	Basic taste sour
Saltiness	Tsaltiness	Basic taste salt
Umami	Tumami	Basic taste umami
Roasted Sesame Seed	Froastedsesameseed	Flavour of roasted sesame seed
Popcorn	Fpopcorn	Flavour of fresh popcorn
Texture (in mouth)		
Oiliness	Teoiliness	Fatty, oily sensation
Grittiness	Tegrittiness	Gritty perception from peel/shell

3.6. Instrumental Measurements

Instrumental measurements (colour, viscosity and rancidity) were performed on all 28 samples (four samples at 7 different time points) plus replicates, see Section 3.4.

Colour measurements were performed using a spectrophotometer (Konica Minolta CM-700d) by measuring L*-, a*-, and b*-values, where L* corresponds to the lightness, a* to the red-green scale (red is positive and green negative) and b* to the yellow-blue scale (yellow is positive and blue is negative) [27].

Rheology is the study of the deformation and flow of materials and can describe the properties of materials, ranging from liquids to elastic solids. One common rheological method is the measurement of viscosity. In this study, the viscosities of the samples were measured using a rotational viscometer (Brookfield DVII+, AMETEK Brookfield Inc., Middleborough, MA, USA) equipped with an S1 spider.

The measurements of rancidity were performed using Thiobarbituric acid reactive substances (TBAR). For these analyses, a Lipid Peroxidation (MDA, Malonaldehyde) Assay Kit (catalogue number MAK085, Sigma-Aldrich, St. Louis, MO, USA) was used for colorimetric assay at λ = 532 nm. The frozen samples stored at −80 °C were thawed before analysis. Since the samples were a mixture of shell parts and soft parts, a stainless-steel garlic press was used to obtain a more homogeneous material. 1.2 mg of the material was weighed into a microcentrifuge vial (three replicates per sample) and centrifuged at 16,000 g for 30 min. 10 μL of the supernatant was transferred into a microcentrifuge vial and gently mixed with 500 μL of 12 mM sulfuric acid. 125 μL of phosphotungstic acid solution was then added and mixed by vortexing. The vial was then incubated at room temperature for 5 min before centrifuging at 13,000 g for 3 min. The pellet was resuspended on ice with water/Butylated hydroxy toluene (100 μL/2 μL) solution and the volume was adjusted to 200 μL with water. A 0.1 M MDA-solution was prepared for the standard calibration curve. This was further diluted into a 0.2 mM MDA standard solution and 0, 2, 4, 6, 8 and 10 μL of the 0.2 mM MDA standard solution were added to separate microcentrifuge tubes. The volume in each vial was adjusted by the addition of water to 200 μL. This generated 0 (blank), 4, 8, 12, 16 and 20 nmole standards. To form the MBA-TBA adduct, 600 μL of a TBA solution was added to each vial containing standard and sample followed by incubation at 95 °C for 60 min. After this, the vials were cooled in an ice bath for 10 min. 200 μL from each vial was pipetted into 96-well plates for analysis in a plate reader for colorimetric assay at λ = 532 nm.

3.7. Statistical Analyses

Data were analysed by calculating mean values and standard deviations. The data were further subjected to two-way analysis of variance (ANOVA) with samples and measurements as fixed effects. Significant differences ($p < 0.05$) between samples were calculated via the Tukey's Post Hoc pairwise comparison test. Results were correlated by Pearson correlation. Regression analysis was performed where the design variables were independent and the measurements were dependent factors (IBM SPSS, version 26). Finally, principal component analysis (PCA; Panel Check V 1.4.2, Nofima, Norway) was performed to give an overview of the results.

4. Results

All the insect food samples were analysed during the same week. Figure 2 shows the sensory attributes of (a) fresh samples and (b) after 6 months of storage. In both fresh and stored samples, the particle size in the attribute "appearance of roughness" was detected, where samples 3 and 4 were perceived as having a statistically significantly higher roughness and also contained the coarsely chopped mealworms. The two samples with higher salt content, samples 1 and 2, were also perceived as significantly saltier. Samples 2 and 3 were perceived as having a significantly higher degree of oiliness, which became more apparent after 6 months of storage. These two samples had a higher oil content. During storage, the odours and flavours of popcorn and roasted sesame seed decreased significantly in all four samples. The decrease was largest at the beginning of the storage.

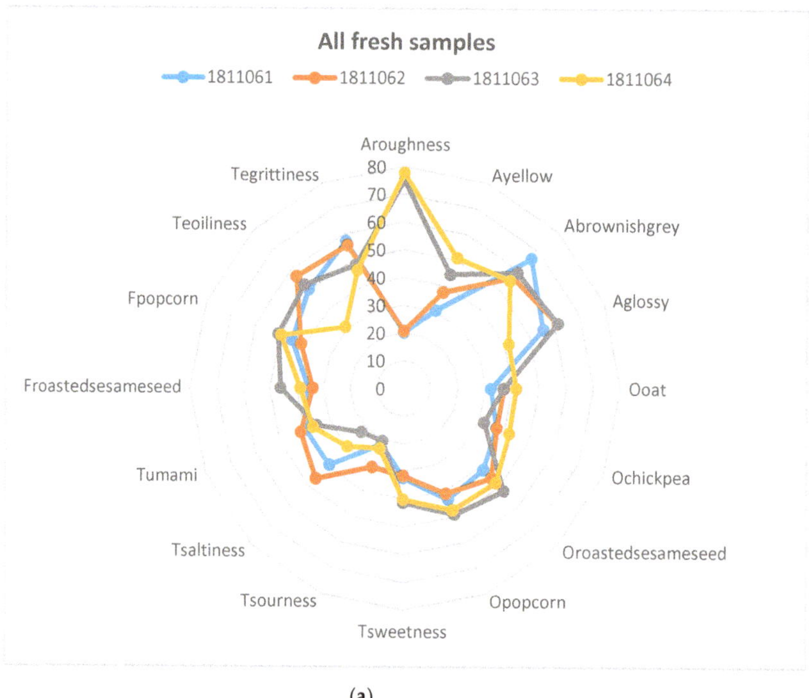

(a)

Figure 2. *Cont.*

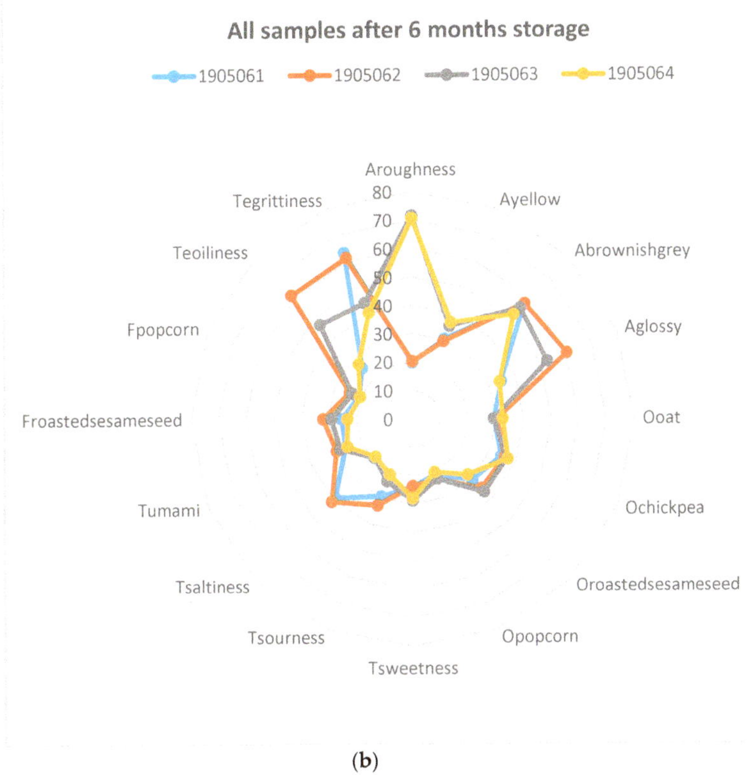

(**b**)

Figure 2. Spider diagram of sensory attributes of (**a**) fresh samples and (**b**) after 6 months of storage.

Figure 3 shows the sensory attributes of the four samples during storage. The same pattern was almost maintained during the 6 months of storage. Figure 3 also shows similar patterns for samples 1 and 2 (with high values for glossy, brownish-grey and grittiness) that were also different from samples 3 and 4 (with high values for roughness).

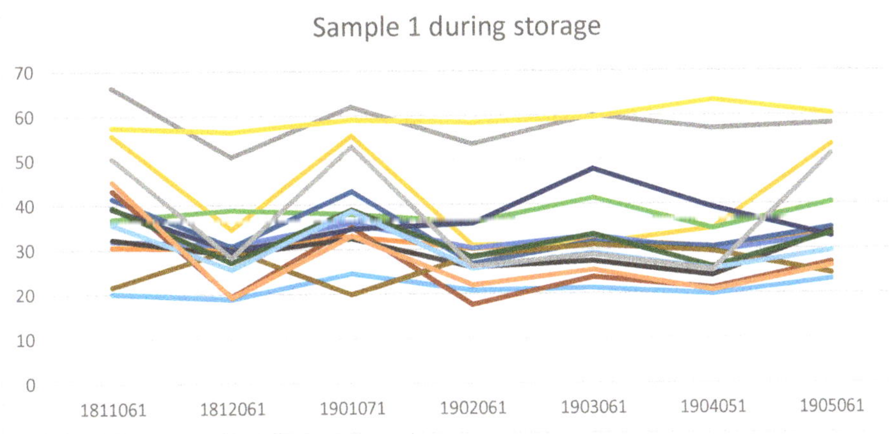

Figure 3. *Cont.*

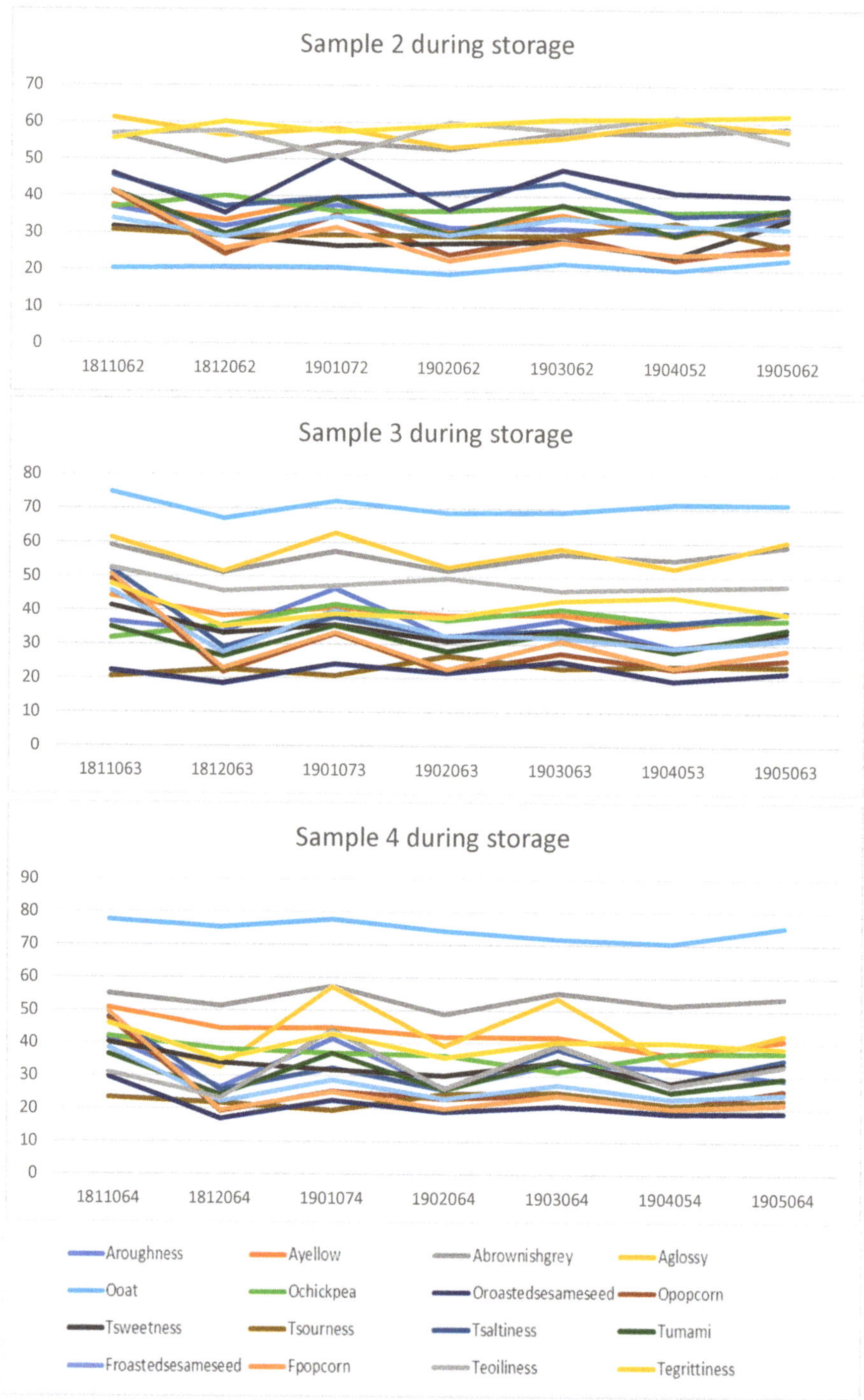

Figure 3. The perceived sensory attributes of all four samples when fresh and after 1, 2, 3, 4, 5 and 6 months of storage.

Table 5 shows the mean values for viscosity, colour and TBAR for the samples when fresh and during six months of storage. The values of viscosity can be seen as an indication of the samples being very inhomogeneous. There were also very few significant differences. Colour measured by L-value seems higher for smaller particles (samples 1 and 2) in Table 5, and all samples differed significantly at the sample level. The difference between particle size was significant for b*-values. The yellow appearance was significantly lower for two samples (1 and 2; Figure 2). Table 5 also shows that there was very little change during storage for up to 6 months in the packages at room temperature, which is supported by the low change in TBAR values over time.

Table 5. Mean values and standard deviations of viscosity, colour (L*, a* and b*) and TBAR.

Sample	Viscosity (mPas)	L*	Colour a*	b*	TBAR µg/g
181106-1	131.2 ± 25.7	51.1 ± 0.4	4.1 ± 0.1	12.6 ± 0.2	5.08 ± 0.97
181206-1	88.0 ± 15.4	52.3 ± 0.8	3.9 ± 0.1	12.5 ± 0.4	6.81 ± 1.41
190107-1	56.8 ± 17.0	52.8 ± 1.3	3.8 ± 0.1	11.6 ± 0.3	6.13 ± 3.08
190206-1	226.4 ± 26.8	51.7 ± 0.3	4.3 ± 0.2	13.1 ± 0.9	6.29 ± 1.80
190306-1	359.2 ± 73.8	51.4 ± 1.1	4.1 ± 0.4	12.8 ± 0.7	7.53 ± 4.18
190405-1	27.2 ± 2.8	54.6 ± 0.3	3.8 ± 0.1	12.0 ± 0.2	6.25 ± 0.79
190506-1	108.0 ± 2.4	53.0 ± 0.4	3.8 ± 0.2	12.3 ± 0.5	6.17 ± 0.42
181106-2	531.2 ± 86.2	48.8 ± 0.5	4.7 ± 0.1	13.2 ± 0.5	3.48 ± 0.60
181206-2	453.6 ± 66.9	48.4 ± 0.2	4.7 ± 0.3	13.1 ± 0.7	5.04 ± 0.32
190107-2	429.6 ± 96.6	47.5 ± 1.0	4.7 ± 0.1	12.3 ± 0.5	4.12 ± 0.90
190206-2	479.2 ± 70.2	47.9 ± 0.6	5.4 ± 0.1	14.3 ± 0.5	3.44 ± 0.57
190306-2	392.8 ± 61.7	46.5 ± 0.2	4.5 ± 0.1	12.0 ± 0.4	3.96 ± 1.16
190405-2	403.2 ± 55.9	47.3 ± 0.1	4.9 ± 0.3	12.9 ± 0.7	3.48 ± 1.25
190506-2	435.2 ± 50.7	47.3 ± 0.5	4.6 ± 0.2	12.5 ± 0.4	4.40 ± 1.20
181106-3	468.0 ± 36.3	42.2 ± 1.1	5.2 ± 0.2	13.8 ± 1.0	2.32 ± 0.30
181206-3	560.8 ± 58.8	39.5 ± 1.9	4.3 ± 0.1	9.6 ± 1.0	3.20 ± 0.84
190107-3	562.4 ± 135.2	42.2 ± 0.5	5.1 ± 0.1	13.2 ± 0.5	3.80 ± 0.90
190206-3	407.2 ± 86.6	40.4 ± 0.1	5.3 ± 0.2	12.2 ± 0.3	3.44 ± 0.54
190306-3	567.2 ± 94.8	41.4 ± 0.1	4.4 ± 0.6	10.9 ± 1.3	6.09 ± 1.32
190405-3	552.8 ± 164.2	40.0 ± 1.7	4.7 ± 0.3	12.0 ± 1.6	4.88 ± 0.85
190506-3	504.0 ± 195.6	40.9 ± 0.2	4.7 ± 0.4	11.2 ± 0.7	4.60 ± 0.18
181106-4	558.4 ± 98.8	44.7 ± 1.4	4.4 ± 0.2	12.4 ± 0.9	2.76 ± 0.83
181206-4	585.6 ± 10.5	44.2 ± 0.2	4.2 ± 0.2	12.8 ± 1.4	4.56 ± 0.24
190107-4	557.6 ± 71.2	43.7 ± 0.4	3.9 ± 1.2	11.4 ± 2.2	4.48 ± 0.07
190206-4	111.2 ± 29.6	45.6 ± 1.2	4.4 ± 0.2	12.8 ± 0.4	4.24 ± 0.66
190306-4	370.4 ± 20.7	40.8 ± 0.5	4.7 ± 0.1	12.0 ± 0.6	2.64 ± 1.07
190405-4	607.2 ± 245.4	40.7 ± 1.6	4.2 ± 0.2	11.1 ± 1.4	3.76 ± 1.00
190506-4	89.6 ± 5.5	45.3 ± 0.4	4.3 ± 0.3	10.8 ± 0.8	4.84 ± 0.28

Figure 4 shows a PCA plot of the mean values over time for the four samples, showing 97.6% of the variance in the resulting data. The four samples are well distributed in each of the four quadrants. Along the first principal component, showing 75.0% of the variance, samples 3 and 4 (coarsely chopped mealworm and low salt content) are plotted close to roughness and opposite samples 1 and 2 (finely grated mealworm and high salt content). The second principal component, representing 22.6% of the variance, was TBAR, with samples 1 and 4 (with low oil content) showing opposite glossy appearance, oiliness and viscosity, which are instead close to samples 2 and 3 (with high oil content). Overall, the results show that the design had a clear impact on the measured parameters.

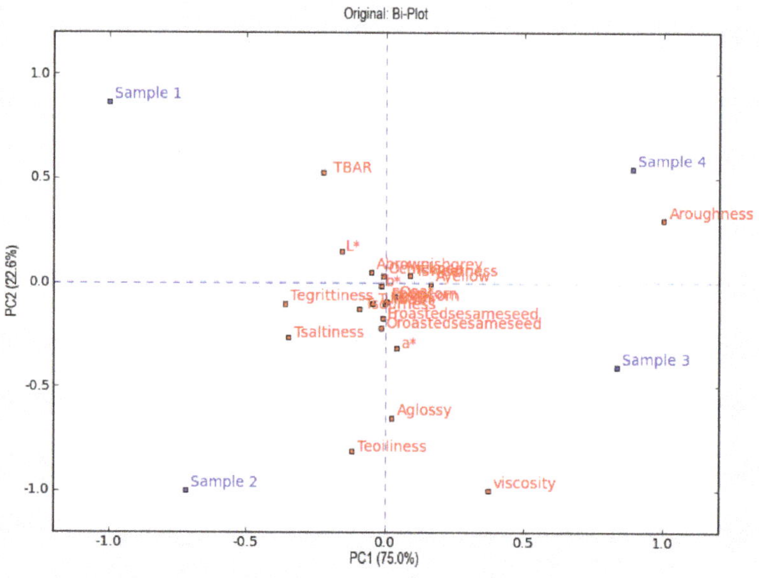

Figure 4. A Principal component analysis (PCA) plot of the four samples.

The results from the PCA are supported by the regression analysis, which showed that all the design parameters had a higher impact than storage time. Oil and salt contents had the largest impact on most attributes, but particle size and rosemary also had significant impacts, mainly on the appearance attributes. The resulting sensorial, rheological, colour and TBAR values were influenced by the design factors to a higher extent than storage time. The only parameters being significantly influenced by the storage time were the odour and flavour attributes of oat, popcorn and roasted sesame seed.

The Pearson correlations also support the PCA results. Table 6 shows that viscosity was negatively correlated to TBAR values. Tsaltiness was positively correlated to Abrownish-grey and Tsourness and negatively correlated to Tsweetness, Ayellow and Aroughness. Further, there was a high correlation between the odour and flavour attributes of popcorn and roasted sesame seed.

Table 6. Significant Pearson correlations between measurements, bold text = |0.7| or larger. Bold text = significant and |0.7| or larger.

	AR	AY	ABG	AG	OO	OC	ORS	OP	TSw	TSo	TSa	TU	FRS	FP	TeO	TeG	Tb	L*	a*	b*	V
AR	x																				
AY	**0.81**	x																			
ABG			x																		
AG			0.39	x																	
OO		0.42		0.49	x																
OC						x															
ORS			0.50	0.63	0.48		x														
OP		0.38	0.54	0.47	0.57		**0.86**	x													
TSw	0.59	**0.72**			0.48		0.47	0.60	x												
TSo	**−0.70**	−0.55							−0.68	x											
Tsa	**−0.88**	−0.52	0.39						−0.47	**0.70**	x										
TU			**0.70**	0.63	0.64		**0.74**	**0.75**			0.50	x									
FRS			0.58	0.66	0.67		**0.85**	**0.84**	0.50			**0.71**	x								
FP			0.56	0.45	0.58		**0.83**	**0.97**	0.61			**0.71**	**0.9**	x							
TeO				**0.90**			0.60					0.53	0.6		x						
TeG	**−0.94**	**−0.71**	0.38						−0.47	0.62	**0.88**					x					
Tb	−0.45	−0.56		−0.44		0.42	−0.43	−0.38	−0.40							0.40	x				
L*	**−0.84**	−0.63							−0.41	0.46	0.69					**0.82**	0.64	x			
a*				0.50			0.39						0.4		0.51		−0.66	−0.50	x		
b*	−0.38									0.39	0.38					0.43				0.44	x
V	0.44	0.46			0.41											−0.39	−0.53	**−0.72**	0.50		x

AR = ARoughness, AY = AYellow, ABG = ABrownishGrey, AG = AGlossy, OO = Ooat, OC = OChickpea, ORS = ORoastedSesameseed, OP = OPopcorn, TSw = TSweet, TSo = TSour, TSa = TSalty, TU = TUmami, FRS = FRoastedSesameseed FP = FPopcorn TeO = TeOiliness, TeG = TeGrittyness, Tb = TBAR, L* = L*, a* = a*, b* = b*, V = Viscosity, Bold text = significant and |0.7| or larger.

5. Discussion

As early as 1975, Meyer-Rochow pointed out that insects could be used as a protein source for humans [28], however even though insects seem to be a sustainable alternative compared to livestock [29], this is not enough for consumers. Food products need to be made with appealing sensory characteristics in order for consumers to buy and eat them; in addition, they must maintain their quality during storage. Sensory determination of shelf life in food products is more or less routine in product development [30]. In this work, we processed yellow mealworms by using traditional food techniques such as chopping, mixing and heat treatment in a retort. The procedures were performed in order to reduce the visibility of the mealworms and to create a model product advised by the results of a focus group in a previous study [6]. The group suggested using ground insects as an ingredient in meat dishes such as sausages and hamburgers. Sausages typically have a fat content between 25% and 30%. Our products contained 12.5% or 37.5% fat, which can be regarded as levels of fat that are commonly used in food products today. The ingredients were included in an experimental design and varied according to input from the focus group. The number of product samples had to be kept low since the total number of samples was large due to the analysis during storage. The study included all 28 samples consisting of the four samples from the experimental design being taken out for analysis at seven different time points during the storage time. A full factorial design would give a larger number of samples from the experimental design [20,31]; however, the number of time points for analysis would then have to decrease.

It was clear from the regression analysis that all the factors in the experimental design influenced the products to a larger extent than the storage time. The results show that the particle size has an impact on the visual appearance of the product. We found that smaller particles, that is finely grated mealworm in samples 1 and 2, had higher values of brownish-grey appearance independent of being fresh or during storage time. However, the samples with larger particles, samples 3 and 4 (coarsely ground mealworm), were perceived as having high values of roughness. The results in this study are, with a few exceptions, in agreement with the results in a previous study [11], where particle size had a large impact on the evaluated parameters, such as brownish-grey appearance. Our results show that the brownish-grey colour was influenced by the particle size and did not change during storage. This is in line with Tonneijick-Srpová et al. [19], who found that browning of minced mealworms could be reduced by blanching in air-tight packaging.

The large change in oil content from 12.5% to 37.5% affected the glossy appearance and the mouthfeel of oiliness. The samples with a higher oil content were perceived to be both more glossy and more oily than the others.

The two samples with higher salt contents were perceived as saltier; here, we cannot detect the effect of particle size as this coincides with the salt levels. Larger liquid volumes in the product could result in a faster release of salt in the mouth, which we were not able to detect in this experiment.

Rosemary level had an impact, mainly on the appearance attributes. This was indicated in a pilot study by Wendin et al. [11], where rosemary seemed to influence both colour and separation. The rosemary extract was added as an antioxidant and was assumed to reduce possible oxidation [32,33]. Nevertheless, no rancidity was perceived by the panel, nor was it detected in the TBAR values. In our previous work, we used the addition of 0%, 0.1% and 0.3% of rosemary extract, where rosemary had a positive impact in preventing rancidity and, as already mentioned, changes in colour and phase separation in samples stored in vacuum-sealed packages in the fridge (5 °C) for 15 days. All four products in this study were very stable during the 6-month storage time, indicating an effective process and well-functioning packaging material.

During storage there was, however, a significant decrease in some of the volatile odour compounds. The sensory panel clearly detected a drop in the odours and flavours of popcorn, oat and roasted sesame seed. This was supported by the regression analysis, which showed that storage time had a significant influence on these sensory attributes. The decrease could probably be related to changes in the chemical composition of volatiles during storage [34].

The results, measured by sensory analysis, TBAR values, colourimetry and viscosity, clearly show that the food products packaged in TRC 200 packages and processed in a retort stayed stable during a storage time of 6 months at room temperature, with the exception of a decrease in the odour and flavour attributes of popcorn and roasted sesame seed.

6. Conclusions

It can be concluded that the measured sensory properties of viscosity, rancidity and colour were impacted mainly by the design parameters of salt and oil content, and to a smaller extent by particle size and the addition of an antioxidant. Storage time of the processed and packaged insect products influenced only the odour and flavour attributes of popcorn and roasted sesame seed. This means that products based on insects are stable during storage when processed in a retort and packaged in TRC packages.

Author Contributions: Conceptualization, K.W. and M.L.; Methodology, K.W., L.M. and H.D.; Validation, M.L.; Formal Analysis, K.W.; Investigation, K.W.; Resources, K.W.; Data Curation, K.W. and M.L.; Writing—Original Draft Preparation, M.L.; Writing—Review and Editing, K.W. and M.L.; Visualization, K.W. and M.L.; Project Administration, K.W.; Funding Acquisition, K.W. All authors have read and agreed to the published version of the manuscript.

Funding: This research was funded by The Knowledge Foundation (KK-stiftelsen) grant number 20170141*. SLU—Swedish University of Agricultural Sciences faculty (to M.L.) is greatly acknowledged for financial support.

Acknowledgments: Thanks to the project group members for discussion, analyses, raw materials, processing, packaging and inspiration: Peter Andersson, Viktoria Olsson, Johan Berg, Karina Birch, Fanny Cedergaardh, Fredrik Davidsson, Sarah Forsberg, Johanna Gerberich, Åsa Josell, Ingemar Jönsson, Susanne Rask, Sofia Stüffe and Per Magnusson.

Conflicts of Interest: The authors declare no conflicts of interest. The funders had no role in the design of the study; in the collection, analyses, or interpretation of data; in the writing of the manuscript, or in the decision to publish the results.

References

1. van Huis, A. Potential of insects as food and feed in assuring food security. *Annu. Rev. Entomol.* **2013**, *58*, 563–583. [CrossRef] [PubMed]
2. VanHuis, A.; Van Itterbeeck, J.; Klunder, H.; Mertens, E.; Halloran, A.; Muir, G.; Vantomme, P. *Edible Insects: Future Prospects for Food and Feed Security (No. 171)*; Food and Agriculture Organization of the United Nations: Rome, Italy, 2013.
3. Zhao, X.; Vázquez-Gutiérrez, J.L.; Johansson, D.P.; Landberg, R.; Langton, M. Yellow mealworm protein for food purposes-extraction and functional properties. *PLoS ONE* **2016**, *11*. [CrossRef] [PubMed]
4. Elhassan, M.; Wendin, K.; Olsson, V.; Langton, M. Quality Aspects of Insects as Food—Nutritional, Sensory, and Related Concepts. *Foods* **2019**, *8*, 95. [CrossRef] [PubMed]
5. Tan, H.S.G.; Fischer, A.R.; Tinchan, P.; Stieger, M.; Steenbekkers, L.P.A.; van Trijp, H.C. Insects as food: Exploring cultural exposure and individual experience as determinants of acceptance. *Food Qual. Prefer.* **2015**, *42*, 78–89. [CrossRef]
6. Wendin, K.; Norman, C.; Forsberg, S.; Langton, M.; Davidsson, F.; Josell, Å.; Prim, M.; Berg, J. Eat'em or not? Insects as a culinary delicacy. In Proceedings of the 10th International Conference on Culinary Arts and Sciences, Copenhagen, Denmark, 6–7 July 2017; Mikkelsen, B., Ofei, K.T., Olsen Tvedebrink, T.D., Quinto Romano, A., Sudzina, F., Eds.; Aalborg University: Copenhagen, Denmark, 2017; pp. 100–106, ISBN978-87-970462-0-3.
7. Videbaek, P.N.; Grunert, K.G. Disgusting or delicious? Examining attitudinal ambivalence towards entomophagy among Danish consumers. *Food Qual. Prefer.* **2020**, *83*, 103913. [CrossRef]
8. Azzollini, D.; Derossi, A.; Fogliano, V.; Lakemond, C.M.M.; Severini, C. Effects of formulation and process conditions on microstructure, texture and digestibility of extruded insect-riched snacks. *Innov. Food Sci. Emerg. Technol.* **2018**, *45*, 344–353. [CrossRef]

9.   Yi, L.; Lakemond, C.M.; Sagis, L.M.; Eisner-Schadler, V.; van Huis, A.; van Boekel, M.A. Extraction and characterisation of protein fractions from five insect species. *Food Chem.* **2013**, *141*, 3341–3348. [CrossRef]

10.   Lawless, H.; Heymann, H. *Sensory Evaluation of Food–Principles and Practices*, 2nd ed.; Springer: New York, NY, USA, 2010.

11.   Wendin, K.; Olsson, V.; Langton, M. Mealworms as Food Ingredient—Sensory Investigation of a Model System. *Foods* **2019**, *8*, 319. [CrossRef]

12.   Chen, N.; Zhao, M.; Chassenieux, C.; Nicolai, T. The effect of adding NaCl on thermal aggregation and gelation of soy protein isolate. *Food Hydrocoll.* **2017**, *70*, 88–95. [CrossRef]

13.   Zeb, A.; Ullah, F. A simple spectrophotometric method for the determination of thiobarbituric acid reactive substances in fried fast foods. *J. Anal. Methods Chem.* **2016**, 1–5. [CrossRef]

14.   Khan, M.A.; Parrish, C.C.; Shahidi, F. Effects of mechanical handling, storage on ice and ascorbic acid treatment on lipid oxidation in cultured Newfoundland blue mussel (*Mytilus edulis*). *Food Chem.* **2006**, *99*, 605–614. [CrossRef]

15.   Morato, P.N.; Rodrigues, J.B.; Moura, C.S.; Silva, F.G.D.; Esmerino, E.A.; Cruz, A.G.; Lollo, P.C.B. Omega-3 enriched chocolate milk: A functional drink to improve health during exhaustive exercise. *J. Funct. Foods* **2015**, *14*, 676–683. [CrossRef]

16.   Dobermann, D.; Swift, J.A.; Field, L.M. Opportunities and hurdles of edible insects for food and feed. *Nutr. Bull.* **2017**, *42*, 293–308. [CrossRef]

17.   Simpson, R.; Jiménez, D.; Almonacid, S.; Nuñez, H.; Pinto, M.; Ramírez, C.; Vega-Castro, O.; Fuentes, L.; Angulo, A. Assessment and outlook of variable retort temperature profiles for the thermal processing of packaged foods: Plant productivity, product quality, and energy consumption. *J. Food Eng.* **2020**, *275*, 109839. [CrossRef]

18.   Kõrge, K.; Laos, K. The influence of different packaging materials and atmospheric conditions on the properties of pork rinds. *J. Appl. Packag. Res.* **2019**, *11*, 1.

19.   Tonneijck-Srpová, L.; Venturini, E.; Humblet-Hua, K.N.P.; Bruins, M.E. Impact of processing on enzymatic browning and texturization of yellow mealworms. *J. Insects Food Feed* **2019**, *5*, 267–276. [CrossRef]

20.   Naes, T.; Nyvold, T.E. Creative design—An efficient tool for product development. *Food Qual. Prefer.* **2004**, *15*, 97–104. [CrossRef]

21.   Teixeira, A.; Almeida, S.; Pereira, E.; Mangachaia, F.; Rodrigues, S. Physicochemical characteristics of sheep and goat pâtés. differences between fat sources and proportions. *Heliyon* **2019**, *5*, e02119. [CrossRef]

22.   Available online: http://orklafoodsrs.se/147585/ (accessed on 3 June 2020).

23.   Available online: https://ods.od.nih.gov/factsheets/Iron-HealthProfessional/ (accessed on 22 April 2020).

24.   Available online: https://ods.od.nih.gov/factsheets/Zinc-HealthProfessional/ (accessed on 22 April 2020).

25.   *Food Processing Technology*; Chapter 12 Heat sterilisation; Elsevier Ltd.: Amsterdam, The Netherlands, 2017. [CrossRef]

26.   ISO Standard 8586-2:2008. *Sensory Analysis-General Guidance for the Selection, Training and Monitoring of Assessors—Part 2: Expert Sensory Assessors*; International Organization for Standardization: Genève, Switzerland, 2008.

27.   CIE. *Technical Report: Colorimetry*; CIE: Washington, DC, USA, 2004.

28.   Meyer-Rochow, V.B. Can insects help to ease the problem of world food shortage. *Search* **1975**, *6*, 261–262.

29.   Rumpold, B.A.; Schlüter, O.K. Potential and challenges of insects as an innovative source for food and feed production. *Innov. Food Sci. Emerg. Technol.* **2013**, *17*, 1–11. [CrossRef]

30.   Freitas, M.A.; Costa, J.C. Shelf life determination using sensory evaluation scores: A general Weibull modeling approach. *Comput. Ind. Eng.* **2006**, *51*, 652–670. [CrossRef]

31.   Box, G.E.P.; Hunter, W.G.; Hunter, J.S. *Statistics for Experimenters, an Introduction to Design, Data Analysis, and Model Building*; John Wiley and Sons: New York, NY, USA, 1986.

32.   Erkan, N.; Ayranci, G.; Ayranci, E. Antioxidant activities of rosemary (*Rosmarinus officinalis* L.) extract, blackseed (*Nigella sativa* L.) essential oil, carnosic acid, rosmarinic acid and sesamol. *Food Chem.* **2008**, *110*, 76–82. [CrossRef] [PubMed]

33. Aruoma, O.I.; Halliwell, B.; Aeschbach, R.; Löligers, J. Antioxidant and pro-oxidant properties of active rosemary constituents: Carnosol and carnosic acid. *Xenobiotica* **1992**, *22*, 257–268. [CrossRef] [PubMed]
34. Jensen, S.; Oestdal, H.; Skibsted, L.H.; Larsen, E.; Thybo, A.K. Chemical changes in wheat pan bread during storage and how it affects the sensory perception of aroma, flavour, and taste. *J. Cereal Sci.* **2011**, *53*, 259–268. [CrossRef]

 *foods*

Article

Traditional Knowledge of the Utilization of Edible Insects in Nagaland, North-East India

Lobeno Mozhui [1,*], L.N. Kakati [1], Patricia Kiewhuo [1] and Sapu Changkija [2]

[1] Department of Zoology, Nagaland University, Lumami, Nagaland 798627, India; lnkakati@nagalanduniversity.ac.in (L.N.K.); patriciakiewhuo16707@gmail.com (P.K.)
[2] Department of Genetics and Plant Breeding, Nagaland University, Medziphema, Nagaland 797106, India; sapu@nagalanduniversity.ac.in
* Correspondence: Lobenommozhui@gmail.com

Received: 2 June 2020; Accepted: 19 June 2020; Published: 30 June 2020

Abstract: Located at the north-eastern part of India, Nagaland is a relatively unexplored area having had only few studies on the faunal diversity, especially concerning insects. Although the practice of entomophagy is widespread in the region, a detailed account regarding the utilization of edible insects is still lacking. The present study documents the existing knowledge of entomophagy in the region, emphasizing the currently most consumed insects in view of their marketing potential as possible future food items. Assessment was done with the help of semi-structured questionnaires, which mentioned a total of 106 insect species representing 32 families and 9 orders that were considered as health foods by the local ethnic groups. While most of the edible insects are consumed boiled, cooked, fried, roasted/toasted, some insects such as *Cossus* sp., larvae and pupae of ants, bees, wasps, and hornets as well as honey, bee comb, bee wax are consumed raw. Certain edible insects are either fully domesticated (e.g., *Antheraea assamensis*, *Apis cerana indica*, and *Samia cynthia ricini*) or semi-domesticated in their natural habitat (e.g., *Vespa mandarinia*, *Vespa soror*, *Vespa tropica tropica*, and *Vespula orbata*), and the potential of commercialization of these insects and some other species as a bio-resource in Nagaland exists.

Keywords: *Antheraea assamensis*; *Apis cerana indica*; entomophagy; food; honey; Nagaland; preparation; *Samia cynthia ricini*; *Vespa mandarinia*; *Vespula orbata*

---

1. Introduction

The world population is estimated to reach 9.8 billion by 2050 with a globally expected increase of 76% in the demand for animal protein [1]. Livestock production worldwide accounts for 70% of the agricultural land and, with the growing demand for meat and the declining availability of agricultural land, there is an urgent need to find additional or alternative sources of protein to provide for the increasing global population. Humans have harvested the eggs, larvae, pupae, and adults of certain insect species from the forest or other suitable habitats for thousands of years as part of regular diets, to stave off famine, to use therapeutically for medicinal and ritual purposes [2], and to try them as novelties [3]. The workshop in Chiang Mai (Thailand) in 2008 on *"Forest insects as food: humans bite back"* [4] emphasized edible insects as a natural food resource. To combat future food insecurity and to develop insects for food and feed, as first suggested by Meyer-Rochow [5], is now considered a viable strategy [6,7]. Insects may have originally been used as a snack or an emergency food item, but even today over 2100 known species of edible insects are still consumed by millions of people of 3071 ethnic groups in 130 countries [8–11]. Entomophagy is advocated as a source to combat future food insecurity mainly because of the insects' abundance, high nutrient composition, high feed conversion efficiency, digestibility, and ease with which they can be bred [6,7,12–14]. Edible insects' nutrient profiles are often very favourable with regard to dietary reference values and daily requirements for normal human growth and health. In marginalized societies, insect consumption has often bridged the gap between the availability and non-availability of conventional food items.

Nagaland is bestowed with natural resources and rich biodiversity supporting various plants and animals including a variety of insects associated with its natural vegetation. While conventional sources of animal-based protein are not always affordable to the rural people and also take a long time to become available, many insects are consumed as an alternative source of protein and as delicacies. The use of insects as food by the Nagas goes way back to those times when they were still called the "Naked people". Insects such as beetles, dragonflies, grubs of white ant, grasshoppers, locusts, crickets, stink bugs, grubs of all sorts of bees and wasps, bee comb, and honey were used as food and played an important role in the diet of the different ethnic groups [15–18]. Most of the studies on entomophagy reported so far from Nagaland [19–22] are preliminary in nature and are restricted to a few insect species available in certain geographical areas of Nagaland. The present study is an extension of an earlier investigation in seven tribal communities of Nagaland by Mozhui et al. [23], providing a more detailed inventory and further documentation of the traditional knowledge of using edible insects as a promising and alternative food source for Nagaland.

2. Materials and Methods

2.1. Study Area

Nagaland is a state located in the north-eastern part of India covering an area of 16,579 km^2. It is situated at 93°20′–95°15′ E and 25°6′–27°4′ N, in the confluence of East Asia, South Asia, and Southeast Asia. Considered one of the biodiversity hotspots (within the Indo-Burma region) of the world, the state has unique geographical location and varied altitudinal range. Out of the total geographical area, 85.43% (14,164 km^2) constitutes the forest cover, of which 5137 km^2 is dense and 9027 km^2 is open forest [24]. Agriculture is the main economy of the state, which includes not only crop raising but all other allied activities such as animal rearing i.e., poultry, horticulture, pisciculture, sericulture, silviculture, livestock i.e., dairy cattle (buffalo, cow, mithun), goats, pigs, etc. Two types of farming systems—jhum or shifting cultivation and terrace or wet cultivation are practiced by the ethnic groups. Jhum cultivation is an extensive method of farming in which the farmers rotate land rather than crops to sustain livelihood [25]. Areas of jhum land are cleared once in five to eight years for better crop production. While in terrace cultivation, the entire hillside is cut into terraces and irrigated by a network of water channels that flows down from one terrace to the other and is easier to maintain than the jhum plots. However, due to the state's wide altitudinal variation, terrace cultivation is found only in some rural pockets and majority of the population are engaged in shifting cultivation. Rice is the dominant crop and the main staple food of the Nagas.

2.2. Data Collection

The documentation is based on a four year field survey from 2014 to 2018 across 53 villages in Nagaland (Figure 1). The methodology followed in this study for data collection and insect identification is similar to that described in detail by Mozhui et al. [23]. Prior to survey, village heads were informed in advance for selection of informants for authentic documentation. Therefore, from each village, 6–8 informants comprising of village heads, traditional knowledge holders, edible insect farmers, edible insect collectors, educated youths, and homemakers were selected for the study. The survey was conducted only after getting ethical clearance from the village heads as well as the informants. Among the 370 informants interviewed, 248 were male and 122 were female, all belonging to the age group 25–104 years (Table 1). The documentation of edible insects in the present study is based on the responses collected from the informants with the help of a semi-structured questionnaire (See Supplementary Material S1). Informants were asked questions on the stages of edible insect consumption, mode of preparation of the insect species, seasonal availability, and preferences for selecting insect species for consumption. All the voucher specimens and photographs of edible insects are deposited at Department of Zoology, Nagaland University, Lumami, Nagaland.

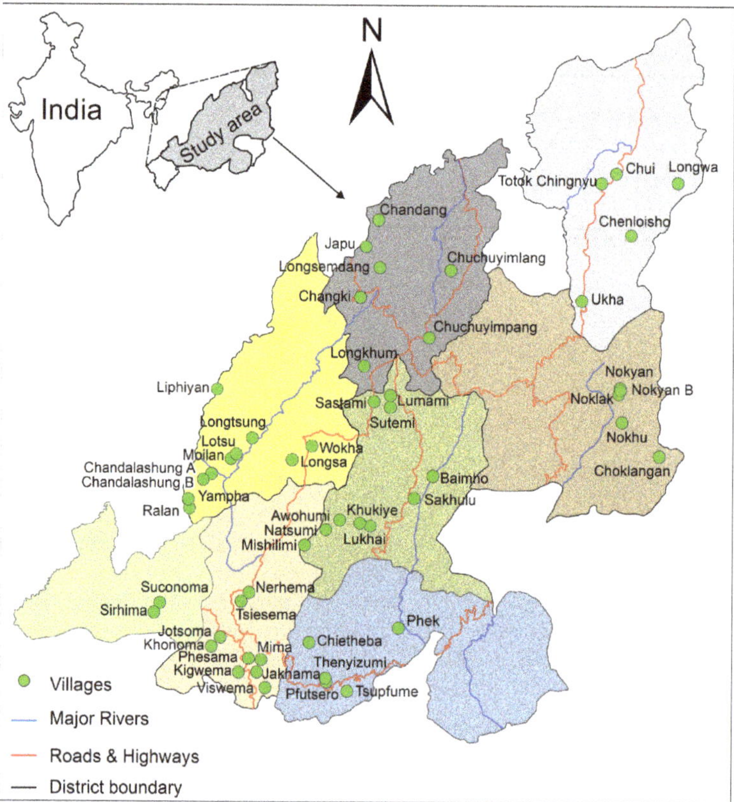

Figure 1. Location map of 53 villages surveyed for the study (Source: Survey of India toposheets).

Table 1. Demographic patterns of informants in the study area.

Gender	Number of Informants
Male	248 (67%)
Female	122 (33%)
Age Group	
25–34	60 (16%)
25–44	58 (16%)
45–54	59 (16%)
55–64	55 (15%)
65–74	57 (15%)
75–84	58 (16%)
85–94	18 (5%)
95–104	5 (1%)
Educational status	
Below high school	230 (62%)
Above high school	140 (38%)
Informant status	
Key informant	198 (54%)
General informant	172 (46%)

3. Results and Discussion

3.1. Edible Insects in Nagaland

The present study contains an inventory of 106 insect species including 82 edible insect species recorded in our earlier work [23] that are regarded as health foods by the ethnic groups in Nagaland (Table 2). Belonging to 32 families and 9 orders, the percentage contribution of 106 edible insect species is: 24% Hymenoptera, 24% Orthoptera, 21% Hemiptera, 13% Coleoptera, 7% Odonata, 7% Lepidoptera, 2% Mantodea, 1% Isoptera, and 1% Diptera (Figure 2). Photographs of certain most frequently abundant and preferred species are depicted in Figure 3.

Table 2. Insect species utilized as food by different ethnic groups in Nagaland (adapted from Mozhui et al., 2017) with new addition of 24 species (*).

Order	Family	Scientific Name	Local Name	Seasonal Availability	Edible Stage	Mode of Consumption
Odonata	Libellulidae	*Diplacodes trivialis* Rambur	Tukhakupu	July–Nov	nymph	B, C
		Orthetrum pruinosum neglectum Rambur	Tukhakupu	July–Nov	nymph	B, C
		Neurothemis fulvia Drury	Tukhakupu	July–Nov	nymph	B, C
		Crocothemis servilia servilia Drury	Tukhakupu	July–Nov	nymph	B, C
		Potamarcha congener Rambur	Tukhakupu	July–Nov	nymph	B, C
		Orthetrum sabina sabina Drury	Tukhakupu	July–Nov	nymph	B, C
		Orthetrum triangulare triangulare Sely	Tukhakupu	July–Nov	nymph	B, C
		Pantala flavescens Fabr.	Tukhakupu	July–Nov	nymph	B, C
Orthoptera	Acrididae	*Acrida exaltata* Walker	Tsütheku	June–Oct	adult	C
		Atractomorpha sp.	Anishe	June–Oct	adult	C
		Choroedocus robustus De Geer	Shenaqhu	June–Oct	adult	C
		Chondracris rosea Serville	Kuprie	June–Oct	adult	RO/TO
		Gastrimargus africanus africanus Saussure	Tluqhu	June–Oct	adult	C
		Hieroglyphus banian Fabr.	Sapathika	June–Oct	adult	C
		Melanoplus sp.	Yeghutukha	June–Oct	adult	C
		Melanoplus bivittatus Say	Shenaqhu	June–Oct	adult	RO
		Oxya hyla Serville	Naghilithika	June–Oct	adult	C
		Oxya fuscovittata Marschall	Naghilithika	June–Oct	adult	C
		Phlaeoba infumata Brunner	Atikha	June–Oct	adult	C
	Gryllidae	*Acheta domesticus* Linn.	Achuqu	Aug–Nov	adult	C
		Gryllus sp.	Ashuko	Sept–Nov	adult	C
		Meloimorpha cincticornis Walker	Petu	Aug–Nov	adult	C
		Teleogryllus sp.	Amishibachuqu	Aug–Nov	adult	C
		Teleogryllus occipitalis Serville	Awusho	Aug–Nov	adult	C
		Tarbinskiellus orientalis Fabr.	Awusho	Aug–Nov	adult	C
		Tarbinskiellus portentosus Lichtenstein	Awusho	Aug–Nov	adult	C
	Gryllotalpidae	*Gryllotalpa orientalis* Burmeister	Alhaqu	Aug–Nov	adult	C
	Tettigoniidae	*Mecopoda nipponensis* Haan	Kaghalapu	June–Oct	adult	RO/TO
		Mecopoda elongata Linn.	Khotsule	June–Oct	adult	RO/TO
		Pseudophyllus titan White	Salhatiti	June–Oct	adult	RO/TO
		Elimaea securigera Brunner	Anishe	June–Oct	adult	C
Orthoptera	Tettigoniidae	*Tettigonia* sp.	Anishe	June–Oct	adult	C
	Schizodactylidae	*Schizodactylus monstrosus* Drury	Awusho	June—Aug	adult	C
Mantodea	Manitidae	*Tenodera sinensis* Saussure	Tsukole	July–Sept	adult	RO
		Hierodula coarctata Saussure	Kupkamichukole	July–Sept	adult	RO
Isoptera	Termitidae	*Macrotermes* sp.	Alho	Nov–Dec	adult	C
Hemiptera	Aetalionidae	*Darthula hardwickii* Gray	Thezü	Sept–Oct	nymph	C, F
	Belostomatidae	*Lethocerus indicus* Lepeltier and Serville	Tsungosho	Nov–Jan	adult	C
	Cicadidae	*Tibicen pruinosa* Say	Akoko	May–Nov	adult	C
		Dundubia intemerata Walker	Akoko	May–Nov	adult	C
		Dundubia oopaga Distant	Akoko	May–Nov	adult	C
		Pomponia sp.	Cievü	May–Nov	adult	C
		Pycna repanda Amyot and Serville	Cievü	May–Nov	adult	C
	Coreidae	*Dalader planiventris* Westwood	Akhane	May–July	adult	C
		Anoplocnemis phasiana Fabr.	Akhane	May–Aug	adult	C
		Notobitus meleagris Fabr.	Akhane	Aug–Sept	adult	C
	Dinidoridae	*Aspongopus nepalensis* Westwood	Akhane	Dec–Jan	adult	C
		Coridius janus Fabr.	Akhane	May–July	adult	C, CH
		Coridius chinensis Dallas	Akhane	May–July	adult	C, CH
		Coridius singhalanus Dist.	Akhane	Feb–Mar	adult	C, CH
	Nepidae	*Laccotrephes ruber* Linn.	Akhane	Oct–Jan	adult	C
	Pentatomidae	*Chinavia hilaris* Say	Akhane	May–July	adult	C
		Cyclopelta siccifolia Westwood	Akhane	May–July	adult	C
		Dolycoris sp.	Akhane	April–May/Nov	adult	C, CH, F
		Erthesina fullo Thunberg	Akhane	Aug–Sept	adult	C
		Eurostus grossipes Dallas	Akhane	May–July	adult	RO

Table 2. *Cont.*

Order	Family	Scientific Name	Local Name	Seasonal Availability	Edible Stage	Mode of Consumption
Hemiptera	Pentatomidae	*Udonga montana Distant	Akhane	April–May/Nov	adult	C, CH, F
	Tessaratomidae	*Tessaratoma javanica* Thunberg	Akhane	May–July	adult	RO
Coleoptera	Cerambycidae	*Anoplophora* sp.	Nukuo	Sept–Jan	all	C, RO
		Batocera rubus Linn.	Akulho	June–Nov	larva, pupa	C
		Batocera parryi Hope	Akulho	June–Nov	larva, pupa	C
		*Batocera rufomaculata De Geer	Akulho	June–Nov	larva, pupa	C
		Orthosoma brunneum Forster	Akulho	July–Nov	all	C, RO
	Chrysomelidae	*Aplosonyx chalybaeus Hope	Akallakulho	Sept–Nov	larva	C
	Curcurlionidae	*Rhynchophorus ferrugineus* Olivier	Akulho	Oct–Nov	all	C
	Dytiscidae	*Cybister limbatus* Fabr.	Azükhanu	July–Nov	adult	C
		Cybister tripunctatus lateralis Fabr.	Azükhanu	July–Nov	adult	C
		Sandracottus manipurensis Vazirani	Dzübolo	July–Nov	adult	C
	Hydropillidae	Hydrophilus caschmirensis Redtenbacher	Azühpüllau	July–Nov	adult	C
	Scarabaeidae	*Xylotrupes gideon* Linn.	Shushu	June–Aug	adult	RO
Coleoptera	Scarabaeidae	*Lepidiota stigma* Fabr.	Kolompvu	Aug–Sept	adult	RO
		Holotricha sp.	Befu	July–Aug	adult	RO
Hymenoptera	Apidae	*Apis cerana indica* Fabr.	Aghui	Whole year	all	R, C
		Apis dorsata dorsata Fabr.	Ati-i	Sept–May	all	R, C
		Apis dorsata laboriosa Smith	Atukhi	Sept–May	all	R, C
		Apis florea Fabr.	Aku-u	Sept–Feb	all	R, C
		Lepidotrigona arcifera Cockerell	Amgho	Sept–Feb	all	R
		Lophotrigona canifrons Smith	Tyita	Sept–Feb	all	R
		Xylocopa violacea Linn.	Khvukhvu	May–Oct	honey	R
	Formicidae	*Oecophylla smaragdina* Fabr.	Satghupu	May–July	all	B, C, CH
	Vespidae	*Parapolybia varia* Fabr.	Pughukhi	May–Sept	all	C, F
		Polistes olivaceus De Geer	Pighikhi	May–Sept	all	C, F
		Polistes stigma Fabr.	Pughukhi	May–Sept	all	C, F
		*Polistes sp.	Pughukhi	May–Sept	all	C, F
		Provespa barthelemyi du Buysson	Akizu	Sept–Feb	all	C, F
		*Ropalida rufoplagiata Cameron	Awopu	Sept–Feb	all	C, F
		*Ropalida sp.	Awopu	Sept–Feb	all	C, F
		Vespa affinis indosinensis Perez	Akhighü	Sept–Feb	all	C, F
		Vespa auraria Smith	Akhibo	Sept–Feb	all	C, F
		Vespa basalis Smith	Akhitsu	May–Nov	all	C, F
		Vespa bicolor Fabr.	Akhitsu	May–Nov	all	C, F
		Vespa ducalis Smith	Angui	May–Nov	all	C, F
		Vespa mandarinia Smith	Akhighü	Sept–Feb	all	C, F
		Vespa tropica tropica Linn.	Akichekhi	May–Nov	all	C, F
		Vespa soror du Buysson	Akhitsü	Sept–Feb	all	C, F
		**Vespula orbata du Buysson	Akhibo	May–Nov	all	C, F
		Vespula sp.	Akhibo	May–Nov	all	C, F
Lepidoptera	Cossidae	*Cossus* sp.	Aphukulho	July–Feb	larva, pupa	B, C
	Crambidae	*Omphisa fuscidentalis* Hampson	Akhaukhulo	Aug–Oct	larva, pupa	C, F, ST
	Hesperiidae	*Erionata torus* Evans	Wuchoninga	Sept–Nov	larva, pupa	B, C
	Lasiocampidae	*Malacosoma* sp.	Michekhane ninga	Nov–Dec	larva, pupa	C
	Noctuidae	*Pericyma cruegeri Butler	Katsulanga	April–May	larva	C, F
	Saturniidae	*Antheraea assamensis Helfer	Muga	Whole year	larva	C
		*Antheraea mylitta Drury	Bokori ra	May–July	larva	C
		Samia cynthia ricini Boisduval	Eri	Whole year	larva, pupa	C, F
Diptera	Tipulidae	*Tipula* sp.	Pochighoh	Sept–Jan	larva	B, C, F

B—boiled; C—cooked; CH—chutney; F—fried; R—raw; RO—roasted; ST—steamed; TO—toasted. * New addition of edible insect in Nagaland. ** New record from India, worker wasp examined, regd. no. ZSI/WGRC/I.R-INV.8972 (P. Girish Kumar and James M. Carpenter, 2018).

3.2. Consumption of Edible Insects

In Nagaland, as elsewhere [26,27] preferences in edible insect consumption are mainly due to six reasons: (a) availability of the insect species, (b) size of the insects, as generally larger insects are preferred for consumption, (c) taboos associated with the insect species, (d) one's own palatability/taste preference, (e) market value of the insect species and, (f) traditional ethno-medicinal knowledge associated with the insect species. With the advancement in technology and availability of western foods, entomophagy practices are discontinued by the locals in many parts of the world [28] in the false belief that they would be more readily accepted as civilized and cultured individuals by representatives of the western world [5]. There are also risks that the cultural and ecological knowledge of entomophagy may get lost while adopting western dietary patterns [7]. However, in Nagaland, even though tribal people are exposed to modern food stuffs, edible insects are still acknowledged and continue to be an important food source providing nutrition and income especially to the rural poor.

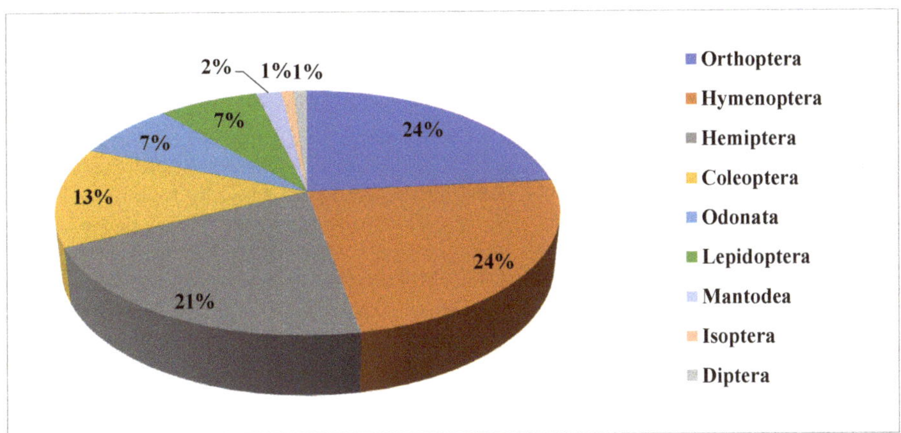

Figure 2. Percentage contribution of edible insects by each insect order.

Figure 3. (**Plate 1**). Certain popular edible insect species of Nagaland. (**a**) Dragon nymphs, (**b**) various aquatic insects, (**c**) katydid *Mecopoda elongata*, (**d**) the sand cricket (*Schizodactylus monstrosus*), (**e**) the dinorid bug *Coridius singhalanus*, (**f**) the red pumpkin bug *Coridius janus*, (**g**) sundried pentatomid bug *Udonga montana*, (**h**) the chrysomelid beetle *Aplosonyx chalybaeus* larvae, and, (**i**) large unidentified wood larvae.

People living in rural areas know very well which species to collect for consumption and this knowledge is acquired by their children, as in other countries with a tradition of consuming insects [29].

It is important to note that edible insects are not only an important source of protein but have ecological advantages over meat [30] and simultaneously aid in maintaining the diversity of habitats for other life forms by sustaining the local environment. The reported benefits of the human consumption of insects as an alternative to conventional food animals are numerous, including comparable levels of protein coupled with lower environmental impact due to lower emissions of greenhouse gases and lower land requirements during production [31,32]. Further, collection of edible insects (those considered as pests) for human consumption has a positive impact on the agricultural crops, being an alternative and efficient biological control method. Thus, consumption coupled with conservation of edible insects can benefit people as well as nature in an area and lead to a sustained use of this important bio-resource [33]. While the use value of different insect species has demonstrated that every Naga tribe has a preference of its own with regard to insect consumption [23], members of the Naga tribes are very specific with the way an insect is prepared for consumption such as boiling in a little amount of water (boiled), cooking with local spices/local ingredients (cooked), frying in hot oil (fried), cooking over fire or over hot charcoal (roasted), as chutney, or eaten raw. Local spices (garlic, ginger), fermented bamboo shoot, dried bamboo shoot, powdered fruits of *Rhus semialata* (Murray) are important local ingredients for preparation. Depending on one's own palatability, the mentioned ingredients are added to enhance the flavour of the insect food, be it as main dish or replacement of conventional meat sources (e.g., beef, chicken, pork). Modes of consumption and preparation of different insects belonging to the nine orders are discussed below.

3.2.1. Order Odonata

Although both adult and nymphal stages of dragonflies are consumed, the nymphs are greatly preferred (Figure 3a). Adult dragonflies are collected in large quantities during the months of July–October, while nymphs are collected mostly during the winter season (December–March). For consumption purpose, the dragon nymphs are prepared by boiling in a little amount of water and cooked until dry, while the adults (wings are removed) are fried and eaten as snacks.

3.2.2. Order Orthoptera

Orthopterans are broadly grouped into (a) grasshopper and katydids, and (b) crickets.

Grasshoppers and Katydids

The Nagas are agriculturalist and look after extensive farmlands, where varieties of locusts and grasshoppers are commonly available. During harvest season (September–October), important edible insect species such as *Melanoplus bivittatus*, *Elimaea securigera*, *Hieroglyphus banian*, *Oxya fuscovittata* (Figure 4a), and *Oxya hyla* are collected in large quantities. The harvested surplus grasshoppers are sundried or smoked and preserved for future use. Grasshoppers are prepared as fried or are cooked with local spices (Figure 4b), while giant katydids (e.g., *Mecopoda nipponensis*, *Mecopoda elongata*, and *Pseudophyllus titan*) are roasted/toasted and eaten as snacks (Figure 4c). Some members of Naga tribe regard *Elimaea securigera* as a "health food" and prepare the insects as cooked or fried to be served as the main dish, replacing conventional meat sources (for e.g., beef, chicken, and pork).

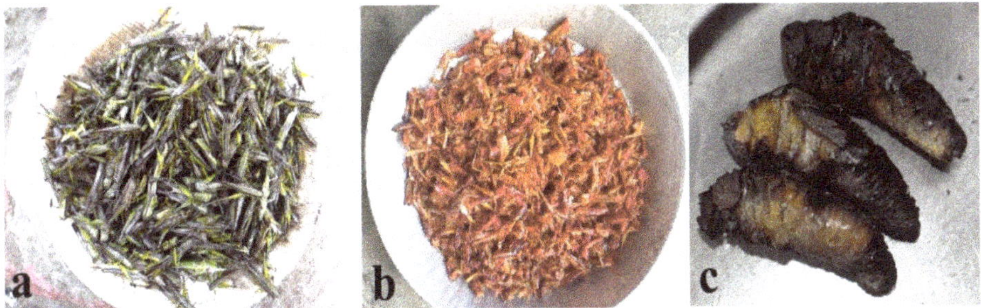

Figure 4. (**Plate 2**). Commonly utilized Orthopterans; (**a**) freshly harvested grasshoppers (*Oxya fuscovittata*), (**b**) grasshoppers cooked with fermented bamboo shoot, (**c**) roasted giant katydid (*Pseudophyllus titan*).

Crickets

Crickets are commonly available from August to November and are generally handpicked or captured by various methods. For instance, field crickets (e.g., *Tarbinskiellus orientalis* and *Tarbinskiellus portentosus*) are dug out after pouring water inside their burrow making them float on the surface. The burrow of the cricket is easily recognisable, as they leave a mound of loose soil at the top of the burrow. Generally, a long stick is put in into the hole so as to keep track of the burrow; after which, the soil is removed slowly and the crickets residing in the burrow are collected. Other crickets such as *Acheta domesticus* and *Teleogryllus occipitalis* are light trapped and collected in large numbers at night. People also spread wheat bran along the paddy fields where crickets are most available. The crickets get attracted to the wheat bran and when they come to feed on it, they are either handpicked or trapped with nets. Crickets are generally preferred in cooked form (with bamboo shoots and local spices). However, some prefer them deep fried or boiled (Figure 5a–c). Besides cooking or frying, the traditional way of preparing crickets such as *Tarbinskiellus orientalis*, *Tarbinskiellus portentosus*, and *Teleogryllus occipitalis* are followed by some members of the Naga tribes. For example, the insects may be properly mixed with salt and dry bamboo shoot, stuffed inside cut bamboo pipes (approximately 30 cm in length) and closed with a banana leaf (the use of banana leaf is to enhance the flavour). The bamboo pipe is placed under the fire, and after 20–30 min the dish prepared is served as the main dish replacing the conventional meat sources.

Figure 5. (**Plate 3**). Preferences in preparation of important cricket species; (**a**) crickets cooked in fresh bamboo shoot, a delightful delicacy, (**b**) fried crickets, a much sought after delicacy, (**c**) boiled cave crickets for consumption.

3.2.3. Order Mantodea and Isoptera

Although, the mantises (*Tenodera sinensis* and *Hierodula coarctata*) are consumed by only some members, the preferred way for consumption is toasting and they are eaten as snacks. Besides toasting, the head and the digestive tract are removed before preparation and then the insects are cooked with local spices until dry. Termite (*Macrotermes* sp. and *Odontotermes* sp.) are popular edible insect species of the Nagas with a high fat content [34] and are collected in plenty during the months of March–May and November–December. While some consumers prefer boiled termites, most prefer cooked or fried termites. When termites are collected in large quantities, they are sun dried to preserve them for longer use.

3.2.4. Order Hemiptera

Stinkbugs

Depending on the species, stink bugs are available all throughout the year. For instance, the most preferred stink bugs such as *Coridius janus*, *Coridius singhalanus*, *Darthula hardwickii*, *Tessaratoma javanica*, and *Udonga montana* are found during the months of May–November. Stink bugs are mostly handpicked during the daytime. The red pumpkin stink bug *Coridius janus* (Fabr.) completes its developmental stages on its host plant (*Cucurbita moschata* Dutch) and is easily handpicked for consumption. They are also available on different bean plants such as string bean: *Phaseolus vulgaris* L.; cowpeas such as *Vigna sinensis* (Savi), *Vigna unguiculata* L.; and Goa bean: *Psophocarpus tetragonolobus* DC, etc.

Most of the stink bugs are preferred in their adult stage; however, the leafhopper (*Darthula hardwickii*) is preferred in its nymphal stage. Preference for consuming only the nymphal stage of *Darthula hardwickii* is mainly because of two reasons: (a) the nymphs are tastier than the adults and, (b) the strong pungent smell of the leafhopper once it gets matured is disliked. While the pentatomid bug *Eurostus grossipes* and the tessaratomid bug *Tessaratoma javanica* are preferred roasted, *Aspongopus nepalensis*, *Coridius janus*, *Coridius chinensis*, and *Coridius singhalanus* are preferred as chutney or cooked for consumption (Figure 6a). With *Udonga montana*, the freshly harvested insects are first boiled (boiling is done to reduce the strong pungent smell), and only then further prepared (with local spices and filtrate of fresh bamboo shoots) as chutney or fried for consumption (Figure 6b). During the month of April and May, *Udonga montana* species are collected in large quantities and people boil the insect and sun dry or smoke them for longer use.

Figure 6. (**Plate 4**). Mode of preparation of various hemipterans: (**a**) roasted stink bugs (*Coridius singhalanus*), (**b**) fried pentatomid bug (*Udonga montana*), (**c**) fried giant water bugs (*Lethocerus indicus*), (**d**) boiled water scorpions (*Laccotrephes ruber*).

Aquatic Hemiptera

Aquatic Hemipterans are commonly available during winter season from October to January. The giant water bug (*Lethocerus indicus*) and water scorpion (*Laccotrephes ruber*) are collected while fishing and are a delicacy among the Naga tribes. Adult *Lethocerus indicus* is popular for its enticing aroma and is consumed boiled, fried, or roasted (Figure 6c). Similarly, adult *Laccotrephes ruber* are cooked with local spices and fermented bamboo shoot and served as the main dish (Figure 6d).

Cicadas

Depending on the species, cicadas are found during the months from May to November. They are usually handpicked or collected with the help of bamboo pole (covered with sticky latex extract) and are trapped on the sticky ends of the bamboo pole [23]. Cicadas (*Tibicen pruinosa, Dundubia intemerata, Dundubia oopaga, Pomponia* sp., and *Pycna repanda*) are equally appreciated by the ethnic groups of Nagaland and are cooked or fried for consumption.

3.2.5. Order Coleoptera

Wood and Timber Larvae

Most of the larval Coleoptera are available for consumption during the period from June to November. All wood or timber larvae are equally appreciated for their food value and are prepared by cooking (with local spices and fermented/dried bamboo shoot) and served as a main dish. While *Batocera rubus, Batocera parryi*, and *Batocera rufomaculata* are consumed only in their larval (Figure 7a,b) and pupal stage, *Anoplophora* sp. and *Orthosoma brunneum* are preferred at all stages (larva, pupa, and adult). Adult *Anoplophora* sp. and *Orthosoma brunneum* are roasted, while their larvae and pupae are cooked for consumption. Corm-borer (*Aplosonyx chalybaeus*) larvae and palm weevil (*Rhynchophorus ferrugineus*) larvae are cooked for consumption.

Figure 7. (Plate 5). Commonly available and consumed Coleopterans; (**a,b**) freshly harvested wood larvae sold at local market, Kohima, Nagaland, (**c**) fried water beetles (*Hydrophilus caschmirensis*).

Aquatic Beetles

Aquatic beetles are available during the months from July to November and are collected during fishing. Predatory diving beetles (e.g., *Cybister limbatus, Cybister tripunctatus lateralis*, and *Sandracottus manipurensis*) and the phytophagous water beetle *Hydrophilus caschmirensis* are considered delicacies among all Naga tribes. For consumption, the elytra and membranous wings of the beetles are removed prior to cooking and then prepared by boiling, cooking, or frying (Figure 7c).

Scarab Beetles

Adult scarab beetles (*Xylotrupes gideon, Lepidiota stigma*, and *Holotricha* sp.) that are found during the months from June to September are consumed by a few tribes and are preferred as roasted/toasted for consumption. Scarab beetles are easily handpicked. For instance, *Lepidiota stigma* is usually found on litchi tree, *Litchi chinensis* Sonn.; the basic mode of collection is to shake the tree branches which makes the beetles fall down into the ground and they are then easily collected.

3.2.6. Order Hymenoptera

Hymenopterans are differentiated as ants, bees, hornets, and wasps. Depending on the species Hymenopterans are available all throughout the year. Ants are mostly available for consumption from May to July; bees, especially *Apis cerana indica* are reared and available in every season, while hornets and wasps are most abundant for consumption from May to February.

Six (6) species of honeybees (Apis cerana indica, Apis dorsata dorsata, Apis dorsata laboriosa, Apis florea, Lepidotrigona arcifera, Lophotrigona canifrons) are commonly used as food in the region. Of all bee species, Apis cerana indica is the most important species as it is commonly available and can be reared or domesticated more easily than the other bee species. All bee stages (larva, pupa, adult) as well as the bees' products are widely used for their food value as well as their medicinal value (Figure 8a,b). Honey is considered to be a strong and effective medical agent for treating cold, conjunctivitis, cough, diarrhoea, and pneumonia.

Figure 8. (**Plate 6**). Widely consumed and marketed Hymenopterans; (**a**) *Apis cerana indica* honey comb, (**b**) honeycomb of *Apis florea*, (**c**) wasp nest of *Vespa soror*, (**d**) large nests of *Vespa tropica tropica*, (**e**) fried wasps (*Provespa barthelemyi*), (**f**) nest of giant hornet (*Vespa mandarinia*), (**g**) healthy hornet grubs, (**h**) freshly harvested *Vespa auraria* by smoking method, and (**i**) worker ants *Oecophylla smaragdina*.

Besides bee and bee products, paper wasps (e.g., *Provespa barthelemyi*) and hornets (e.g., *Vespa soror, Vespa tropica tropica,* and *Vespa mandarinia*) are important edible insects (Figure 8c–h). Preferences in preparation of wasps (larva, pupa, and adult) depend on one's own taste and the preparation is done by frying or cooking (Figure 8c). The giant hornet (*Vespa mandarinia*) is an important edible species mainly because the grubs are a delicacy, but it is difficult to capture and has a high market value

(Figure 8d). For consumption, the larvae, pupae, and the adult are prepared by frying or cooking with local spices replacing the conventional meat sources (beef, chicken, pork etc.).

Different methods are followed to capture/collect the bees, wasps, and hornets. The common honeybee (*Apis cerana indica*) is collected by smoking. In the process, a piece of cloth is burnt, and the smoke is blown into the bees' nest. The adults fly away and do not return until the smoke disappears. The rock honeybees *Apis dorsata dorsata* and *Apis dorsata laboriosa* are collected in two steps. Firstly, the beehive is pierced to allow honey to drip down; after which the honey is applied on the exposed body areas. This way, although the bees remain on the hive they do not sting the collector and are easily collected. The nest entrance of the underground stingless bee *Lophotrigona canifrons* is very small; therefore, a stick is inserted into the entrance so as to keep track of the nest. The bees' nest and the bee products are then collected by digging the soil away.

Giant hornet of the species *Vespa mandariana* are collected by the meat trap method, but for other hornet species the smoking out method is used [23]. With only its honey balls considered edible, the violet carpenter bee (*Xylocopa violacea*) is preferred by only a few Nagas and is eaten raw. The weaver ant (*Oecophylla smaragdina*), which had its chemical composition analysed by Chakravorty et al. [34], is considered as an important food item by all members of the Naga tribes (Figure 8i). While, the immature stages (eggs, larvae, and pupa) of the ant are boiled or cooked (with local spices) for consumption, the adult ants are prepared as chutney in a mixture of chilly, salt, and dry fish ground to powdered form and are also preserved for longer use.

3.2.7. Order Lepidoptera

Lepidopterans are mostly available for consumption from May to December. However, the eri silkworm (*Samia cynthia ricini*), which is reared widely in the region, is available for consumption throughout the year. The carpenter worm (*Cossus* sp.) with high protein and fat content, total phenol, and total antioxidant content [35] is the most popular and valued edible species, mainly because of its enticing aroma, its high market value, and ethno-medicinal properties. Although, some people prefer the larvae raw, most of the consumers prefer the larvae boiled or cooked with local spices served as a main dish on the menu (Figure 9a). The bamboo larvae (*Omphisa fuscidentalis*) are also an important edible insect that is preferred mainly for its crispness after being fried and eaten as snacks (Figure 9b). With *Omphisa fuscidentalis*, besides cooking or frying of the larvae, the traditional way of steaming is followed by some Nagas. For instance, the larvae (properly mixed with local spices) are wrapped with banana leaves that are then placed under warm ash. After 20–30 min, the packages are taken out and the larvae which are perfectly steamed are eaten as snacks.

The eri silkworm (larva and pupa) is relished by all Naga tribes and is prepared fried or cooked with local spices for consumption (Figure 9c,d). Tent caterpillars (*Malacosoma* sp.) are also a delicacy among the ethnic groups and are cooked or fried for consumption. The caterpillars are handpicked, when collected in plenty; the caterpillars are smoked for future use.

3.2.8. Order Diptera

The only dipteran species consumed, i.e., *Tipula* sp. (larva) is an important edible species and is prepared by boiling or cooking (faeces and gut removed) with local spices and served as a main dish on the menu. The larvae are mostly available for consumption during winter season (September–January).

Figure 9. (**Plate 7**). Mode of preparation of important Lepidopterans; (**a**) fried carpenter worms (*Cossus* sp.), (**b**) fried bamboo caterpillars (*Omphisa fuscidentalis*), (**c**) silkworm larvae and pupae cooked with fermented bamboo shoot and local spices, (**d**) fried silkworms.

4. Rearing and Marketing of Edible Insects

Having a high feed conversion and low feed consumption ratio, coupled with fast production of protein as compared to conventional animal and plant sources [36], rearing of edible insects is found to be more advantageous in providing unconventional protein to humans. While most of the insects documented from Nagaland are collected from the wild, the eri silkworm, muga silkworm (*Antheraea assamensis*), and *Apis cerana indica* are reared for personal consumption as well as for marketing (Figure 10a,c). However, wasps (e.g., *Provespa barthelemyi*) and hornets (*Vespa affinis indosinensis, Vespa auraria, Vespa basalis, Vespa bicolor, Vespa soror, Vespa tropica tropica, Vespa mandarinia,* and *Vespula otrbata*) are semi-domesticated. All throughout the year, different kinds of edible insects and insect products are sold in the markets (Figure 10b,c). While most of the edible insect species that are sold in the markets are freshly harvested ones, different honeybee products are now packaged and sold at specialized stores. The Nagaland Beekeeping and Honey Mission (NBHM) set up by the Government of Nagaland implements programs and policies for the promotion and development of beekeeping in the state. NBHM-brand processed honey is commercially available; besides many local entrepreneurs are also putting in an effort to commercialize different bee products of Nagaland. Apart from honey, all the other edible insects documented in the present study are not packaged and cannot be kept for longer durations. Therefore, even though there is potentiality of edible insects to become a staple food, ways for packaging edible insects and increasing their shelf-life to promote them as food nationally and internationally is the need of the hour.

Figure 10. (Plate 8). Local insect rearing and various marketed edible insects; (**a**) rearing of eri silkworm (*Samia cynthia ricini*) at Socūnoma village, Dimapur district, (**b**) different species of wood larvae sold at local market, Kohima district, (**c**) a woman selling eri silkworms at local market, Dimapur district, (**d**) sundried termites sold at local market, Mokokchung district, (**e**) sundried grasshoppers sold at local market, Dimapur district, (**f**) smoked tent caterpillars sold at local market, Mokokchung district.

Traditional way of increasing their shelf-life is practiced by the local insect sellers by simply smoking or sun drying the insect species (Figure 10d–f). The traditional method is nothing but minimizing the moisture content of the insect to reduce the rate of deterioration. The pentatomid bug *Udonga montana* that is boiled and further smoked/sundried (maintains its crispness) is also available in the local markets. Boiling and steaming an insect (larva and adult) not only maintains its size and keeps the desired aroma and flavour but also gives a higher score in terms of crispness [37]. Food dehydration is one way for the commercialization of edible insects, while the other way is to turn them into pickles. All over the world, the main concern with edible insect consumption, collection, and marketing is overexploitation. Entomophagy is an age-old practice in Nagaland and the different ethnic groups believe in sustainable utilization of the edible insect species. For instance, while collecting different kinds of bees and wasps, the queen along with some workers is left behind in a brood to enable them for further development and production. This effort for sustainability has been followed through generations and till today apiculture is successful in the region.

5. Conclusions and Recommendations

For the different ethnic groups in Nagaland, edible insects are a natural, renewable resource that plays an important role in providing nutrition. Currently, edible insects are mainly harvested from the wild and sometimes from agricultural crops where they occur as pests. However overharvesting edible insects from the wild to supply to the urban market for improving livelihood is not sustainable and is a strain to the rural biodiversity. Therefore, while edible insects have the potential to improve rural livelihood, semi-cultivation and farming of insects has to be seen as a priority. Looking into the Indian scenario, Nagaland has the potentiality of commercializing insects as a bio-resource. At present, only bees (especially *Apis cerana indica*), eri silkworm, and muga silkworm are mass produced in Nagaland. However, besides the mentioned three edible insects, there are groups of insects e.g., crickets, giant water bug, water beetles, palm weevil larvae, grasshoppers, etc. that are successfully reared in other parts of the world [38,39]. Nagaland harbours large number of edible insects such as aquatic bugs and beetles (*Cybister limbatus, Cybister tripunctatus lateralis, Lethoceruus indicus, Hydrophilus cashmirensis*) bees (*Apis cerana indica, Apis dorsata dorsata, Apis dorsata laboriosa*, etc.), crickets (*Tarbinskiellus orientalis, Tarbinskiellus portentosus, Teleogryllus occipitalis*), grasshoppers (*Oxya fuscovittata* and *Oxya hyla*), palm weevil (*Rhynchophorus ferrugineus*) and wasps (*Vespa mandarinia, Vespa soror, Vespa tropica tropica, Vespula orbata*, etc.), which have the potential for mass production in the region.

A study undertaken by the International Livestock Research Institute (ILRI), reported an increasing demand for livestock products and the present state production is not enough to feed the growing population. As per sample survey report of 2017–2018, the state produced 45.23% of total requirement of livestock worth 1206.15 crore Indian rupees but leaving a shortfall of 54.77%. Out of this shortfall, the state imported animal husbandry products worth 212.05 crore Indian rupees in monetary terms. In the backdrop of the growing demand for meat and the declining availability of agricultural land in the state, edible insects could be the alternative source of protein. Therefore, the present study, not only raises awareness among the tribal communities but also develops deep interest amongst policy makers and stake holders of the potential growth of the edible insect sector and thereby to promote funding into edible insect research and development. We believe that mass production, proper commercialization, and marketing strategies can improve livelihoods of tribal communities (especially the womenfolk) living in remote villages.

Supplementary Materials: The following are available online at http://www.mdpi.com/2304-8158/9/7/852/s1, Supplementary Material S1: Questionnaire format.

Author Contributions: Conceptualization, L.M. and L.N.K.; methodology, L.M., L.N.K., P.K., and S.C.; validation, L.N.K. and S.C.; formal analysis, L.M.; investigation, L.M., L.N.K., P.K., and S.C.; resources, L.M., P.K., and S.C.; data curation: L.M., L.N.K., and S.C.; writing—original draft preparation, L.M.; writing—review and editing, L.N.K. and S.C. All authors have read and agreed to the published version of the manuscript.

Funding: This research received no external funding.

Acknowledgments: The authors would like to acknowledge the field guides, local interpreters, and the different ethnic groups without whom the study would not have been possible. Lobeno Mozhui would like to thank National Fellowship for Higher Education of Scheduled Tribe Students (NFST) for financial assistance to pursue Ph.D.

Conflicts of Interest: The authors declare no conflict of interest.

References

1. Tilman, D.; Balzer, C.; Hill, J.; Befort, B.L. Global food demand and the sustainable intensification of agriculture. *Proc. Natl. Acad. Sci. USA* **2011**, *108*, 20260–20264. [CrossRef] [PubMed]
2. Meyer-Rochow, V.B. Therapeutic arthropods and other, largely terrestrial, folk-medicinally important invertebrates: A comparative survey and review. *J. Ethnobiol. Ethnomed.* **2017**, *13*, 9. [CrossRef] [PubMed]

3. Shelomi, M. Why we still don't eat insects: Assessing entomophagy promotion through a diffusion of innovations framework. *Trends Food Sci. Technol.* **2015**, *45*, 311–318. [CrossRef]

4. Durst, P.B.; Johnson, D.V.; Leslie, R.; Shono, K. *Edible Forest Insects-Humans Bite Back*; FAO Publication: Bangkok, Thailand, 2010.

5. Meyer-Rochow, V.B. Can insects help to ease the problem of world food shortage? *Search* **1975**, *6*, 261–262.

6. Van Huis, A.; Van Itterbeeck, J.; Klunder, H.; Mertens, E.; Halloran, A.; Muir, G.; Vantomme, P. *Edible Insects: Future Prospects for Food and Feed Security*; Food and Agricultural Organisation of the United Nations: Rome, Italy, 2013; pp. 1–190.

7. Van Huis, A.; Dicke, M.; Van Loon, J.J.A. Insects to feed the world. *J. Insects Food Feed* **2015**, *1*, 3–5. [CrossRef]

8. Costa-Neto, E.M.; Dunkel, F.V. Insects as food: History, culture and modern use around the world. In *Insects as Sustainable Food Ingredients*; Dossey, A.T., Morales-Ramos, J.L., Rojas, M.G., Eds.; Elsevier: San Diego, CA, USA, 2016; pp. 29–60.

9. Jongema, Y. *List of Edible Insects of the World*; Wageningen UR: Wageningen, The Netherlands, 2015. Available online: https://tinyurl.com/mestm6p (accessed on 10 May 2020).

10. Mitsuhashi, J. *Edible Insects of the World*; CRC Press: Boca Raton, FL, USA, 2016. [CrossRef]

11. Bernard, T.; Womeni, H.M. Entomophagy: Insects as food. In *Insect Physiology and Ecology*; Shields, V.D.C., Ed.; InTechOpen, Agricultural and Biological Sciences: London, UK, 2017; pp. 233–249. [CrossRef]

12. Seni, A. Edible insects: Future prospects for dietary regimen. *Int. J. Curr. Microbiol. Appl. Sci.* **2017**, *6*, 1302–1314. [CrossRef]

13. Alexander, P.; Brown, C.; Arneth, A.; Dias, C.; Finnigan, J.; Moran, D.; Rounsevell, M. Could consumption of insects, cultured meat or imitation meat reduce global agricultural land use? *Glob. Food Secur.* **2017**, *15*, 22–32. [CrossRef]

14. Sere, A.; Bougma, A.; Ouilly, J.T.; Traore, M.; Sangare, H.; Lykke, A.M.; Ouedraogo, A.; Gnankine, O.; Bassole, I.H.N. Traditional knowledge regarding edible insects in Burkina Faso. *J. Ethnobiol. Ethnomed.* **2018**, *14*, 1–11. [CrossRef]

15. Hutton, J.H. *The Angami Nagas: With Some Notes on Neighbouring Tribes*; Macmillan: London, UK, 1921.

16. Hutton, J.H. *The Sema Nagas*; Macmillan: London, UK, 1921.

17. Mills, J.P. *The Lhota Nagas*; Macmillan: London, UK, 1922.

18. Mills, J.P. *The Ao Nagas*; Macmillan: London, UK, 1926.

19. Meyer-Rochow, V.B.; Changkija, S. Uses of insects as human food in Papua New Guinea, Australia and North-East India: Cross-cultural considerations and cautious conclusions. *Ecol. Food Nutr.* **1997**, *36*, 159–185. [CrossRef]

20. Ao, A.; Singh, H.K. Utilization of insects as human food in Nagaland, India. *J. Entomol.* **2004**, *66*, 308–310.

21. Singson, L.; Das, A.N.; Ahmed, R. Edible insects of Nagaland and its nutritional benefit. *Periodic Res.* **2016**, *5*, 1–7.

22. Pongener, A.; Ao, B.; Yenisetti, S.C.; Lal, P. Ethnozoology and entomophagy of ao tribe in the district of Mokokchung, Nagaland. *Ind. J. Tradit. Knowl.* **2019**, *18*, 508–515.

23. Mozhui, L.; Kakati, L.N.; Changkija, S. A study on the use of insects as food in seven tribal communities in Nagaland, Northeast India. *J. Human Ecol.* **2017**, *60*, 42–53. [CrossRef]

24. Changkija, S. *Biodiversity of Nagaland*; Department of Forest, Ecology, Environment and Wildlife: Nagaland, India, 2014.

25. Kuotsuo, R.; Chatterjee, D.; Deka, B.C.; Kumar, R.; Ao, M.; Vikramjeet, K. Shifting cultivation: An 'organic like' farming in Nagaland. *Indian J. Hill Farm.* **2014**, *27*, 23–28.

26. Ghosh, S.; Jung, C.; Meyer-Rochow, V.B. What governs selection and acceptance of edible insect species? In *Edible Insects in Sustainable Food Systems*; Halloran, A., Vantomme, P., Roos, N., Eds.; Springer: Cham, Switzerland, 2018; pp. 331–351.

27. Meyer-Rochow, V.B. Food taboos: Their origins and purposes. *J. Ethnobiol. Ethnomed.* **2009**, *5*, 18. [CrossRef]

28. Müller, A. Insects as food in Laos and Thailand: A case of "Westernisation"? *Asian J. Soc. Sci.* **2019**, *47*, 204–223. [CrossRef]

29. Ramos-Elorduy, J. Anthropo-entomophagy: Culture, evolution and sustainability. *Entomol. Res.* **2009**, *39*, 271–288. [CrossRef]

30. Belluco, S.; Losasso, C.; Maggioletti, M.; Alonzi, C.C.; Paoletti, M.G.; Ricci, A. Edible insects in a food safety and nutritional perspective: A critical review. *Compr. Rev. Food Sci. Food Saf.* **2013**, *12*, 296–313. [CrossRef]

Foods **2020**, *9*, 852

31. Testa, M.; Stillo, M.; Maffei, G.; Andriolo, G.; Gardois, P.; Zotti, C.M. Ugly but tasty: A systematic review of possible human and animal health risks related to entomophagy. *Crit. Rev. Food Sci. Nutr.* **2016**, *57*, 3734–3759. [CrossRef]

32. Oonincx, D.G.A.B.; de Boer, J.M. Environmental impact of the production of mealworms as a protein source for humans—A lifecycle assessment. *PLoS ONE* **2012**, *7*, e51145. [CrossRef]

33. Gahukar, R.T. Edible insects collected from forests for family livelihood and wellness of rural communities—A review. *Glob. Food Sec.* **2020**. [CrossRef]

34. Chakravorty, J.; Ghosh, S.; Megu, K.; Jung, C.; Meyer-Rochow, V.B. Nutritional and anti-nutritional composition of *Oecophylla smaragdina* (Hymenoptera: Formicidae) and *Odontotermes* sp. (Isoptera: Termitidae): Two preferred edible insects of Arunachal Pradesh, India. *J. Asia-Pacif. Entomol.* **2016**, *19*, 711–720. [CrossRef]

35. Aochen, C.; Krishnappa, R.; Firake, D.M.; Pyngrope, S.; Aochen, S.; Ningombam, A.; Behere, G.T.; Ngachan, S.V. Loungu (Carpenter worm): Indigenous delicious insects with immense dietary potential in Nagaland state, India. *Ind. J. Trad. Knowl.* **2020**, *19*, 145–151.

36. Nagagaki, B.J.; Defoliart, G.R. Comparison of diets for mass-rearing *Acheta domesticus* (Orthoptera: Gryllidae) as a novelty food, and comparison of food conversion efficiency with values reported for livestock. *J. Econ. Entomol.* **1991**, *84*, 891–896. [CrossRef]

37. Baek, M.; Yoon, Y.I.; Kim, M.; Hwang, J.S.; Goo, T.W.; Yun, E.Y. Physical and sensory evaluation of Tenebrio molitor larvae coked by various cooking methods. *Korean J. Food Cook. Sci.* **2015**, *31*, 534–543. [CrossRef]

38. Halloran, A.; Caparros Megido, R.; Oloo, J.; Weigel, T.; Nsevolo, P.; Francis, F. Comparative aspect of cricket framing in Thailand, Cambodia, Lao People's Democratic Republic, Democratic Republic of the Congo, and Kenya. *J. Insects Food Feed* **2018**, *4*, 101–114. [CrossRef]

39. Hanboonsong, Y.; Durst, P.B. Edible insects. In *Lao PDR: Building on Tradition to Enhance Food Security*; FAO: Bangkok, Thailand, 2014; pp. 1–55.

 *foods*

Article

Effect of Thermal Processing on Physico-Chemical and Antioxidant Properties in Mulberry Silkworm (*Bombyx mori* L.) Powder

Artorn Anuduang [1,2], Yuet Ying Loo [1,*], Somchai Jomduang [2,3], Seng Joe Lim [1] and Wan Aida Wan Mustapha [1,*]

[1] Department of Food Sciences, Faculty of Science and Technology, Universiti Kebangsaan Malaysia, UKM Bangi 43600, Malaysia; a.anuduang@gmail.com (A.A.); joe@ukm.edu.my (S.J.L.)
[2] Division of Food Science and Technology, Faculty of Agro-Industry, Chiang Mai University, Chiang Mai 50100, Thailand; somchai.j@cmu.ac.th
[3] Biosafe Holding Partnership Limited, 353 Moo 9, Tambol Sanklang, Sanpatong District, Chiang Mai 50120, Thailand
* Correspondence: wanaidawm@ukm.edu.my (W.A.W.M.); yuetying88@gmail.com (Y.Y.L.); Tel.: +603-8921-3870 (ext. 5963) (W.A.W.M.)

Received: 26 April 2020; Accepted: 30 June 2020; Published: 3 July 2020

Abstract: The mulberry silkworm (*Bombyx mori* L.) is a common edible insect in many countries. However, the impact of thermal processing, especially regarding Thai silkworm powder, is poorly known. We, therefore, determined the optimum time for treatment in hot water and subsequent drying temperatures in the production of silkworm powder. The silkworms exposed to 90 °C water for 0, 5, 10, 15, and 20 min showed values of Total Phenolic Compounds (TPCs), 2,2-Diphenyl-1-picrylhydrazyl free radical scavenging (DPPH) assay, 2,2′-Azino-bis(3-ethylbenzothiazoline-6-sulfonic acid (ABTS) assay, and Ferric Reducing Antioxidant Power (FRAP) assay that were significantly ($p < 0.05$) higher at the 5 min exposure time compared with the other times. The reduction of microorganisms based on log CFU/g counts was ≥3 log CFU/g (99%) at the 5 min treatment. To determine the optimum drying temperature, the silkworms exposed to 90 °C water for 5 min were subjected to a hot-air dryer at 80, 100, 120, and 140 °C. The TPC value was the highest ($p < 0.05$) at 80 °C. The silkworm powder possessed significantly ($p < 0.05$) higher DPPH, ABTS radical scavenging ability, and ferric ion reducing capability (FRAP assay) at 80 °C compared with other drying temperatures. This study indicates that shorter exposure times to hot water and a low drying temperature preserve the antioxidant activities. High antioxidant activities (in addition to its known protein and fat content) suggest that silkworms and silkworm powder can make a valuable contribution to human health.

Keywords: silkworm; edible insects; thermal processing; antioxidant activities; silkworm powder

---

1. Introduction

The custom of consuming insects as a food source (entomophagy) has been widely practiced by many people from all around the world for thousands of years [1]. Global population growth has led to the increase of food consumption, especially protein. Due to the high protein content, short life cycle, and rapid growth cycle, edible insects may serve as an alternative source of protein for humans, as suggested by Meyer-Rochow [2]. Food security is seen as one of the major global concerns, and therefore, the benefits of edible insect have been re-evaluated, as insects are rich in protein, fats, minerals, and essential vitamins, and contain low amount of carbohydrates. Moreover, insects are abundant and easily bred in large quantities [3]. The mulberry silkworm (*Bombyx mori* L.) is an edible insect that has been bred for a long time and can be produced throughout the year.

Silkworms produce silk from plant proteins. In the sericulture industry, the after-silking silkworms are the main by-product and have become a popular food among the 194 species of edible insects in Thailand. In many Asian countries, silkworm has been utilized as food, medicine, and animal feed due to its nutritional profile [4]. Meyer-Rochow [5] mentioned that the larvae, pupae, and adults of *B. mori* could serve as medicine to treat wounds, sore throats, fever, bleeding, brain hemorrhage, hemorrhoids, and others. Besides phenolic compounds, silkworms also exhibit antioxidant activity [6]. Studies revealed that some pigments found on the yellow Nangnoi silk (Thai silk) are associated with the presence of carotenoids and flavonoids [7,8]. The phenolic compounds are well-known for antioxidant activity as well as some biological functions such as anti-hypertensive [9], antiviral [10], antioxidant [11], antibacterial, and anti-inflammatory activities [12]. Dutta et al. [13] reported that the antioxidant potential of the edible insect *Vespa affinis* L. could mediate its therapeutic activities in oxidative stress-associated health disorders. Dutta et al. [14] also suggested that the antioxidant potential of the edible insect, *Brachytrupes orientalis* extract in hydro-alcoholic (AEBO) has a significant role against cellular oxidative impairment.

Silkworm powder, also known as pury, is reported to have a high content of nutrients and tends to be an alternative functional food [15]. The yellowish silkworm powder was reported to possess a high content of nutrients such as protein, essential amino acids, polyunsaturated fatty acid, carbohydrates, vitamins, and minerals. Suk et al. [16] reported that the incorporation of silkworm powder in wheat flour noodles could reduce postprandial glucose response and act as a potential carbohydrate staple food for glycemic control. Park et al. [17] suggested that silkworm powder could be an alternative additive for meat products, as it improved the meat product's physicochemical properties such as protein, fat, and ash contents. Biró et al. [18] concluded that the addition of silkworm powder in buckwheat pasta could increase its nutritional value.

A thermal process is usually employed in the food industry to reduce microbial activity and prolong the shelf life of silkworm products [19]. In addition, thermal processing of food has resulted in physical or chemical changes [20] to ensure the quality standards of food products. Baek et al. [21] stated that hot air drying and oven broiling methods gave the highest score in sensory evaluation in terms of the hardness, crispiness, and aroma in processed mealworm larvae. Meanwhile, Jensen et al. [22] reported that different processing methods were available to alter the taste, aroma, and texture of the insects. In this study, silkworm was processed into a yellowish powder by treatment in hot water, dried, and ground. Thermal processing may destroy some of the bioactive compounds and affect the antioxidant activity in food [23]. Norafida and Aminah [24] revealed that heat treatments (blanching, steaming, or boiling) changed the quality of food products such as texture, taste, color, and nutritional content. Hence, this study aimed to determine the optimal times for treatment in hot water and subsequent drying in the production of silkworm powder.

2. Materials and Methods

2.1. Sample Preparation

Fresh after-silking silkworms (FASSs) were collected from Natural Thai Golden Silk Ltd. (skincare manufacturing industry, Payao Province, Thailand). Fifth instar silkworms are used to produce silk sheets in the process of developing a skincare product. The silkworms are collected and disposed of as waste after silking. Thus, FASS was chosen in this study to transform the waste into a value-added product. The samples were transferred immediately to the laboratory in the Faculty of Agro-Industry, Chiang Mai University under cold conditions. The silkworm samples were then kept at −18 °C until further use.

2.2. Thermal Processing

2.2.1. Hot Water Treatment (Pre-Treatment of Silkworms)

A total of 4 kg of frozen FASS was used in this study. FASSs were exposed to 90 °C water (ratio of 1:2) for 5, 10, 15, and 20 min. The treatment time started when the water temperature reached the desired temperature (90 °C). After each treatment time, the samples were blotted with tissue paper to remove the excess water on the surface of silkworms. The silkworm samples were further analyzed for their physicochemical properties and antioxidant activities. The experiments were carried out in triplicate.

2.2.2. Drying Process

The silkworms that were exposed to 90 °C water for 5 min were then dried. The treated FASS was centrifuged at $1500\times g$ rpm for 1 min to remove the excess water from the silkworm. FASS was dried in a hot-air dryer (Kluynamthai, Bangkok, Thailand) at four different temperatures (80, 100, 110, and 120 °C). The moisture content of the sample was measured every 30 min by recording the weight. Weight changes in the sample were used to calculate the moisture content of the sample. The drying process ended when less than 15% of moisture content was achieved for each drying temperature. The dried silkworms were ground into powder using a blender at 25,000 rpm (Nanotech, Beijing, China) and filtered using a 1.0 mm mesh sieve to obtain fine silkworm powder. The powder was then further analyzed for its physicochemical properties and antioxidant activities.

2.3. Physicochemical Properties

The color, size, and weight of the silkworm were determined in this study. The surface color of silkworms was measured using a Hunter Lab colorimeter (model Color Quest XE, Reston, VA, USA). The color scale gave the measurement of color (L*, a*, and b*) in units of approximate visual uniformity throughout the solid color. L* value determined lightness, where the score of 100 showed perfect white color and 0 for black color. Moreover, a* measured the redness when positive, whereas b* measured the yellowness when positive. The length and width of silkworms were measured by a Vernier caliper. The weight of the silkworm was recorded, and the average weight was calculated.

2.4. Proximate Analysis

Moisture content (method 934.01), ash (method 923.03), and total fat (method 991.36) were analyzed using the protocols of the Association of the Official Analytical Chemists (AOAC). Protein content (N × 6.25) was examined using the AOAC Kjeldahl method (984.13). Carbohydrate was obtained by the formula of 100 − (the sum of moisture, protein, fat, and ash).

2.5. Total Microorganism

The total microorganisms in the silkworm were determined using the aerobic plate count method, as described in the FDA's Bacteriological Analytical Manual (Chapter 3) [25]. The silkworms (10 g) were pre-enriched in tryptic soy broth (90 mL; Merck, Darmstadt, Germany) in a 1:9 sample-to-broth ratio. The pre-enriched sample was diluted in broth by a 10-fold serial dilution (10^{-2}, 10^{-3}, and 10^{-4}). Then, 1 mL of each dilution was pipetted and spread onto the plate count agar (PCA; Merck, Darmstadt, Germany). The agar plates were incubated at 37 °C for 24 h. The experiments were carried out in triplicate, and the results were expressed as log colony-forming units per gram (log CFU/g).

2.6. Antioxidant Assays

Extraction was performed before the antioxidant assays. Fresh silkworm, hot water-treated silkworm, and silkworm powder were weighed at 10 g each (dry weight equivalent) and mixed with 50 mL of 85% (*v/v*) ethanol (Merck, Darmstadt, Germany). The mixture of each sample was left at the ambient temperature for 1 h. The mixture was then filtered using Whatman® Grade 4 filter paper.

2.7. Total Phenolic Compounds (TPCs)

Total phenolic compounds (TPCs) of the silkworms were determined by the Folin–Ciocalteu method with slight modification from Li et al. [26]. Approximately 100 µL of silkworm extract was mixed with 2 mL of 10% (*v/v*) Folin–Ciocalteu reagent (Merck, Darmstadt, Germany), and the mixture was left for 3 min at the ambient temperature. Then, 2.5 mL of 7.5% (*w/v*) sodium carbonate (Sigma, Darmstadt, Germany) was added to the mixture and incubated in the dark for 2 h. Absorbance was recorded using a spectrophotometer (BMG Labtech, Ortenberg, Germany) at 760 nm. Gallic acid was used as the standard for calibration. TPC was expressed in terms of µmol gallic acid equivalents per 1 g dry weight (DW) sample (µmol GAE/g of DW sample).

2.8. The 2,2-Diphenyl-1-picrylhydrazyl (DPPH) Free Radical Scavenging Assay

The determination of the scavenging activity of the silkworm extracts on the stable free radical DPPH was done as described in Phongthai et al. [27] with slight modification. The DPPH stock solution was prepared by dissolving 40 mg of DPPH (Sigma, Germany) in 100 mL methanol (Merck, Germany) to yield an absorbance of 0.70 ± 0.01 at 517 nm. Silkworm extract (100 µL) or blank were mixed with 0.1 mM of DPPH reagent (2.9 mL). The mixture was then left in the dark for 30 min at ambient temperature. The absorbance of the mixture was measured by spectrophotometer (Thermo Spectronic, Model Genesys 10UV-Scanning, CE, Wisconsin, WI, USA) at 517 nm. All samples and controls were done in triplicate. A standard curve was prepared using standard Trolox at concentrations of 0.01–0.40 mg/mL. Ethanol served as negative control (without samples). The values were expressed in terms of µmol equivalent of Trolox/dry basis (µmol TE/g DW).

2.9. The 2,2′-Azino-bis(3-ethylbenzothiazoline-6-sulfonic Acid (ABTS) Assay

Antioxidant activities of silkworm extract were determined using the method as described by Phongthai et al. [27] with slight modification. The stock solution was prepared by mixing 7 mM 2, 2-azinobis-3-ethyl-benzothiazoline-6-sulfonic acid (ABTS) reagent with 2.45 mM potassium persulfate ($K_8S_2O_8$) at a ratio of 1:1. The mixture was left in the dark for 12 h at ambient temperature. After 12 h, the mixture was diluted with deionized water until the absorbance at 734 nm wavelength reached 0.70 ± 0.05 using the spectrophotometer. An aliquot of silkworm extract (150 µL) was mixed with 2.85 mL of ABTS radical cation stock solution before 2 h incubation in the dark. The absorbance of the mixture was measured at 734 nm. Ethanol served as negative control (without samples). The value was expressed in terms of µmol Trolox equivalent/g (µmol TE/g DW).

2.10. Ferric Reducing Antioxidant Power (FRAP) Assay

The antioxidant activity of silkworm extract was determined according to Phongthai et al. [27] with some modifications. The stock solution of FRAP reagent was prepared using 300 mM acetate buffer, pH 3.6 (3.1 g sodium acetate trihydrate and 16 mL glacial acid made up to 1:1 with distilled water), 10 mM TPTZ (2,4,6-tris [2-pyridyl]-s-triazine) in 40 mM HCl, and 20 mM $FeCl_3.6H_2O$ in a ratio of 10:1:1. Approximately 0.8 mL of silkworm extract was added to 4 mL of FRAP reagent and left at ambient temperature for 10 min. The absorbance was measured at 595 nm using a spectrophotometer after 10 min. A standard calibration curve of ferrous sulphate ($FeSO_4$) was plotted to estimate the activity capacity of samples. Ethanol served as negative control (without samples). The results were reported in terms of µmol Fe^{2+}/g dry basis sample.

2.11. Statistical Analysis

The results were expressed as the mean ± standard deviation (SD) of three independent experiments. Statistical comparisons of the results were subjected to one-way ANOVA using SPSS version 20. The differences in antioxidant activity and TPC among the different exposure times to hot water at the

$p < 0.05$ level of significance, and the differences in antioxidant activity and TPC among the different drying temperature at the $p < 0.05$ level of significance were analyzed by Duncan's triplicates range test.

3. Results and Discussion

3.1. Physicochemical Properties of FASS

The physicochemical properties of FASS are shown in Table 1. FASS showed a high lightness with an L* value of 58.11 ± 0.63 and b* value of 24.45 ± 0.98, indicating that FASS in this study was yellowish. According to Taba and Gogoi [28], fifth instar silkworm exists in three forms, i.e., bluish-white, yellow, and bluish-green. FASS had a length of 33.19 ± 1.72 mm and a breadth of 6.97 ± 0.22 mm. Meanwhile, its weight was 1.04 ± 0.06 g. The size of FASS in this study was almost similar to the size of fifth instar larvae in the study of Gurjar et al. [29] with a length of 57.86 ± 7.06 mm and breadth of 7.15 ± 0.43 mm.

Table 1. Physicochemical properties, total microorganism count and antioxidant activity of fresh Thai silkworm after silking (FASS).

Properties	Value
Physicochemical Properties	
Color L*	58.11 ± 0.63
a*	1.14 ± 0.99
b*	24.45 ± 0.98
Size (mm) length	33.19 ± 1.72
Width	6.97 ± 0.22
Weight (g)	1.04 ± 0.06
Proximate Analysis	
Moisture content (%)	75.83 ± 0.63
Protein (% DM basis)	66.66 ± 1.23
Fat (% DM basis)	20.64 ± 2.24
Ash (% DM basis)	1.32 ± 0.13
Carbohydrate (% DM basis)	2.59 ± 0.43
Total Microorganism (log CFU/g)	6.54 ± 0.13
Antioxidant Activities	
DPPH (μmol TE/g DW)	11.57 ± 0.80
ABTS (μmol TE/g DW)	32.75 ± 4.57
FRAP (μmol of Fe^{2+}/g DW)	54.28 ± 2.86

3.2. Proximate Analysis

The results of the proximate analysis are shown in Table 1. Protein has the highest percentage (66.66% of dry matter (DM)) among the components in silkworms. Karthick Raja et al. [30] reported that the protein content in dried silkworm ranged from 52 to 72%. The high content of protein in silkworm was due to the intake of mulberry leaves. Mulberry leaves are the only food source of the silkworm, and the leaves are reported to have high protein content [31].

Fats are identified as the second-highest percentage of components in silkworms. In this study, FASS contained 20.64% DM basis of fat content, which falls in the range of 18.9–36%, as reported by Kim et al. [32]. Meanwhile, Chieco et al. [33] mentioned that the silkworm contains essential fatty acids (EFA), mainly in the form of polyunsaturated fatty acids (PUFA). In addition, the silkworms also possess a higher n-3 to n-6 fatty acid ratio, which differs from other insects.

In this study, FASS exhibited high moisture content (75.83%), and this may have been due to the sole food source of the silkworm, the mulberry leaves [34]. The ash content of FASS was 1.32% on a dry matter basis. The ash content indicated the amount of minerals present in the insect. Omotoso [35] reported that the silkworm is rich in minerals such as sodium (Na), potassium (K), calcium (Ca), iron (Fe), magnesium (Mg), and zinc (Zn). In addition, the carbohydrate content was found to have a 2.59% DM basis in FASS. Carbohydrate is identified as the primary food source in silkworms.

3.3. Total Microorganisms

Table 1 shows that the microorganisms count of the silkworm was 6.54 log CFU/g. The incidence of a high amount of microorganisms is reasonable because the fresh silkworms were rich in nutrients with high moisture content. Li et al. [36] revealed that microorganisms could be found in the intestine of lepidopteran insects such as silkworm, which play essential roles in promoting growth and development.

3.4. Antioxidant Activities

A linear calibration curve of Trolox was obtained with good linearity of correlation coefficient ($R^2 = 0.979$) and a linear equation of y = 185.3x + 18.97, which was used as positive control for DPPH scavenging activity. The value of the DPPH assay for FASS was 11.57 ± 0.80 µmol TE/g DW. In the FRAP assay, the antioxidant activity of FASS was determined by the ability of antioxidant compounds in silkworm to reduce Fe^{3+} to Fe^{2+} in the FRAP reagent. A linear calibration curve of ferrous sulphate ($FeSO_4$) was obtained with good linearity of the correlation coefficient ($R^2 = 0.996$) and a linear equation of y = 12.66x + 0.031, which was used as a positive control for the FRAP assay. The FRAP value of FASS was 54.28 ± 2.86 µmol Fe^{2+}/g on a dry basis, indicating that the fresh silkworm exhibited a strong ferric ion reducing capacity. A linear calibration curve of Trolox was obtained with good linearity of correlation coefficient ($R^2 = 0.982$) and a linear equation of y = 771.1x + 9.971, which was used as a positive control for ABTS scavenging activity. The value of the ABTS assay of FASS in this study was 32.75 ± 4.57 µmol TE/g DW. A higher ABTS radical scavenging capability was observed in FASS compared with DPPH measurement. This could be due to the high protein content in fresh silkworm, as reported by Zheng et al. [37], whereby the ABTS assay was more sensitive in evaluating the radical scavenging activities of amino acids and peptides compared with the DPPH assay.

3.5. Effect of Hot Water Treatment

FASS was exposed to 90 °C water before the process of producing silkworm powder. The exposure of FASS to hot water was to kill them thoroughly and for sterilization purposes. As shown in Figure 1A, the TPC of FASS (18.21 ± 1.31 µmol GAE/g DW) was significantly ($p < 0.05$) higher following 5 min of exposure compared with the other time points. The values of DPPH free radical scavenging assay (22.43 ± 2.85 µmol TE/g DW), ABTS assay (35.37 ± 2.99 µmol TE/g DW), and FRAP assay (55.26 ± 4.05 µmol of Fe^{2+}/g DW) were significantly ($p < 0.05$) higher following 5 min exposure compared with the other time points (Figure 1B–D). Therefore, the samples exposed to 90 °C water for 5 min were used for subsequent drying treatment.

The TPC, DPPH free radical scavenging assay, and FRAP assay of FASS showed a significant reduction ($p < 0.05$) when the exposure time was extended to 10, 15, and 20 min. The reduction of TPC indicated the loss of antioxidant compound, and this was due to the large surface area of FASS in contact with water [38]. In addition, cooking in the boiling water had an adverse effect on the polyphenol level, as it could diffuse into the boiling water, resulting in the loss of phenolic compounds [39]. On the other hand, it was interesting to note that TPC and antioxidant activity of FASS were significantly ($p < 0.05$) increased at the 5 min exposure time and gradually decreased after exposure to hot water for 10, 15, and 20 min. The hot water treatment caused the destruction of cell and sub-cellular components in the silkworms, resulting in the release of potent radical-scavenging antioxidants [40]. Furthermore, a study suggested that the thermal inactivation of oxidative enzymes during the heating process led to the suppressed oxidation by antioxidants [39].

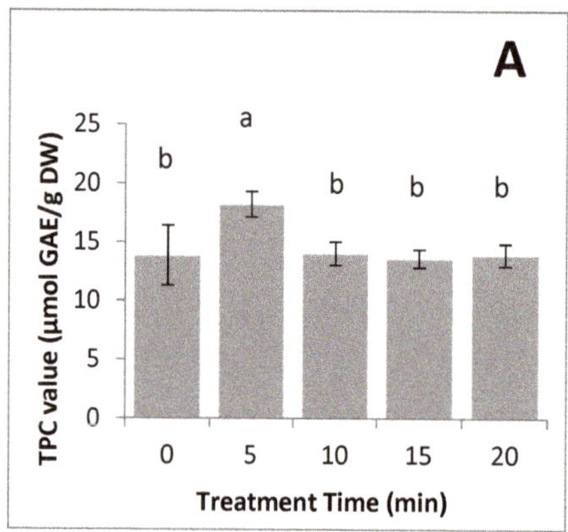
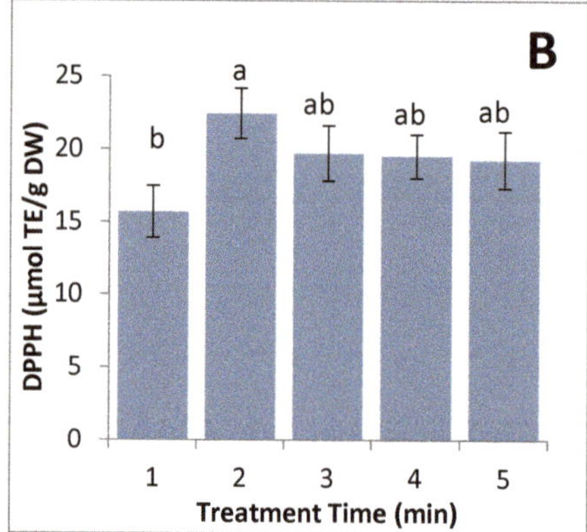
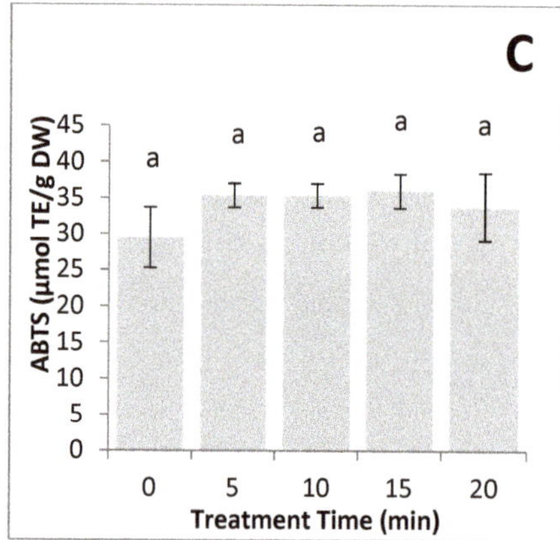
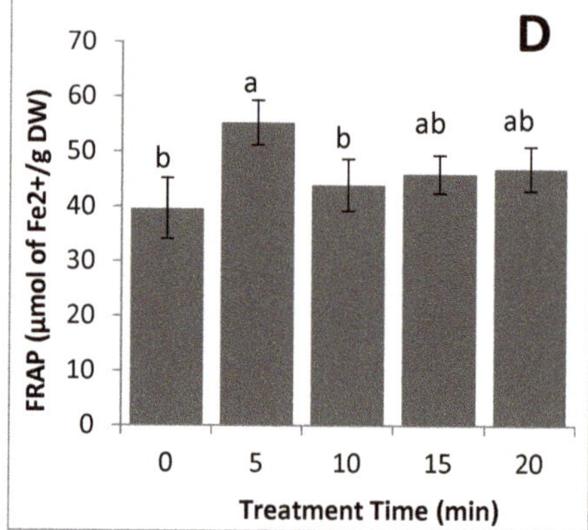

Figure 1. Effect of different treatment time of hot water on antioxidant activity in silkworms. (**A**) Total phenolic compound (TPC), (**B**) DPPH free radical scavenging assay, (**C**) ABTS assay, and (**D**) ferric reducing antioxidant power (FRAP). The results were expressed as mean ± SD of triplicate measurements. The different letter on the bar indicates significantly difference among treatment ($p < 0.05$).

Table 2 shows that the total microorganism load at 0 min of exposure time was 6.69 log CFU/g. Following exposure for 5 min, the microorganism count was 1.30 log CFU/g, while the total microorganism count was less than 1.0 log CFU/g for exposure times of 10, 15, and 20 min. It was observed that the reduction in the number of CFU/g was ≥3 log CFU/g (99%). The whole body of the silkworm, including the intestine, was used in the production of silkworm powder. The intestine of silkworm contains a high amount of microorganisms, which affect the quality and shelf life of silkworm powder [41]. Hence, a pre-treatment of the fresh silkworm is an essential step before the production of silkworm powder.

Table 2. Effect of water heat treatment time on total microorganism load in fresh after silking silkworms.

Treatment Time (min)	Total Microorganisms Load (log CFU/g)
0	6.69 ± 0.39
5	1.30 ± 0.38
10	<1
15	<1
20	<1

3.6. Effect of Drying Temperature

The silkworms exposed to 90 °C water for 5 min were subjected to a hot air dryer. Table 3 shows the effect of different drying temperatures on the physicochemical properties and antioxidant activities of silkworms. For the color measurement of silkworm powder, L* and b* values were significantly ($p < 0.05$) decreased when the drying temperature increased. The color fading in the silkworm powder may have been due to the sensitivity of the silkworm color pigment to high temperature [42]. In addition, Dorouzi et al. [43] reported the formation of browning pigments by the non-enzymatic browning or Maillard reactions at high drying temperature, leading to a decrease in lightness. The moisture content of the silkworm powder was measured until the final moisture content was less than 15%, as the growth of microorganism could be prevented when the moisture content was reduced to less than 15% [44,45].

Table 3. Chemical properties and antioxidant activities of dried silkworms in a hot-air oven at different drying temperatures.

Properties	Drying Temperature (°C)			
	80	100	120	140
Color L*	47.07 [a] ± 0.27	44.80 [c] ± 0.24	46.13 [b] ± 0.40	44.42 [c] ± 0.44
a*	4.68 [c] ± 0.10	5.83 [b] ± 0.24	6.61 [a] ± 0.22	6.67 [a] ± 0.08
b*	17.01 [a] ± 0.38	14.36 [c] ± 0.68	15.96 [b] ± 0.50	13.75 [c] ± 0.28
Moisture content (%)	8.08 [a] ± 0.37	4.30 [b] ± 0.30	2.20 [c] ± 0.59	0.70 [d] ± 0.18
Total phenolic compounds (μmol GAE/g DW)	51.81 [c] ± 0.84	47.12 [a] ± 0.77	48.96 [ab] ± 1.53	50.29 [bc] ± 2.12
Antioxidant Activities				
DPPH (μmol TE/g DW)	69.09 [b] ± 0.58	68.59 [ab] ± 0.31	68.71 [b] ± 0.15	67.94 [a] ± 0.37
ABTS (μmol TE/g DW)	69.27 [b] ± 2.85	68.90 [b] ± 0.56	68.29 [b] ± 0.87	57.63 [a] ± 0.91
FRAP (μmol of Fe^{2+}/g DW)	46.73 [b] ± 4.54	43.02 [ab] ± 2.23	41.71 [ab] ± 2.61	39.89 [a] ± 2.17

The results were expressed as mean ± SD ($n = 3$) of triplicate measurements. Mean values with the different superscript lowercase letter are significantly different ($p < 0.05$).

The TPC of silkworm powder was significantly ($p < 0.05$) decreased at 100 °C and increased at 120 and 140 °C. This was due to the release of potent radical-scavenging antioxidants, as reported by Kao et al. [40]. However, the results showed that the TPC of silkworm powder was significantly higher (51.81 ± 0.84 μmol GAE/g DW) at a low drying temperature (80 °C). The DPPH free radical scavenging assay showed a significant ($p < 0.05$) decease as the drying temperature increased and the highest value of DPPH free radical scavenging (69.09 ± 0.58 μmol TE/g DW) was obtained at 80 °C. The results of ABTS and FRAP assays showed a significant ($p < 0.05$) reduction in silkworm powder as the drying temperature increased. In the ABTS assay, silkworm powder possessed the highest ABTS radical scavenging capability (69.27 ± 2.85 μmol TE/g DW) of antioxidant at 80 °C. The FRAP assay of silkworm powder showed a significantly ($p < 0.05$) higher value at 80 °C compared with the other drying temperatures. The decreased antioxidant activity upon exposure to high temperature was associated with the degradation of antioxidant compounds [46]. Furthermore, the decrease in antioxidant activity in silkworm powder was due to the formation of pro-oxidants by the Maillard reaction when exposed to high temperature [47].

The presence of antioxidant activity in fifth instar silkworms is probably because of the production of sericin. Silkworms produce large quantity of sericin at the fifth instar stage larvae. Sericin was reported to have high antioxidant properties [48]. The high consumption of mulberry leaves during late larval stage contributes to the antioxidant activity, as mulberry leaves have been identified to contain large quantities of phenolic compounds such as flavonoid and its derivatives [49]. Thermal processing, especially water heat treatment, may cause the loss of phenolic compounds, as reports showed that phenolic are highly soluble in water. Phenolics may be lost by leaching. The high temperature may destroy the cell wall and other subcellular compartments, thus facilitates the release of the compound into the water [50]. The reduction of antioxidant compounds may due to certain factors such as temperature, air velocity, heat exposure time, and the content of antioxidant compound [51].

4. Conclusions

Silkworm and silkworm powder contain antioxidant activities. The results showed that thermal processing could affect the antioxidant activity and some physicochemical properties during the production of silkworm powder. Shorter exposure time to hot water and a lower drying temperature should be applied to prevent the loss of bioactive compounds in silkworm powder. Silkworm powder could be a potential protein source for animal products, and the application may lead to valuable findings in various fields such as pharmaceutical and cosmetic industries.

Author Contributions: A.A. and S.J. developed the study design. A.A. performed the experiments. A.A. and Y.Y.L. interpreted the data, drafted the manuscript, and revised the manuscript. S.J., S.J.L., and W.A.W.M. supervised the project and checked on the manuscript. All authors read and approved the final version of the manuscript.

Funding: This research received no external funding.

Acknowledgments: This research was funded by the Dana Modal Insan (MI-2020-007) research grant provided by Universiti Kebangsaan Malaysia, Tech Enterprise Service Network of Ministry of Science and Technology, Thailand and private grant from NTGS Co. Ltd., Thailand.

Conflicts of Interest: The authors declare no conflict of interest.

References

1. Barsics, F.; Megido, R.C.; Brostaux, Y.; Barsics, C.; Blecker, C.; Haubruge, E.; Francis, F. Could new information influence attitudes to foods supplemented with edible insects? *Br. Food J.* **2017**, *119*, 2027–2039. [CrossRef]
2. Meyer-Rochow, V.B. Can insects help to ease the problem of world food shortage. *Search* **1975**, *6*, 261–262.
3. Ghosh, S.; Jung, C.; Meyer-Rochow, V.B. What governs selection and acceptance of edible insect species? In *Edible Insects in Sustainable Food Systems*; Springer: Berlin/Heidelberg, Germany, 2018; pp. 331–351.
4. Dong, H.-L.; Zhang, S.-X.; Tao, H.; Chen, Z.-H.; Li, X.; Qiu, J.-F.; Cui, W.-Z.; Sima, Y.-H.; Cui, W.-Z.; Xu, S.-Q. Metabolomics differences between silkworms (*Bombyx mori*) reared on fresh mulberry (Morus) leaves or artificial diets. *Sci. Rep.* **2017**, *7*, 1–16. [CrossRef] [PubMed]
5. Meyer-Rochow, V.B. Therapeutic arthropods and other, largely terrestrial, folk-medicinally important invertebrates: A comparative survey and review. *J. Ethnobiol. Ethnomed.* **2017**, *13*, 9. [CrossRef]
6. Butkhup, L.; Jeenphakdee, M.; Jorjong, S.; Samappito, S.; Samappito, W.; Butimal, J. Phenolic composition and antioxidant activity of Thai and Eri silk sericins. *Food Sci. Biotechnol.* **2012**, *21*, 389–398. [CrossRef]
7. Tamura, Y.; Nakajima, K.-I.; Nagayasu, K.-I.; Takabayashi, C. Flavonoid 5-glucosides from the cocoon shell of the silkworm, *Bombyx Mori*. *Phytochemistry* **2002**, *59*, 275–278. [CrossRef]
8. Tabunoki, H.; Higurashi, S.; Ninagi, O.; Fujii, H.; Banno, Y.; Nozaki, M.; Kitajima, M.; Miura, N.; Atsumi, S.; Tsuchida, K. A carotenoid-binding protein (CBP) plays a crucial role in cocoon pigmentation of silkworm (*Bombyx mori*) larvae. *FEBS Lett.* **2004**, *567*, 175–178. [CrossRef]
9. Tao, M.; Wang, C.; Liao, D.; Liu, H.; Zhao, Z.; Zhao, Z. Purification, modification and inhibition mechanism of angiotensin I-converting enzyme inhibitory peptide from silkworm pupa (*Bombyx mori*) protein hydrolysate. *Process. Biochem.* **2017**, *54*, 172–179. [CrossRef]
10. Zhao, P.; Xia, F.; Jiang, L.; Guo, H.; Xu, G.; Sun, Q.; Wang, B.; Wang, Y.; Lu, Z.; Xia, Q. Enhanced antiviral immunity against *Bombyx mori* cytoplasmic polyhedrosis virus via overexpression of peptidoglycan recognition protein S2 in transgenic silkworms. *Dev. Comp. Immunol.* **2018**, *87*, 84–89. [CrossRef]
11. Rattana, S.; Katisart, T.; Sungthong, B.; Butiman, C. Acute and sub-acute toxicities of Thai Silkworm Powder (*Bombyx mori* Linn.) from three races in male Wistar rats and in vitro antioxidant activities. *Pharmacogn. J.* **2017**, *9*, 541–545. [CrossRef]
12. Xu, S.; Wang, F.; Wang, Y.; Wang, R.; Hou, K.; Tian, C.; Ji, Y.; Yang, Q.; Zhao, P.; Xia, Q. A silkworm based silk gland bioreactor for high-efficiency production of recombinant human lactoferrin with antibacterial and anti-inflammatory activities. *J. Biol. Eng.* **2019**, *13*, 61. [CrossRef] [PubMed]
13. Dutta, P.; Dey, T.; Manna, P.; Kalita, J. Antioxidant potential of *Vespa affinis* L., a traditional edible insect species of North East India. *PLoS ONE* **2016**, *11*, e0156107. [CrossRef] [PubMed]
14. Dutta, P.; Dey, T.; Dihingia, A.; Manna, P.; Kalita, J. Antioxidant and glucose metabolizing potential of edible insect, Brachytrupes orientalis via modulating Nrf2/AMPK/GLUT4 signaling pathway. *Biomed. Pharmacother.* **2017**, *95*, 556–563. [CrossRef] [PubMed]

15. Lee, J.H.; Kim, S.; Jo, Y.-Y.; Kweon, H.; Jeon, J.Y.; Ju, W.-T.; Kim, H.-B.; Kim, K.-Y.; Kim, S.-W.; Kim, S.-B. Effect of humidity on the quality characteristics of the 3 rd day of 5 th instar silkworm powder. *Int. J. Ind. Entomol.* **2019**, *39*, 74–81.

16. Suk, W.; Kim, J.; Kim, D.-Y.; Lim, H.; Choue, R. Effect of Wheat Flour Noodles with *Bombyx mori* Powder on Glycemic Response in Healthy Subjects. *Prev. Nutr. Food Sci.* **2016**, *21*, 165. [CrossRef]

17. Park, Y.-S.; Choi, Y.-S.; Hwang, K.-E.; Kim, T.-K.; Lee, C.-W.; Shin, D.-M.; Han, S.G. Physicochemical properties of meat batter added with edible silkworm pupae (*Bombyx mori*) and transglutaminase. *Korean J. Food Sci. Anim. Resour.* **2017**, *37*, 351. [CrossRef]

18. Biró, B.; Fodor, R.; Szedljak, I.; Pásztor-Huszár, K.; Gere, A. Buckwheat-pasta enriched with silkworm powder: Technological analysis and sensory evaluation. *LWT* **2019**, *116*, 108542. [CrossRef]

19. Hashim, M.A.; Yahya, F.; Mustapha, W.A.W. Effect of different drying methods on the morphological structure, colour profile and citral concentration of Lemongrass (*Cymbopogon citratus*) powder. *Asian J. Agric. Biol.* **2019**, *7*, 93–102.

20. Nur Tantiyani, A.O.; Razalib, M.E.F.M. Drying of Instant Coffee in a Spray Dryer. *J. Kejuruter.* **2019**, *31*, 295–301.

21. Baek, M.; Yoon, Y.-I.; Kim, M.; Hwang, J.-S.; Goo, T.-W.; Yun, E.-Y. Physical and sensory evaluation of *Tenebrio molitor* larvae cooked by various cooking methods. *Korean J. Food Cook. Sci.* **2015**, *31*, 534–543. [CrossRef]

22. Jensen, A.B.; Evans, J.; Jonas-Levi, A.; Benjamin, O.; Martinez, I.; Dahle, B.; Roos, N.; Lecocq, A.; Foley, K. Standard methods for *Apis mellifera* brood as human food. *J. Apic. Res.* **2019**, *58*, 1–28. [CrossRef]

23. Hashim, H.; Embong, F.A.; Yee, C.K.; Daud, N.M.; Rahman, H.A. Effect of different drying methods and ethanol concentrations on antioxidant and anti-hyperglycaemic properties of gynura procumbens leaves extract. *Malays. Appl. Biol.* **2018**, *47*, 157–164.

24. Norafida, A.; Aminah, A. Effect of blanching treatments on antioxidant activity of frozen green Capsicum (*Capsicum annuum* L. var bell pepper). *Int. Food Res. J.* **2018**, *25*, 1427–1434.

25. Maturin, L.; Peeler, J. BAM aerobic plate count. In *Bacteriological Analytical Manual*; US Food and Drug Administration: Silver Spring, MD, USA, 2001.

26. Li, H.-B.; Cheng, K.-W.; Wong, C.-C.; Fan, K.-W.; Chen, F.; Jiang, Y. Evaluation of antioxidant capacity and total phenolic content of different fractions of selected microalgae. *Food Chem.* **2007**, *102*, 771–776. [CrossRef]

27. Phongthai, S.; D'Amico, S.; Schoenlechner, R.; Homthawornchoo, W.; Rawdkuen, S. Fractionation and antioxidant properties of rice bran protein hydrolysates stimulated by in vitro gastrointestinal digestion. *Food Chem.* **2018**, *240*, 156–164. [CrossRef]

28. Taba, M.; Gogoi, H. Report on Wild Eri Silkworm *Samia canningii* Hutton (Lepidoptera: Saturniidae) from Arunachal Pradesh, India. *Natl. Acad. Sci. Lett.* **2019**, *42*, 147–150. [CrossRef]

29. Gurjar, T.S.; Siddhapara, M.R.; Surani, P.M. Biology of mulberry silkworm, *Bombyx mori* L. on mulberry, *Morus alba* L. *J. Entomol.* **2018**, *6*, 276–280.

30. Karthick Raja, P.; Aanand, S.; Stephen Sampathkumar, J.; Padmavathy, P. Silkworm pupae meal as alternative source of protein in fish feed. *J. Entomol. Zool. Stud.* **2019**, *7*, 78–85.

31. Yu, Y.; Li, H.; Zhang, B.; Wang, J.; Shi, X.; Huang, J.; Yang, J.; Zhang, Y.; Deng, Z. Nutritional and functional components of mulberry leaves from different varieties: Evaluation of their potential as food materials. *Int. J. Food Prop.* **2018**, *21*, 1495–1507. [CrossRef]

32. Kim, H.-W.; Setyabrata, D.; Lee, Y.J.; Jones, O.G.; Kim, Y.H.B. Pre-treated mealworm larvae and silkworm pupae as a novel protein ingredient in emulsion sausages. *Innov. Food Sci. Emerg. Technol.* **2016**, *38*, 116–123. [CrossRef]

33. Chieco, C.; Morrone, L.; Bertazza, G.; Cappellozza, S.; Saviane, A.; Gai, F.; Di Virgilio, N.; Rossi, F. The Effect of Strain and Rearing Medium on the Chemical Composition, Fatty Acid Profile and Carotenoid Content in Silkworm (*Bombyx mori*) Pupae. *Animals* **2019**, *9*, 103. [CrossRef] [PubMed]

34. Rahmathulla, V.; Tilak, R.; Rajan, R. Influence of moisture content of mulberry leaf on growth and silk production in *Bombyx mori* L. *Caspian J. Environ. Sci.* **2006**, *4*, 25–30.

35. Omotoso, O.T. An Evaluation of the Nutrients and Some Anti-nutrients in Silkworm, *Bombyx mori* L. (Bombycidae: Lepidoptera). *Jordan J. Biol. Sci.* **2015**, *8*, 45–50. [CrossRef]

36. Li, M.; Li, F.; Lu, Z.; Fang, Y.; Qu, J.; Mao, T.; Wang, H.; Chen, J.; Li, B. Effects of TiO_2 nanoparticles on intestinal microbial composition of silkworm. *Bombyx Mori. Sci. Total Environ.* **2020**, *704*, 135273. [CrossRef] [PubMed]

37. Zheng, L.; Lin, L.; Su, G.; Zhao, Q.; Zhao, M. Pitfalls of using 1, 1-diphenyl-2-picrylhydrazyl (DPPH) assay to assess the radical scavenging activity of peptides: Its susceptibility to interference and low reactivity towards peptides. *Food Res. Int.* **2015**, *76*, 359–365. [CrossRef]

38. Chuah, A.M.; Lee, Y.-C.; Yamaguchi, T.; Takamura, H.; Yin, L.-J.; Matoba, T. Effect of cooking on the antioxidant properties of coloured peppers. *Food Chem.* **2008**, *111*, 20–28. [CrossRef]

39. Gunathilake, K.; Ranaweera, K.; Rupasinghe, H. Effect of different cooking methods on polyphenols, carotenoids and antioxidant activities of selected edible leaves. *Antioxidants* **2018**, *7*, 117. [CrossRef]

40. Kao, F.-J.; Chiu, Y.-S.; Chiang, W.-D. Effect of water cooking on antioxidant capacity of carotenoid-rich vegetables in Taiwan. *J. Food Drug Anal.* **2014**, *22*, 202–209. [CrossRef]

41. Garofalo, C.; Osimani, A.; Milanović, V.; Taccari, M.; Cardinali, F.; Aquilanti, L.; Riolo, P.; Ruschioni, S.; Isidoro, N.; Clementi, F. The microbiota of marketed processed edible insects as revealed by high-throughput sequencing. *Food Microbiol.* **2017**, *62*, 15–22. [CrossRef]

42. Ma, M.; Hussain, M.; Dong, S.; Zhou, W. Characterization of the pigment in naturally yellow-colored domestic silk. *Dyes Pigment.* **2016**, *124*, 6–11. [CrossRef]

43. Dorouzi, M.; Mortezapour, H.; Akhavan, H.-R.; Moghaddam, A.G. Tomato slices drying in a liquid desiccant-assisted solar dryer coupled with a photovoltaic-thermal regeneration system. *Solar Energy* **2018**, *162*, 364–371. [CrossRef]

44. Mujuru, F.M.; Kwiri, R.; Nyambi, C.; Winini, C.; Moyo, D.N. Microbiological quality of Gonimbrasia belina processed under different traditional practices in Gwanda, Zimbabwe. *Int. J. Curr. Microbiol. Appl. Sci.* **2014**, *3*, 1085–1094.

45. Ho, L.-H.; Ramli, N.F.; Tan, T.-C.; Muhamad, N.; Haron, M.N. Effect of extraction solvents and drying conditions on total phenolic content and antioxidant properties of watermelon rind powder. *Sains Malays.* **2018**, *47*, 99–107. [CrossRef]

46. Correia, P.; Guiné, R.P.; Correia, A.C.; Gonçalves, F.; Brito, M.; Ribeiro, J. Physical, chemical and sensorial properties of kiwi as influenced by drying conditions. *Agric. Eng. Int. CIGR J.* **2017**, *19*, 203–212.

47. De Carvalho Tavares, I.M.; de Castilhos, M.B.M.; Mauro, M.A.; Ramos, A.M.; de Souza, R.T.; Gómez-Alonso, S.; Gomes, E.; Da-Silva, R.; Hermosín-Gutiérrez, I.; Lago-Vanzela, E.S. BRS Violeta (BRS Rúbea× IAC 1398-21) grape juice powder produced by foam mat drying. Part I: Effect of drying temperature on phenolic compounds and antioxidant activity. *Food Chem.* **2019**, *298*, 124971. [CrossRef] [PubMed]

48. Kunz, R.I.; Brancalhão, R.M.C.; Ribeiro, L.D.F.C.; Natali, M.R.M. Silkworm sericin: Properties and biomedical applications. *BioMed Res. Int.* **2016**, *2016*, 1–19. [CrossRef]

49. Taufik, Y.; Widiantara, T.; Garnida, Y. The effect of drying temperature on the antioxidant activity of black mulberry leaf tea (Morus nigra). *Rasayan J. Chem.* **2016**, *9*, 889–895.

50. Minatel, I.O.; Borges, C.V.; Ferreira, M.I.; Gomez, H.A.G.; Chen, C.-Y.O.; Lima, G.P.P. Phenolic compounds: Functional properties, impact of processing and bioavailability. In *Phenolic Compounds Biological Activity*; InTech: Rijeka, Croatia, 2017; pp. 1–24.

51. Vega-Gálvez, A.; Ah-Hen, K.; Chacana, M.; Vergara, J.; Martínez-Monzó, J.; García-Segovia, P.; Lemus-Mondaca, R.; Di Scala, K. Effect of temperature and air velocity on drying kinetics, antioxidant capacity, total phenolic content, colour, texture and microstructure of apple (var. Granny Smith) slices. *Food Chem.* **2012**, *132*, 51–59.

Article

Nutritional Value of the Larvae of the Alien Invasive Wasp *Vespa velutina nigrithorax* and Amino Acid Composition of the Larval Saliva

Hyeyoon Jeong [1], Ja Min Kim [1,2], Beomsu Kim [1], Ju-Ock Nam [1,2], Dongyup Hahn [1,2,3,*] and Moon Bo Choi [2,4,*]

[1] School of Food Science and Biotechnology, College of Agriculture and Life Sciences, Kyungpook National University, Daegu 41566, Korea; ddi02084@naver.com (H.J.); kimjamin1987@naver.com (J.M.K.); vincent1580@naver.com (B.K.); namjo@knu.ac.kr (J.-O.N.)
[2] Institute of Agricultural Science and Technology, Kyungpook National University, Daegu 41566, Korea
[3] Department of Integrative Biology, Kyungpook National University, Daegu 41566, Korea
[4] School of Applied Biosciences, College of Agriculture and Life Sciences, Kyungpook National University, Daegu 41566, Korea
* Correspondence: dohahn@knu.ac.kr (D.H.); kosinchoi@hanmail.net (M.B.C.)

Received: 30 May 2020; Accepted: 30 June 2020; Published: 6 July 2020

Abstract: The systematic investigations on the value of social wasps as a food resource are deficient, in spite of the long history of the utilization of social wasps as food and pharmaceutical bioresources. *Vespa velutina nigrithorax* is an invasive alien wasp species that is currently dominating in East Asia and Europe, bringing huge economic damages. As a control over alien species is made when the valuable utilization of the invasive species as a potential resource are discovered, investigations on the potential of *V. v. nigrithorax* as a useful bioresource are also in demand. Nutritional and heavy metal analyses of the larvae revealed their balanced and rich nutritional value and safety as a food resource. The larval saliva amino acid composition was investigated for further study on amino acid supplementation and exercise enhancement.

Keywords: edible insects; alternative food resource; wasp larva; *Vespa velutina nigrithorax*

---

1. Introduction

Social wasps are recognized as harmful insects. They are a threat to humans due to their stinging behaviors that might lead to death [1,2] and their aggressive nature to hunt for honeybees brings huge economic losses in the beekeeping industry [1,3,4]. However, especially in East Asia, social wasps are bioresources that have been utilized for medicinal and food purpose for a long time. In Korea and China, nests of social wasps (*Nidus vespae*) have been used as an ingredient of crude drugs [5,6], or an ingredient of health-promoting traditional liquors [5]. The entomophagic records about wasps are mostly related to consumption of larvae, which are available in Papua New Guinea [7], China [8] and Laos [9]. In Japan, larvae of wasps, 'hachinoko' in Japanese, have long been consumed, and even canned products of the larvae are available at markets [10]. However, wasp larvae have been rather delicacies—not part of daily diets—due to their limited supply from nature. Particularly in China and Japan, where wasp materials provided by wasp hunters are traded at high prices and where wasps were traditionally used as food or pharmacological resources, there have been attempts to develop wasp breeding techniques to ensure sufficient supply of materials [8,11]. Fortunately, in some cases, embryo nest breeding was often successful, but none of the reliable systematic breeding technology has been established yet [8]. To establish wasp larvae as a regular food resource, techniques for mass rearing to ensure sufficient quantities of materials is required [12]. However, prior to the supply issue, it is necessary to evaluate if wasp larvae are worth development for a regular food resource in regards to their nutritional value and food safety. The investigation on *Vespa mandarinia* larvae by Kim and Jung [13] in 2013 is the only attempt to evaluate the nutritional value of wasp larvae to investigate potential as a food material including taxonomic identification.

Invasion of *V. v. nigrithorax* is a momentous ecological issue to Europe and East Asia now. Since the first invasion of *V. v. nigrithorax* into South Korea in 2003, it has spread throughout the country [1,5,14–16], and invaded Europe [17] and Japan [18]. *V. v. nigrithorax* have a serious economic impact as they hunt honeybees in apiaries [2,19] and a public health impact as they sting people in urban areas [1]. This wasp is also causing ecosystem disturbances through competition and interference among *Vespa* species within the ecosystem [19–21]. The southern region of Korea is densely populated by *V. v. nigrithorax*, and this wasp is the most prevalent in urban and forest areas [22]. In this context of momentous significance, *V. v. nigrithorax* was selected for investigation.

In this study, we collected larvae from *V. v. nigrithorax* nests and examined the amino acid composition of their saliva, and nutritional analyses of larvae were performed. The nutritional value of *V. v. nigrithorax* larvae have not been revealed, though *V. v. nigrithorax* is now one of the frequently encountered species in East Asia.

2. Materials and Methods

2.1. Wasp Larvae Collection and Preparation

Two *V. v. nigrithorax* nests were collected from tree branches at heights of approximately 10–15 m in October 2018 in Daegu, South Korea (Figure 1). One of the authors (M.B.C.) identified the builders of these nests as *V. v. nigrithorax*. The nests contained mature colonies with approximately 1200–1400 adults. The adults were removed from the nests and the combs were separated. Healthy larvae of the fourth or fifth stage were selected for saliva collection. The rest of the larvae were preserved at −80 °C for nutritional analysis.

Figure 1. The nest of *V. v. nigrithorax* that hovers larvae. (**A**) Larvae of *V. v. nigrithorax* secreting saliva; (**B**) larvae of *V. v. nigrithorax* in cells; (**C**) the nest of *V. v. nigrithorax* hung on a tree; (**D**) the nest of *V. mandarinia*.

2.2. Larval Saliva Preparation

To induce salivation, capillary pipette tips were used to gently tap the mouth parts of the larvae and collect saliva. A 50–200 μL volume of clear saliva was collected from every larva. The tubes containing saliva were immediately frozen and kept at −80 °C until analysis.

2.3. Amino Acid Analysis of the Larvae and Larval Saliva

Qualitative and quantitative analysis for the composition of amino acids in larval saliva was performed through the amino acid automatic analyzer (L-8900, Hitachi, Tokyo, Japan) equipped with an ion exchange column (No. 2622 SCPF, 4.6 mm × 60 mm). Buffer solutions PF-1, 2, 3, 4, 6, PF-RG, R-3, and C-1 (Wako, Osaka, Japan) were used as the mobile phase with a flow rate of 0.35 mL/min, and the column temperature was maintained at 50 °C. The reaction liquid flow rate was 0.3 mL/min, monitoring the wavelengths of 440 nm from 570 nm, and the injection volume was 20 μL. The concentration of amino acids in larval saliva samples was determined by the peak area of standard samples. For hydrolyzed amino acid analysis, 1 g of freeze-dried larvae and 40 mL of 6 N HCl were placed in a round flask and mixed, followed by hydrolysis by injecting nitrogen gas at 110 °C for 24 h. The hydrochloric acid was concentrated under reduced pressure at 50 °C, and the concentrated sample was diluted with 50 mL of 0.2 N sodium citrate buffer (pH 2.2) and filtered with filter paper (0.45 μm). The filtered sample (30 μL) was analyzed using an amino acid analyzer (L-8900, Hitachi, Tokyo, Japan).

2.4. Nutirtional Analysis of the Larvae

The nutritional composition (ash, moisture, lipid, protein and carbohydrate) of lyophilized larvae was determined by following the standard method of the Korean Food Standard Codex [23]. Briefly, the crude protein content was determined by the Kjeldahl method using a crude protein analyzer (Kjeltec™ 8400, Foss, Hilleroed, Denmark), and was calculated by multiplying the nitrogen content using a coefficient of 6.25. Crude fat was extracted from larvae in a Soxhlet apparatus with ethyl ether as the solvent. Moisture content was determined by drying the sample in an oven at 105 °C until a constant weight was obtained. Ash content was determined by dry-ashing in a furnace (J-FM3, Jisico, Seoul, Korea) at 550 °C for 12 h. Carbohydrate content had been determined by subtracting the sum of the weights including ash, moisture, protein and lipid.

2.5. Hazardous Heavy Metals Ananlyses of the Larvae

The contents of hazardous metals, arsenic (As), cadmium (Cd), mercury (Hg) and lead (Pb) were determined using an inductively coupled plasma atomic emission spectroscopy instrument (ICP/OES, Avio 500, Perkin-Elmer, Waltham, MA, USA). For analysis, 140 mg of the lyophilized larvae underwent a nitric acid-pretreatment and were digested with a microwave digestion system (UltraWAVE, Milestone, Sorisole, Italy). The digested samples were diluted with 2% nitric acid (w/w) and injected for instrumental analysis. The analytical condition used for ICP/OES is described in Table 1.

Table 1. Analytical condition used for inductively coupled plasma atomic emission spectroscopy instrument (ICP/OES).

Condition	
Parameter	**Specification**
Power	1500 W
Plasma gas flow	10 L/min
Auxiliary gas flow	0.2 L/min
Nebulizer gas flow	0.6 L/min
Nebulizer	Concentric nebulizer, glass

Table 1. *Cont.*

Condition	
Parameter	**Specification**
Torch	Quartz
Spray chamber	Cylconic spary chamber
Injector	Alumina, inner diameter 2 mm
Pump tubing	Polyvinyl Chloride (PVC)
Sample pump flow	1.5 mL/min
Rinse/read delay	60 s
Integration time	3 (3 replicates)
Plasma view	Axial

3. Results and Discussion

3.1. Amino Acid Composition of Larval Saliva

One of a few examples for the application of wasp materials to food is the development of a supplementary beverage for physical exercise inspired from the amino acid composition of *V. mandarinia* larval saliva [24]. The beverage products do not contain the actual larval saliva of *V. mandarinia*, but an amino acid mixture composed of the same amino acid composition of *V. mandarinia* larval saliva. The products were developed on the basis of a series of investigations on exercise endurance enhancement effects by supplementation of the amino acid mixture of which the composition is the same as the larval saliva of *V. mandarinia* [25–27]. This idea has come from trophallaxis, a unique way of survival observed in social wasps. Trophallaxis is common among wasps; carnivorous larvae which feed on meat pellets prepared by adult workers, while larvae produce and pass an oral exudate to adult workers in response [27,28]. For social wasps, larval saliva is considered a source of energy and motivation for adult workers that fly a distance of about 100 km daily only to prepare meat pellets for larvae [29]. Takashi and colleagues reported the composition of amino acids in the larval saliva from five species of hornets including *Vespa mandarinia*, *Vespa crabo*, *Vespa tropica*, *Vespa analis* and *Vespa xanthoptera*, collected in central Japan, which has a temperate climate [27]. Based on this research, other related researches were performed subsequently, which were focused on observations of the metabolism of animals administered an amino acid mixture of an identical composition to the larval saliva of *V. mandarinia* [25,26,29–32]. The vespa amino acid mixture (VAAM) was coined, based on the free amino acid composition of the larval saliva of *V. mandarina*, which may be related to aggressiveness and the longest distance flown for hunting [25,27,29]. Studies on the metabolism using of exercising animals revealed that amino acids in larval saliva may act as energy sources and enhance exercise performance through controlling the metabolism [25,26,29–31]. *Vespa mandarinia*, the most dominant hornet species in central Japan, has been used as a reference, as it boasts the longest flight distance and the most extensive hunting domains [25,27,29].

Reports on *V. v. nigrithorax* invasions to temperate regions sparked an increasing interest in them as their distribution is expanding [1,16,33,34]. However, studies on trophallaxis and the larval saliva of *V. v. nigrithorax* has not been performed despite the significance of these topics. According to previous research by Takashi et al. [27], each wasp species has a peculiar larval saliva amino acid composition and larval saliva is the principal component of trophallaxis in hornets. There may be a number of factors that influenced the expansion of the *V. v. nigrithorax* territory beyond the tropical region. In addition, *V. v. nigrithorax* are found in numbers of 1500–2500 individuals per nest, and their nests are 2–3 times larger than those of other *Vespa* spp [2,35,36]. The amino acid contents in the larval saliva of *V. v. nigrithorax* might be correlated with the energy required for exercise and metabolism and could have influenced the growth of their population, which might inspire us to coin a supplementary diet formula. Thus, the amino acid composition of the larval saliva of *V. v. nigrithorax* was investigated.

As *V. mandarinia* has been the standard for the physiological studies involved in vespa amino acid mixture (VAAM) supplementation, our results on the amino acid composition of the larval saliva of *V. v. nigrithorax* were also presented alongside the amino acid composition of the larval saliva of *V. mandarinia* reported by Takashi et al. [27] (Table 2). A notable feature of the amino acid composition of the larval saliva of *V. v. nigrithorax* was the high proportion of the physiological amino acids compared to the total amount of amino acids measured from the hydrolyzed product of larval saliva. The total amount of physiological amino acids in the larval saliva of *V. v. nigrithorax* was 38.29 μmol/mL and the total amount of amino acids deduced from the hydrolyzed product was 67.24 μmol/mL; these results suggest that the larval saliva of *V. v. nigrithorax* has a higher ratio of free amino acids than that of *V. mandarinia*. Physiological amino acids are valuable nutrients for adult wasp workers; adult wasps are able to consume only liquid foods and hardly digest polymer-type nutrients, owing to their short and simplified gut structure and narrow waists [37]. Worker wasps in particular, spend most of their time flying for hunting purposes; therefore, free amino acids in larval saliva could help them prolong their flight time, as has been reported in previous studies on animal exercise physiology and VAAM supplementation [25,29]. The amino acids that composed the physiological amino acid mixture of the larval saliva of *V. v. nigrithorax* were similar to those of saliva from other wasps, as revealed by Takashi et al. [27]. A notable feature of the physiological amino acid composition of the larval saliva of *V. v. nigrithorax* was the high proportion of proline. The content of proline is believed to be correlated with the daily flight distance and hunting domain size of *Vespa* spp. [27], as proline is metabolized by insects to produce energy for hunting [38]. Branched chain amino acids (BCAAs) such as Ile, Val, and Leu were detected in substantial amounts (3.28, 5.45, and 3.75 mol%, respectively). In animals, BCAAs are directly metabolized in muscles as energy sources and the intake of BCAAs helps prevent muscular loss caused by endurance exercise [39]. The other essential amino acids (threonine, phenylalanine, lysine, and histidine) were detected in fair amounts (5.48, 5.34, 9.88, and 4.28 mol%, respectively), except for tryptophan and methionine.

Table 2. Physiological amino acid compositions of larval saliva of *V. v. nigrithorax* compared to *V. mandarinia*.

Amino Acids	*V. mandarinia* (μmol/mL) [a]	*V. v. nigrithorax* (μmol/mL)	*V. mandarinia* (mol%) [a]	*V. velutina* (mol%)
phosphoserine	-	0.08	-	0.22
Tau	0.39	0.82	0.51	2.15
NH$_3$	0.87	2.70	1.13	7.04
Asp	0.12	0.06	0.16	0.16
Thr	5.40	2.10	7.03	5.48
Ser	1.86	0.96	2.42	2.50
Glu	2.40	1.71	3.13	4.46
sarcosine	-	0.98	-	2.57
Gly	14.37	3.30	18.72	8.63
Ala	4.54	0.74	5.91	1.93
Val	4.40	2.09	5.73	5.45
Cys	-	0.36	-	0.94
Ile	3.40	1.26	4.43	3.28
Leu	4.62	1.44	6.02	3.75
Tyr	4.48	1.06	5.84	2.76
Phe	2.89	2.04	3.76	5.34
β-Ala	0.15	0.04	0.20	0.10
β-amino isobutyric acid	-	0.20	-	0.52
GABA	0.26	0.07	0.34	0.17

Table 2. *Cont.*

Amino Acids	*V. mandarinia* (μmol/mL) [a]	*V. v. nigrithorax* (μmol/mL)	*V. mandarinia* (mol%) [a]	*V. velutina* (mol%)
EtAm	0.82	0.64	1.07	1.67
Orn	0.82	0.57	1.07	1.50
Lys	6.48	3.78	8.44	9.88
His	1.94	1.64	2.53	4.28
3-MeHis	0.36	0.12	0.47	0.31
Arg	2.64	0.74	3.44	1.92
hydroxy proline	-	0.16	-	0.41
Pro	13.55	8.65	17.65	22.58
Total	76.76	38.29	100.00	100.00

[a] Amino acid contents in larval saliva of *V. mandarinia* is quoted from the previous research by Takashi et al. [27].

The amino acid content of the hydrolyzed product of the larval saliva of *V. v. nigrithorax* revealed that a few amino acids were present as peptide constituents (Table 3). Acidic amino acids aspartate and glutamate were detected in significantly different amounts in the physiological and hydrolyzed amino acid mixtures. Aspartate and glutamate were detected in amounts of 0.06 and 1.71 μmol/mL, respectively, in the physiological amino acid mixture, while they were detected in amounts of 3.21 and 10.71 μmol/mL, respectively, in the hydrolyzed amino acid mixture. Other amino acids such as glycine, alanine, isoleucine, and lysine were detected in the hydrolyzed amino acid mixture in quantities that were about 2–3 times higher than those in the physiological amino acid mixture. The larval saliva of *V. v. nigrithorax* did not contain higher amino acid and protein amounts than that of *V. mandarinia*. Takashi et al. [27] suggested that high amino acid and protein contents in larval saliva may be correlated with the body weight and the ecological dominance of wasps in nature. However, the larval saliva of the invasive alien species *V. v. nigrithorax* has not been subjected to amino acid analysis up to this point.

Table 3. Hydrolyzed amino acid compositions of larval saliva of *V. v. nigrithorax* compared to that of *V. mandarinia*.

Amino Acids	*V. mandarinia* μmol/mL [a]	*V. v. nigrithorax* μmol/mL	*V. mandarinia* mol% [a]	*V. v. nigrithorax* mol%
Asp	12.29	3.21	5.62	4.77
Thr	6.71	3.18	3.07	4.74
Ser	8.47	1.59	3.87	2.36
Glu	32.17	10.71	14.71	15.93
Gly	29.85	8.24	13.65	12.25
Ala	8.90	2.14	4.07	3.18
Cys	0.67	0.57	0.31	0.85
Val	11.96	4.26	5.47	6.34
Met	3.15	0.33	1.44	0.49
Ile	15.27	3.29	6.98	4.90
Leu	15.98	3.44	7.31	5.11
Tyr	7.74	0.76	3.54	1.13
Phe	6.12	3.45	2.80	5.13
Lys	22.81	7.29	10.43	10.85
His	14.13	4.00	6.46	5.96
Arg	12.75	1.60	5.83	2.38
Pro	9.72	9.17	4.44	13.64
Total	218.69	67.24	100.00	100.00

[a] Amino acid contents in larval saliva of *V. mandarinia* is quoted from the previous research by Takashi et al. [27].

3.2. Nutritional Composition of the Larvae

Table 4 represents the nutrient composition of the lyophilized *V. v. nigrithorax* larvae compared to other representative edible insect larvae, *Protaetia brevitarsis seulensis* and *Tenebrio molitor*. The nutrient compositions of *V. mandarinia, P. b. seulensis* and *T. molitor* larvae were quoted from previous reports [35–37]. Among carbohydrates in the larvae, the sugar content was 6.08 g/100 g. And the content of saturated fat was 4.31 g/100 g, while the content of trans fat was detected under the limit of quantification. Total cholesterol content was 31.97 mg/100 g. Sodium content was 60.91 mg/100 g.

Table 4. Nutritional composition of *V. v. nigrithorax* larvae, compared to those of *V. mandarinia, P. b. seulensis* and *T. molitor* larvae (g/100 g).

Nutrients	*Vespa velutina nigrithorax*	*Vespa mandarinia*[a]	*Protaetia brevitarsis seulensis*[a]	*Tenebrio molitor*[a]
Moisture	2.78	-	6.66	5.33
Crude protein	48.64	59.7	57.86	46.44
Crude fat	13.23	20.6	16.57	32.70
Crude ash	3.04	4.1	8.36	2.86
Carbohydrate	32.31	15.6	10.56	12.67

[a] Nutritional composition of *V. mandarinia* [13], *P. b. seulensis* [40] and *T. molitor* [41] larvae are quoted from previous research.

The content of protein in *V. v. nigrithorax* larvae was similar to that of *T. molitor* larvae, and less than in *P. b. seulensis* larvae. The content of fat in *V. v. nigrithorax* larvae was slightly less than in *P. b. seulensis* larvae and far less than in *T. molitor* larvae. Interestingly, carbohydrate content in *V. v. nigrithorax* larvae was nearly three times of the content in other larvae. Overall nutrient contents in *V. v. nigrithorax* larvae imply the larvae may provide a nutritionally balanced diet, as they contain sufficient and balanced amounts of essential nutrients compared to other representative edible insect larvae, *P. b. seulensis* and *T. molitor*. Compared to dry whole milk, the protein content in *V. v. nigrithorax* larvae (48.64%) was an overwhelmingly great amount, as protein takes 26.3% of dry milk weight [42]. The protein content in *V. v. nigrithorax* larvae was even more than commercial dried whey protein concentrate products (protein contents 35–39%) [43].

In 2010, the South Korean government enacted the "Act of Fosterage and Support of the Insect Industry" to establish a legal basis for supporting the growth of the insect industry [44]. Along with the approved insect food ingredients (crickets, *T. molitor* larvae, *P. b. seulensis* larvae, and silkworm), wasps are often ingested in Korea even if they are not approved as food ingredients yet [13,45]. However, the nutritional analysis may provide firm grounds for approval of wasp larvae as a food ingredient along with the entomophagic records of wasp larvae in Japan [46], Papua New Guinea [7] and Laos [9].

3.3. Amino Acid Composition in Larvae

Composition of free amino acids in the larvae and amino acids in the hydrolyzed products of the larvae (Table 5.) demonstrated that the larvae of *V. v. nigrithorax* might be a balanced source for essential amino acids. Total content of free amino acids was nearly a quarter of constitutional amino acids of the larvae, which reflected that the larvae might be good resources for amino acid supplement. Twenty-seven amino acids were found in the free amino acid mixture, and sixteen amino acids were detected in hydrolyzed products of the larvae. In particular, essential amino acid contents, except Met and Trp (Arg, His, Ile, Leu, Lys, Phe, Thr and Val), were detected in substantial amounts in hydrolyzed products of the larvae (7.74–24.86 mg/g).

Table 5. Amino acids composition in the larvae.

Amino Acids	Free Amino Acids in The Larvae (mg/g)	Free Amino Acids in the Larvae (mol%)	Hydrolyzed Amino Acids in the Larvae (mg/g)	Hydrolyzed Amino Acids in the Larvae (mol%)
phosphoserine	0.63	0.53		
Tau	8.14	10.17		
NH$_3$	0.30	2.76		
Asp	0.45	0.53	24.86	7.38
Thr	3.17	4.16	13.35	4.43
Ser	1.59	2.36	11.37	4.28
Glu	2.87	3.05	44.71	12.02
Gly	10.26	21.37	17.51	9.22
Ala	2.85	5.00	12.12	5.38
Val	2.79	3.72	18.08	6.10
Cys	-	-	3.31	1.08
Ile	2.68	3.19	13.61	4.10
Leu	3.31	3.95	21.81	6.57
Tyr	3.62	3.12	12.07	2.63
Phe	2.68	2.54	12.90	3.09
β-Ala	0.56	0.99		
β-amino isobutyric acid	0.30	0.46		
EtAm	0.15	0.51		
Orn	0.62	0.73		
Lys	5.17	5.53	21.79	5.89
His	1.29	1.30	7.74	3.23
3-MeHis	0.10	0.10		
Arg	6.43	5.77	14.21	1.97
hydroxy proline	0.37	0.44		
Pro	12.24	16.62	24.58	13.21
Total	74.41	100	277.68	100

The amount of each hydrolyzed amino acid reflects that *V. v. nigrithorax* larvae could be a good supplementation for essential amino acids. Compared to the guideline of the World Health Organization (WHO) for daily adult essential amino acid requirement [47], the essential amino acids that might be provided by *V. v. nigrithorax* larvae are balanced and sufficient. Table 6 represents daily adult amino acid requirement suggested by the WHO, the amount of each essential amino acid required for an adult of 70 kg weight, and the amount of each essential amino acid equivalent to amino acid contained in 100 g of the larvae. On the basis of the guideline from the WHO, the essential amino acid requirements of a 70 kg adult was deduced, and could be mostly fulfilled by supplementation of amino acids provided from *V. v. nigrithorax* larvae. The amounts of Thr, Cys, Phe, Tyr, Lys, and His contained in 100 g of the larvae were more than the daily amino acid requirements for a 70 kg adult (104–143% of the requirement). Val and Ile provided by 100 g of the larvae were estimated to account for 99% and 97% of the requirements, respectively. The Leu content in 100 g of the larvae was 80% of the requirement. Except Trp and Met which were not detected in the larvae, the majority of the essential amino acids in the larvae are sufficient to fulfil the daily requirements of amino acids in adults.

Table 6. Comparison of the essential amino acids amounts required for an adult per day and amino acids contained in 100 g of *V. v. nigrithorax* larvae.

Amino Acids	Daily Adult Amino Acid Requirements (mg/kg) [a]	Daily Amino Acid Requirements for a 70-kg Adult (mg)	Amino Acid Provided by 100 g of *V. v. nigrithorax* Larvae (mg) [b]	Ratio of Amino Acid Provided to Requirements (%)
Thr	15.0	1050	1335	127
Val	39.0	1820	1808	99
Cys	4.1	287	331	115
Ile	20.0	1400	1361	97
Leu	39.0	2730	2181	80
Phe + Tyr	25.0	1750	2497	143
Lys	30.0	2100	2179	104
His	10.0	700	774	111
Trp	4.0	280	-	
Met	10.4	728	-	

[a] Based on the daily amino acids requirements for an adult suggested by the WHO (World Health Organization) [47]; [b] deduced from the amino acid contents in the hydrolyzed product of the larvae in Table 5.

3.4. Hazardous Heavy Metal Analysis

By the analysis using the ICP/OES instrument, contents of As, Pb and Cd were not detected, while 23.56 ppb of Hg was detected in the lyophilized larvae. Compared to other edible insects, wasp larvae have an extraordinary carnivorous nature; they are fed on smaller insects hunted by adult wasps and adults are fed on saliva of the larvae. In nature, it is common to contain more heavy metals in the body if an animal species is a higher predator. However, hazardous heavy metal contents in the larvae were not as considerable as a carnivorous animal which might accumulate heavy metals from prey. There is no regulation legislated by the Ministry of Food and Drug Safety (formerly Korea Food & Drug Administration, KFDA)for edible insect products concerning heavy metal contents yet. For fishery products, up to 500 ppb of Hg is allowed to be detected, therefore, the amount of Hg in the larvae would not harm consumers.

3.5. General Discussion

The habitat of *V. v. nigrithorax* is quickly expanding northward, particularly in Korea [48], since its first reported appearance in 2003 [16]. *V. v. nigrithorax* has serious negative public health, economic, and ecological impacts [1]. In recent years, invasion of alien species has gradually increased in Korea. However, in most of the cases, the controls over the invasive species were not successful, and various damages are rapidly increasing through the nationwide spread [49–52]. However, in some cases, the control of alien species is made when the valuable utilization of the invasive species as a potential resource are discovered [53]. *V. v. nigrithorax* was designated an ecological disturbance species in 2019, causing comprehensive impact in Korea, but the attempts to discover the utilization as a useful bioresource from this species are rare [54].

Owing to environmental pressures, food and feed insecurity, and a growing demand for protein, the consumption of insects to ease the problem of global food shortages has been advocated since 1975 [55]. Along with the approved insect food ingredients (crickets, mealworm, *Protaetia brevitarsis* larvae, and silkworm), wasps are also often ingested in Korea even if they are not yet approved as food ingredients by KFDA. The history of utilizing the social wasps as a food and pharmaceutical resources suggests that *V. v. nigrithorax* could be a valuable bioresource.

4. Conclusions

From this study, the potential of *V. v. nigrithorax* larvae as a food resource has been exhibited. They contain decent amount of nutrients, while hazardous metal contents are scarce. However, additional investigations in sanitary issues and sustainable production technology are necessary to accept the wasp larvae as an alternative food source. The larval saliva of *V. v. nigrithorax* will not provide original food sources, but the amino acid contents of it can be mimicked and might inspire the development of amino acid supplement formulas for exercise enhancement. This study will initiate future systematic investigations on the potential of *V. v. nigrithorax* as a valuable bioresource.

Author Contributions: Conceptualization, M.B.C. and D.H.; methodology, D.H., H.J. and B.K.; validation, J.M.K. and M.B.C.; formal analysis, M.B.C. and J.-O.N.; investigation, H.J. and J.M.K.; resources, M.B.C.; data curation, J.M.K. and J.-O.N.; writing—original draft preparation, H.J., D.H. and M.B.C.; writing—review and editing, D.H. and M.B.C.; visualization, H.J. and B.K.; supervision, D.H. and M.B.C.; project administration, M.B.C.; funding acquisition, M.B.C. All authors have read and agreed to the published version of the manuscript.

Funding: This study was supported by the National Research Foundation of Korea (NRF) grant funded by the Korean government (Ministry of Science and ICT, No. 2018R1D1A1B07050065).

Conflicts of Interest: The authors declare no conflict of interest.

References

1. Choi, M.B.; Kim, J.K.; Lee, J.W. Increase trend of social hymenoptera (wasps and honeybees) in urban areas, inferred from moving-out case by 119 rescue services in Seoul of South Korea. *Entomol. Res.* **2012**, *42*, 308–319. [CrossRef]

2. Choi, M.B.; Kwon, O. Occurrence of Hymenoptera (wasps and bees) and their foraging in the southwestern part of Jirisan National Park, South Korea. *J. Ecol. Environ.* **2015**, *38*, 367–374. [CrossRef]

3. Abrol, D. Ecology, behaviour and management of social wasp *Vespa velutina* Smith (Hymenoptera: Vespidae), attacking honeybee colonies. *Korean J. Apic.* **1994**, *9*, 5–10.

4. Shah, F.; Shah, T. Vespa velutina, a serious pest of honey bees in Kashmir. *Bee World* **1991**, *72*, 161–164. [CrossRef]

5. Kwon, H.-O.; Kim, C.-S.; Lee, Y.-S.; Choi, M.B. Abundance of diet-derived polychlorinated dibenzo-p-dioxins and polychlorinated dibenzofurans in the bodies and nests of the yellow-legged hornet *Vespa velutina nigrithorax* and risks to human health in South Korea. *Sci. Total Environ.* **2019**, *654*, 1033–1039. [CrossRef] [PubMed]

6. Wang, B.; Zhang, C.; Gao, P.; Wu, X.; Zhao, Y. Research progress on Nidus vespae, a traditional Chinese medicine derived from insents. *J. Pharm. Sci. Innov.* **2013**, *2*, 1–9. [CrossRef]

7. Mercer, C. Sago grub production in Labu swamp near Lae, Papua New Guinea. *Klinkii* **1994**, *5*, 30–34.

8. Feng, Y.; Zhao, M.; Ding, W.; Chen, X. Overview of edible insect resources and common species utilisation in China. *J. Insects Food Feed* **2020**, *6*, 13–25. [CrossRef]

9. Boulidam, S. Edible insects in a Lao market economy. *Edible For. Insects* **2010**, 131–141.

10. Mitsuhashi, J. Insects as traditional foods in Japan. *Ecol. Food Nutr.* **1997**, *36*, 187–199. [CrossRef]

11. Nonaka, K.; Yanagihara, H. Reviving the consumption of insects in Japan: A promising case of hebo (*Vespula* spp., wasps) by high school club activities. *J. Insects Food Feed* **2020**, *6*, 45–50. [CrossRef]

12. Feng, Y.; Zhao, M.; He, Z.; Chen, Z.; Sun, L. Research and utilization of medicinal insects in China. *Entomol. Res.* **2009**, *39*, 313–316. [CrossRef]

13. Kim, H.; Jung, C. Nutritional characteristics of edible insects as potential food materials. *Korean J. Apic.* **2013**, *28*, 1–8.

14. Choi, M.B.; Kim, T.G.; Kwon, O. Recent trends in wasp nest removal and Hymenoptera stings in South Korea. *J. Med Entomol.* **2019**, *56*, 254–260. [CrossRef] [PubMed]

15. Do, Y.; Kim, J.B.; Shim, J.; Kim, C.J.; Kwon, O.; Choi, M.B. Quantitative analysis of research topics and public concern on V. velutina as invasive species in Asian and European countries. *Entomol. Res.* **2019**, *49*, 456–461. [CrossRef]

16. Kim, J.-K.; Choi, M.; Moon, T.-Y. Occurrence of *Vespa velutina* Lepeletier from Korea, and a revised key for Korean Vespa species (Hymenoptera: Vespidae). *Entomol. Res.* **2006**, *36*, 112–115. [CrossRef]

17. Villemant, C.; Barbet-Massin, M.; Perrard, A.; Muller, F.; Gargominy, O.; Jiguet, F.; Rome, Q. Predicting the invasion risk by the alien bee-hawking Yellow-legged hornet *Vespa velutina nigrithorax* across Europe and other continents with niche models. *Biol. Conserv.* **2011**, *144*, 2142–2150. [CrossRef]

18. Ueno, T. Establishment of the invasive hornet *Vespa velutina* (Hymenoptera: Vespidae) in Japan. *Int. J. Chem. Environ. Biol. Sci.* **2014**, *2*, 3.

19. Monceau, K.; Maher, N.; Bonnard, O.; Thiéry, D. Evaluation of competition between a native and an invasive hornet species: Do seasonal phenologies overlap? *Bull. Entomol. Res.* **2015**, *105*, 462–469. [CrossRef]

20. Cini, A.; Cappa, F.; Petrocelli, I.; Pepiciello, I.; Bortolotti, L.; Cervo, R. Competition between the native and the introduced hornets *Vespa crabro* and *Vespa velutina*: A comparison of potentially relevant life-history traits. *Ecol. Entomol.* **2018**, *43*, 351–362. [CrossRef]

21. Yamasaki, K.; Takahashi, R.; Harada, R.; Matsuo, Y.; Nakamura, M.; Takahashi, J.-I. Reproductive interference by alien hornet *Vespa velutina* threatens the native populations of Vespa simillima in Japan. *Sci. Nat.* **2019**, *106*, 1–5. [CrossRef] [PubMed]

22. Park, J.; Jung, C. Risk prediction of the distribution of invasive hornet, *Vespa velutina nigrothorax* in Korea using CLIMEX model. *Korean J. Apic.* **2018**, *31*, 293–303. [CrossRef]

23. Codex, K.F.S. *KFDA, 10. General Test Methods*; Korea Food Drug Administration: Seoul, Korea, 2010.

24. Abe, T.; Tsuchita, H.; Iida, K. Amino Acid Composition. U.S. Patent No. 6,224,861, 2001.

25. Abe, T.; Takiguchi, Y.; Tamura, M.; Shimura, J.; Yamazaki, K.-I. Effects of Vespa Amino Acid Mixture (VAAM) lsolated from Hornet Larval Saliva and Modified VAAM Nutrients on Endurance Exercise in Swimming Mice. *Jpn. J. Phys. Fit. Sports Med.* **1995**, *44*, 225–237. [CrossRef]

26. Sasai, H.; Matsuo, T.; Fujita, M.; Saito, M.; Tanaka, K. Effects of regular exercise combined with ingestion of vespa amino acid mixture on aerobic fitness and cardiovascular disease risk factors in sedentary older women: A preliminary study. *Geriatr. Gerontol. Int.* **2011**, *11*, 24–31. [CrossRef] [PubMed]

27. Takashi, A.; Yoshiya, T.; Hiromitsu, M.; Yasuko, Y.K. Comparative study of the composition of hornet larval saliva, its effect on behaviour and role of trophallaxis. *Comp. Biochem. Physiol. Part C Comp. Pharmacol.* **1991**, *99*, 79–84. [CrossRef]

28. Matsuura, M.; Yamane, S. *Biology of the Vespine Wasps*; Springer-Verlag Berlin Heidelberg: Berlin, Germany, 1990.

29. Tsuchita, H.; ShiraiMorishita, Y.; Shimizu, T.; Abe, T. Effects of a Vespa amino acid mixture identical to hornet larval saliva on the blood biochemical indices of running rats. *Nutr. Res.* **1997**, *17*, 999–1012. [CrossRef]

30. Ogasawara, J.; Abe, T. Amino acid mixture identical to vespa larval saliva increases both leptin secretion and basal lipolysis in rat adipocytes. *Food Sci. Technol. Res.* **2008**, *14*, 95–98. [CrossRef]

31. Shinozaki, F.; Abe, T. Synergistic effect of Vespa amino acid mixture on lipolysis in rat adipocytes. *Biosci Biotechnol. Biochem.* **2008**, *72*, 1860–1868. [CrossRef]

32. Shinozaki, F.; Abe, T.; Kamei, A.; Watanabe, Y.; Yasuoka, A.; Shimada, K.; Kondo, K.; Arai, S.; Kumagai, K.; Kondo, T.; et al. Coordinated regulation of hepatic and adipose tissue transcriptomes by the oral administration of an amino acid mixture simulating the larval saliva of Vespa species. *Genes Nutr.* **2016**, *11*, 21. [CrossRef]

33. Franklin, D.N.; Datta, S.; Keeling, M.J.; Franklin, D.N.; Keeling, M.J.; Brown, M.A.; Datta, S.; Keeling, M.J.; Cuthbertson, A.G.S.; Budge, G.E. Invasion dynamics of Asian hornet, *Vespa velutina* (Hymenoptera: Vespidae): A case study of a commune in south-west France. *Appl. Entomol. Zool.* **2017**, *52*, 221–229. [CrossRef]

34. Takahashi, J.; Okuyama, H.; Kiyoshi, T.; Takeuchi, T.; Martin, S.J. Origins of *Vespa velutina* hornets that recently invaded Iki Island, Japan and Jersey Island, UK. *Mitochondrial DNA Part A* **2019**, *30*, 434–439. [CrossRef]

35. Choi, M.B.; Martin, S.J.; Lee, J.W. Distribution, spread, and impact of the invasive hornet *Vespa velutina* in South Korea. *J. Asia-Pac. Entomol.* **2012**, *15*, 473–477. [CrossRef]

36. Rome, Q.; Muller, F.; Touret-Alby, A.; Darrouzet, E.; Perrard, A.; Villemant, C. Caste differentiation and seasonal changes in *Vespa velutina* (Hym.: Vespidae) colonies in its introduced range. *J. Appl. Entomol.* **2015**, *139*, 771–782. [CrossRef]

37. Morimoto, R. Experimental study on the trophallactic behavior in *Polistes* (Hymenoptera, Vespidae). *Acta Hym. Fukuoka* **1960**, *1*, 99–103.

38. Beenakkers, A.; Van Der Horst, D.J.; Van Marrewijk, W.J.A. Biochemical process directed to flight muscle metabolism. In *Comprehensive Insect Physiology, Biochemistry and Pharmacology*; Pergamon Press: Oxford, UK, 1985; Volume 10.

39. Assenza, A.; Bergero, D.; Tarantola, M.; Piccione, G.; Caola, G. Blood serum branched chain amino acids and tryptophan modifications in horses competing in long-distance rides of different length. *J. Anim. Physiol. Anim. Nutr.* **2004**, *88*, 172–177. [CrossRef] [PubMed]

40. Chung, M.Y.; Gwon, E.-Y.; Hwang, J.-S.; Goo, T.-W.; Yun, E.-Y. Analysis of general composition and harmful material of *Protaetia brevitarsis*. *J. Life Sci.* **2013**, *23*, 664–668. [CrossRef]

41. Ravzanaadii, N.; Kim, S.-H.; Choi, W.-H.; Hong, S.-J.; Kim, N.-J. Nutritional value of mealworm, *Tenebrio molitor* as food source. *Int. J. Ind. Entomol.* **2012**, *25*, 93–98. [CrossRef]

42. Haytowitz, D.; Ahuja, J.; Thomas, R.; Nickle, M.; Roseland, J.; Williams, J.; Showell, B.; Somanchi, M.; Khan, M.; Nguyen, Q. *USDA National Nutrient Database for Standard Reference, Release 24*; US Department of Agriculture: Washington, DC, USA, 2011.

43. Evans, J.; Zulewska, J.; Newbold, M.; Drake, M.; Barbano, D. Comparison of composition, sensory, and volatile components of thirty-four percent whey protein and milk serum protein concentrates. *J. Dairy Sci.* **2009**, *92*, 4773–4791. [CrossRef]

44. Kim, Y.-S.; Park, C.G.; Kim, T.; Choi, J.W. A Study on the Legal Status of Insect Industry. *Korean J. Appl. Entomol.* **2018**, *57*, 401–408.

45. Jung, C. Prospects of insect food commercialization: A mini review. *Korean J. Soil Zool.* **2013**, *17*, 5–8.

46. Payne, C.L.; Evans, J.D. Nested houses: Domestication dynamics of human–wasp relations in contemporary rural Japan. *J. Ethnobiol. Ethnomed.* **2017**, *13*, 13. [CrossRef] [PubMed]

47. World Health Organization; United Nations University. *Protein and amino acid requirements in human nutrition*; World Health Organization: Geneva, Switzerland, 2007; Volume 935.

48. Choi, M.B.; Lee, S.A.; Suk, H.Y.; Lee, J.W. Microsatellite variation in colonizing populations of yellow-legged Asian hornet, *Vespa velutina nigrithorax*, in South Korea. *Entomol. Res.* **2013**, *43*, 208–214. [CrossRef]

49. Choi, M.B.; Lee, S.Y.; Yoo, J.S.; Jun, J.; Kwon, O. First record of the western black widow spider *Latrodectus hesperus* Chamberlin & Ivie, 1935 (Araneae: Theridiidae) in South Korea. *Entomol. Res.* **2019**, *49*, 141–146.

50. Lee, H.; Kim, M.J.; Bae, Y.S.; Kim, D.E. New occurrence of the invasive alien leaf-footed bug *Leptoglossus gonagra* (Hemiptera: Coreidae) in South Korea. *Entomol. Res.* **2020**, *50*, 14–22. [CrossRef]

51. Lee, W.; Lee, Y.; Kim, S.; Lee, J.-H.; Lee, H.; Lee, S.; Hong, K.-J. Current status of exotic insect pests in Korea: Comparing border interception and incursion during 1996–2014. *J. Asia-Pac. Entomol.* **2016**, *19*, 1095–1101. [CrossRef]

52. Sung, S.; Kwon, Y.S.; Lee, D.K.; Cho, Y. Predicting the potential distribution of an invasive species, *Solenopsis invicta* Buren (Hymenoptera: Formicidae), under climate change using species distribution models. *Entomol. Res.* **2018**, *48*, 505–513. [CrossRef]

53. Kong, J.-Y.; Yeon, S.-C.; Lee, H.J.; Kang, C.; Park, J.-K.; Jeong, K.-S.; Hong, I.-H. Protective Effects of Nutria Bile against Thioacetamide-Induced Liver Injury in Mice. *Evid.-Based Complement. Altern. Med.* **2019**, *2019*, 6059317. [CrossRef]

54. Kim, J.; Kim, M.; Lee, M.; Lee, Y.J.; Kim, H.R.; Nam, J.O.; Choi, M.B.; Hahn, D. Antibacterial potential of *Nidus vespae* built by invasive alien hornet, *Vespa velutina nigrithorax*, against food-borne pathogenic bacteria. *Entomol. Res.* **2020**, *50*, 28–33. [CrossRef]

55. Meyer-Rochow, V.B. Can insects help to ease the problem of world food shortage. *Search* **1975**, *6*, 261–262.

MDPI

St. Alban-Anlage 66

4052 Basel

Switzerland

Tel. +41 61 683 77 34

Fax +41 61 302 89 18

www.mdpi.com

Foods Editorial Office

E-mail: foods@mdpi.com

www.mdpi.com/journal/foods